Organic Solar Cells

Devices, Circuits, and Systems

Series Editor
Krzysztof Iniewski
CMOS Emerging Technologies Research Inc.,
Vancouver, British Columbia, Canada

PUBLISHED TITLES:

Atomic Nanoscale Technology in the Nuclear Industry
Taeho Woo

Biological and Medical Sensor Technologies
Krzysztof Iniewski

Building Sensor Networks: From Design to Applications
Ioanis Nikolaidis and Krzysztof Iniewski

Circuits at the Nanoscale: Communications, Imaging, and Sensing
Krzysztof Iniewski

Design of 3D Integrated Circuits and Systems
Rohit Sharma

Electrical Solitons: Theory, Design, and Applications
David Ricketts and Donhee Ham

Electronics for Radiation Detection
Krzysztof Iniewski

**Embedded and Networking Systems:
Design, Software, and Implementation**
Gul N. Khan and Krzysztof Iniewski

Energy Harvesting with Functional Materials and Microsystems
Madhu Bhaskaran, Sharath Sriram, and Krzysztof Iniewski

**Graphene, Carbon Nanotubes, and Nanostuctures:
Techniques and Applications**
James E. Morris and Krzysztof Iniewski

High-Speed Devices and Circuits with THz Applications
Jung Han Choi

High-Speed Photonics Interconnects
Lukas Chrostowski and Krzysztof Iniewski

**High Frequency Communication and Sensing:
Traveling-Wave Techniques**
Ahmet Tekin and Ahmed Emira

Integrated Microsystems: Electronics, Photonics, and Biotechnology
Krzysztof Iniewski

PUBLISHED TITLES:

Integrated Power Devices and TCAD Simulation
Yue Fu, Zhanming Li, Wai Tung Ng, and Johnny K.O. Sin

Internet Networks: Wired, Wireless, and Optical Technologies
Krzysztof Iniewski

Labs on Chip: Principles, Design, and Technology
Eugenio Iannone

Low Power Emerging Wireless Technologies
Reza Mahmoudi and Krzysztof Iniewski

Medical Imaging: Technology and Applications
Troy Farncombe and Krzysztof Iniewski

Metallic Spintronic Devices
Xiaobin Wang

MEMS: Fundamental Technology and Applications
Vikas Choudhary and Krzysztof Iniewski

Micro- and Nanoelectronics: Emerging Device Challenges and Solutions
Tomasz Brozek

Microfluidics and Nanotechnology: Biosensing to the Single Molecule Limit
Eric Lagally

MIMO Power Line Communications: Narrow and Broadband Standards, EMC, and Advanced Processing
Lars Torsten Berger, Andreas Schwager, Pascal Pagani, and Daniel Schneider

Mobile Point-of-Care Monitors and Diagnostic Device Design
Walter Karlen

Nano-Semiconductors: Devices and Technology
Krzysztof Iniewski

Nanoelectronic Device Applications Handbook
James E. Morris and Krzysztof Iniewski

Nanopatterning and Nanoscale Devices for Biological Applications
Šeila Selimović

Nanoplasmonics: Advanced Device Applications
James W. M. Chon and Krzysztof Iniewski

Nanoscale Semiconductor Memories: Technology and Applications
Santosh K. Kurinec and Krzysztof Iniewski

Novel Advances in Microsystems Technologies and Their Applications
Laurent A. Francis and Krzysztof Iniewski

Optical, Acoustic, Magnetic, and Mechanical Sensor Technologies
Krzysztof Iniewski

Optical Fiber Sensors: Advanced Techniques and Applications
Ginu Rajan

PUBLISHED TITLES:

Organic Solar Cells: Materials, Devices, Interfaces, and Modeling
Qiquan Qiao

Radiation Effects in Semiconductors
Krzysztof Iniewski

Semiconductor Radiation Detection Systems
Krzysztof Iniewski

Smart Grids: Clouds, Communications, Open Source, and Automation
David Bakken

Smart Sensors for Industrial Applications
Krzysztof Iniewski

Solid-State Radiation Detectors: Technology and Applications
Salah Awadalla

Technologies for Smart Sensors and Sensor Fusion
Kevin Yallup and Krzysztof Iniewski

Telecommunication Networks
Eugenio Iannone

Testing for Small-Delay Defects in Nanoscale CMOS Integrated Circuits
Sandeep K. Goel and Krishnendu Chakrabarty

VLSI: Circuits for Emerging Applications
Tomasz Wojcicki

Wireless Technologies: Circuits, Systems, and Devices
Krzysztof Iniewski

Wireless Transceiver Circuits: System Perspectives and Design Aspects
Woogeun Rhee

FORTHCOMING TITLES:

Advances in Imaging and Sensing
Shuo Tang, Dileepan Joseph, and Krzysztof Iniewski

Analog Electronics for Radiation Detection
Renato Turchetta

Cell and Material Interface: Advances in Tissue Engineering, Biosensor, Implant, and Imaging Technologies
Nihal Engin Vrana

Circuits and Systems for Security and Privacy
Farhana Sheikh and Leonel Sousa

CMOS: Front-End Electronics for Radiation Sensors
Angelo Rivetti

CMOS Time-Mode Circuits and Systems: Fundamentals and Applications
Fei Yuan

FORTHCOMING TITLES:

Electrostatic Discharge Protection of Semiconductor Devices and Integrated Circuits
Juin J. Liou

Gallium Nitride (GaN): Physics, Devices, and Technology
Farid Medjdoub and Krzysztof Iniewski

Laser-Based Optical Detection of Explosives
Paul M. Pellegrino, Ellen L. Holthoff, and Mikella E. Farrell

Mixed-Signal Circuits
Thomas Noulis and Mani Soma

Magnetic Sensors: Technologies and Applications
Simone Gambini and Kirill Poletkin

MRI: Physics, Image Reconstruction, and Analysis
Angshul Majumdar and Rabab Ward

Multisensor Data Fusion: From Algorithm and Architecture Design to Applications
Hassen Fourati

Nanoelectronics: Devices, Circuits, and Systems
Nikos Konofaos

Nanomaterials: A Guide to Fabrication and Applications
Gordon Harling, Krzysztof Iniewski, and Sivashankar Krishnamoorthy

Optical Imaging Devices: New Technologies and Applications
Dongsoo Kim and Ajit Khosla

Physical Design for 3D Integrated Circuits
Aida Todri-Sanial and Chuan Seng Tan

Power Management Integrated Circuits and Technologies
Mona M. Hella and Patrick Mercier

Radiation Detectors for Medical Imaging
Jan S. Iwanczyk

Radio Frequency Integrated Circuit Design
Sebastian Magierowski

Reconfigurable Logic: Architecture, Tools, and Applications
Pierre-Emmanuel Gaillardon

Soft Errors: From Particles to Circuits
Jean-Luc Autran and Daniela Munteanu

Terahertz Sensing and Imaging: Technology and Devices
Daryoosh Saeedkia and Wojciech Knap

Tunable RF Components and Circuits: Applications in Mobile Handsets
Jeffrey L. Hilbert

Wireless Medical Systems and Algorithms: Design and Applications
Pietro Salvo and Miguel Hernandez-Silveira

Organic Solar Cells

Materials, Devices, Interfaces, and Modeling

EDITED BY **QIQUAN QIAO**
SOUTH DAKOTA STATE UNIVERSITY
BROOKINGS, SOUTH DAKOTA

KRZYSZTOF INIEWSKI MANAGING EDITOR
CMOS EMERGING TECHNOLOGIES RESEARCH INC.
VANCOUVER, BRITISH COLUMBIA, CANADA

CRC Press is an imprint of the
Taylor & Francis Group, an **informa** business

CRC Press
Taylor & Francis Group
6000 Broken Sound Parkway NW, Suite 300
Boca Raton, FL 33487-2742

© 2015 by Taylor & Francis Group, LLC
CRC Press is an imprint of Taylor & Francis Group, an Informa business

No claim to original U.S. Government works

Printed on acid-free paper
Version Date: 20141230

International Standard Book Number-13: 978-1-4822-2983-7 (Hardback)

This book contains information obtained from authentic and highly regarded sources. Reasonable efforts have been made to publish reliable data and information, but the author and publisher cannot assume responsibility for the validity of all materials or the consequences of their use. The authors and publishers have attempted to trace the copyright holders of all material reproduced in this publication and apologize to copyright holders if permission to publish in this form has not been obtained. If any copyright material has not been acknowledged please write and let us know so we may rectify in any future reprint.

Except as permitted under U.S. Copyright Law, no part of this book may be reprinted, reproduced, transmitted, or utilized in any form by any electronic, mechanical, or other means, now known or hereafter invented, including photocopying, microfilming, and recording, or in any information storage or retrieval system, without written permission from the publishers.

For permission to photocopy or use material electronically from this work, please access www.copyright.com (http://www.copyright.com/) or contact the Copyright Clearance Center, Inc. (CCC), 222 Rosewood Drive, Danvers, MA 01923, 978-750-8400. CCC is a not-for-profit organization that provides licenses and registration for a variety of users. For organizations that have been granted a photocopy license by the CCC, a separate system of payment has been arranged.

Trademark Notice: Product or corporate names may be trademarks or registered trademarks, and are used only for identification and explanation without intent to infringe.

Visit the Taylor & Francis Web site at
http://www.taylorandfrancis.com

and the CRC Press Web site at
http://www.crcpress.com

Contents

Preface xiii
Editors xv
Contributors xvii

SECTION I Materials

Chapter 1 Conjugated Polymers as Electron Donors in Organic Solar Cells 3

Yan Zhou and Siyi Wang

Chapter 2 Donor and Acceptor Functionalized Silsesquioxane Nanostructures for Organic-Based Photovoltaic Devices 19

Hemali Rathnayake and John Ferguson

Chapter 3 Next-Generation Transparent Electrode Materials for Organic Solar Cells 43

Kuan Sun, Yijie Xia, and Jianyong Ouyang

SECTION II Modeling

Chapter 4 Relating Synthesis Parameters to the Morphology of the Photoactive Layer in Organic Photovoltaic Solar Cells Using Molecular Dynamics Simulations 89

S. M. Mortuza and Soumik Banerjee

Chapter 5 Insights Obtained from Modeling of Organic Photovoltaics: Morphology, Interfaces, and Coupling with Charge Transport 113

Rajeev Kumar, Jan-Michael Carrillo, Monojoy Goswami, and Bobby G. Sumpter

SECTION III Morphology, Interface, Charge Transport, and Defect States

Chapter 6 Photoexcited Carrier Dynamics in Organic Solar Cells 143

Sou Ryuzaki and Jun Onoe

Chapter 7 Defect States in Organic Photovoltaic Materials, Thin Films, and Devices 167

John A. Carr and Sumit Chaudhary

Chapter 8 Interfacial Materials toward Efficiency Enhancement of Polymer Solar Cells 201

Zhiqiang Zhao, Wenfeng Zhang, Xuemei Zhao, and Shangfeng Yang

Chapter 9 Nanophase Separation in Organic Solar Cells 247

Wei Chen, Feng Liu, Ondrej E. Dyck, Gerd Duscher, Huipeng Chen, Mark D. Dadmun, Wei You, Qiquan Qiao, Zhengguo Xiao, Jinsong Huang, Wei Ma, Jong K. Keum, Adam J. Rondinone, Karren L. More, and Jihua Chen

Chapter 10 Engineering of Active Layer Nanomorphology via Fullerene Ratios and Solvent Additives for Improved Charge Transport in Polymer Solar Cells 281

Swaminathan Venkatesan, Evan Ngo, and Qiquan Qiao

SECTION IV Devices

Chapter 11 Inorganic–Organic Nanocomposites and Their Assemblies for Solar Energy Conversion 307

Jaehan Jung, Ming He, and Zhiqun Lin

Chapter 12 Organic Tandem Solar Cells 337

Ning Li, Tayebeh Ameri, and Christoph J. Brabec

Chapter 13 Graphene-Based Polymer and Organic Solar Cells 379
Reg Bauld, Faranak Sharifi, and Giovanni Fanchini

Index 395

Preface

Current energy consumption mainly depends on fossil fuels that are limited and can cause environmental issues such as greenhouse gas emission and global warming. This has stimulated people to search for alternate, clean, and renewable energy sources. Solar cells are one of the most promising clean and readily available energy sources. Successful utilization of solar energy can help to reduce the dependence on fossil fuels. Organic solar cells have provided a unique opportunity and gained extensive attention as a next-generation photovoltaic technology due to their lightweight, mechanical flexibility, and solution-based cost-effective processing. However, organic solar cells still suffer from low efficiency and short lifetime. Significant progress has been achieved in improving device efficiencies. Recently, organic solar cell efficiencies above 10% have been reached.

Organic solar cells are a very broad field that covers materials, modeling, morphology, interface, and device engineering. This book aims to provide a reader with a comprehensive knowledge of organic solar cell technology ranging from materials, simulation and modeling, device physics, and engineering. With this book available, a reader can gain a broad understanding on the most recent and frontier developments in organic solar cells. Specifically, this book will deliver the following four sections: (1) materials that will cover active layer, interfacial, and transparent electrode materials; (2) modeling that will include approaches to relate synthesis parameters to morphology of the photoactive layer using molecular dynamics simulations and obtaining insights to coupling morphology and interfaces with charge transport in organic solar cells; (3) device physics that will consist of photoexcited carrier dynamics, defect states, interface engineering, and nanophase separation and (4) device engineering that will consist of inorganic–organic hybrids, tandem structure, and graphene-based polymer solar cells.

We hope this book will provide in-depth knowledge of organic solar cells to researchers and scientists in both academic institutions and industry sectors. The chapters are organized in sections of materials, modeling, device physics, and device engineering in a coherent manner that will be easily understood. This book will also benefit young investigators including students and postdoctoral researchers who have just entered the organic solar cell field from broad disciplines including chemistry, material science and engineering, physics, nanotechnology, nanoscience, and electrical engineering.

Qiquan Qiao, PhD

Editors

Dr. Qiquan Qiao is an associate professor of electrical engineering at South Dakota State University (SDSU). His current research focuses on polymer photovoltaic, dye-sensitized solar cell and lithium ion battery materials and devices. Dr. Qiao has published more than 60 peer-reviewed papers in leading journals, including the *Journal of the American Chemical Society*, *Advanced Materials*, *Advanced Functional Materials*, *Energy and Environmental Science*, *Nanoscale*, and *Nano Energy*. In addition to his recent receipt of the 2014 F O Butler Award for Excellence in Research at SDSU, he also received the 2012 3M Faculty Award and the College of Engineering Young Investigator Award. In 2010, Dr. Qiao received the US NSF CAREER, and in 2009 the Bergmann Memorial Award from the US–Israel Bi-national Science Foundation. During his graduate study, Dr. Qiao received the 2006 American Society of Mechanical Engineers Solar Energy Division Graduate Student Research Award and the 2006 Chinese Government Award for Outstanding Students Abroad.

Krzysztof (Kris) Iniewski is an R&D manager at Redlen Technologies, Inc., a startup company in Vancouver, Canada. Redlen's revolutionary production process for advanced semiconductor materials enables a new generation of more accurate, all-digital, radiation-based imaging solutions. Kris is also a president of CMOS Emerging Technologies Research, Inc. (www.cmosetr.com), an organization of high-tech events covering communications, microsystems, optoelectronics, and sensors. In his career, Dr. Iniewski held numerous faculty and management positions at the University of Toronto, University of Alberta, SFU, and PMC-Sierra, Inc. He has published more than 100 research papers in international journals and conferences. He holds 18 international patents granted in the United States, Canada, France, Germany, and Japan. He is a frequently invited speaker and has consulted for multiple organizations internationally. He has written and edited several books for CRC Press, Cambridge University Press, IEEE Press, Wiley, McGraw Hill, Artech House, and Springer. His personal goal is to contribute to healthy living and sustainability through innovative engineering solutions. In his leisure time, Kris can be found hiking, sailing, skiing, or biking in beautiful British Columbia. He can be reached at kris.iniewski@gmail.com.

Contributors

Tayebeh Ameri
Institute of Materials for Electronics and Energy Technology (i-MEET)
Department of Materials Science and Engineering
Friedrich-Alexander University Erlangen-Nuremberg
Erlangen, Germany

Soumik Banerjee
School of Mechanical and Materials Engineering
Washington State University
Pullman, Washington

Reg Bauld
Department of Physics and Astronomy
University of Western Ontario
London, Ontario, Canada

Christoph J. Brabec
Institute of Materials for Electronics and Energy Technology (i-MEET)
Department of Materials Science and Engineering
Friedrich-Alexander University Erlangen-Nuremberg
and
Bavarian Center for Applied Energy Research (ZAE Bayern)
Erlangen, Germany

John A. Carr
Electrical and Computer Engineering
Iowa State University
Iowa City, Iowa

Jan-Michael Carrillo
National Center for Computational Sciences
Oak Ridge National Laboratory
Oak Ridge, Tennessee

Sumit Chaudhary
Electrical and Computer Engineering
Iowa State University
Iowa City, Iowa

Huipeng Chen
Department of Chemistry
University of Tennessee-Knoxville
Knoxville, Tennessee

Jihua Chen
Center for Nanophase Materials Sciences
Oak Ridge National Laboratory
Oak Ridge, Tennessee

Wei Chen
Materials Science Division
Argonne National Laboratory
and
Institute for Molecular Engineering
University of Chicago
Argonne, Illinois

Mark D. Dadmun
Department of Chemistry
University of Tennessee-Knoxville
Knoxville, Tennessee

Gerd Duscher
Department of Materials Science and Engineering
University of Tennessee-Knoxville
Knoxville, Tennessee

Ondrej E. Dyck
Department of Materials Science and Engineering
University of Tennessee-Knoxville
Knoxville, Tennessee

Giovanni Fanchini
Department of Physics and Astronomy
University of Western Ontario
London, Ontario, Canada

John Ferguson
Department of Chemistry
Western Kentucky University
Bowling Green, Kentucky

Monojoy Goswami
Center for Nanophase Materials Sciences
and
Computer Science and Mathematics Division
Oak Ridge National Laboratory
Oak Ridge, Tennessee

Ming He
School of Materials Science and Engineering
Georgia Institute of Technology
Atlanta, Georgia

Jinsong Huang
Department of Mechanical and Materials Engineering
University of Nebraska-Lincoln
Lincoln, Nebraska

Jaehan Jung
School of Materials Science and Engineering
Georgia Institute of Technology
Atlanta, Georgia

Jong K. Keum
Center for Nanophase Materials Sciences
and
Neutron Science Division
Oak Ridge National Laboratory
Oak Ridge, Tennessee

Rajeev Kumar
Center for Nanophase Materials Sciences
and
Computer Science and Mathematics Division
Oak Ridge National Laboratory
Oak Ridge, Tennessee

Ning Li
Institute of Materials for Electronics and Energy Technology (i-MEET)
Department of Materials Science and Engineering
Friedrich-Alexander University Erlangen-Nuremberg
Erlangen, Germany

Zhiqun Lin
School of Materials Science and Engineering
Georgia Institute of Technology
Atlanta, Georgia

Feng Liu
Department of Polymer Science and Engineering
University of Massachusetts-Amherst
Amherst, Massachusetts

Wei Ma
Department of Physics
North Carolina State University
Raleigh, North Carolina

Contributors

Karren L. More
Center for Nanophase Materials
  Sciences
and
Materials Science and Technology
  Division
Oak Ridge National Laboratory
Oak Ridge, Tennessee

S. M. Mortuza
School of Mechanical and Materials
  Engineering
Washington State University
Pullman, Washington

Evan Ngo
Center for Advanced Photovoltaics
Department of Electrical Engineering
South Dakota State University
Brookings, South Dakota

Jun Onoe
Department of Physical Science and
  Engineering
Nagoya University
Nagoya, Japan

Jianyong Ouyang
Department of Materials Science and
  Engineering
National University of Singapore
Singapore

Qiquan Qiao
Center for Advanced Photovoltaics
Department of Electrical Engineering
  and Computer Science
South Dakota State University
Brookings, South Dakota

Hemali Rathnayake
Department of Chemistry
Western Kentucky University
Bowling Green, Kentucky

Adam J. Rondinone
Center for Nanophase Materials
  Sciences
Oak Ridge National Laboratory
Oak Ridge, Tennessee

Sou Ryuzaki
Institute for Materials Chemistry and
  Engineering
Kyushu University
Fukuoka, Japan

Faranak Sharifi
Department of Physics and Astronomy
University of Western Ontario
London, Ontario, Canada

Bobby G. Sumpter
Center for Nanophase Materials
  Sciences
and
Computer Science and Mathematics
  Division
Oak Ridge National Laboratory
Oak Ridge, Tennessee

Kuan Sun
Department of Materials Science and
  Engineering
National University of Singapore
Singapore

and

School of Power Engineering
Chongqing University
Chongqing, China

Swaminathan Venkatesan
Center for Advanced Photovoltaics
Department of Electrical Engineering
South Dakota State University
Brookings, South Dakota

Siyi Wang
Department of Chemistry
University of Southern California
Los Angeles, California

Yijie Xia
Institute of Materials Research and
  Engineering
Singapore

Zhengguo Xiao
Department of Mechanical and
  Materials Engineering
University of Nebraska-Lincoln
Lincoln, Nebraska

Shangfeng Yang
Hefei National Laboratory for Physical
  Science at Microscale
CAS Key Laboratory of Materials
  for Energy Conversion and
  Department of Materials Science and
  Engineering
University of Science and Technology
  of China (USTC)
Hefei, China

Wei You
Department of Chemistry
University of North Carolina-Chapel
  Hill
Chapel Hill, North Carolina

Wenfeng Zhang
School of Engineering
Anhui Agricultural University
Hefei, China

Xuemei Zhao
Hefei National Laboratory for Physical
  Science at Microscale
CAS Key Laboratory of Materials
  for Energy Conversion and
  Department of Materials Science and
  Engineering
University of Science and Technology
  of China (USTC)
Hefei, China

Zhiqiang Zhao
Hefei National Laboratory for Physical
  Science at Microscale
CAS Key Laboratory of Materials
  for Energy Conversion and
  Department of Materials Science and
  Engineering
University of Science and Technology
  of China (USTC)
Hefei, China

Yan Zhou
Department of Chemical Engineering
Stanford University
Stanford, California

Section I

Materials

1 Conjugated Polymers as Electron Donors in Organic Solar Cells

Yan Zhou and Siyi Wang

CONTENTS

1.1 Polymer/Polymer Bulk Heterojunction Solar Cells.. 10
References... 13

Organic π-conjugated polymers and oligomers, due to their potential in the development of plastic solar cells that are lightweight, mechanically flexible, and low cost, have drawn significant attention as active materials in organic photovoltaic (OPV) devices.[1–3] Because of the discovery of rapid and efficient photoinduced electron transfer from poly(2-methoxy-5-(2′-ethyl-hexyloxy)-*para*-phenylenevinylene (MEH-PPV) to C_{60} in the early 1990s,[4,5] the use of fullerene derivatives as electron acceptors in OPVs has become ubiquitous. The so-called bulk-heterojunction (BHJ) OPVs have been the most studied, using a single-layer blend of a conjugated polymer and a fullerene derivative, such as [6,6]-phenyl–C_{61}–butyric acid methyl ester ($PC_{61}BM$) or [6,6]-phenyl–C_{71}–butyric acid methyl ester ($PC_{71}BM$). In this type of device, appropriate phase segregation enables the formation of a BHJ at the interface of the donor and acceptor components. Current state-of-the-art single-layer conjugated polymer-based OPVs achieve power conversion efficiencies in the range of ~7%, and the highest reported efficiency of 6.5% is for a tandem solar cell.[6–15] Solar cells with optimized performance have been developed using P3HT/$PC_{61}BM$ blends,[16] and these cells feature nearly ideal photon-to-current quantum efficiency (IPCE) in the mid-visible region. However, to produce highly efficient OPVs, light absorption by the active layer must be extended into the near-infrared region while, at the same time, preserving the high IPCE and open circuit voltage. In principle, the overall power conversion efficiency of the device is mainly determined by several individual efficiencies: light-harvesting efficiency across the visible and near-infrared regions, exciton diffusion to the donor–acceptor interface, photoinduced charge separation, and the mobility of the charge carriers produced by photoinduced charge separation and charge collection.[1]

Sariciftci et al. reported efficient photoinduced electron transfer (PET) in a conjugated polymer-fullerene (Buckminsterfullerene, C_{60}) composite in 1992.[4] In this report, a time scale of 45 fs PET was observed from a conjugated polymer to fullerene, which is several orders of magnitude faster than any photoexcitation radiative decay or back electron transfer in the process. Consequently, the quantum efficiency of charge separation in such a composite can approach unity. The tendency for fullerene to crystallize in organic solvents and on surfaces, however, leads to unfavorable phase separation in the composite. This implies that charge carriers do not have the necessary channels to reach the electrodes. Hence, efficient solar cells are still not achievable in this context.

The conceptually new photovoltaic device, namely the BHJ solar cell, was demonstrated by Wudl, Heeger, and coworkers in 1995, simply using the blend of poly(2-methoxy-5-(2'-ethyl-hexyloxy)-1,4-phenylene vinylene) (MEH-PPV) and a fullerene derivative ([6,6]-phenyl-C61-butyric acid methyl ester, $PC_{61}BM$).[17] The replacement of fullerene with more soluble and less symmetrical PCBM decreases the formation of large fullerene clusters and instead increases the possibility to form a D-A interpenetrating network in the composite. Thus, this network, with a large conjugated polymer–fullerene interfacial area and the appropriate phase domain size, at least in one dimension comparable to the exciton diffusion length, enables the required compromise between optical length and exciton diffusion length as mentioned in the discussion of bilayer heterojunction solar cells. In other words, the absorbing sites in the blend are most likely within the exciton drifting length to the D-A interface. In the mean time, efficient hole and electron transport can also be realized with the likelihood of the formation of a bi-continuous donor and acceptor network in such a blend. It is also noted that the fast PET can effectively improve the photostability of the conjugated polymers, due to the fast quenching of the highly reactive excited states and, thus, the reduction of any possible photooxidation associated with oxygen and water, etc. In addition, the use of a single active layer greatly simplifies the solution processing, a great advantage over solution-processed bilayer devices.

Since then, many efforts have been put forth to design new soluble π-conjugated polymers as donor material for BHJ solar cells when fullerene derivatives are primarily used as an electron acceptor, for example, $PC_{61}BM$ and $PC_{71}BM$. Figure 1.1 shows a list of conjugated polymers from which PCE over 3% has been achieved in blend with either $PC_{61}BM$ or $PC_{71}BM$.

Regioregular Poly(3-hexylthiophene) (RR-P3HT) is perhaps the most investigated conjugated polymer in the field of BHJ solar cells. RR-P3HT has HOMO and LUMO levels at −5.2 and −3.2 eV, respectively, with an optical bandgap of ~2.0 eV. Using RR-P3HT as the donor and $PC_{61}BM$ as the acceptor with a device structure of Glass/ITO/P3HT:$PC_{61}BM$/TiO_x/Alumina, BHJ solar cells have been able to exhibit EQE of 75% and PCE up to 5%.[9] Unfortunately, the success of RR-P3HT has not been repeated in other homo-conjugated polymers. The high efficiency of RR-P3HT/$PC_{61}BM$ devices may result from a unique microcrystalline lamellar stacking in the blends. Sariciftci et al. have studied a series of regioregular poly(3-alkylthiophenes), with butyl, hexyl, octyl, decyl, and dedecyl as solubilizing groups. They observed that chain length longer than eight carbons facilitates diffusion rates of $PC_{61}BM$ in the blend during the thermal annealing.[18] This leads to unfavored phase separation and thereby lowers the device performance. From

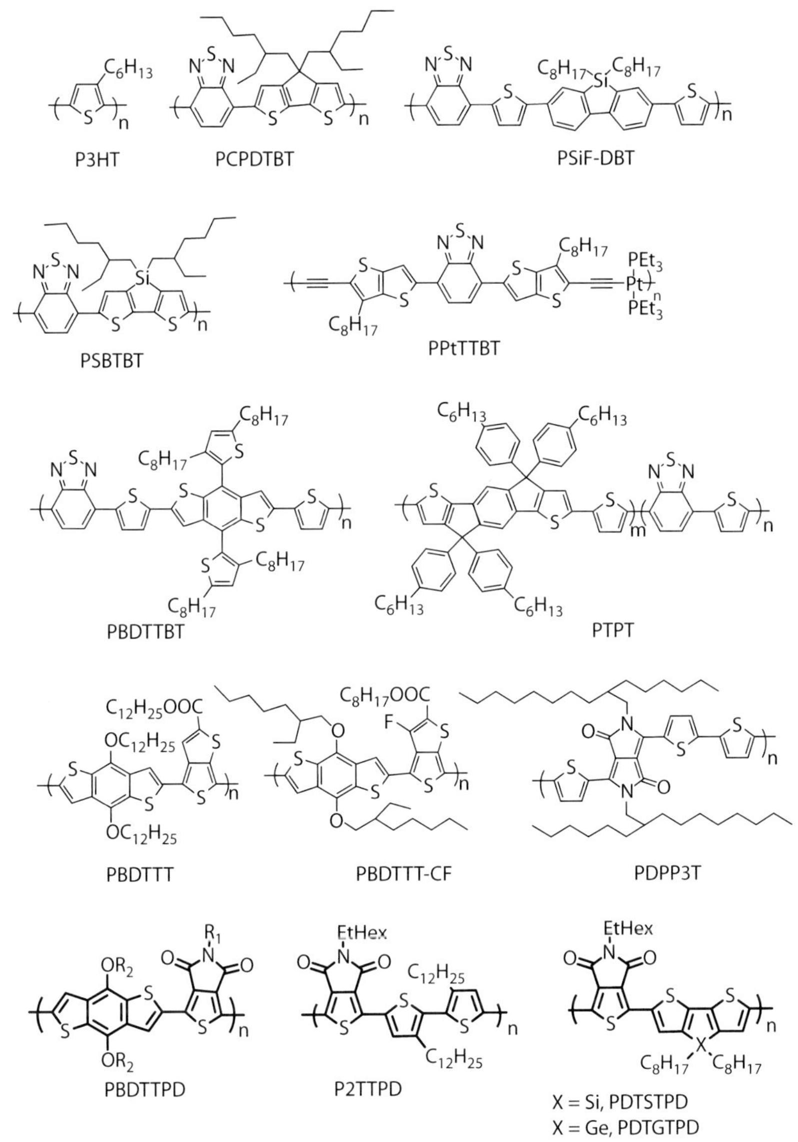

FIGURE 1.1 Representative p-type conjugated polymers with power conversion efficiency.

this study, it is reasonable to conclude that the passive solubilizing chains also affect the optical and electronic properties of a conjugated material and, hence, its performance in a device.

In the course of studying RR-P3HT/PC$_{61}$BM BHJ solar cells, people have learned many useful processing techniques to improve device performance. It has been demonstrated that device performance based on RR-P3HT/PCBM can be dramatically enhanced by careful selection of processing solvent,[19] solvent vapor annealing,[7] thermal annealing,[20] and the addition of high-boiling point additives.[21] (Note: The additive effect was first reported in the blend of PCPDTBT.[22]) These strategies are now widely applied in other conjugated polymer-based BHJ solar cells.

It is arguable that the further improvement of PCEs in RR-P3HT/PCBM cells is hindered by its relatively large bandgap (~1.9–2.0 eV). That is to say a RR-P3HT/PCBM blend only absorbs lights with a wavelength shorter than 650 nm, at best about 22.4% of the total amount of photons under AM 1.5 G. It is therefore important to design polymers that can harvest more photons from the available sunlight. It also needs to consider that narrowing the bandgap will consequently cause a decrease in open circuit voltage and thus a decrease in PCE. Through a careful estimation, an optical bandgap of 1.4 eV will be ideal for polymer/PCBM BHJ solar cells provided the appropriate energy offset is present.[3,23] We are proud to point out the Reynolds group was among the first to provide an analysis on this issue.[23] In practice, an optimal bandgap of 1.3 to 1.8 eV has been reported for conjugated polymers as the active absorbing materials blended with PCBM in high-performance BHJ solar cells. A few examples are listed to illuminate the common features that these polymers share.

PCPDTBT is a donor and acceptor alternating conjugated polymer. It possesses an optical bandgap (E_g^{opt}) of 1.4 eV (absorption onset 890 nm) and an electrochemical bandgap (E_g^{echem}) of 1.7 eV with a HOMO level of −5.3 eV and a LUMO level of −3.6 eV.[24] This polymer is the first low bandgap polymer with a highly efficient photovoltaic response in the near-IR region and has a PCE of 3.2% blended with $PC_{71}BM$ (a PCE of 2.7% blended with $PC_{61}BM$). The V_{oc} is typically at 0.65 V; the highest observed values approach 0.7 eV. With the addition of a small amount of alkanedithiols in the solvent, PCPDTBT/$PC_{71}BM$ devices have shown PCEs up to 5.5% with V_{oc} of ~0.62, J_{sc} of ~16.2 mA cm^{-2}, and fill factor (FF) of ~0.55.[22] The role of dithiols is to alter the BHJ morphology.[25,26]

PSiF-DBT is a 2,7- silafluorene (SiF) and 4,7-di(2′-thienyl)-2,1,3-benzothiadiazole (DBT) alternating polymer with an optical bandgap of 1.8 eV. The HOMO level from electrochemical measurements was −5.4 eV. In a blend with $PC_{60}BM$, the optimized PSiF-DBT BHJ solar cells exhibited a PCE up to 5.4%, a large V_{oc} of 0.9 V, a J_{sc} of 9.5 mA cm^{-2}, and a FF of 0.51.[27] Noticeably, no thermal or solvent annealing was performed on these devices. In addition, this combination is also the first low bandgap polymer with a PCE over 5%.

PSPTPB is a combination of PCPDTBT and PSiF-DBT from the material design point of view. From the study of Psi-DBT, it has been shown that the Si atom has a pronounced effect on the device performance. Yang et al. therefore replaced the bridge carbon atom in cylcopentadithiophene (CPDT) with silicon atom and obtained poly[(4,4′-bis(2-ethylhexyl)dithieno[3,2-b:2′,3′-d]silole)-2,6-diyl-alt-(2,1,3-benzothiadiazole)-4,7-diyl] (PSPTPB).[28] The HOMO and LUMO levels are −5.1 and −3.3 eV measured by cyclic voltammetry. (Note: These values are calculated using −4.8 eV as a ferrocene/ferrocenium standard versus vacuum level. It can be converted to −5.4 and −3.6 eV for a comparison reason because −5.1 eV as the ferrocene/ferrocenium standard versus vacuum level is used in this dissertation.) The optical bandgap (E_g^{opt}) of 1.45 eV is very similar to that of PCPDTBT. The photovoltaic device with a structure of ITO/

PEDOT-PSS/PSBTBT: PCBM/Ca/Al showed a maximum PEC up to 5.1% with a V_{oc} of 0.68V, a J_{sc} of 12.7 mA cm^{-2}, and a FF of 0.55.

PPtTTBT is a metallated polymer with an E_g^{opt} of 1.8 eV. HOMO and LUMO levels are −5.1 and −3.3 eV, respectively.[29] The best solar cell performance is based on the PPtTTBT and PC$_{71}$BM blend and yields an open circuit voltage (V_{oc}) of 0.79, a short circuit current (J_{sc}) of 10.1 mA cm^{-2}, a FF of 0.51, and a PCE of 4.13% under simulated AM 1.5 G illumination. In an early report, Wong et al., using a similar polymer as a donor and PC$_{61}$BM as an acceptor, demonstrated PCEs up to ~5% with EQEs as high as 87% at 570 nm.[30] Serious doubts, however, have been raised, suggesting that the reported efficiencies are significantly overestimated (in Chapter 3, more details will be given).[31] Nevertheless, the concept of designing materials that demonstrate triplet excitons is still interesting as demonstrated in PPtTTBT.

PBDTTBT is different from most of linear donor–acceptor type conjugated polymers in that it possesses a cross-conjugation segment in its donor. This polymer has three absorption bands, a distinction from a two-band donor–acceptor system, and therefore absorbs broadly from 300 to 700 nm with an E_g^{opt} of 1.75 eV. PBDTTBT presents HOMO and LUMO levels of −5.6 and −3.7 eV (after correction), respectively. The HOMO level is about 0.3 eV deeper than PSPTPB and PCPDTBT.[13] The deeper HOMO level leads to a higher V_{oc} of 0.9 V, an expected result, considering the difference between the HOMO level of a donor and the LUMO level of PCBM. PCEs up to 5.66% have been obtained from the PBDTTBT-based device with a V_{oc} of 0.92 V, a J_{sc} of 10.7 mA cm^{-2}, and a FF of 0.58, which is one of the highest PCEs for single active layer OPVs.

PTPT is perhaps the only existing random donor–acceptor copolymer that exhibits efficient photovoltaic response, with EQE as high as 63% at 540 nm and over 50% for a broad range. The PTPT/PC$_{71}$BM cell has the highest AM1.5G PCE of 4.4%, with a V_{oc} of 0.81 V, a J_{sc} of 10.2 mA cm^{-2}, and a FF of 0.53.[32] The devices also showed encouraging stability after encapsulation under ambient conditions, in which only 15% loss was observed after two months storage in air.

It is interesting to notice that all the high-performance D-A conjugated polymers mentioned contain benzothiadiazole as an acceptor. Thieno[3,4-b]thiophene (TT) and diketopyrrolopyrrole (DPP) are the two other known acceptors being used in D-A conjugated polymers that demonstrate their applicability in high-performance solar cells.

PBDTTT and its derivatives are benzo[1,2-b:4,5-b']dithiophene (BDT) and Thieno[3,4-b]thiophene (TT) alternating polymers with HOMO levels from −5.20 to −5.5 eV and LUMO levels from −3.6 to −3.8 eV.[14,15,33] The PBDTTT-CF$_3$/PC$_{71}$BM-blend BHJ solar cells prepared from chlorobenzene (CB) show a V_{oc} of 0.76 V, a J_{sc} of 10.2 mA cm^{-2}, and a FF of 0.51, which corresponds to a PCE of 3.92%. Upon using a mixture of chlorobenzene (CB) and 1,8-diiodoctane (DIO) (97:3 by volume) as a cosolvent, the PCE of this blend is almost doubled to 7.4% with an increase in J_{sc} to 14.5 mA cm^{-2}, and a boost in FF to 0.69. This is the highest PCE in a polymeric solar cell reported to date.[33]

PDPP3T is an ambipolar conjugated polymer. Field-effect transistors based on PDPP3T exhibit nearly balanced electron and hole mobilities of 0.01 and 0.04 cm^2

$V^{-1} s^{-1}$, respectively. The best cells obtained for PDPP3T/PC$_{61}$BM in a 1:2 weight ratio have a V_{oc} of 0.68 V, a J_{sc} of 8.3 mA cm^{-2}, and a FF of 0.67, yielding a PCE of 3.8%.[34] The PCEs of the devices increase to ~4.7% due to a jump in photocurrent to 11.8 mA cm^{-2}, switching PC$_{61}$BM to PC$_{71}$BM. In addition, it needs to be pointed out that the EQE of the optimized device is still relatively low, about 35% on average. It can be ascribed to inefficient electron transfer, due to approaching the minimum offset of $eV_{oc} = E_g^{opt} - 0.6$ eV with a V_{oc} of 0.65~0.68 V and a small optical bandgap of 1.3 eV.[35] An interesting observation for PDPP3T is that its molecular weights affect the performance of photovoltaic devices but have little or no influence on charge mobilities.

Isoindigo-based polymers have been developed as the donors in organic BHJ solar cells recently. With different aromatic spacers, which are copolymerized with isoindigo, the performance of solar cells are varied a lot. With 2,5-thiophenyl, 2,2′-bithiophenyl, and terthiophyl, J_{sc} are gradually increased from 9.1 mA cm^2 to 13.1 mA cm^{-2}, and V_{oc} decreased from 0.91 V to 0.70 V. FF of the devices are also increased from 0.54 to 0.69, and PCEs are increased from 4.5% to 6.3%.[36] The spacers vary from thiophenyl to terthiophenyl, and the LUMO levels of the donors are increased gradually. The energy offset of the LUMO levels between the donor and PCBM are increased, which results in higher J_{sc} and FF and lower V_{oc}. Higher driving force is achieved with the highest LUMO energy (terthiophenyl), and the best PCE is obtained.[37]

Thieno[3,4-c]pyrrole-4,6-dione (TPD) is another important acceptor unit in high-efficiency donor polymers. Because of the strong electron-withdrawing effect of TPD, although copolymerized with electron-donating aromatic structures, the HOMO level of the polymer is relevantly low compared to other D-A polymers. The typical HOMO levels of these class polymers are around 5.4 eV to 5.6 eV.[38] Such deep HOMO levels ensure high V_{oc} (around or above 0.9 V) of the polymer/fullerene solar cell. Spacers, such as BDT, oligothiophene, DTS, and DTGe, are applied in donor polymers.[39,40]

With different substitution in PBDTTPD polymers, there are dramatic differences in performances of the solar cells with PC$_{71}$BM as the acceptor. R1 as the substitution on the BDT is believed to be crucially important for the solubility of the polymer donor. With a bulky β-branch side chain, such as EH (2-ethylhexyl) on BDT, the PCEs of the devices are all higher than the ones with linear chains. However, on the TPD part, a linear chain is preferred for better device performance. PCE up to 8.5% is achieved with EH on BDT and C7 on TPD. J_{sc} as high as 12.7 mA cm^{-2} is obtained. Very high V_{oc} (0.97 V) and FF (0.70) is achieved in this polymer system. 5% v/v of 1-chloronaphthlene (CN) is used to optimize the morphology of the BHJ films. Notably, with the exact same donors, only 5% of the additive is changed from DIO to CN, the J_{sc} is increased from 11.5 to 12.5 mA cm^{-2}, and V_{oc} is increased from 0.89 V to 0.93 V. PCE is enhanced from 6.8% to 7.5%.[41]

With bithiophene as the spacer in TPD polymers (PBTTPD), PCE up to 7.3% is achieved with J_{sc} of 13.1 mA cm^{-2} and V_{oc} of 0.92 V. Different additives are screened and 1,7-diiodoheptane is the best one in this system. A 0.5 vol% of additive is believed to "dissolve" the PCBM cluster, reduce the grain boundary, and induce high crystalline donor domains. The optimized phase separation structure is identified by X-ray scatter experiment.[42]

With terthiophene as the spacer (PTPD3T), PCE is even higher: 7.7% is achieved with very high FF of 0.79. Such high FF is very rare in organic BHJ solar cells. Interaction between oxygen atoms and sulfate atoms results in a highly ordered, closely packed structure of donor polymer. Optimized vertical and horizontal phase separation structures and high charge carrier mobilities suppressed both bulk and interfacial recombination, which result in the high FF.[43]

With dithienosilole (DTS) as the spacer, solar cells with PDTSTPD/PC$_{71}$BM give the PCE as high as 7.3%. V_{oc} of 0.88 V, J_{sc} of 12.2 mA cm^{-2} and FF of 0.68; 3 vol% of DIO is required. In this work, solar cells with a thick active layer (>200 nm) are tested, and 6.1% of PCE are achieved. J_{sc} are increased from 12.2 mA cm^{-2} to 13.3 mA cm^{-2}, and V_{oc} and FF are reduced to 0.85 V and 0.54, respectively. By changing the atom of Si to Ge, higher performance is achieved. In the first report of this donor, PCE up to 7.3% is achieved. J_{sc} as high as 12.6 mA cm^{-2}, V_{oc} of 0.85 V, and FF of 0.68 is reported. By optimizing the fabrication process of the ZnO layer, 8.5% PCE is achieved with the same recipe. Much higher J_{sc} (14.0 mA cm^{-2}) is obtained by treating the ZnO with UV-ozone.[38,44]

DPP-based polymers play a very important role in the polymer-fullerene solar cell. The low bandgap property realizes absorption up to 900 nm, which covers the near-IR energy in the solar spectrum. Numerous aromatic spacers are copolymerized with the DPP unit, and excellent solar cell performances are achieved. V_{oc} around 0.7~0.8 V was always found in DPP/fullerene-based BHJ solar cells. Strong near-IR absorption enabled high J_{sc} of DPP solar cells. A current up to 19 mA cm^{-2} can be achieved by a single junction BHJ cell with DPP as a donor. Relatively high hole mobilities of DPP polymer also suppressed the carrier recombination, which ensured a high FF. The absorption of DPP polymers were complementary with P3HT, which is strongly absorbed at 600 nm. With this property, DPP polymers were well applied in the tandem organic BHJ solar cell, in which the highest PCE was up to 10.6%.[45,46]

The photovoltaic data of some representative high-performance OPVs are summarized in Table 1.1. As one can see, the efficiencies for OPVs are steadily approaching the predicted values of 10%~12%. Designing new polymers will still be the focus in realizing such a grand goal.

TABLE 1.1
Photovoltaic Data of Representative High-Performance OPVs

	V_{oc}/V	J_{sc}/mA cm^{-2}	FF	E_g^{opt}	PCE (%)
RR-P3HT	0.61	11.1	0.66	1.85	5.0
PCPDTBT	0.62	16.1	0.55	1.40	5.5
PSiF-DBT	0.90	9.5	0.51	1.82	5.4
PSPTPB	0.68	12.7	0.55	1.45	5.1
PPtTTBT	0.79	10.1	0.51	1.80	4.1
PTPT	0.81	10.2	0.53	1.76	4.4
PBDTTBT	0.74	14.5	0.69	1.62	7.4
PDPP3T	0.65	11.8	0.60	1.30	4.7

1.1 POLYMER/POLYMER BULK HETEROJUNCTION SOLAR CELLS

Friend and Holmes reported the first all-polymer BHJ solar cells in 1995, blending MEH-PPV as a donor with CN-PPV as an acceptor.[47] Compared to polymer/fullerene photovoltaic devices, all-polymer BHJ solar cells have several foreseeable advantages. In order to maximize the PCEs of solar cells, it is necessary to absorb across the whole visible and near-IR solar spectrum. In a polymer/polymer blend, both donor and acceptor components contribute to harvest light. It is possible to make a complementary polymer pair that absorb across the field. Although $PC_{71}BM$ provides enhanced absorption in the blue region (peak at *ca.* 500 nm) with respect to $PC_{61}BM$ (peak at *ca.* 349 nm), it contributes little in the red and near-IR regions in the case of polymer/fullerene systems. In addition, polymer synthesis offers much higher versatility and flexibility than fullerene chemistry as demonstrated in the development of polymer/fullerene solar cells. Thus, it enables the control of the energy levels to a great extent. The energy offset between the LUMO level of a donor and the LUMO level of an acceptor can be precustomized to obtain the highest possible V_{oc} that is dependent on the difference between the LUMO level of an acceptor and the HOMO level of a donor without sacrificing the driving force for charge transfer and separation.

Figure 1.2 shows some donor and acceptor polymer combinations used in all-polymer BHJ solar cells. The state-of-the-art all-polymer cells in the last decade exhibit PCEs of 1.5%–1.9%, using the combination ranging from POPT/MEH-CN/PPV,[48,49] P3HT/F8BT,[50] and M3EH-PPV/CN-ether-PPV to TVPT/PDIDTT.[27,51] These efficiencies are significantly less than those of polymer/PCBM BHJ solar cells. The efficiency discrepancy has tentatively been attributed to the lower electron mobility of most conjugated polymers compared to fullerene derivatives, which is consequently led by the poor phase segregation resulting from binary polymer demixing and the limited availability of n-type conjugated polymers.

Polymer blends have an intrinsic proclivity toward demixing (phase separate) into their individual pure phases due to their low entropy of mixing. There is no exception for rigid rod-like conjugated polymers. Binary random coil-like polymers and their demixing have been extensively studied and are relatively well understood.[52,53] Phase separation in conjugated polymers is much less understood even though some demixing mechanisms, for example, nucleation and growth of one phase in a surrounding phase or the process of spinodal decompositions, have been proposed based on what has been learned from random coil polymer systems. It also needs to be noted that these models are suitable for bulk phase separation but may not be applicable to thin-film phase separation.[54] Nevertheless, people have applied these principles to control the phase separation process in order to obtain the characteristic domain size of 10–20 nm for photovoltaics. Certain progress has been achieved through techniques such as thermal annealing and the use of mixed solvents.[55,56]

The PCEs of all polymer solar cells have been busted to more than 4% during the past three years; this rapid increase in performance started the new development cycle in this area. Thermal annealing with a higher temperature and higher molecular weight batch raises the PCE of PFBT/P3HT from 1.8% to 2.7%.[57,58] Optimal phase separation is achieved in this system and determined by AFM. A further

Conjugated Polymers as Electron Donors in Organic Solar Cells

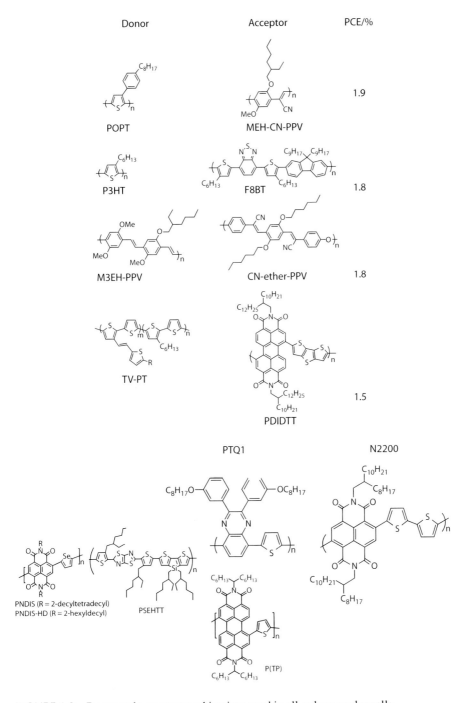

FIGURE 1.2 Donor and acceptor combinations used in all-polymer solar cells.

improvement of PCE is achieved by conjugation of the donor polymer (P3HT) and acceptor (PFBT) with a covalent bond. Three percent of PCE is obtained by this single component polymer with a nanosized phase separation structure.

By replacing a polythiophene-based donor to a donor–acceptor polymer with a narrow bandgap, which can harvest more light and gain a higher V_{oc}, the PCE of all polymer solar cells with PDI or NDI polymers as acceptors starts to increase very quickly.[59–63]

NDI-2T, the most famous electron-transporting polymer, is well studied in all polymer solar cells for its high electron mobility. However, with P3HT as the donor, the PCE is lower than 0.3%.[64] PTB7, with lower crystallinity and narrower bandgap compared to P3HT, is applied as the donor. By mixing PTB7 with NDI-2T in o-xylene, an optimized phase separation structure is obtained, and PCE as high as 2.66% is achieved with a V_{oc} of 0.80 V. O-xylene plays an important role here to achieve the 3-D continuance phase separation structure. With chlorobenzene as the solvent, phase separation size is too large for efficient exciton diffusion. Meanwhile, with chloroform as the solvent, phase separation size is too small so that there is a dead island in the BHJ film, which will lead to charge recombination.[60]

All polymer solar cells with NDI-2T as the electron acceptor and an amorphous low bandgap polymer (PTQ1) as the donor shows a PCE up to 4%.[62] Different D-A weight ratio is screened in this work, and J_{sc} up to 8.85 mA cm^{-2} and V_{oc} as high as 0.84 V are achieved with a D-A ratio of 7/3. Relevantly low FF (0.55) is observed. Different D-A blending ratios result in different PL quenching efficiency and charge carrier mobility. Balanced charge transportation and exciton separation is reached at a D-A ratio of 7/3, which results in the highest J_{sc} and PCE.

NDI copolymerized with a carbazole derivative results in an electron-accepting polymer (PC-NDI). Blends with P3HT show higher efficiency than those with NDI-2T. PCE up to 1.3% is achieved while it blends with PTB7. Almost three times higher PCE (3.7%) is achieved by blending PC-NDI with an engineered PTB7, with which a tris(thienylenevinylene) (TTV) is grafted as a side chain. One percent volume ratio of DIO is used to tune the morphology. High V_{oc} of 0.88 V is achieved in this system.[61]

PCE as high as 3.3% is achieved by blending NDI-Se with a novel donor named PSEHTT. Better electron and hole mobilities are observed in the optimized system. V_{oc} as high as 0.76 V and J_{sc} as high as 7.78 mA cm^{-2} are obtained, respectively.[63]

PDI polymers with perylenetetracarboxylic diimide as the acceptor have also shown much better performance in the past two years by blending with narrow bandgap donors.

A PCE around 2% is achieved by blending PDI with TV-PT in the literature. TV as the side chain is believed to be important for optimal phase separation.[65] This exact acceptor gives much lower performance while blending the high crystalline RR-P3HT. With the "swallow tail" modified PDI polymer, PCE up to 2.1%, which is nearly one order of magnitude higher than those results reported before, is achieved by blending with RR-P3HT.[66]

Binary additives of both a high boiling point solvent (DIO) and a nonvolatile small molecule (PDI-2DTT) are added to the BHJ film. The PDI-2DTT is the repeating unit of the acceptor polymer, which results in de-aggregation of the acceptor and a better mixing with the donor. Better-balanced electron/hole mobility is also observed when both additives are added. PCE up to 3.45% along with V_{oc} of 0.75 and J_{sc} of 8.55 mA cm^{-2} is achieved. The FF of this system is only 0.52.[59]

PDI-T with the "swallow tail" blended with isoindigo-based donors result in PCE as high as 3.6%. Different aromatic spacers are screened for the isoindigo donor polymers, and dramatic changes of phase separation sizes of the all-polymer blends are observed. The author is able to determine that the PCE is only related to the phase separation size of the all-polymer blend in this materials system. Five percent of the polystyrene (PS) side chain is grafted in the donor, and more than 20% enhancement of the J_{sc} and PCE are obtained for both the "best" and "worst" donors, due to reduction of phase separation size.

Many efforts have been devoted to design p-type conjugated polymers, resulting in a large pool of p-type conjugated polymers. In contrast, the number of solution-processable n-type conjugated polymers is still limited, as mentioned in Section 1.1.2. The available n-channel polymeric materials are still dominated by perylene diimide and naphthalene diimide-containing polymers.[67–69] To design and prepare other n-type polymers with variable electron affinity, high electron mobility, and good ambient stability is thereby becoming the biggest challenge for all-polymer BHJ solar cells.

With the increased synthetic efforts and the better understanding of phase-separation for rigid-rod type polymers, all-polymer BHJ solar cells have the potential to push PCEs to higher levels in OPVs.

REFERENCES

1. Brabec, C. J.; Sariciftci, N. S.; Hummelen, J. C., Plastic solar cells. *Adv. Funct. Mater.* **2001**, *11* (1), 15–26.
2. Brabec, C. J., Organic photovoltaics: Technology and market. *Sol. Energy Mat. Sol. C* **2004**, *83* (2–3), 273–292.
3. Scharber, M. C.; Mühlbacher, D.; Koppe, M.; Denk, P.; Waldauf, C.; Heeger, A. J.; Brabec, C. J., Design rules for donors in bulk-heterojunction solar cells—Towards 10% energy-conversion efficiency. *Adv. Mater.* **2006**, *18* (6), 789–794.
4. Sariciftci, N. S.; Smilowitz, L.; Heeger, A. J.; Wudl, F., Photoinduced electron transfer from a conducting polymer to buckminsterfullerene. *Science* **1992**, *258* (5087), 1474–1476.
5. Brabec, C. J.; Zerza, G.; Cerullo, G.; De Silvestri, S.; Luzzati, S.; Hummelen, J. C.; Sariciftci, S., Tracing photoinduced electron transfer process in conjugated polymer/fullerene bulk heterojunctions in real time. *Chem. Phys. Lett.* **2001**, *340* (3–4), 232–236.
6. Reyes-Reyes, M.; Kim, K.; Carroll, D. L., High-efficiency photovoltaic devices based on annealed poly(3-hexylthiophene) and 1-(3-methoxycarbonyl)-propyl-1-phenyl-(6,6) C-61 blends. *Appl. Phys. Lett.* **2005**, *87* (8), 083506.
7. Li, G.; Shrotriya, V.; Huang, J.; Yao, Y.; Moriarty, T.; Emery, K.; Yang, Y., High-efficiency solution processable polymer photovoltaic cells by self-organization of polymer blends. *Nat. Mater.* **2005**, *4* (11), 864–868.
8. Kim, Y.; Cook, S.; Tuladhar, S. M.; Choulis, S. A.; Nelson, J.; Durrant, J. R.; Bradley, D. D. C.; Giles, M.; Mcculloch, I.; Ha, C. S.; Ree, M., A strong regioregularity effect in self-organizing conjugated polymer films and high-efficiency polythiophene: Fullerene solar cells. *Nat. Mater.* **2006**, *5* (3), 197–203.

9. Kim, J. Y.; Kim, S. H.; Lee, H.-H.; Lee, K.; Ma, W.; Gong, X.; Heeger, A. J., New architecture for high-efficiency polymer photovoltaic cells using solution-based titanium oxide as an optical spacer. *Adv. Mater.* **2006**, *18* (5), 572–576.
10. Kim, K.; Liu, J.; Namboothiry, M. A. G.; Carroll, D. L., Roles of donor and acceptor nanodomains in 6% efficient thermally annealed polymer photovoltaics. *Appl. Phys. Lett.* **2007**, *90* (16), 163511.
11. Kim, J. Y.; Lee, K.; Coates, N. E.; Moses, D.; Nguyen, T. Q.; Dante, M.; Heeger, A. J., Efficient tandem polymer solar cells fabricated by all-solution processing. *Science* **2007**, *317* (5835), 222–225.
12. Green, M. A.; Emery, K.; Hishikawa, Y.; Warta, W., Solar cell efficiency tables (version 31). *Prog. Photovolt.* **2008**, *16* (1), 61–67.
13. Huo, L.; Hou, J.; Zhang, S.; Chen, H.-Y.; Yang, Y., A polybenzo[1,2-b:4,5-b']dithiophene derivative with deep HOMO level and its application in high-performance polymer solar cells. *Angew. Chem. Int. Ed.* **2010**, *49* (8), 1500–1503.
14. Chen, H.-Y.; Hou, J.; Zhang, S.; Liang, Y.; Yang, G.; Yang, Y.; Yu, L.; Wu, Y.; Li, G., Polymer solar cells with enhanced open-circuit voltage and efficiency. *Nat. Photon.* **2009**, *3* (11), 649–653.
15. Liang, Y.; Wu, Y.; Feng, D.; Tsai, S.-T.; Son, H.-J.; Li, G.; Yu, L., Development of new semiconducting polymers for high performance solar cells. *J. Am. Chem. Soc.* **2009**, *131* (1), 56–57.
16. Padinger, F.; Rittberger, R. S.; Sariciftci, N. S., Effects of postproduction treatment on plastic solar cells. *Adv. Funct. Mater.* **2003**, *13* (1), 85–88.
17. Yu, G.; Gao, J.; Hummelen, J. C.; Wudl, F.; Heeger, A. J., Polymer photovoltaic cells: Enhanced efficiencies via a network of internal donor-acceptor heterojunctions. *Science* **1995**, *270* (5243), 1789–1791.
18. Nguyen, L. H.; Hoppe, H.; Erb, T.; Günes, S.; Gobsch, G.; Sariciftci, N. S., Effects of annealing on the nanomorphology and performance of poly(alkylthiophene): Fullerene bulk-heterojunction solar cells. *Adv. Funct. Mater.* **2007**, *17* (7), 1071–1078.
19. Moulé, A. J.; Meerholz, K., Controlling morphology in polymer-fullerene mixtures. *Adv. Mater.* **2008**, *20* (2), 240–245.
20. Ma, W.; Yang, C.; Gong, X.; Lee, K.; Heeger, A. J., Thermally stable, efficient polymer solar cells with nanoscale control of the interpenetrating network morphology. *Adv. Funct. Mater.* **2005**, *15* (10), 1617–1622.
21. Yao, Y.; Hou, J.; Xu, Z.; Li, G.; Yang, Y., Effects of solvent mixtures on the nanoscale phase separation in polymer solar cells. *Adv. Funct. Mater.* **2008**, *18* (12), 1783–1789.
22. Peet, J.; Kim, J. Y.; Coates, N. E.; Ma, W. L.; Moses, D.; Heeger, A. J.; Bazan, G. C., Efficiency enhancement in low-bandgap polymer solar cells by processing with alkane dithiols. *Nat. Mater.* **2007**, *6* (7), 497–500.
23. Thompson, B. C.; Kim, Y.-G.; Reynolds, J. R., Spectral broadening in MEH-PPV: PCBM-based photovoltaic devices via blending with a narrow band gap cyanovinylene–dioxythiophene polymer. *Macromolecules* **2005**, *38* (13), 5359–5362.
24. Mühlbacher, D.; Scharber, M.; Morana, M.; Zhu, Z.; Waller, D.; Gaudiana, R.; Brabec, C., High photovoltaic performance of a low-bandgap polymer. *Adv. Mater.* **2006**, *18* (21), 2884–2889.
25. Lee, J. K.; Ma, W. L.; Brabec, C. J.; Yuen, J.; Moon, J. S.; Kim, J. Y.; Lee, K.; Bazan, G. C.; Heeger, A. J., Processing additives for improved efficiency from bulk heterojunction solar cells. *J. Am. Chem. Soc.* **2008**, *130* (11), 3619–3623.
26. Hoven, C. V.; Dang, X.-D.; Coffin, R. C.; Peet, J.; Nguyen, T.-Q.; Bazan, G. C., Improved performance of polymer bulk heterojunction solar cells through the reduction of phase separation via solvent additives. *Adv. Mater.* **2010**, *22* (8), E63–E66.

27. Tan, Z. A.; Zhou, E.; Zhan, X.; Wang, X.; Li, Y.; Barlow, S.; Marder, S. R., Efficient all-polymer solar cells based on blend of tris(thienylenevinylene)-substituted polythiophene and poly[perylene diimide-alt-bis(dithienothiophene)]. *Appl. Phys. Lett.* **2008**, *93* (7), 073309.
28. Hou, J.; Chen, H.-Y.; Zhang, S.; Li, G.; Yang, Y., Synthesis, characterization, and photovoltaic properties of a low band gap polymer based on silole-containing polythiophenes and 2,1,3-benzothiadiazole. *J. Am. Chem. Soc.* **2008**, *130* (48), 16144–16145.
29. Baek, N. S.; Hau, S. K.; Yip, H.-L.; Acton, O.; Chen, K.-S.; Jen, A. K. Y., High performance amorphous metallated π-conjugated polymers for field-effect transistors and polymer solar cells. *Chem. Mater.* **2008**, *20* (18), 5734–5736.
30. Wong, W.-Y.; Wang, X.-Z.; He, Z.; Djurisic, A. B.; Yip, C.-T.; Cheung, K.-Y.; Wang, H.; Mak, C. S. K.; Chan, W.-K., Metallated conjugated polymers as a new avenue towards high-efficiency polymer solar cells. *Nat. Mater.* **2007**, *6* (7), 521–527.
31. Gilot, J.; Wienk, M. M.; Janssen, R. A. J., On the efficiency of polymer solar cells. *Nat. Mater.* **2007**, *6* (10), 704–705.
32. Chen, C.-P.; Chan, S.-H.; Chao, T.-C.; Ting, C.; Ko, B.-T., Low-bandgap poly(thiophene-phenylene-thiophene) derivatives with broaden absorption spectra for use in high-performance bulk-heterojunction polymer solar cells. *J. Am. Chem. Soc.* **2008**, *130* (38), 12828–12833.
33. Liang, Y.; Xu, Z.; Xia, J.; Tsai, S.-T.; Wu, Y.; Li, G.; Ray, C.; Yu, L., For the bright future—Bulk heterojunction polymer solar cells with power conversion efficiency of 7.4%. *Adv. Mater.* **2011**, *22* (20), E135–E138.
34. Bijleveld, J. C.; Zoombelt, A. P.; Mathijssen, S. G. J.; Wienk, M. M.; Turbiez, M.; de Leeuw, D. M.; Janssen, R. A. J., Poly(diketopyrrolopyrrole–terthiophene) for ambipolar logic and photovoltaics. *J. Am. Chem. Soc.* **2009**, *131* (46), 16616–16617.
35. Veldman, D.; Meskers, S. C. J.; Janssen, R. A. J., The energy of charge-transfer states in electron donor-acceptor blends: Insight into the energy losses in organic solar cells. *Adv. Funct. Mater.* **2009**, *19* (12), 1939–1948.
36. Wang, E.; Ma, Z.; Zhang, Z.; Vandewal, K.; Henriksson, P.; Inganäs, O.; Zhang, F.; Andersson, M. R., An easily accessible isoindigo-based polymer for high-performance polymer solar cells. *J. Am. Chem. Soc.* **2011**, *133* (36), 14244–14247.
37. Ma, Z.; Sun, W.; Himmelberger, S.; Vandewal, K.; Tang, Z.; Bergqvist, J.; Salleo, A.; Andreasen, J. W.; Inganas, O.; Andersson, M. R.; Muller, C.; Zhang, F.; Wang, E., Structure-property relationships of oligothiophene-isoindigo polymers for efficient bulk-heterojunction solar cells. *Energy Environ. Sci.* **2014**, *7*, 361–369.
38. Chu, T. Y.; Lu, J.; Beaupre, S.; Zhang, Y.; Pouliot, J. R.; Wakim, S.; Zhou, J.; Leclerc, M.; Li, Z.; Ding, J.; Tao, Y., Bulk heterojunction solar cells using thieno[3,4-c]pyrrole-4,6-dione and dithieno[3,2-b:2',3'-d]silole copolymer with a power conversion efficiency of 7.3%. *J. Am. Chem. Soc.* **2011**, *133* (12), 4250–4253.
39. Wang, D. H.; Kim do, Y.; Choi, K. W.; Seo, J. H.; Im, S. H.; Park, J. H.; Park, O. O.; Heeger, A. J., Enhancement of donor-acceptor polymer bulk heterojunction solar cell power conversion efficiencies by addition of Au nanoparticles. *Angew. Chem. Int. Ed.* **2011**, *50* (24), 5519–5523.
40. Amb, C. M.; Chen, S.; Graham, K. R.; Subbiah, J.; Small, C. E.; So, F.; Reynolds, J. R., Dithienogermole as a fused electron donor in bulk heterojunction solar cells. *J. Am. Chem. Soc.* **2011**, *133* (26), 10062–10065.
41. Cabanetos, C.; El Labban, A.; Bartelt, J. A.; Douglas, J. D.; Mateker, W. R.; Fréchet, J. M. J.; McGehee, M. D.; Beaujuge, P. M., Linear side chains in benzo[1,2-b:4,5-b'] dithiophene–Thieno[3,4-c]pyrrole-4,6-dione polymers direct self-assembly and solar cell performance. *J. Am. Chem. Soc.* **2013**, *135* (12), 4656–4659.

42. Su, M. S.; Kuo, C. Y.; Yuan, M. C.; Jeng, U. S.; Su, C. J.; Wei, K. H., Improving device efficiency of polymer/fullerene bulk heterojunction solar cells through enhanced crystallinity and reduced grain boundaries induced by solvent additives. *Adv. Mater.* **2011**, *23* (29), 3315–3319.
43. Guo, X.; Zhou, N.; Lou, S. J.; Smith, J.; Tice, D. B.; Hennek, J. W.; Ortiz, R. P.; Navarrete. J. T. L.; Li, S.; Strzalka, J.; Chen, L. X.; Chang, R. P. H.; Facchetti, A.; Marks, T. J., Polymer solar cells with enhanced fill factors. *Nat. Photon.* **2013**, *7* (10), 825–833.
44. Small, C. E.; Chen, S.; Subbiah, J.; Amb, C. M.; Tsang, S.-W.; Lai, T.-H.; Reynolds, J. R.; So, F., High-efficiency inverted dithienogermole-thienopyrrolodione-based polymer solar cells. *Nat. Photon.* **2012**, *6* (2), 115–120.
45. Dou, L.; Chang, W.-H.; Gao, J.; Chen, C.-C.; You, J.; Yang, Y., A selenium-substituted low-bandgap polymer with versatile photovoltaic applications. *Adv. Mater.* **2013**, *25*, 825–831.
46. You, J.; Dou, L.; Hong, Z.; Li, G.; Yang, Y., Recent trends in polymer tandem solar cells research. *Prog. Polym. Sci.* **2013**, *38* (12), 1909–1928.
47. Halls, J. J. M.; Walsh, C. A.; Greenham, N. C.; Marseglia, E. A.; Friend, R. H.; Moratti, S. C.; Holmes, A. B., Efficient photodiodes from interpenetrating polymer networks. *Nature* **1995**, *376* (6540), 498–500.
48. Granstrom, M.; Petritsch, K.; Arias, A. C.; Lux, A.; Andersson, M. R.; Friend, R. H., Laminated fabrication of polymeric photovoltaic diodes. *Nature* **1998**, *395* (6699), 257–260.
49. Holcombe, T. W.; Woo, C. H.; Kavulak, D. F. J.; Thompson, B. C.; Frechet, J. M. J., All-polymer photovoltaic devices of poly(3-(4-n-octyl)-phenylthiophene) from grignard metathesis (GRIM) polymerization. *J. Am. Chem. Soc.* **2009**, *131* (40), 14160–14161.
50. McNeill, C. R.; Abrusci, A.; Zaumseil, J.; Wilson, R.; McKiernan, M. J.; Burroughes, J. H.; Halls, J. J. M.; Greenham, N. C.; Friend, R. H., Dual electron donor/electron acceptor character of a conjugated polymer in efficient photovoltaic diodes. *Appl. Phys. Lett.* **2007**, *90* (19), 193506.
51. Zhan, X.; Tan, Z. A.; Domercq, B.; An, Z.; Zhang, X.; Barlow, S.; Li, Y.; Zhu, D.; Kippelen, B.; Marder, S. R., A high-mobility electron-transport polymer with broad absorption and its use in field-effect transistors and all-polymer solar cells. *J. Am. Chem. Soc.* **2007**, *129* (23), 7246–7247.
52. Pincus, P., Dynamics of fluctuations and spinodal decomposition in polymer blends. II. *J. Chem. Phys.* **1981**, *75* (4), 1996–2000.
53. Walheim, S.; Boltau, M.; Mlynek, J.; Krausch, G.; Steiner, U., Structure formation via polymer demixing in spin-cast films. *Macromolecules* **1997**, *30* (17), 4995–5003.
54. Kim, J.-S.; Ho, P. K. H.; Murphy, C. E.; Friend, R. H., Phase separation in polyfluorene-based conjugated polymer blends: Lateral and vertical analysis of blend spin-cast thin films. *Macromolecules* **2004**, *37* (8), 2861–2871.
55. McNeill, C. R.; Greenham, N. C., Conjugated-polymer blends for optoelectronics. *Adv. Mater.* **2009**, *21* (38–39), 3840–3850.
56. McNeill, C. R.; Westenhoff, S.; Groves, C.; Friend, R. H.; Greenham, N. C., Influence of nanoscale phase separation on the charge generation dynamics and photovoltaic performance of conjugated polymer blends: Balancing charge generation and separation. *J. Phys. Chem. C* **2007**, *111* (51), 19153–19160.
57. Mori, D.; Benten, H.; Kosaka, J.; Ohkita, H.; Ito, S.; Miyake, K., Polymer/polymer blend solar cells with 2.0% efficiency developed by thermal purification of nanoscale-phase-separated morphology. *ACS Appl. Mater. Interfaces* **2011**, *3* (8), 2924–2927.
58. Mori, D.; Benten, H.; Ohkita, H.; Ito, S.; Miyake, K., Polymer/polymer blend solar cells improved by using high-molecular-weight fluorene-based copolymer as electron acceptor. *ACS Appl. Mater. Interfaces* **2012**, *4* (7), 3325–3329.

59. Cheng, P.; Ye, L.; Zhao, X.; Hou, J.; Li, Y.; Zhan, X., Binary additives synergistically boost the efficiency of all-polymer solar cells up to 3.45%. *Energy Environ. Sci.* **2014**, *7*, 1351–1356.
60. Zhou, N.; Lin, H.; Lou, S. J.; Yu, X.; Guo, P.; Manley, E. F.; Loser, S.; Hartnett, P.; Huang, H.; Wasielewski, M. R.; Chen, L. X.; Chang, R. P. H.; Facchetti, A.; Marks, T. J., Morphology-performance relationships in high-efficiency all-polymer solar cells. *Adv. Energy Mater.* **2014**, 4:1300785. doi: 10.1002/aenm.201300785.
61. Zhou, E.; Cong, J.; Hashimoto, K.; Tajima, K., Control of miscibility and aggregation via the material design and coating process for high-performance polymer blend solar cells. *Adv. Mater.* **2013**, *25*, 6991–6996.
62. Mori, D.; Benten, H.; Okada, I.; Ohkita, H.; Ito, S., Low-bandgap donor/acceptor polymer blend solar cells with efficiency exceeding 4%. *Adv. Energy Mater.* **2014**, 4:1301006. doi: 10.1002/aenm.201301006.
63. Earmme, T.; Hwang, Y.-J.; Murari, N. M.; Subramaniyan, S.; Jenekhe, S. A., All-polymer solar cells with 3.3% efficiency based on naphthalene diimide-selenophene copolymer acceptor. *J. Am. Chem. Soc.* **2013**, *135* (40), 14960–14963.
64. Yan, H.; Collins, B. A.; Gann, E.; Wang, C.; Ade, H.; McNeill, C. R., Correlating the efficiency and nanomorphology of polymer blend solar cells utilizing resonant soft X-ray scattering. *ACS Nano* **2011**, *6* (1), 677–688.
65. Zhou, E.; Cong, J.; Wei, Q.; Tajima, K.; Yang, C.; Hashimoto, K., All-polymer solar cells from perylene diimide based copolymers: Material design and phase separation control. *Angew. Chem. Int. Ed.* **2011**, *50* (12), 2799–2803.
66. Zhou, Y.; Yan, Q.; Zheng, Y.-Q.; Wang, J.-Y.; Zhao, D.; Pei, J., New polymer acceptors for organic solar cells: The effect of regio-regularity and device configuration. *J. Mater. Chem. A* **2013**, *1* (22), 6609–6613.
67. Guo, X.; Watson, M. D., Conjugated polymers from naphthalene bisimide. *Org. Lett.* **2008**, *10* (23), 5333–5336.
68. Yan, H.; Chen, Z.; Zheng, Y.; Newman, C.; Quinn, J. R.; Dotz, F.; Kastler, M.; Facchetti, A., A high-mobility electron-transporting polymer for printed transistors. *Nature* **2009**, *457* (7230), 679–686.
69. Chen, Z.; Zheng, Y.; Yan, H.; Facchetti, A., Naphthalenedicarboximide- vs perylene-dicarboximide-based copolymers. Synthesis and semiconducting properties in bottom-gate N-channel organic transistors. *J. Am. Chem. Soc.* **2008**, *131*, 8–9.

2 Donor and Acceptor Functionalized Silsesquioxane Nanostructures for Organic-Based Photovoltaic Devices

Hemali Rathnayake and John Ferguson

CONTENTS

2.1 Introduction 19
2.2 Functionalized Polysilsesquioxanes 20
    2.2.1 Overview 20
    2.2.2 Synthesis of Polysilsesquioxanes by Sol-Gel Method 23
2.3 Functional Silsesquioxanes/Bridged Silsesquioxanes Nanocomposites 25
2.4 Functionalized Silsesquioxanes Nanostructures for Organic-Based Photovoltaic Devices 26
    2.4.1 Overview 26
    2.4.2 Perylenediimide- and Poly(3-hexylthiophene)-Functionalized Silsesquioxane Nanoparticles: Synthesis, Particle Morphology, and Photovoltaic Performance 27
        2.4.2.1 Synthesis and Particle Morphology 28
        2.4.2.2 Photovoltaic Performance 31
2.5 Summary 33
References 33

2.1 INTRODUCTION

Organic semiconductors functionalized nanostructures play a major role in optoelectronic device applications. Reliable synthesis of such nanostructures with narrow average size distributions or specific aspect ratios provides opportunities to study size-dependent optical and electrical properties. Much advancement has been made investigating novel functionalized nanostructures for electronic device applications.[1–12]

Recent research has provided many examples of ligand-functionalized hybrid nanostructures developed initially from planar metal and metal oxide surface monolayers.[1,2,7,8,10] Some of these nanostructures are now commercially available, and significant advances have been made toward their use in light-emitting displays, photovoltaic devices,[13–16] biodiagnostics,[17–18] and as probes of interfacial phenomena.[19–22] Related ligand-functionalization chemistry also applies to carbon nanotubes,[23–26] silica nanoparticles,[27–30] magnetic nanoparticles,[31–34] clay sheets,[35–37] and numerous other nanostructures.

The hierarchical assembly of nanoscale building blocks with tunable dimensions and structural complexities is essential for controlling the distinctive geometry of 1-D structures. These particular dimensional nanomaterials are utilized for a variety of nanostructure applications, including high-strength nanocomposites, field-emitting surfaces, sensors, nanotransistors, biomaterial delivery tools, optical devices, electrode materials, energy storage devices, and catalysts.[38–47] Many 1-D nanomaterials, such as cerium oxide nanotubes, zinc oxide nanotubes, gold nanorods, carbon nanotubes (CNT), and silicon nanowires, have been recently fabricated and reported.[48–52] The anisotropic nature of cylindrical polymer brushes has been applied to template 1-D nanostructures, including metal, semiconductors, or magnetic nanowires.[53–58] Great efforts have been made to develop techniques to produce unique 1-D structures, such as laser-assisted catalytic growth,[59] vapor–liquid–solid,[60] solution–liquid–solution techniques,[61] template-directed synthesis,[62] and others.[63] However, achieving distinct geometries at the extremely small size of 1-D structures remains a great challenge in this field.

Despite the many advances, a number of challenges lie ahead in the tailored ligand chemistry of nanoparticles and 1-D nanomaterials, including (1) expanding the scope of ligand chemistry in terms of ligand variety (ranging from small molecules to functionalized molecules and polymers), (2) exploiting new functionality on nanostructures to enable dispersion in specific polymer matrices, (3) delicate balancing over directed assembly of nanoparticles in polymer/copolymer assemblies, (4) improving the utility of both nanoparticles and 1-D nanostructures in devices by replacing conventional insulating molecules/polymers with electronically active ligands, and (5) controlling the aspect ratios of 1-D nanostructures to produce smaller feature size.

As a part of continuing development for preparing novel materials with enhanced properties, we discuss here the recent development of organic–inorganic hybrid nanostructures derived from a silsesquioxane core structure with regards to their synthesis, characterization, properties, and applications as donors and acceptors for organic-based photovoltaics.

2.2 FUNCTIONALIZED POLYSILSESQUIOXANES

2.2.1 Overview

The polysilsesquioxanes are a common class of organic–inorganic hybrid materials in which organic components are bonded to a silsesquioxane backbone to make either silsesquioxanes with the empirical formula of $R\text{-}SiO_{1.5}$ or functionalized-bridged silsesquioxanes with the general formula of $R\text{-}Si_2O_3$, in which R is an organic functional group.

The organic groups covalently attach to the trifunctional silicon groups through Si-C bonds and can be varied in length, rigidity, geometry of substitution, and functionality. Because the organic group remains an integral component of the material, the structural geometry and functionality provide an opportunity to modulate bulk properties, such as porosity, thermal stability, refractive index, optical clarity, chemical resistance, hydrophobicity, and dielectric constant of bulk materials. The fine degree of control over bulk chemical and physical properties has made these hybrid materials excellent candidates for a variety of applications ranging from optoelectronic devices[64–67] to catalyst supports,[68–69] ceramics,[70] and membrane applications.

Polyhedral oligomeric silsesquioxanes (POSS) belong to the silsesquioxane family having the empirical formula of $RSiO_{1.5}$, and R may be a hydrogen atom or an organic functional group, for example, an alkyl, alkylene, acrylate, hydroxyl, or epoxide unit.[71–74] The most common POSS compounds are cage structures consisting of a silica cage core and eight organic functional groups attached to the corners of the cage (Figure 2.1). Depending on the number of organic functional groups, these molecular silica can be divided into monofunctional POSS and multifunctional POSS (Figure 2.1).[71] When all the organic groups are nonreactive, they are referred to as molecular silica. POSS nanostructures have diameters in the range of 1–3 nm and, hence, may be considered as the smallest existing silica particles.[71]

Over the past decade, POSS have been used as molecular building blocks for many applications, especially for electronic device applications.[75–79] These hybrid structures have been incorporated into a variety of materials to enhance their properties, such as thermal stability, optical transparency, toughness, and solubility. Many of these properties are critical for applications in electronic devices, especially for organic-based optoelectronics. Significant advances have been made in incorporating POSS cages into organic light emitting materials.[77–81] Conjugated polymers functionalized with silsesquioxane cages have shown improved thermal and color stability, higher brightness, and improved quantum efficiencies compared to polymers alone.[77] Moreover, tailoring silsesquioxane cages onto light-emitting conjugated polymers and macromolecules minimizes energy trap and aggregate formation while improving the lifetime of the devices.[77] It is evidenced that incorporating light-emitting

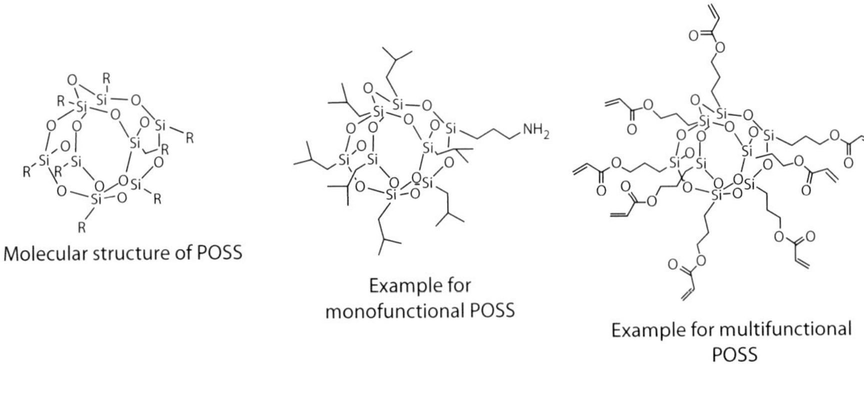

Molecular structure of POSS

Example for monofunctional POSS

Example for multifunctional POSS

FIGURE 2.1 Representative molecular structures of polyhedral oligomeric silsesquioxanes (POSS).

units into a ladder polysiloxane backbone reduces the electron delocalization of conjugated polymer and thus suppress the formation of aggregation.[80–82]

On the other hand, bridged silsesquioxanes are silsesquioxane networks formed by bridged polysilsesquioxanes, which contain a very high content of organic groups in networks with a wide range of desirable physical properties. These networks are able to demonstrate the full potential of the organic groups utilized with the structural stability, toughness, and tensile strength greater than either an organic or inorganic component alone.[83] The organic group chosen to be incorporated into the silsesquioxane network can retain many of its chemical properties inside of the network while also playing a significant role in the physical properties of the network.[84] Bridged polysilsesquioxanes are prepared in a single step from molecular precursors having variety of organic fragments. The organic fragments in the building blocks range from rigid arylenic,[85–90] acetylenic,[85,89–97] and olefinic[98–101] bridging groups to flexible alkylenes ranging from one to 14 methylene groups in length.[89–103] A variety of functionalized groups, such as amines,[104] ethers,[105] sulfides,[106] phosphines, amides, ureas, carbamates, and carbonates, also include bridging groups.[107–112] The chemical structures of a selected few examples of such bridging silane precursors are shown in Figure 2.2.

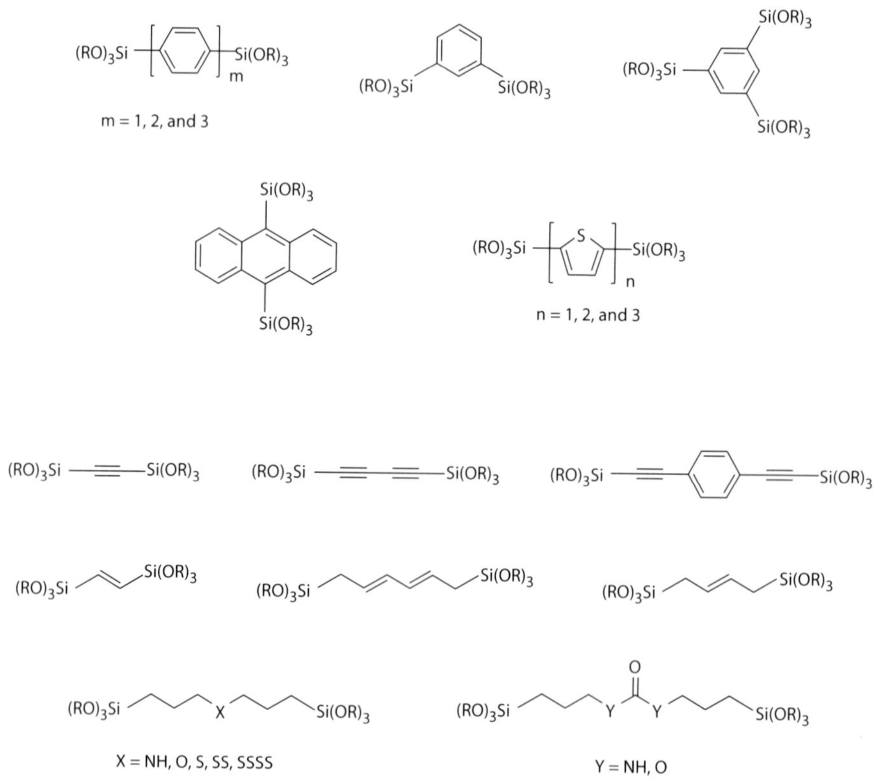

FIGURE 2.2 Chemical structures of bridging silane precursors with a variety of bridging groups.

2.2.2 Synthesis of Polysilsesquioxanes by Sol-Gel Method

The common method of preparing silsesquioxanes is sol-gel polymerization of organosilane monomers containing two or more trichlorosilyl or trialkoxysilyl groups. Sol-gel polymerization of unmodified tetraalkoxysilanes forms an amorphous silica gel in the solution when catalyzed with either an acid or a base.[83] The alkoxy groups in the presence of water and an acid or base undergo hydrolysis to form silanol groups and release a corresponding alcohol into the solution.[83] Without any change to reaction conditions, condensation occurs between silanol groups and alkoxy groups to form the silica network with by-products of water and alcohol, respectively.[83] As these silane bonds form, the silsesquioxane network forms as a gel in solution until all monomers are polymerized. The choice of solvents, catalysts, and processing conditions of this sol-gel polymerization influence the physical properties of the resulting silica network.[83–85]

As depicted in Scheme 2.1, bridged polysilsesquioxanes undergo sol-gel polymerization in the same fashion; the bridging organic group does not form bonds during polymerization as it is already incorporated into the silane network by its bonds at the monomer level.[83–85] The length and flexibility of the bridging organic group also influences the physical properties of tensile strength and porosity of the resulting silsesquioxane network.[83,84]

The sol-gel process of creating the silsesquioxane network in solution results in the creation of many different types of materials depending upon the processing techniques used to extract the network from the solution.[83] For example, if the solution is centrifuged or filtered before complete gelation has occurred, either due to a lack of time or necessary concentration of one or more reagents, silsesquioxane nanoparticles can be collected from the precipitate with their size and morphology controlled by the reaction conditions.[113] Partially formed silsesquioxane networks in the solution can also be coated onto surfaces to form films as the solvent evaporates or removed from the solution as fibers.[84]

After gelation has occurred, the method by which these gels are processed differentiates between two different types of resulting materials. Allowing the gels to air dry, possibly after washing with another solvent or introducing additives to slow the drying process, will result in xerogels.[114,115] Varying with the degree of porosity and the drying conditions, silsesquioxane networks will undergo 80%–95% shrinkage as they dry to form xerogels.[84] This dramatic change in the volume of the gels leads to a reduction in their porosities and can cause structural cracks if additives are not used. Crushing and washing the gel with water before drying under vacuum at high temperatures can create xerogel powders with comparable porosities to monolithic xerogels.[83]

Aerogels created from the gelled solution require more careful processing techniques but more closely resemble the original structure of the gelled solution. By adjusting the temperature and pressure, the gel can be dried at the supercritical point of the solvent contained within its networks. Because of safety concerns of working with supercritical alcohols, the solvent is often replaced with supercritical carbon dioxide.[83] Aerogels undergo less shrinkage than xerogels during the drying process and retain much higher porosities as a direct result.[83]

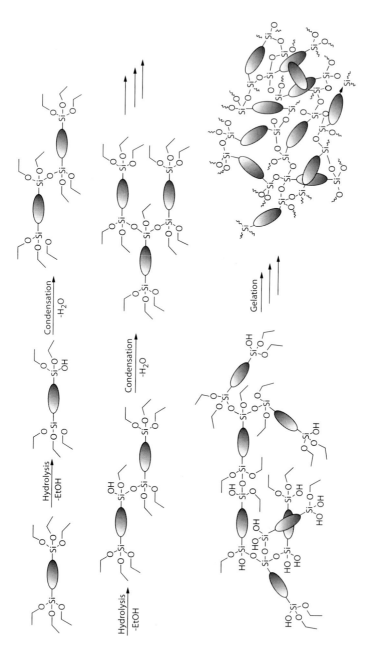

SCHEME 2.1 Schematic representation for the formation of bridged polysilsesquioxanes bulk materials.

Although a wide range of ormosil materials has been prepared in bulk form,[113,116–118] the use of the sol-gel process as a route to make silsesquioxane nanoparticles and nanocomposites has been limited until quite recently.

2.3 FUNCTIONAL SILSESQUIOXANES/BRIDGED SILSESQUIOXANES NANOCOMPOSITES

Recent research on organosilica hybrids, ormosils, or silsesquioxanes and bridged silsesquioxanes has provided examples of tailoring various types of organic moieties onto particles for a wide variety of applications. There are three known methods to produce functionalized silicon-based particles: (1) surface functionalized silica particles by the Stöber method, (2) organically modified silica particles (ormosils) by cocondensation of tetraethoxysilane (TEOS) and organotrialkoxysilane precursor, and (3) functionalized silsesquioxane particles by base-catalyzed direct hydrolysis and condensation of organotrialkoxysilanes. The schematic diagram for the preparation of these three types of silica-based particles is shown in Figure 2.3.

The first approach involved surface functionalization of organic moieties onto *presynthesized* Stöber silica particles, which was originally developed by Stöber and coworkers.[119] In the Stöber method, monodisperse silica particles were prepared from tetraalkoxysilane precursors, such as TEOS, in a mixture of water, ammonia, and alcohol. However, the capability of chemically tailoring the surface of silica nanoparticles made from the Stöber method is limited due to lesser nanoparticle surface coverage with ligands (less residual silanol groups) as well as by difficulties in using some specific ligands. The second approach involves the Stöber synthesis of ormosil particles starting from mixtures of TEOS and organotrialkoxysilane precursors with the general formula RSi(OR')$_3$, where R = methyl, phenyl, octyl, mercaptopropyl, aminopropyl, and R' = methyl or ethyl.[120–125] However, it has been also shown that this method gives less ligand density, which is less than 20%. The

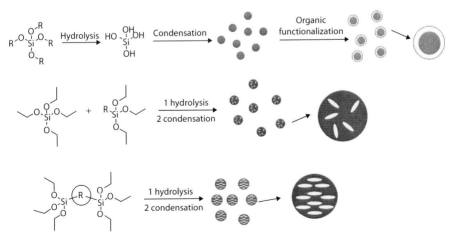

FIGURE 2.3 Schematic diagram for the preparation of functionalized silica-based particles from three known methods.

direct hydrolysis and condensation of organotrialkoxy silanes is found to be the most effective and efficient method to make functionalized silica-based particles that gives higher load of organic groups with silsesquioxane core structure. This method also provides more organic character to the inorganic core with retained three-dimensional architecture in solution.

As described in Section 2.2.2, the well-known method to make functionalized silsesquioxanes is the hydrolytic sol-gel route, which involves base- or acid-catalyzed hydrolysis and condensation reactions of monomeric alkoxy silane precursors in aqueous solvent systems. In addition to the hydrolytic sol-gel route, some other approaches have been developed in recent years by a few other research groups to produce organosilicas, including a modified Stöber sol-gel route, a nonhydrolytic sol-gel route, and miniemulsion polymerization.[113–118] Recently Noda et al. introduced miniemulsion polymerization in the presence of a nonionic emulsifier to synthesize spherical methylsilsesquioxane particles with an average diameter of 0.2–2.0 μm.[125,126] Later, this method was modified to yield a spherical organosilica network with controlled particle size from 3–15 nm, using benzethonium chloride emulsifier in aqueous sodium hydroxide solution.[127] To date, a range of organosilicas with the general formula [R_2SiO_3]n, (R is methyl, phenyl, ethyl, octyl, mercaptopropyl, vinyl, acrylic, aminopropyl, and isocyanate) has been synthesized using the size-controlled emulsion polymerization method and/or modified Stöber sol-gel routes.[118–128] To extend the method available for synthesis of silsesquioxane nanoparticles, emulsion polymerization offers an additional attractive option. Arkhireeva and Hay reported the synthesis of methyl, ethyl, phenyl, and vinyl silsesquioxane nanoparticles with an average diameter of 3.5–320 nm, which were prepared in aqueous media using a cationic and nonionic surfactant as the emulsifier and sodium hydroxide as the catalyst at room temperature.[129] However, there is less effort devoted to make organic semiconductors-functionalized silsesquioxane nanostructures that have potential applicability in organic-based photovoltaic devices.[130,131] The recent developments in the preparation of such silsesquioxane nanostructures are discussed in Section 2.4.

2.4 FUNCTIONALIZED SILSESQUIOXANES NANOSTRUCTURES FOR ORGANIC-BASED PHOTOVOLTAIC DEVICES

2.4.1 Overview

Conjugated polymers functionalized with silsesquioxane POSS cages have shown improved thermal and color stability, higher brightness, and improved quantum efficiencies compared to polymers alone.[79] Indeed, only a few reports describe surface functionalized electronically active silica nanoparticles[132] and donor and acceptor functionalized silsesquioxane nanoparticles[130,131] for photovoltaic applications. Senkovskyy et al. recently introduced a novel surface-initiated grafting method to graft regioregular poly(3-hexylthiophene), P3HT, from surface-modified silica particles and evaluated the resulting photovoltaic properties by blending with a derivative of fullerene C_{60} (PCBM).[132] These *nano*-P3HT particles in combination with PCBM showed a reasonably good photovoltaic performance with overall power conversion efficiency of 1.8%–2.3%.[132] Rathnayake et al. developed a modified Stöber method to prepare perylenediimide (PDI) and P3HT-functionalized silsesquioxane nanoparticles and evaluated the photovoltaic performance for organic-based solar cells.[130,131] The method developed to make these dye-functionalized silsesquioxane nanoparticles has been shown to be very useful in incorporating higher content of organic dyes into the silsesquioxane network to obtain highly fluorescent acceptor functionalized nanoparticles. This approach is also clearly not limited to specific ligands but could be applied broadly across any appropriately functionalized silane precursor to make hybrid nanostructures of various sizes and compositions. Section 2.4.2 describes a detailed discussion of preparing PDI and P3HT functionalized silsesquioxane and their applicability as a light-absorbing layer for photovoltaic devices.

2.4.2 Perylenediimide- and Poly(3-hexylthiophene)-Functionalized Silsesquioxane Nanoparticles: Synthesis, Particle Morphology, and Photovoltaic Performance

Poly(3-hexylthiophene) (P3HT) has been the most used donor material in polymer-based OPVs along with either a fullerene derivative (PCBM) or a perylenediimide derivative (PDI) as an acceptor.[133,134] Power conversion efficiency up to ~10% has been achieved for P3HT/PCBM solar cells,[133] making them a prime candidate for future low-cost solar power generation. However, bulk heterojunction solar cells are strongly dependent on the polymer morphology and their molecular interactions with the substrates and other components in the blends.[135–138] Linear conjugated polymers (LCPs) exhibit very complex self-assembly behavior due to their structural flexibility, longer chain length, and wide molecular weight distribution. It is essential to develop LCPs having both improved optoelectronic properties and organizable self-assembly properties. To improve the progress of organic-based devices, synthetic methods need to be developed to make well-defined three-dimensional structures with a controlled size, shape, and delicately organized self-assembly properties.

Significant research efforts have been focused on incorporating LCPs onto a variety of nanoparticle surfaces to enhance not only their optical responses, but also to improve guidable self-assembly properties.[132,136] Among such nanostructures, functionalized silicas, silsesquioxanes, and ormosils have provided rich platforms for the covalent attachment of polymers, macromolecules, and dendrimers for a variety of applications in microelectronics.[75,76,79] For example, these organic/inorganic hybrid structures have been incorporated into organic light-emitting materials to improve thermal and color stabilities, brightness, and quantum efficiencies.

The progress of the synthetic efforts for OPV has been placed on the development of novel donor materials. However, progress with alternative acceptor materials has not been as visible as for donor polymers. So far, acceptors, such as conjugated polymers, carbon nanotubes, perylenes, and inorganic semiconducting nanoparticles, have not satisfied expectations compared to PCBM and related fullerenes.[133,134] For example, solar cells comprised of blends of perylenediimide derivatives (PDIs) and P3HT exhibit power conversion efficiencies below 0.2% due to the formation of micrometer-sized PDI crystals upon annealing. The micrometer-sized crystals act as barriers to separate holes and electrons generated at the donor–acceptor interface resulting low power conversion efficiencies.[134]

In recent years, substituted perylenediimides (PDIs) have been studied extensively as a major class of organic semiconductors, which serve as both acceptors and n-type semiconductors.[139–142] Functionalized PDI derivatives show unique optoelectronic properties due to the formation of supramolecular architectures via noncovalent intermolecular interactions between the extended π-conjugation.[143,144] Novel PDI derivative-substituted donor systems synthesized for efficient energy and charge transfer studies have been reported for oligo(p-phenylenevinylene)s,[145–148] oligo(p-phenylene)s,[149] oligothiophenes,[150–154] oligopyrroles,[155] oligofluorenes,[156–158] porphyrin,[159–161] and phthalocyanine.[162] The inherent optoelectronic characteristics and high thermal and photostability of PDI derivatives make them promising alternatives to PCBM. The creation of supramolecular architecture with well-defined shape and size would be an elegant approach to avoid the formation of microcrystalline domains upon annealing. Therefore, tailoring PDI derivatives to a silsesquioxane networks would be a way to improve power conversion efficiency of the P3HT/PDI donor/acceptor system. However, a greater understanding of structure–property relationships is vital to reaching this goal.

2.4.2.1 Synthesis and Particle Morphology

In the typical procedure introduced by Rathnayake et al.,[130] PDI-functionalized bridged silsesquioxane nanoparticles were prepared using base-catalyzed condensation of PDI-triethoxysilane precursor either presence of catalytic amount of tetraethoxysilane (TEOS) or without TEOS (Scheme 2.2).

Although the particle size was controlled by the molar ratios of silane precursor to NH$_4$OH to TEOS, slow drop feeding the monomer into the reaction and the addition of TEOS controlled the particle regularity. The reaction conditions were evaluated to produce a best-suited synthetic procedure for the production of a wide size range of nanoparticles with narrow size distribution. These nanoparticles are soluble in

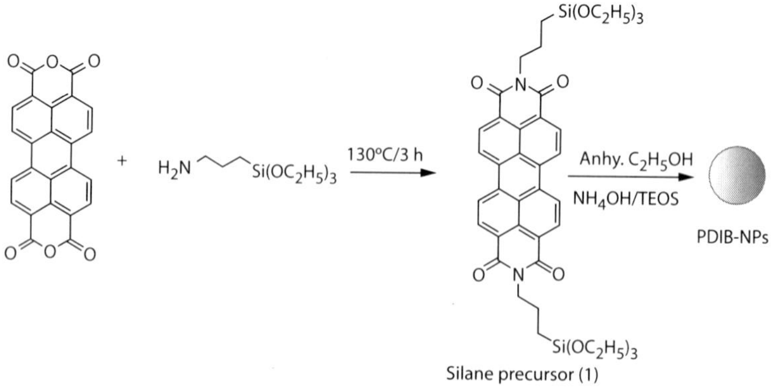

SCHEME 2.2 Synthesis scheme for the preparation of PDIB-NPs. (From Rathnayake, H. et al., *Nanoscale*, 4, 4631–4640, 2012. Reproduced by permission of The Royal Society of Chemistry.)

most organic solvents to give homogeneous solutions from clear to opaque, depending upon the particle size. In this manner, the particle size was controlled ranging from 30 to 200 nm. The size and the shape of the nanoparticles were visualized using transmission electron microscopy (TEM). The morphology of these particles is shown in Figure 2.4 and further confirmed the successful preparation of these novel electronically active nanoparticles.

It has been shown that the modified Stöber method developed here is not limited to specific ligands. The range of materials achievable in this fashion is extremely diverse due to the capability of incorporating a wide range of ligand compositions, from reactive functional groups to fluorescent molecules to semiconducting polymers. The same research group later adapted this modified Stöber method to make P3HT-functionalized siloxane nanoparticles and evaluated their photovoltaic parameters as a promising donor nanomaterial for OPVs.[131]

The application of this modified Stöber method to prepare P3HT-functionalized siloxane nanoparticles is depicted in Scheme 2.3. The silane precursor of P3HT, 3 was prepared via subsequent synthetic steps starting from 2,5-dibromo-3-hexylthiophene. In a typical procedure, P3HT-NPs were prepared by condensation of P3HT-silane 3 with ammonium hydroxide solution in the presence of 20%–40% tetraethoxysilane (TEOS) in a solvent mixture of anhydrous ethanol and THF (4:1).

The reaction conditions and the effect of base concentration on the particle size distribution were reported. The particle size was controlled by the amount of base concentration, and it has shown that the smaller-size particles were yielded at a higher NH_4OH concentration. As the amount of TEOS used in this method was in the range of 20%–40%, which is higher than the usual catalytic amount of TEOS, the resulting silica core is a siloxane network. The size and shape of these functionalized hybrids are shown in Figure 2.5. According to TGA analysis, the percent weight of the polymer fraction attached to the siloxane core is in the range of 55% to 70%, which also confirms the incorporation of both silicas (SiO_2) and silsesquioxanes ($SiO_{1.5}$) into the core structure. The degree of condensation in typical base-catalyzed polysilsesquioxanes is in the range of 80%–90%. However, the ligand density of functionalized siloxanes is still higher over the conventional methods, such as ligand functionalization from surface modified silica and organically modified silica particles.

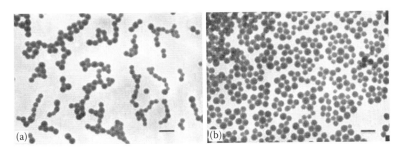

FIGURE 2.4 Transmission electron microscopy images of PDIB-NPs in two different sizes; (a) ~40 nm and (b) ~60 nm (scale bars: 100 nm). (From Rathnayake, H. et al., *Nanoscale*, 4, 4631–4640, 2012. Reproduced by permission of The Royal Society of Chemistry.)

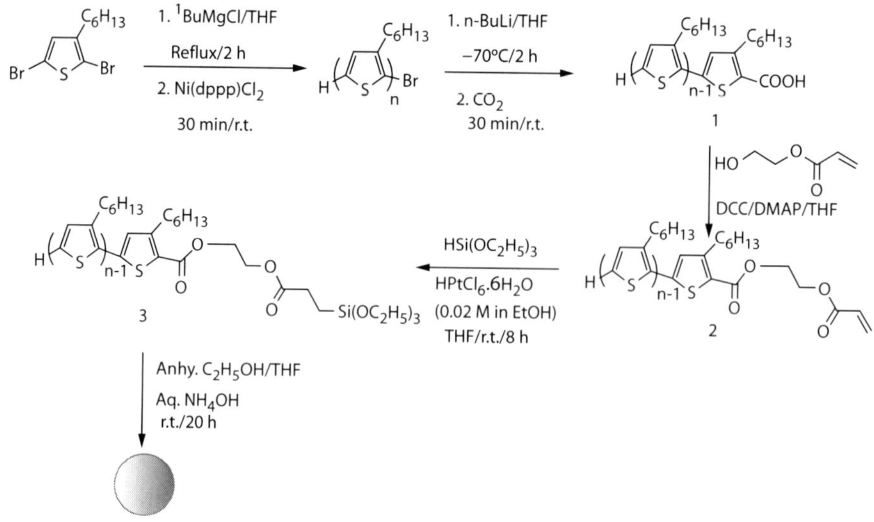

SCHEME 2.3 Preparation of P3HT-SSQ nanoparticles by the direct hydrolysis and condensation method. (From Rathnayake, H., *Nanoscale*, 5, 3212–3215, 2013. Reproduced by permission of The Royal Society of Chemistry.)

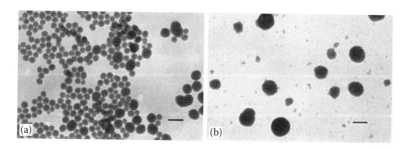

FIGURE 2.5 Transmission electron microscopy images of P3HT-SSQ NPs (sizes: [a] ~55 nm and [b] 120 nm; scale bars: 100 nm in both images). (From Rathnayake, H., *Nanoscale*, 5, 3212–3215, 2013. Reproduced by permission of The Royal Society of Chemistry.)

The direct hydrolysis and condensation process described above is clearly not limited to specific ligands but could be applied broadly across any appropriately functionalized silane precursor to make hybrid nanostructures of various sizes, shapes, and compositions. As shown in Figure 2.6, different morphologies of functionalized silsesquioxane nanostructures would be feasible to produce through this method. For example, the hydrolysis and condensation of organomonoethoxy silanes with

Donor and Acceptor Functionalized Silsesquioxane Nanostructures

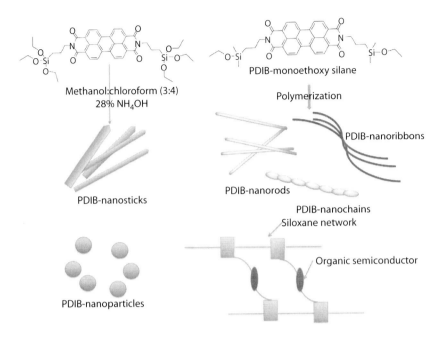

FIGURE 2.6 Morphologies of a variety of PDI-functionalized nanostructures prepared from the modified Stöber method.

different base concentrations yields functionalized nanostructures varied from nanoribbons to nanorods to nanochains.

2.4.2.2 Photovoltaic Performance

The photovoltaic performances of PDI-silsesquioxane nanoparticles and P3HT-siloxane nanoparticles for the bulk-heterojunction solar cells were reported with configuration of ITO/PEDOT:PSS/Active layer/Ca (or LiF)/Al. The various device parameters, including annealing conditions and compositions of the active layer blends, were optimized (Figure 2.7). The reported photovoltaic parameters for the devices made with PDI-silsesquioxane nanoparticles as an acceptor are a V_{oc} of 0.53 V, J_{sc} of 6.88 mA/cm^2, and a fill factor (FF) of 0.42 ± 0.02, when the devices were annealed at 150°C. The current density of P3HT/PDIB-NPs blends was improved fivefold with a power conversion efficiency of 1.56% ± 0.06%, which represents a nearly threefold improvement in performance over previously reported P3HT-block-PDI copolymer solar cells.[134] These results demonstrate the beneficial effects on device performance by having nanosized crystalline particles of PDI with a silsesquioxane structure. In the case of P3HT-siloxane nanoparticles, the power conversion efficiency of 1.58% with a V_{oc} of 0.55 V, J_{sc} of 9.61 mA/cm^2, and a FF of 0.30 ± 0.02 at room temperature was reported (Figure 2.7b). These results demonstrate that the siloxane core structure at least does not affect the optoelectronic properties of P3HTs.

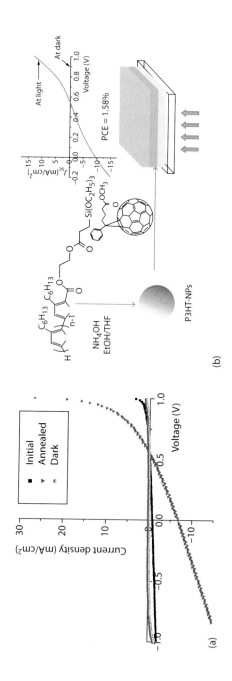

FIGURE 2.7 Current density/voltage curves for PDI-nanoparticles as an acceptor (a) and P3HT-siloxane nanoparticles as a donor (b). (From Rathnayake, H. et al., *Nanoscale*, 5, 3212–3215, 2013; Rathnayake, H. et al., *Nanoscale*, 4, 4631–4640, 2012. Reproduced by permission of The Royal Society of Chemistry.)

2.5 SUMMARY

Hybrid organic–inorganic nanostructures derived from polysilsesquioxanes continue to show a wide range of desirable chemical, physical, and optoelectronic properties. These properties are an integral part of the inorganic network, which allows novel morphologies, such as spherical nanoparticles, nanoribbons, nanorods, and nanochains. The base- or acid-catalyzed hydrolysis and condensation of organoalkoxysilanes through either sol-gel polymerization or the modified Stöber sol-gel method enables the formation of a silsesquioxane network with variety of morphologies. The recent advancement on incorporating organic semiconductors to silsesquioxane networks and their applicability as promising active layer materials for organic solar cells suggests that these hybrid nanostructures will continue to expand the horizons of materials science.

REFERENCES

1. Calhoun, M. F.; Sanchez, J.; Olaya, D.; Gershenson, M. E.; Podzorov, V. Electronic Functionalization of the Surface of Organic Semiconductors with Self-Assembled Monolayers. *Nat. Mater.* 2008, 7, 84–89.
2. Taratula, O.; Galoppini, E.; Wang, D.; Chu, D.; Zhang, Z.; Chen, H.; Saraf, G.; Lu, Y. Binding Studies of Molecular Linkers to ZnO and MgZnO Nanotip Films. *J. Phys. Chem. B* 2006, 110, 6506–6515.
3. Taratula, O.; Galoppini, E.; Mendelsohn, R.; Reyes, P. I.; Zhang, Z.; Duan, Z.; Zhong, J.; Lu, Y. Stepwise Functionalization of ZnO Nanotips with DNA. *Langmuir* 2009, 25, 2107–2113.
4. Kimling, J.; Maier, M.; Okenve, B.; Kotaidis, V.; Ballot, H.; Plech, A. Turkevich Method for Gold Nanoparticle Synthesis Revisited. *J. Phys. Chem. B* 2006, 110, 15700–15707.
5. Brust, M.; Walker, M.; Bethell, D.; Schiffrin, D. J.; Whyman, R. Synthesis of Thiol-Derivatised Gold Nanoparticles in a Two-Phase Liquid–Liquid System. *J. Chem. Soc., Chem. Commun.* 1994, 801–802.
6. Nam, J. M.; Thaxton, C. S.; Mirkin, C. A. Nanoparticle-Based Bio-Bar Codes for the Ultrasensitive Detection of Proteins. *Science* 2003, 301, 1884–1886.
7. Goodman, C. M.; McCusker, C. D.; Yilmaz, T.; Rotello, V. M. Toxicity of Gold Nanoparticles Functionalized with Cationic and Anionic Side Chains. *Bioconjug. Chem.* 2004, 15, 897–900.
8. Fan, C.; Wang, S.; Hong, J. W.; Bazan, G. C.; Plaxco, K. W.; Heeger, A. J. Beyond Superquenching: Hyper-Efficient Energy Transfer from Conjugated Polymers to Gold Nanoparticles. *Proc. Natl. Acad. Sci. U.S.A.* 2003, 100, 6297–6301.
9. Oyelere, A. K.; Chen, P. C.; Huang, X.; El-Sayed, I. H.; El-Sayed, M. A. Peptide-Conjugated Gold Nanorods for Nuclear Targeting. *Bioconjug. Chem.* 2007, 18, 1490–1497.
10. Zweifel, D. A.; Wei, A. Sulfide-Arrested Growth of Gold Nanorods. *Chem. Mater.* 2005, 17, 4256–4261.
11. Tang, J.; Redl, F.; Zhu, Y.; Siegrist, T.; Brus, L. E.; Steigerwald, M. L. An Organometallic Synthesis of TiO_2 Nanoparticles. *Nano Lett.* 2005, 5, 543–548.
12. Littau, K. A.; Szajowski, P. J.; Muller, A. J.; Kortan, A. R.; Brus, L. E. A Luminescent Silicon Nanocrystal Colloid via a High-Temperature Aerosol Reaction. *J. Phys. Chem.* 1993, 97, 1224–1230.
13. Koleilat, G. I.; Levina, L.; Shukla, H.; Myrskog, S. H.; Hinds, S.; Pattantyus-Abraham, A. G.; Sargent, E. H. Efficient, Stable Infrared Photovoltaics Based on Solution-Cast Colloidal Quantum Dots. *ACS Nano* 2008, 2, 833–840.

14. Yan, X.; Cui, X.; Li, B.; Li, L. Large, Solution-Processable Graphene Quantum Dots as Light Absorbers for Photovoltaics. *Nano Lett.* 2010, 10, 1869–1873.
15. Noone, K. M.; Strein, E.; Anderson, N. C.; Wu, P.-T.; Jenekhe, S. A.; Ginger, D. S. Broadband Absorbing Bulk Heterojunction Photovoltaics Using Low-Bandgap Solution-Processed Quantum Dots. *Nano Lett.* 2010, 10, 2635–2639.
16. Arango, A. C.; Oertel, D. C.; Xu, Y.; Bawendi, M. G.; Bulović, V. Heterojunction Photovoltaics Using Printed Colloidal Quantum Dots as a Photosensitive Layer. *Nano Lett.* 2009, 9, 860–863.
17. Earhart, C.; Jana, N. R.; Erathodiyil, N.; Ying, J. Y. Synthesis of Carbohydrate-Conjugated Nanoparticles and Quantum Dots. *Langmuir* 2008, 24, 6215–6219.
18. Kim, K. S.; Hur, W.; Park, S.-J.; Hong, S. W.; Choi, J. E.; Goh, E. J.; Yoon, S. K.; Hahn, S. K. Bioimaging for Targeted Delivery of Hyaluronic Acid Derivatives to the Livers in Cirrhotic Mice Using Quantum Dots. *ACS Nano* 2010, 4, 3005–3014.
19. Vaia, R. A.; Maguire, J. F. Polymer Nanocomposites with Prescribed Morphology: Going beyond Nanoparticle-Filled Polymers. *Chem. Mater.* 2007, 19, 2736–2751.
20. Tangirala, R.; Hu, Y.; Joralemon, M.; Zhang, Q.; He, J.; Russell, T. P.; Emrick, T. Connecting Quantum Dots and Bionanoparticles in Hybrid Nanoscale Ultra-Thin Films. *Soft Matter* 2009, 5, 1048–1054.
21. Böker, A.; He, J.; Emrick, T.; Russell, T. P. Self-Assembly of Nanoparticles at Interfaces. *Soft Matter* 2007, 3, 1231–1248.
22. Balazs, A. C.; Emrick, T.; Russell, T. P. Nanoparticle Polymer Composites: Where Two Small Worlds Meet. *Science* 2006, 314, 1107–1110.
23. Erlanger, B. F.; Chen, B. X.; Zhu, M.; Brus, L. Binding of an Anti-Fullerene IgG Monoclonal Antibody to Single Wall Carbon Nanotubes. *Nano Lett.* 2001, 1, 465–467.
24. Liu, I.-C.; Huang, H. M.; Chang, C. Y.; Tsai, H.-C.; Hsu, C. H.; Tsiang, R. C. C. Preparing a Styrenic Polymer Composite Containing Well-Dispersed Carbon Nanotubes: Anionic Polymerization of a Nanotube-Bound P-Methylstyrene. *Macromolecules* 2004, 37, 283–287.
25. Liu, Y. L.; Chen, W. H. Modification of Multiwall Carbon Nanotubes with Initiators and Macroinitiators of Atom Transfer Radical Polymerization. *Macromolecules* 2007, 40, 8881–8886.
26. Banerjee, S.; Wong, S. S. Synthesis and Characterization of Carbon Nanotube-Nanocrystal Heterostructures. *Nano Lett.* 2002, 2, 195–200.
27. Ling, X. Y.; Reinhoudt, D. N.; Huskens, J. Ferrocenyl-Functionalized Silica Nanoparticles: Preparation, Characterization, and Molecular Recognition at Interfaces. *Langmuir* 2006, 22, 8777–8783.
28. Cauda, V.; Schlossbauer, A.; Kecht, J.; Zürner, A.; Bein, T. Multiple Core–Shell Functionalized Colloidal Mesoporous Silica Nanoparticles. *J. Am. Chem. Soc.* 2009, 131, 11361–11370.
29. Shin, J. H.; Metzger, S. K.; Schoenfisch, M. H. Synthesis of Nitric Oxide-Releasing Silica Nanoparticles. *J. Am. Chem. Soc.* 2007, 129, 4612–4619.
30. Zhou, Q.; Wang, S.; Fan, X.; Advincula, R.; Mays, J. Living Anionic Surface-Initiated Polymerization (LASIP) of a Polymer on Silica Nanoparticles. *Langmuir* 2002, 18, 3324–3331.
31. Kang, K.; Choi, J.; Nam, J. H.; Lee, S. C.; Kim, K. J.; Lee, S.-W.; Chang, J. H. Preparation and Characterization of Chemically Functionalized Silica-Coated Magnetic Nanoparticles as a DNA Separator. *J. Phys. Chem. B* 2009, 113, 536–543.
32. Zhou, Z.; Liu, G.; Han, D. Coating and Structural Locking of Dipolar Chains of Cobalt Nanoparticles. *ACS Nano* 2009, 3, 165–172.
33. Gao, J.; Gu, H.; Xu, B. Multifunctional Magnetic Nanoparticles: Design, Synthesis, and Biomedical Applications. *Acc. Chem. Res.* 2009, 42, 1097–1107.

34. Keng, P. Y.; Kim, B. Y.; Shim, I.-B.; Sahoo, R.; Veneman, P. E.; Armstrong, N. R.; Yoo, H.; Pemberton, J. E.; Bull, M. M.; Griebel, J. J.; Ratcliff, E. L.; Nebesny, K. G.; Pyun, J. Colloidal Polymerization of Polymer-Coated Ferromagnetic Nanoparticles into Cobalt Oxide Nanowires. *ACS Nano* 2009, 3, 3143–3157.
35. Wheeler, P. A.; Wang, J.; Baker, J.; Mathias, L. J. Synthesis and Characterization of Covalently Functionalized Laponite Clay. *Chem. Mater.* 2005, 17, 3012–3018.
36. Wheeler, P. A.; Wang, J.; Mathias, L. J. Poly(methyl methacrylate)/Laponite Nanocomposites: Exploring Covalent and Ionic Clay Modifications. *Chem. Mater.* 2006, 18, 3937–3945.
37. Tchinda, A. J.; Ngameni, E.; Kenfack, I. T.; Walcarius, A. One-Step Preparation of Thiol-Functionalized Porous Clay Heterostructures: Application to Hg(II) Binding and Characterization of Mass Transport Issues. *Chem. Mater.* 2009, 21, 4111–4121.
38. Guo, S.; Li, J.; Ren, W.; Wen, D.; Dong, S.; Wang, E. Carbon Nanotube/Silica Coaxial Nanocable as a Three-Dimensional Support for Loading Diverse Ultra-High-Density Metal Nanostructures: Facile Preparation and Use as Enhanced Materials for Electrochemical Devices and SERS. *Chem. Mater.* 2009, 21, 2247–2257.
39. Ji, Q.; Iwaura, R.; Kogiso, M.; Jung, J. H.; Yoshida, K.; Shimizu, T. Direct Sol–Gel Replication without Catalyst in an Aqueous Gel System: From a Lipid Nanotube with a Single Bilayer Wall to a Uniform Silica Hollow Cylinder with an Ultrathin Wall. *Chem. Mater.* 2004, 16, 250–254.
40. Wei, B. Q.; Vajtai, R.; Jung, Y.; Ward, J.; Zhang, R.; Ramanath, G.; Ajayan, P. M. Assembly of Highly Organized Carbon Nanotube Architectures by Chemical Vapor Deposition. *Chem. Mater.* 2003, 15, 1598–1606.
41. Satishkumar, B. C.; Doorn, S. K.; Baker, G. A.; Dattelbaum, A. M. Fluorescent Single Walled Carbon Nanotube/Silica Composite Materials. *ACS Nano* 2008, 2, 2283–2290.
42. Norman, R. S.; Stone, J. W.; Gole, A.; Murphy, C. J.; Sabo-Attwood, T. L. Targeted Photothermal Lysis of the Pathogenic Bacteria, Pseudomonas Aeruginosa, with Gold Nanorods. *Nano Lett.* 2008, 8, 302–306.
43. Sun, B.; Sirringhaus, H. Surface Tension and Fluid Flow Driven Self-Assembly of Ordered ZnO Nanorod Films for High-Performance Field Effect Transistors. *J. Am. Chem. Soc.* 2006, 128, 16231–16237.
44. Kim, S.; Kim, S. K.; Park, S. Bimetallic Gold–Silver Nanorods Produce Multiple Surface Plasmon Bands. *J. Am. Chem. Soc.* 2009, 131, 8380–8381.
45. Fuhrer, M. S.; Kim, B. M.; Dürkop, T.; Brintlinger, T. High-Mobility Nanotube Transistor Memory. *Nano Lett.* 2002, 2, 755–759.
46. Züttel, A.; Sudan, P.; Mauron, P.; Kiyobayashi, T.; Emmenegger, C.; Schlapbach, L. Hydrogen Storage in Carbon Nanostructures. *Int. J. Hydrogen Energy* 2002, 27, 203–212.
47. Huang, P. X.; Wu, F.; Zhu, B. L.; Gao, X. P.; Zhu, H. Y.; Yan, T. Y.; Huang, W. P.; Wu, S. H.; Song, D. Y. CeO$_2$ Nanorods and Gold Nanocrystals Supported on CeO$_2$ Nanorods as Catalyst. *J. Phys. Chem. B* 2005, 109, 19169–19174.
48. Tang, C. C.; Bando, Y.; Liu, B. D.; Golberg, D. Cerium Oxide Nanotubes Prepared from Cerium Hydroxide Nanotubes. *Adv. Mater.* 2005, 17, 3005–3009.
49. Wu, H. Q.; Wei, X. W.; Shao, M. W.; Gu, J. S. Synthesis of Zinc Oxide Nanorods Using Carbon Nanotubes as Templates. *J. Cryst. Growth* 2004, 265, 184–189.
50. Yu, M. F.; Files, B. S.; Arepalli, S.; Ruoff, R. S. Tensile Loading of Ropes of Single Wall Carbon Nanotubes and Their Mechanical Properties. *Phys. Rev. Lett.* 2000, 84, 5552–5555.
51. Nishioka, K.; Niidome, Y.; Yamada, S. Photochemical Reactions of Ketones to Synthesize Gold Nanorods. *Langmuir* 2007, 23, 10353–10356.
52. Hochbaum, A. I.; Gargas, D.; Hwang, Y. J.; Yang, P. Single Crystalline Mesoporous Silicon Nanowires. *Nano Lett.* 2009, 9, 3550–3554.

53. Johnson, J. C.; Choi, H.-J.; Knutsen, K. P.; Schaller, R. D.; Yang, P.; Saykally, R. J. Single Gallium Nitride Nanowire Lasers. *Nat. Mater.* 2002, 1, 106–110.
54. Wang, Z. L.; Song, J. Piezoelectric Nanogenerators Based on Zinc Oxide Nanowire Arrays. *Science* 2006, 312, 242–246.
55. Djalali, R.; Li, S. Y.; Schmidt, M. Amphipolar Core–Shell Cylindrical Brushes as Templates for the Formation of Gold Clusters and Nanowires. *Macromolecules* 2002, 35, 4282–4288.
56. Zhang, M.; Drechsler, M.; Müller, A. H. E. Template-Controlled Synthesis of Wire-Like Cadmium Sulfide Nanoparticle Assemblies within Core–Shell Cylindrical Polymer Brushes. *Chem. Mater.* 2004, 16, 537–543.
57. Zhang, M.; Estournès, C.; Bietsch, W.; Müller, A. H. E. Superparamagnetic Hybrid Nanocylinders. *Adv. Funct. Mater.* 2004, 14, 871–882.
58. Yuan, J.; Drechsler, M.; Xu, Y.; Zhang, M.; Müller, A. H. E. Cadmium Selenide Nanowires within Core–Shell Cylindrical Polymer Brushes: Synthesis, Characterization and the Double-Loading Process. *Polymer* 2008, 49, 1547–1554.
59. Duan, X.; Lieber, C. M. General Synthesis of Compound Semiconductor Nanowires. *Adv. Mater.* 2000, 12, 298–302.
60. Ma, C.; Wang, Z. L. Road Map for the Controlled Synthesis of CdSe Nanowires, Nanobelts, and Nanosaws—A Step towards Nanomanufacturing. *Adv. Mater.* 2005, 17, 2635–2639.
61. Grebinski, J. W.; Richter, K. L.; Zhang, J.; Kosel, T. H.; Kuno, M. Synthesis and Characterization of Au/Bi Core/Shell Nanocrystals: A Precursor toward II–VI Nanowires. *J. Phys. Chem. B* 2004, 108, 9745–9751.
62. Adelung, R.; Aktas, O. C.; Franc, J.; Biswas, A.; Kunz, R.; Elbahri, M.; Kanzow, J.; Schürmann, U.; Faupel, F. Strain-Controlled Growth of Nanowires within Thin-Film Cracks. *Nat. Mater.* 2004, 3, 375–379.
63. Milenkovic, S.; Hassel, A. W.; Schneider, A. Effect of the Growth Conditions on the Spatial Features of Re Nanowires Produced by Directional Solidification. *Nano Lett.* 2006, 6, 794–799.
64. Cerveau, G.; Corriu, R. J. P. Some Recent Developments of Polysilsesquioxanes Chemistry for Material Science. *Coord. Chem. Rev.* 1998, 178–180, 1051–1071.
65. Corriu, R. A New Trend in Metal-Alkoxide Chemistry: The Elaboration of Monophasic Organic–Inorganic Hybrid Materials. *Polyhedron* 1998, 17, 925–934.
66. Cerveau, G.; Corriu, R. J. P.; Framery, E. Sol–Gel Process: Influence of the Temperature on the Textural Properties of Organosilsesquioxane Materials. *J. Mater. Chem.* 2000, 10, 1617–1622.
67. Cerveau, G.; Corriu, R. J. P.; Framery, E. Influence of the Nature of the Catalyst on the Textural Properties of Organosilsesquioxane Materials. *Polyhedron* 2000, 19, 307–313.
68. Lindner, E.; Schneller, T.; Auer, F.; Mayer, H. A. Chemistry in Interphases—A New Approach to Organometallic Syntheses and Catalysis. *Angew. Chem., Int. Ed.* 1999, 38, 2154–2174.
69. Schubert, U. Catalyst Made of Organic-Inorganic Hybrid Material. *New J. Chem.* 1994, 18, 1049–1058.
70. Corriu, R. J. P. Ceramics and Nanostructures from Molecular Precursors. *Angew. Chem., Int. Ed.* 2000, 39, 1376–1398.
71. Kuo, S.-W.; Chang, F.-C. POSS Related Polymer Nanocomposites. *Prog. Polym. Sci.* 2011, 36, 1649–1696.
72. Cordes, D. B.; Lickiss, P. D.; Rataboul, F. Recent Developments in the Chemistry of Cubic Polyhedral Oligosilsesquioxanes. *Chem. Rev.* 2010, 110, 2081–2173.
73. Shea, K. J.; Loy, D. A. Bridged Polysilsesquioxanes. Molecular-Engineered Hybrid Organic–Inorganic Materials. *Chem. Mater.* 2001, 13, 3306–3319.

74. Lickiss, P. D.; Rataboul, F. Fully Condensed Polyhedral Oligosilsesquioxanes (POSS): From Synthesis to Application. In *Advances in Organometallic Chemistry*; Hill, A. F.; Fink, M. J.; Eds.; Elsevier Academic Press Inc.: San Diego, CA, 2008; Vol. 57; pp. 1–116.
75. Leu, C. M.; Chang, Y. T.; Wei, K. H. Polyimide-Side-Chain Tethered Polyhedral Oligomeric Silsesquioxane Nanocomposites for Low-Dielectric Film Applications. *Chem. Mater.* 2003, 15, 3721–3727.
76. Tegou, E.; Bellas, V.; Gogolides, E.; Argitis, P.; Eon, D.; Cartry, G.; Cardinaud, C. Polyhedral Oligomeric Silsesquioxane (POSS) Based Resists: Material Design Challenges and Lithographic Evaluation at 157 nm. *Chem. Mater.* 2004, 16, 2567–2577.
77. Vourdas, N.; Bellas, V.; Tegou, E.; Brani, O.; Constantoudis, V.; Argitis, P.; Tserepi, A.; Gogolides, E.; Eon, D.; Cartry, G.; Cardinaud, C. *Plasma Process and Polymers*; Wiley-VCH: Weinheim, Germany, 2005; p. 281.
78. Sellinger, A.; Laine, R. M. Organic-Inorganic Hybrid Light Emitting Devices (HLED). U.S. Patent 6,517,958 B1, February 11, 2003.
79. Xiao, S.; Nguyen, M.; Gong, X.; Cao, Y.; Wu, H.; Moses, D.; Heeger, A. J. Stabilization of Semiconducting Polymers with Silsesquioxane. *Adv. Funct. Mater.* 2003, 13, 25–29.
80. Arias, A. C.; MacKenzie, J. D.; McCulloch, I.; Rivnay, J.; Salleo, A. Materials and Applications for Large Area Electronics: Solution-Based Approaches. *Chem. Rev.* 2010, 110, 3–24.
81. Tipnis, R.; Bernkopf, J.; Jia, S.; Krieg, J.; Li, S.; Storch, M.; Laird, D. Large-Area Organic Photovoltaic Module—Fabrication and Performance. *Sol. Energy Mat. Sol. C* 2009, 93, 442–446.
82. Hoth, C. N.; Choulis, S. A.; Schilinsky, P.; Brabec, C. J. High Photovoltaic Performance of Inkjet Printed Polymer: Fullerene Blends. *Adv. Mater.* 2007, 19, 3973–3978.
83. Sharp, K. G. Inorganic/Organic Hybrid Materials. *Adv. Mater.* 1998, 10, 1243–1248.
84. Loy, D. A.; Shea, K. J. Bridged Polysilsesquioxanes. Highly Porous Hybrid Organic–Inorganic Materials. *Chem. Rev.* 1995, 95, 1431–1442.
85. Shea, K. J.; Loy, D. A.; Webster, O. Arylsilsesquioxane Gels and Related Materials. New Hybrids of Organic and Inorganic Networks. *J. Am. Chem. Soc.* 1992, 114, 6700–6710.
86. Shea, K. J.; Loy, D. A.; Webster, O. W. Aryl Bridged Polysilsesquioxanes: A New Class of Microporous Materials. *Polym. Mater. Sci. Eng.* 1990, 63, 281.
87. Shea, K. J.; Loy, D. A.; Webster, O. W. Aryl-Bridged Polysilsesquioxanes—New Microporous Materials. *Chem. Mater.* 1989, 1, 572–574.
88. Shea, K. J.; Webster, O.; Loy, D. A. Aryl-Bridged Polysilsesquioxanes—New Microporous Materials. *Mater. Res. Soc. Symp. Proc.* 1990, 180, 975.
89. Small, J. H.; Shea, K. J.; Loy, D. A. Arylene- and Alkylene-Bridged Polysilsesquioxanes. *J. Non-Cryst. Solids* 1993, 160, 234–246.
90. Corriu, R. J. P.; Moreau, J. J. E.; Thepot, P.; Man, M. W. C. New Mixed Organic-Inorganic Polymers: Hydrolysis and Polycondensation of Bis(trimethoxysilyl)organometallic Precursors. *Chem. Mater.* 1992, 4, 1217–1224.
91. Boury, B.; Corriu, R. J. P. Adjusting the Porosity of a Silica-Based Hybrid Material. *Adv. Mater.* 2000, 12, 989–992.
92. Boury, B.; Corriu, R. J. P.; Delord, P.; Le Strat, V. Structure of Silica-Based Organic–Inorganic Hybrid Xerogel. *J. Non-Cryst. Solids* 2000, 265, 41–50.
93. Boury, B.; Corriu, R. J. P.; Le Strat, V. Generation of Microporosity in a Hybrid Material. Access to Pillared Amorphous Organosilicate. *Chem. Mater.* 1999, 11, 2796–2803.
94. Boury, B.; Chevalier, P.; Corriu, R. J. P.; Delord, P.; Moreau, J. J. E.; Chiman, M. W. Hybrid Organic–Inorganic Xerogel Access to Meso- and Microporous Silica by Thermal and Chemical Treatment. *Chem. Mater.* 1999, 11, 281–291.
95. Boury, B.; Corriu, R. J. P.; Strat, V. L.; Delord, P. Generation of Porosity in a Hybrid Organic–Inorganic Xerogel by Chemical Treatment. *New J. Chem.* 1999, 23, 531–538.

96. Chevalier, P.; Corriu, R. J. P.; Delord, P.; Moreau, J. J. E.; Man, M. W. C. Design of Porous Silica from Hybrid Organic–Inorganic Precursors. *New J. Chem.* 1998, 22, 423–433.
97. Chevalier, P. M.; Corriu, R. J. P.; Moreau, J. J. E.; Man, M. W. C. Chemistry of Hybrid Organic–Inorganic. Access to Silica Materials through Chemical Selectivity. *J. Sol-Gel Sci. Technol.* 1997, 8, 603–607.
98. Loy, D. A.; Carpenter, J. P.; Yamanaka, S. A.; McClain, M. D.; Greaves, J.; Hobson, S.; Shea, K. J. Polymerization of Bis(triethoxysilyl)ethenes. Impact of Substitution Geometry on the Formation of Ethenylene- and Vinylidene-Bridged Polysilsesquioxanes. *Chem. Mater.* 1998, 10, 4129–4140.
99. Carpenter, J. P.; Yamanaka, S. A.; McClain, M. D.; Loy, D. A.; Greaves, J.; Shea, K. J. The Impact of Substitution Geometry on the Formation of Ethenylene-Bridged Polysilsesquioxanes. *Polym. Prepr. (Am. Chem. Soc., Div. Polym. Chem.)* 1998, 39, 589.
100. Shaltout, R. M.; Loy, D. A.; Carpenter, J. P.; Dorhout, P. K.; Shea, K. J. Polymerization of the Z and E isomers of bis-(triethoxysilyl)-2-butene. *Polym. Prepr. (Am. Chem. Soc., Div. Polym. Chem.)* 1999, 40, 906.
101. Corriu, R. J. P.; Moreau, J. J. E.; Thépot, P.; Man, M. W. C. Evidence for a Template Effect in Control of Organic-Inorganic Hybrid Solid Formation by Intermolecular Interactions. *J. Mater. Chem.* 1994, 4, 987–989.
102. Oviatt, H. W.; Shea, K. J.; Small, J. H. Alkylene-Bridged Silsesquioxane Sol-Gel Synthesis and Xerogel Characterization. Molecular Requirements for Porosity. *Chem. Mater.* 1993, 5, 943–950.
103. Loy, D. A.; Jamison, G. M.; Baugher, B. M.; Russick, E. M.; Assink, R. A.; Prabakar, S.; Shea, K. J. Alkylene-Bridged Polysilsesquioxane Aerogels: Highly Porous Hybrid Organic–Inorganic Materials. *J. Non-Cryst. Solids* 1995, 186, 44–53.
104. Li, C.; Glass, T.; Wilkes, G. L. NMR Studies of Sol-Gel Derived Hybrid Materials Based on Triethoxysilylated Diethylenetriamine and Tetramethoxysilane. *J. Inorg. Organomet. Polym.* 1999, 9, 79–106.
105. Loy, D. A.; Beach, J. V.; Baugher, B. M.; Assink, R. A.; Shea, K. J.; Tran, J.; Small, J. H. Dialkylene Carbonate-Bridged Polysilsesquioxanes. Hybrid Organic–Inorganic Sol–Gels with a Thermally Labile Bridging Group. *Chem. Mater.* 1999, 11, 3333–3341.
106. Kohjiya, S.; Ikeda, Y. Reinforcement of General-Purpose Grade Rubbers by Silica Generated In Situ. *Rubber Chem. Technol.* 2000, 73, 534–550.
107. Bezombes, J.-P.; Chuit, C.; Corriu, R. J.; Reyé, C. Preparation and Characterization of Organic-Inorganic Hybrid Materials Incorporating Diphosphino Moieties. Study of the Accessibility of the Phosphorus Atoms Included into the Material. *Can. J. Chem.* 2000, 78, 1519–1525.
108. Corriu, R. J. P.; Hoarau, C.; Mehdi, A.; Reyé, C. Study of the Accessibility of Phosphorus Centres Incorporated within Ordered Mesoporous Organic–Inorganic Hybrid Materials. *Chem. Commun.* 2000, 71–72.
109. Bezombes, J.-P.; Chuit, C.; Corriu, R. J. P.; Reyé, C. Organic–Inorganic Hybrid Materials Incorporating Phosphorus Centres. Evidence for Some Factors Controlling the Accessibility of Phosphorus Centres. *J. Mater. Chem.* 1999, 9, 1727–1734.
110. Morris, L. J.; Downs, A. J.; Greene, T. M.; McGrady, G. S.; Herrmann, W. A.; Sirsch, P.; Gropen, O.; Scherer, W. Photo-Induced Tautomerisation of methyltrioxorhenium(VII): The Intermediate in Olefinmetathesis? *Chem. Commun.* 2000, 1, 67–68.
111. Aliev, A.; Ou, D. L.; Ormsby, B.; Sullivan, A. C. Porous Silica and Polysilsesquioxane with Covalently Linked Phosphonates and Phosphonic Acids. *J. Mater. Chem.* 2000, 10, 2758–2764.
112. Loy, D. A.; Beach, J. V.; Baugher, B. M.; Assink, R. A.; Shea, K. J.; Tran, J.; Small, J. H. Dialkylenecarbonate-bridged polysilsesquioxanes. Hybrid organic sol-gels with a thermally labile bridging group. *Mater. Res. Soc. Symp. Proc. (Org./Inorg. Hybrid Mater. II)* 1999, 576, 99–104.

113. Arkhireeva, A.; Hay, J. N. Synthesis of Organically Modified Silica Particles for Use as Nanofillers in Polymer Systems. *Polym. Polym. Compos. Rapra.* 2004, 12, 101–110.
114. Brinker, C. J.; Clark, D. E.; Ulrich, D. R.; Eds. *Better Ceramics through Chemistry.* North-Hoolan: New York, 1984; pp. 79–84.
115. Gesser, H. D.; Goswami, P. C. Aerogels and Related Porous Materials. *Chem. Rev.* 1989, 89, 765–788.
116. Iwamoto, T.; Mackenzie, J. D. Hard Ormosils Prepared with Ultrasonic Irradiation. *J. Sol-Gel Sci. Technol.* 1995, 4, 141–150.
117. Ou, D. L.; Seddon, A. B. Near- and Mid-Infrared Spectroscopy of Sol–Gel Derived Ormosils: Vinyl and Phenyl Silicates. *J. Non-Cryst. Solids* 1997, 210, 187–203.
118. Prabakar, S.; Assink, R. A. Hydrolysis and Condensation Kinetics of Two Component Organically Modified Silica Sols. *J. Non-Cryst. Solids* 1997, 211, 39–48.
119. Stöber, W.; Fink, A.; Bohn, E. Controlled Growth of Monodisperse Silica Spheres in the Micron Size Range. *J. Colloid Interface Sci.* 1968, 26, 62–69.
120. Van Blaaderen, A.; Vrij, A. Synthesis and Characterization of Monodisperse Colloidal Organo-Silica Spheres. *J. Colloid Interface Sci.* 1993, 156, 1–18.
121. Hatakeyama, F.; Kanzaki, S. Synthesis of Monodispersed Spherical B-Silicon Carbide Powder by a Sol-Gel Process. *J. Am. Ceram. Soc.* 1990, 73, 2107–2110.
122. Yacoub-George, E.; Bratz, E.; Tiltscher, H. Preparation of Functionalized Polyorganosiloxane Spheres for the Immobilization of Catalytically Active Compounds. *J. Non-Cryst. Solids* 1994, 167, 9–15.
123. Silva, C. R.; Airoldi, C. Acid and Base Catalysts in the Hybrid Silica Sol–Gel Process. *J. Colloid Interface Sci.* 1997, 195, 381–387.
124. Reynolds, K. J.; Colón, L. A. Submicron Sized Organo-Silica Spheres for Capillary. *J. Liq. Chromatogr. Relat. Technol.* 2000, 23, 161–173.
125. Etienne, M.; Lebeau, B.; Walcarius, A. Organically Modified Mesoporous Silica Spheres with MCM-41 Architecture. *New J. Chem.* 2002, 26, 384–386.
126. Noda, I.; Isikawa, M.; Yamawaki, M.; Sasaki, Y. A Facile Preparation of Spherical Methylsilsesquioxane Particles by Emulsion Polymerization. *Inorg. Chim. Acta* 1997, 263, 149–152.
127. Noda, I.; Kamoto, T.; Yamada, M. Size-Controlling Synthesis of Narrowly Distributed Particles of Methylsilsesquioxane Derivatives. *Chem. Mater.* 2000, 12, 1708–1714.
128. Noda, I.; Kamoto, T.; Sasaki, Y.; Yamada, M. Topological Morphology of Methylsilsesquioxane Derivatives. *Chem. Mater.* 1999, 11, 3693–3701.
129. Arkhireeva, A.; Hay, J. N.; Oware, W. A Versatile Route to Silsesquioxane Nanoparticles from Organically Modified Silane Precursors. *J. Non-Cryst. Solids* 2005, 351, 1688–1695.
130. Rathnayake, H.; Binion, J.; McKee, A.; Scardino, D. J.; Hammer, N. I. Perylenediimide Functionalized Bridged-Siloxane Nanoparticles for Bulk Heterojunction Organic Photovoltaics. *Nanoscale* 2012, 4, 4631–4640.
131. Rathnayake, H.; Wright, N.; Patel, A.; Binion, J.; McNamara, L. E.; Scardino, D. J.; Hammer, N. I. Synthesis and Characterization of Poly(3-Hexylthiophene)-Functionalized Siloxane Nanoparticles. *Nanoscale* 2013, 5, 3212–3215.
132. Senkovskyy, V.; Tkachov, R.; Beryozkina, T.; Komber, H.; Oertel, U.; Horecha, M.; Bocharova, V.; Stamm, M.; Gevorgyan, S. A.; Krebs, F. C.; Kiriy, A. "Hairy" Poly(3-Hexylthiophene) Particles Prepared via Surface-Initiated Kumada Catalyst-Transfer Polycondensation. *J. Am. Chem. Soc.* 2009, 131, 16445–16453.
133. Lee, J. U.; Kim, Y. D.; Jo, J. W.; Kim, J. P.; Jo, W. H. Efficiency Enhancement of P3HT/PCBM Bulk Heterojunction Solar Cells by Attaching Zinc Phthalocyanine to the Chain-End of P3HT. *J. Mater. Chem.* 2011, 21, 17209–17218.
134. Rajaram, S.; Armstrong, P. B.; Kim, B. J.; Fréchet, J. M. J. Effect of Addition of a Diblock Copolymer on Blend Morphology and Performance of Poly(3-Hexylthiophene):Perylene Diimide Solar Cells. *Chem. Mater.* 2009, 21, 1775–1777.

135. Hoppe, H.; Sariciftci, N. S. Morphology of Polymer/Fullerene Bulk Heterojunction Solar Cells. *J. Mater. Chem.* 2006, 16, 45–61.
136. Thompson, B. C.; Fréchet, J. M. J. Polymer–Fullerene Composite Solar Cells. *Angew. Chem., Int. Ed.* 2008, 47, 58–77.
137. Kroon, R.; Lenes, M.; Hummelen, J. C.; Blom, P. W. M.; de Boer, B. Small Bandgap Polymers for Organic Solar Cells (Polymer Material Development in the Last 5 Years). *Polym. Rev.* 2008, 48, 531–582.
138. Kiriy, N.; Bocharova, V.; Kiriy, A.; Stamm, M.; Krebs, F. C.; Adler, H.-J. Designing Thiophene-Based Azomethine Oligomers with Tailored Properties: Self-Assembly and Charge Carrier Mobility. *Chem. Mater.* 2004, 16, 4765–4771.
139. Horowitz, G.; Kouki, F.; Spearman, P.; Fichou, D.; Nogues, C.; Pan, X.; Garnier, F. Evidence for N-Type Conduction in a Perylene Tetracarboxylic Diimide Derivative. *Adv. Mater.* 1996, 8, 242–245.
140. Conboy, J. C.; Olson, E. J. C.; Adams, D. M.; Kerimo, J.; Zaban, A.; Gregg, B. A.; Barbara, P. F. Impact of Solvent Vapor Annealing on the Morphology and Photophysics of Molecular Semiconductor Thin Films. *J. Phys. Chem. B* 1998, 102, 4516–4525.
141. Malenfant, P. R. L.; Dimitrakopoulos, C. D.; Gelorme, J. D.; Kosbar, L. L.; Graham, T. O.; Curioni, A.; Andreoni, W. N-Type Organic Thin-Film Transistor with High Field-Effect Mobility Based on a N,N′-Dialkyl-3,4,9,10-Perylene Tetracarboxylic Diimide Derivative. *Apply. Phys. Lett.* 2002, 80, 2517–2519.
142. Chesterfield, R. J.; McKeen, J. C.; Newman, C. R.; Ewbank, P. C.; da Silva Filho, D. A.; Brédas, J.-L.; Miller, L. L.; Mann, K. R.; Frisbie, C. D. Organic Thin Film Transistors Based on N-Alkyl Perylene Diimides: Charge Transport Kinetics as a Function of Gate Voltage and Temperature. *J. Phys. Chem. B* 2004, 108, 19281–19292.
143. Sugiyasu, K.; Fujita, N.; Shinkai, S. Visible-Light-Harvesting Organogel Composed of Cholesterol-Based Perylene Derivatives. *Angew. Chem., Int. Ed.* 2004, 43, 1229–1233.
144. Chen, Z.; Stepanenko, V.; Dehm, V.; Prins, P.; Siebbeles, L. D. A.; Seibt, J.; Marquetand, P.; Engel, V.; Würthner, F. Photoluminescence and Conductivity of Self-Assembled π–π Stacks of Perylene Bisimide Dyes. *Chem.-Eur. J.* 2007, 13, 436–449.
145. Peeters, E.; van Hal, P. A.; Meskers, S. C. J.; Janssen, R. A. J.; Meijer, E. W. Photoinduced Electron Transfer in a Mesogenic Donor–Acceptor–Donor System. *Chem.-Eur. J.* 2002, 8, 4470–4474.
146. Schenning, A. P. H. J.; v. Herrikhuyzen, J.; Jonkheijm, P.; Chen, Z.; Würthner, F.; Meijer, E. W. Photoinduced Electron Transfer in Hydrogen-Bonded Oligo(p-Phenylene vinylene)–Perylene Bisimide Chiral Assemblies. *J. Am. Chem. Soc.* 2002, 124, 10252–10253.
147. Neuteboom, E. E.; Meskers, S. C. J.; van Hal, P. A.; van Duren, J. K. J.; Meijer, E. W.; Janssen, R. A. J.; Dupin, H.; Pourtois, G.; Cornil, J.; Lazzaroni, R.; Brédas, J.-L.; Beljonne, D. Alternating Oligo(p-Phenylene vinylene)–Perylene Bisimide Copolymers: Synthesis, Photophysics, and Photovoltaic Properties of a New Class of Donor–Acceptor Materials. *J. Am. Chem. Soc.* 2003, 125, 8625–8638.
148. Jonkheijm, P.; Stutzmann, N.; Chen, Z.; de Leeuw, D. M.; Meijer, E. W.; Schenning, A. P. H. J.; Würthner, F. Control of Ambipolar Thin Film Architectures by Co-Self-Assembling Oligo(p-Phenylene vinylene)s and Perylene Bisimides. *J. Am. Chem. Soc.* 2006, 128, 9535–9540.
149. Holman, M. W.; Liu, R.; Zang, L.; Yan, P.; DiBenedetto, S. A.; Bowers, R. D.; Adams, D. M. Studying and Switching Electron Transfer: From the Ensemble to the Single Molecule. *J. Am. Chem. Soc.* 2004, 126, 16126–16133.
150. You, C.-C.; Saha-Möller, C. R.; Würthner, F. Synthesis and Electropolymerization of Novel Oligothiophene-Functionalized Perylene Bisimides. *Chem. Commun.* 2004, 2030–2031.
151. Chen, S.; Liu, Y.; Qiu, W.; Sun, X.; Ma, Y.; Zhu, D. Oligothiophene-Functionalized Perylene Bisimide System: Synthesis, Characterization, and Electrochemical Polymerization Properties. *Chem. Mater.* 2005, 17, 2208–2215.

152. Cremer, J.; Mena-Osteritz, E.; Pschierer, N. G.; Müllen, K.; Bäuerle, P. Dye-Functionalized Head-to-Tail Coupled Oligo(3-Hexylthiophenes)—Perylene–Oligothiophene Dyads for Photovoltaic Applications. *Org. Biomol. Chem.* 2005, 3, 985–995.
153. Chen, L. X.; Xiao, S.; Yu, L. Dynamics of Photoinduced Electron Transfer in a Molecular Donor–Acceptor Quartet. *J. Phys. Chem. B* 2006, 110, 11730–11738.
154. Zhan, X.; Tan, Z.; Domercq, B.; An, Z.; Zhang, X.; Barlow, S.; Li, Y.; Zhu, D.; Kippelen, B.; Marder, S. R. A High-Mobility Electron-Transport Polymer with Broad Absorption and Its Use in Field-Effect Transistors and All-Polymer Solar Cells. *J. Am. Chem. Soc.* 2007, 129, 7246–7247.
155. Ko, H. C.; Kim, S.; Choi, W.; Moon, B.; Lee, H. Formation of Insoluble Perylene-tetracarboxylic Diimide Films by Electro- or Photo-Crosslinking of Pyrrole Units. *Chem. Commun.* 2006, 69–71.
156. Becker, S.; Ego, C.; Grimsdale, A. C.; List, E. J. W.; Marsitzky, D.; Pogantsch, A.; Setayesh, S.; Leising, G.; Müllen, K. Optimisation of Polyfluorenes for Light Emitting Applications. *Synth. Met.* 2001, 125, 73–80.
157. Ego, C.; Marsitzky, D.; Becker, S.; Zhang, J.; Grimsdale, A. C.; Müllen, K.; MacKenzie, J. D.; Silva, C.; Friend, R. H. Attaching Perylene Dyes to Polyfluorene: Three Simple, Efficient Methods for Facile Color Tuning of Light-Emitting Polymers. *J. Am. Chem. Soc.* 2003, 125, 437–443.
158. Gómez, R.; Veldman, D.; Blanco, R.; Seoane, C.; Segura, J. L.; Janssen, R. A. J. Energy and Electron Transfer in a Poly(Fluorene-Alt-Phenylene) Bearing Perylenediimides as Pendant Electron Acceptor Groups. *Macromolecules* 2007, 40, 2760–2772.
159. Van der Boom, T.; Hayes, R. T.; Zhao, Y.; Bushard, P. J.; Weiss, E. A.; Wasielewski, M. R. Charge Transport in Photofunctional Nanoparticles Self-Assembled from Zinc 5,10,15,20-Tetrakis(perylenediimide)porphyrin Building Blocks. *J. Am. Chem. Soc.* 2002, 124, 9582–9590.
160. Hayes, R. T.; Walsh, C. J.; Wasielewski, M. R. Using Three-Pulse Femtosecond Spectroscopy to Probe Ultrafast Triplet Energy Transfer in Zinc Meso-Tetraarylporphyrin–Perylene-3,4-Dicarboximide Dyads. *J. Phys. Chem. A* 2004, 108, 3253–3260.
161. Ahrens, M. J.; Kelley, R. F.; Dance, Z. E. X.; Wasielewski, M. R. Photoinduced Charge Separation in Self-Assembled Cofacial Pentamers of Zinc-5,10,15,20-Tetrakis(perylenediimide)porphyrin. *Phys. Chem. Chem. Phys.* 2007, 9, 1469–1478.
162. Prodi, A.; Chiorboli, C.; Scandola, F.; Iengo, E.; Alessio, E.; Dobrawa, R.; Würthner, F. Wavelength-Dependent Electron and Energy Transfer Pathways in a Side-to-Face Ruthenium Porphyrin/Perylene Bisimide Assembly. *J. Am. Chem. Soc.* 2005, 127, 1454–1462.

3 Next-Generation Transparent Electrode Materials for Organic Solar Cells

Kuan Sun, Yijie Xia, and Jianyong Ouyang

CONTENTS

3.1 Introduction 44
3.2 Transparent Conducting Polymers...44
    3.2.1 Structure, Properties, and Synthesis of Transparent Conductive Polymers 44
    3.2.2 Conductivity Enhancement of PEDOT:PSS 47
    3.2.3 Application of PEDOT:PSS as the Transparent Electrode of OSCs.....52
3.3 Carbon-Based Nanomaterial 56
    3.3.1 Carbon Nanotubes 56
        3.3.1.1 Structure and Properties of Carbon Nanotubes 56
        3.3.1.2 Application of Carbon Nanotubes as the Transparent Electrode of OSCs..58
    3.3.2 Graphene..61
        3.3.2.1 Preparation, Structure, and Properties of Graphene.............61
        3.3.2.2 Application of Graphene as the Transparent Electrode of OSCs 61
3.4 Metal Nanostructures 64
    3.4.1 Metal Nanomeshes...64
    3.4.2 Metal Nanowire Grids 67
    3.4.3 Metal Ultrathin Films..69
3.5 Transparent Conducting Oxide rather than ITO...70
3.6 Multilayer Thin Films...72
3.7 Summary and Outlook 74
Acknowledgment 74
References..74

3.1 INTRODUCTION

Solar cells have been attracting great attention due to energy and environmental concerns. Although the crystalline silicon solar cells are still the main player in practical application today, they have a problem of too long payback time because of the high fabrication cost. Organic solar cells (OSCs) are regarded as the next-generation solar cells due to their merits, including low fabrication cost, light weight, high mechanical flexibility, tunable chemical and physical properties via molecular design, and fast module installation.[1,2] An OSC usually has a photoactive layer 100–200 nm thick sandwiched between two electrodes. At least one electrode must be transparent to harvest light. OSCs can be semitransparent or transparent in the visible range when both electrodes are transparent.[3–6] A transparent electrode is needed not only for solar cells, but also for other optoelectronic devices. Indium tin oxide (ITO) is the most popular material for transparent electrodes now. ITO is a wide bandgap degenerate semiconductor. It has superior optical and electrical properties. For example, a 150-nm-thick ITO thin film can have a sheet resistance of below 15 Ω/sq and an optical transmittance of above 90% in the visible range. Nevertheless, there are several severe problems for ITO.[7,8] Indium is a scarce element on Earth, which has caused the skyrocketing of the indium price. ITO has poor mechanical flexibility so that it cannot be used for flexible electronic devices. Flexible electronic devices are regarded as the next-generation electronic devices.

To develop new transparent electrode materials is important for solar cells and other optoelectronic devices. The requirements for new transparent electrode materials include at least high transparency and low sheet resistance. In order to be used in flexible electronics, they should have high mechanical flexibility as well. Several types of materials, including transparent conducting polymers, carbon nanotubes, graphene, nanostructured metals, and transparent conductive oxides (TCOs) rather than ITO have been proposed as the substitute of ITO. This chapter reviews the fabrication, structure, and properties of these materials and their application as the transparent electrode of OSCs.

3.2 TRANSPARENT CONDUCTING POLYMERS

3.2.1 Structure, Properties, and Synthesis of Transparent Conductive Polymers

Since the report on conductive polyacetylene in the 1970s, conducting polymers have been attracting great attention.[9] Conducting polymers are conjugated polymers in an oxidized or reduced state. There are alternative σ and π bonds along the backbone for conjugated polymers. The π electrons are delocalized along the whole polymer chains, giving rise to the formation of valence and conduction bands. But the bandwidth of the conduction band and valence band is very narrow for neutral conjugated polymers so that the lowest unoccupied molecular orbital (LUMO) and highest occupied molecular orbital (HOMO) are more popular to describe their electronic structure. The optical properties of neutral conjugated polymers are determined by the electron transition from HOMO to LUMO. The oxidation or reduction doping of conjugated polymers turns them highly conductive. The dramatic increase in the conductivity after doping is related to the appearance of new energy levels. As shown in Figure 3.1, two polaron levels are generated between the conduction and valence bands when a conjugated polymer is lightly doped. After the conjugated polymer is further doped, the coupling turns two polarons into a bipolaron. Accordingly, the two polaron levels become two bipolaron levels. When the conjugated polymers are heavily doped, the bipolarons interact, causing the bipolaron levels to become two bipolaron bands. Finally, the two bipolaron bands overlap and merge as one band. A heavily doped conjugated polymer behaves as a metal.

The change in the electronic structure gives rise to the significant change in the optical properties during the process from neutral to a doped state. There are remarkable color changes for conjugated polymers during this process. Figure 3.2 shows the UV-Vis-NIR absorption spectra of poly(3,4-ethylenedioxythiophene):poly(styrenesulfonate) (PEDOT:PSS, chemical structure shown in Scheme 3.1) during the reduction process from an oxidative doping state at 0.5 V to the neutral state at −0.8 V.[10] The polymer film is transparent in the visible range from 400 to 700 nm when in the doped state. The

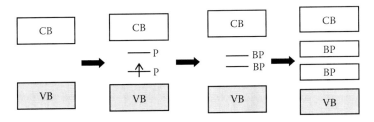

FIGURE 3.1 Change of the electronic structure of conjugated polymers during oxidation doping. VB: valence band, CB: conduction band, P: polaron, and BP: bipolaron.

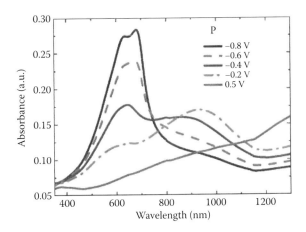

FIGURE 3.2 In situ UV-Vis-NIR absorption spectra of a PEDOT:PSS film during the electrochemical reduction from 0.5 V versus Ag/AgCl to −0.8 V versus Ag/AgCl. The polymer is in a doped state at 0.5 V versus Ag/AgCl and neutral state at −0.8 V versus Ag/AgCl. (Reproduced from Y. Xia, J. Ouyang, *ACS Appl. Mater. Interf.*, 4, 4131, 2012.)

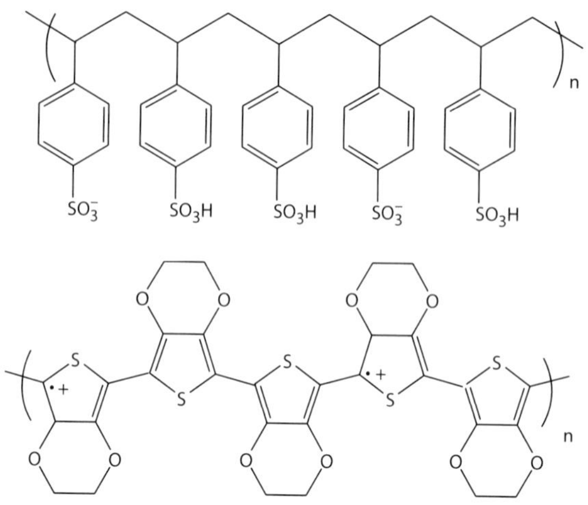

SCHEME 3.1 Chemical structure of PEDOT:PSS.

absorption in the visible range gradually increases during the dedoping process. It has a blue color with an absorption maximum close to 700 nm at −0.8 V.

Polymers with high conductivity and high transparency can have important application in many areas, particularly for optoelectronic devices. Most conducting polymers are not transparent. But PEDOT, its derivatives, and analogues exhibit high transparency in the visible range. PEDOT is thus the most important transparent conducting polymer. In addition, PEDOT has other merits in processability and thermal stability.

PEDOTs are synthesized through the cation radical polymerization of its monomer, ethylenedioxythiophene (EDOT). EDOT becomes a positively charged radical after it loses an electron. The radical structure makes the monomer chemically active, and the monomer molecules can connect into a polymer chain. The polymerization can take place by applying an electrochemical potential or using an oxidizing agent.

Pei et al. were the first to report the electrochemical polymerization of EDOT.[11] The oxidation of EDOT in acetonitrile solution of 0.1 M Bu_4NClO_4 starts at 1.04 V. Oxidation at this potential leads to the growth of a PEDOT on the anode. They observed a conductivity of about 200 S/cm for the PEDOT doped with ClO_4^-. The electropolymerization of EDOT can also proceed in an aqueous solution. Surfactant is used for the dispersion of the monomer in water.[12]

PEDOT can also be synthesized in a solution through the polymerization of its monomer with an oxidizing agent, such as persulfate salts, iron (III) salts or bromine.[13] PEDOT doped with tosylate (PEDOT:TsO) exhibited the highest conductivity among the PEDOTs by solution polymerization. Ha et al. carried out a comprehensive investigation on the chemical polymerization of EDOT with iron tosylate as the oxidizing agent.[14] By optimizing experimental conditions, they obtained a conductivity of 750 S/cm for PEDOT:TsO and a conductivity of 900 S/cm for methanol-substituted EDOT. PEDOTs doped with small counter anions are insoluble in any solvent. Conductive PEDOT thin films can be prepared by spin-coating a solution consisting of EDOT and an oxidizing agent prior to the polymerization.

When polystyrenesulfonate (PSS⁻) is used as the counter anion of PEDOT, the polymer can be dispersed with good stability in water and some polar organic solvents.[15] Positively charged PEDOT chains are stabilized by negatively charged PSS⁻ chains in solvent. PEDOT is in fact an oligomer with the molecular weight of less than 1000 to 2500 Da (about 6 to 18 repeating units). The PEDOT:PSS aqueous solution can be readily processed into conductive polymer films on various substrates. The conductivity of PEDOT:PSS depends on the ratio of PEDOT to PSS. Increasing the PSS content reduces the electrical conductivity of PEDOT:PSS. The conductivity is about 1 S/cm when the molar ratio of PEDOT to PSS is 1:2.5, and it decreases to 10^{-3} S/cm when the molar ratio turns to 1:6.

A variation for the chemical polymerization of EDOT in solution is the vapor-phase polymerization (VPP).[16–21] An oxidizing agent, typically iron(III) tosylate, is coated onto a substrate. It forms a solid thin film on the substrate after drying. Exposure of this substrate to EDOT vapor produces a PEDOT:TsO film on the substrate. The conductivity of the PEDOT:TsO films can be as high as 1500 S/cm by this method. Recently, Evans et al. modified the VPP method by blending the oxidizing agent with a tri-block polymer, poly(ethylene glycol–propylene glycol–ethylene glycol) (PEG–PPG–PEG).[22] They observed a conductivity of about 2500 S/cm.

3.2.2 Conductivity Enhancement of PEDOT:PSS

Among the PEDOT polymers, solution-processable PEDOT:PSS has gained the most attention. High-quality PEDOT:PSS films can be readily prepared on substrates through the conventional solution-processing techniques, such as coating, printing, and painting.[23,24] However, an as-prepared PEDOT:PSS film from its aqueous solution usually has a conductivity below 1 S/cm, which is remarkably lower than ITO.

The conductivity of PEDOT:PSS is determined not only by its doping degree but also by the amount of PSS and its microstructure. In order to stabilize PEDOT in water and polar organic solvents, PSS must be excessive. PSS serves as not only the counter anions for the charge balance of the conducting polymer, but also as a surfactant to stabilize PEDOT in solvents. But PSS is an insulator, and the presence of more PSS can decrease the conductivity of PEDOT:PSS.

PEDOT:PSS aqueous solutions are commercially available from both Heraeus and Agfa. Clevios P, Clevios PH 500, Cleivios PH750, and Clevios PH 1000 are four grades supplied by Heraeus. In these aqueous solutions, PEDOT has a molecular weight of about 1000–2500 g/mol whereas the molecular weight of PSS is approximately 400,000 g/mol.[25] The PEDOT chains are much shorter than the PSS chains. The short PEDOT chains attach to the long PSS chains due to the Coulombic attraction among them. PEDOT:PSS is a polyelectrolyte with the hydrophobic PEDOT and hydrophilic PSS so that it has a necklace structure in water. There are principally two types of conformations for the PSS segments owing to the different Coulombic interactions and the interactions between the polymer and water medium. For the PSS segments attached with PEDOT, the Coulombic repulsions among the PSS anions are screened by the positive charges on PEDOT, leading to a coil conformation for these PSS segments together with attached PEDOT chains (Scheme 3.2).[26] In order to reduce the interactions between PEDOT and water, these PSS segments form

SCHEME 3.2 Conformations of PEDOT:PSS. The thin and thick curves stand for PSS and PEDOT chains, respectively. (Reproduced from Y. Xia, J. Ouyang, *J. Mater. Chem.*, 21, 4927, 2011.)

blobs to prevent PEDOT from interacting with water. The blobs thus have a core/shell structure with PEDOT in the core and PSS in the shell. On the other hand, the PSSH segments without PEDOT attached (excess PSS segment) dissociate into PSS anions and protons in water. These PSS segments adopt a linear conformation to minimize the Coulombic repulsions among the PSS anions. They act as the strings between the blobs. As a result, all the blobs try to separate as far as possible. This necklace structure in the solution is conserved in the as-prepared PEDOT:PSS films. The shell richness of the insulating PSS forms an energy barrier for the charge transport across the PEDOT chains. Moreover, the coil PEDOT conformation leads to the localization of the charge carriers on the PEDOT chains. The excess PSS and coil PEDOT conformation are the two reasons for the low conductivity of as-prepared PEDOT:PSS films from an aqueous solution.

As-prepared PEDOT:PSS films have a large sheet resistance because of their low conductivity. To greatly improve the conductivity of PEDOT:PSS is important for its application as the transparent electrode of optoelectronic devices. Kim et al. were the first to report the conductivity enhancement of PEDOT:PSS by adding dimethyl sulfoxide (DMSO) or dimethylformamide (DMF) into PEDOT:PSS aqueous solution in 2002.[27] The conductivity at room temperature was enhanced by about two orders of magnitude when DMSO was added. The temperature dependences of the resistances of the pristine and treated PEDPOT:PSS films indicate that the charge transport across the PEDOT chains by charge hopping. Since then, many other organic solvents or organic solids were reported to enhance the conductivity of PEDOT:PSS.[10,26,28–50] Polar organic solvents with a high boiling point, such as ethylene glycol (EG), nitromethanol, glycerol, and other organic solvents with multiple hydroxyl groups, were added into the PEDOT:PSS aqueous solution, and they can significantly enhance the conductivity to about 200 S/cm for PEDOT:PSS prepared from Clevios P aqueous solution. Organic solids, such as D-sorbitol, were also added in the PEDOT:PSS aqueous solution, and they can have a similar effect on the conductivity of PEDOT:PSS as the organic solvents with high boiling points. The conductivity can be enhanced to 600–700 S/cm for PEDOT:PSS prepared from the Clevios PH1000 aqueous solution after the addition of EG or DMSO.[41] Untreated Clevios P and Clevios PH1000 have almost the same conductivity of about 0.3 S/cm, but their conductivities become saliently different after a treatment. The different conductivities are attributed to the different PEDOT molecular weights for Clevios P and Clevios PH1000.[10]

Ionic liquids were also added into PEDOT:PSS aqueous solution for the conductivity enhancement. The conductivity of PEDOT:PSS depends on the structure and concentration of the ionic liquids.[32] The optimal conductivity was higher than 100 S/

cm when 1-butyl-3-methylimidazolium tetrafluoroborate tetrafluoroborate [(BMIm)BF$_4$] or 1-butyl-3-methylimidazolium bromide [(BMIm)Br] was added into Clevios P aqueous solution. Recently, Badre et al. added ionic liquids into Clevios PH1000 aqueous solution.[51] They observed a conductivity of 2084 S/cm when 1-ethyl-3-methylimidazolium tetracyanoborate (EMIM-TCB) was added.

Surfactants were also added into PEDOT:PSS aqueous solution to enhance the conductivity of PEDOT:PSS. Ouyang et al. added the nonionic surfactant polyoxyethylene(12) tridecyl ether into PEDOT:PSS aqueous solution in 2002.[52] This nonionic surfactant can enhance the conductivity of PEDOT:PSS by a factor of 20.[33] The conductivity of PEDOT:PSS can be enhanced by more than two orders of magnitude when an anionic surfactant is used. Anionic surfactants, particularly fluoro-surfactants, were also studied to enhance the conductivity of Clevios PH1000 by Bao et al.[53-55]

Apart from adding compounds into the PEDOT:PSS aqueous solution, the conductivity of PEDOT:PSS film can be enhanced by a posttreatment. Ouyang et al. observed significant conductivity enhancement when the PEDOT:PSS films were treated with EG or DMSO.[29] The conductivity enhancement is similar with that by adding EG and DMSO into PEDOT:PSS aqueous solution. Significant enhancement in the conductivity of PEDOT:PSS films was also observed through a postcoating treatment with salts, zwitterions, cosolvents, and acids.[37-41] The posttreatment does not affect the coating process of the PEDOT:PSS aqueous solution. In addition, the compounds used for the posttreatment can be easily removed from the final PEDOT:PSS films.

The conductivity of PEDOT:PSS films can be significantly improved through a treatment with organic solvents having multiple hydroxyl groups, such as EG or glycerol. It is surmised that geminal diols, which have two hydroxyl groups on one carbon atom, can also enhance the conductivity of PEDOT:PSS. Hexafluoroacetone (HFA) can hydrolyze into a geminal diol with water. It was investigated for the posttreatment of PEDOT:PSS films prepared from Clevios PH1000 aqueous solution.[41] It can improve the conductivity of the PEDOT:PSS films to 1164 S/cm. The conductivity could be further improved to be higher than 1300 S/cm when the treatment was repeated three times.

Chu et al. found that the conductivity of PEDOT:PSS films prepared from Clevios PH1000 aqueous solution can be significantly improved through a posttreatment with alcohols.[44] The conductivity enhancement is quite sensitive to the chemical structure of alcohols. It is 1015 S/cm after the PEDOT:PSS films are treated with methanol whereas it is only 286 S/cm after treatment with iso-butanol. In addition, the alcohol-treated PEDOT:PSS films have a better stability in the conductivity than EG-treated and pristine PEDOT:PSS films.

Kim et al. combined the solvent addition and posttreatment to enhance the conductivity of PEDOT:PSS.[40] The conductivity of PEDOT:PSS is improved to more than 600 S/cm after the addition of an appropriate amount of EG into Clevios PH1000 aqueous solution. The PEDOT:PSS films are then treated with EG vapor, and the conductivity is further enhanced to more than 1400 S/cm.

The two components, PEDOT and PSS, of PEDOT:PSS have different hydrophilicities. They can have different interactions with solvents of different polarities. Xia et al. treated PEDOT:PSS prepared from Clevios P solution with cosolvents of water with several organic solvents, including methanol, ethanol, iso-propyl alcohol

(IPA), actonitrile (ACN), acetone, and THF.[26] The conductivity of PEDOT:PSS was enhanced to almost 100 S/cm, and the optimal concentration of the organic solvent in the cosolvent is about 80% by volume. The conductivity enhancement is attributed to the preferential solvations of PEDOT and PSS with the cosolvent of organic solvent and water.

The conductivity enhancement of PEDOT:PSS is related to the phase segregation between PEDOT and PSS during the treatment. When the Coulombic attraction between PEDOT and PSS is reduced, phase segregation can occur because PEDOT and PSS have different hydrophilicities. The Coulombic attraction can be screened by various compounds like polar organic solvents. Another strategy to induce the phase segregation is through the protonation of PSS, $PSS^- + H^+ \rightarrow PSSH$. The absence of Coulombic attraction between the positively charged PEDOT chains and neutral PSSH chains can lead to the phase segregation between PEDOT and PSS. This suggests that the conductivity of PEDOT:PSS can be enhanced through a treatment with acids.[37] The conductivity of the PEDOT:PSS films prepared from Clevios P aqueous solution is enhanced to higher than 100 or 200 S/cm when oxalic acid or sulfurous acid is used for the treatment. Howden et al. also observed conductivity enhancement of PEDOT prepared by vapor deposition after a treatment with acids.[56]

The conductivity enhancement becomes very significant when PEDOT:PSS films are treated with sulfuric acid.[57] The conductivity of PEDOT:PSS films prepared from Clevios PH1000 aqueous solution reached about 2400 S/cm after a treatment with 1.5 M H_2SO_4 (Figure 3.3). The conductivity further increased to 3065 S/cm when the PEDOT:PSS films were treated with 1 M H_2SO_4 solution at 160°C three times. This conductivity is higher than that (~2000 S/cm) of ITO on plastic and comparable to that (3000–6000 S/cm) of ITO on glass.

Moreover, the H_2SO_4 treatment changes the conduction mechanism of the PEDOT:PSS films as revealed by the temperature dependence of the resistances. As illustrated in Figure 3.4, the resistance of the H_2SO_4-treated PEDOT:PSS films is almost constant from 320 K down to 240 K. It then increases with the lowering

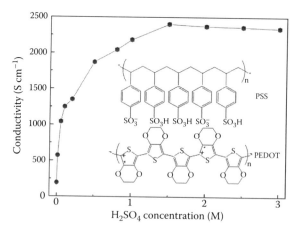

FIGURE 3.3 Conductivities of PEDOT:PSS films treated with H_2SO_4 solutions of various concentrations at 160°C. (Reproduced from Y. J. Xia et al., *Adv. Mater.*, 24, 2436, 2012.)

Next-Generation Transparent Electrode Materials for Organic Solar Cells 51

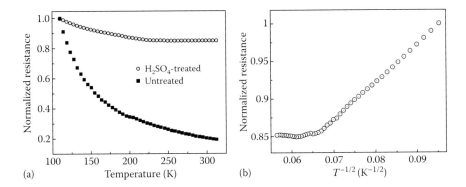

FIGURE 3.4 Temperature dependencies of the normalized resistances of untreated and H$_2$SO$_4$-treated PEDOT:PSS films. (a) Normalized resistances versus temperature. (b) Analysis of resistance-temperature relationship of the H$_2$SO$_4$-treated PEDOT:PSS film with the VRH model. The PEDOT:PSS film was treated with 1 M H$_2$SO$_4$ for three times. (Reproduced from Y. J. Xia et al., *Adv. Mater.*, 24, 2436, 2012.)

temperature. This is saliently different from the temperature dependence of the untreated PEDOT:PSS films, which show continuous resistance increases with the lowering temperature. The temperature dependence of the untreated PEDOT:PSS films was analyzed by a one-dimensional variable range hopping (VRH) model[58]

$$R(T) = R_0 \exp\left[\left(\frac{T_0}{T}\right)^{1/2}\right].$$

The temperature dependence of the H$_2$SO$_4$-treated PEDOT:PSS films conforms with the hopping model only at the low temperature range. The H$_2$SO$_4$-treatment PEDOT:PSS becomes metallic or semimetallic at room temperature. Metallic behavior has been rarely observed on solution-processed conducting polymers.

The treatments did not affect the transmittance of the PEDOT:PSS films in the visible range. Figure 3.5 presents the transmittance spectra in the visible range of three H$_2$SO$_4$-treated PEDOT:PSS films with different thicknesses. The transmittance of the 66-nm-thick PEDOT:PSS film, which has a sheet resistance of 67 Ω^{-1}, is 87% at 550 nm. The transmittance is higher than 90% in the wavelength range below 500 nm and decreases a little but is still higher than 80% at the wavelength above 550 nm. The transmittance of the 109-nm-thick PEDOT:PSS film, which has a sheet resistance of 39 Ω/sq, is more than 80% at 550 nm. The thickness of the ITO layer on ITO/PET is about 100 nm, quite close to the thickness of the thickest PEDOT:PSS film.

The PEDOT:PSS films treated with sulfuric acid can have a conductivity comparable to ITO on glass and high transparency in the visible range. But sulfuric acid is a strong acid and highly corrosive. Recently, it was discovered that similar conductivity enhancement could be achieved by using mild organic acids to treat PEDOT:PSS. The conductivity of PEDOT:PSS films can be higher than 3300 S/cm after being treated with methanesulfonic acid.[59] Methanesulfonic acid is an organic acid and not corrosive. The

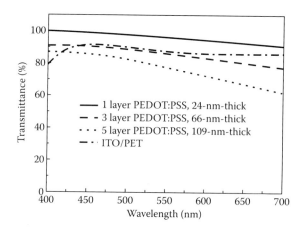

FIGURE 3.5 Transmittance spectra of H$_2$SO$_4$-treated PEDOT:PSS films. The transmittance spectrum of ITO/PET, which has a sheet resistance of 50 Ω/sq, is present for comparison. (Reproduced from Y. J. Xia et al., *Adv. Mater.*, 24, 2436, 2012.)

mechanism for the conductivity enhancement by the methanesulfonic acid treatment is similar to that by the sulfuric acid treatment. Although methanesulfonic acid has almost the same acidity as PSSH, it can protonate PSS$^-$ into PSSH and cause the phase segregation of PSSH from PEDOT. This is related to the hydrophilicity of PSSH and PEDOT. The different hydrophilicities lead to the segregation of PSSH from PEDOT. This phase segregation is the driving force for the protonation of PSS$^-$ by methanesulfonic acid.

3.2.3 Application of PEDOT:PSS as the Transparent Electrode of OSCs

PEDOT:PSS can have high conductivity and high transparency in the visible range after a treatment while its work function (ca. 5.0 eV) is hardly affected by the treatments. Highly conductive PEDOT:PSS films were investigated as the transparent electrode of OSCs.

Zhang et al. made the first attempt to use PEDOT:PSS as the transparent electrode for OSCs in 2002.[60] They added glycerol or D-sorbitol into Clevios P aqueous solution. The conductivity of the PEDOT:PSS film was ~10 S/cm. They used PEDOT:PSS films with a thicknesses of 150–200 nm and transmittance of over 80% in the wavelength range of 350–600 nm as the transparent electrode of OSCs. Poly(2-methoxy-5-(2-ethylhexyloxy)-1,4-phenylenevinylene) (MEH-PPV) and [6,6]-phenyl-C61-butyric acid methyl ester (PC$_{61}$BM) were used as the active materials. Their best OSCs achieved a power conversion efficiency (PCE) of 0.36% under light of 78 mW/cm^2. Table 3.1 lists the photovoltaic performance of OSCs with PEDOT:PSS as the transparent electrode. The conductivity enhancement of PEDOT:PSS was carried out by adding organic solvents into the PEDOT:PSS aqueous solution.

PEDOT:PSS films through a posttreatment were also studied as the transparent electrode of OSCs. Figure 3.6 presents the architecture of an OSC with a H$_2$SO$_4$-treated PEDOT:PSS film replacing ITO as the transparent electrode and the chemical structure of poly(3-hexylthiophene) (P3HT) and PC$_{61}$BM. The H$_2$SO$_4$-treated

TABLE 3.1
Photovoltaic Performance of OSCs with PEDOT:PSS as the Transparent Electrode

Transparent Electrode	Thickness (nm)	R_s (Ω/sq)	T (%)	Active Layer	J_{sc} (mA/cm²)	V_{oc} (V)	FF	PCE (%)	Year	Ref.
PH500	100	100,000	—	P3HT:PC$_{61}$BM	0.184	0.58	0.25	0.03	2008	[61]
PH500 + 5%DMSO	100	213	>90	P3HT:PC$_{61}$BM	9.73	0.63	0.54	3.27	2008	[61]
PH500 + 5%DMSO	100	213	>90	P3HT:PC$_{61}$BM	9.16	0.61	0.5	2.8	2008	[61]
PH500 + DMSO + DEG + sorbital: Surfynol/PH500 + DMSO + DEG[a]	110–450	250	>77	P3HT:PC$_{61}$BM	8.5	0.58	0.53	2.6	2008	[62]
PH500 + 5%DMSO + zonyl	200	230	>75	APFO3:PC$_{61}$BM	4.15	0.86	0.51	1.82	2008	[63]
PH500 + 5%DMSO + zonyl	250	—	—	APFO3:PC$_{61}$BM	4.06	0.98	0.56	2.23	2008	[63]
PH500 + 5%DMSO + sorbital[b]	2100	900	—	P3HT:PC$_{61}$BM	7.03	0.60	0.48	2.0	2008	[64]
PH500 + 5%DMSO/ZnO/C$_{60}$-SAM	130	370	>80	P3HT:PC$_{61}$BM	9.41	0.61	0.53	3.08	2009	[65]
PH500 + 5%DMSO/ZnO/C60-SAM	170	220	>75	P3HT:PC$_{61}$BM	8.42	0.62	0.58	2.99	2009	[65]
PH500 + 5%DMSO	—	370	—	P3HT:PC$_{61}$BM	7.86	0.61	0.52	2.51	2009	[65]
PH750 + 5%DMSO + 13%IPA	168	102	95	P3HT:bisPC$_{61}$BM	8.7	0.66	0.61	3.5	2009	[66]
PH510 + 7%DMSO	300	>63	>77	P3HT:PC$_{61}$BM	9.47	0.57	0.64	3.48	2009	[67]
BPhen:Li/Au/PH750 + 5%DMSO/air	120	570	—	P3HT:PC$_{61}$BM	—	0.51	0.30	0.6	2010	[68]
PH500	200	—	70–80	P3HT:PC$_{61}$BM	6.68	0.59	0.47	1.86	2010	[69]
PH500 + 5%DMSO[b]	180–340	>63	>67	P3HT:PC$_{61}$BM	6.62	0.61	0.54	2.17	2010	[70]
PH1000 + 5%DMSO	160	115	85–90	P3HT:PC$_{61}$BM	7.20	0.55	0.45	1.8	2010	[71]
PH1000 + 5%DMSO	140–200	420	—	P3HT:PC$_{61}$BM	7.20	0.55	0.58	2.4	2011	[72]
PH1000 + 5%DMSO + 1%Zonyl	100	750	96	P3HT:PC$_{61}$BM	5.9	0.59	0.58	2.0	2011	[73]
PH1000 + 10%DMSO + 10%IPA[b]	1000–2000	>25	—	P3HT:PC$_{61}$BM	6.90	—	—	1.9	2012	[74]
PH1000 + 5%DMSO + 0.1%Zonyl	—	46	82	P3HT:PC$_{61}$BM	6.69	0.58	0.56	2.16	2012	[54]
PH1000 + 73% IPA + 1.3%EG[b]	—	84	91	P3HT:PC$_{61}$BM	8.8	0.59	0.58	3.2	2013	[75]

Note: Compounds were added into PEDOT:PSS aqueous solutions for conductivity enhancement.
[a] The second layer was formed by spray coating.
[b] The layer was formed by spray coating.

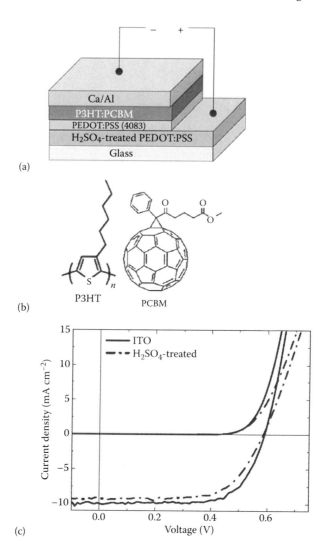

FIGURE 3.6 (a) Architecture of an OSC with H_2SO_4-treated PEDOT:PSS as the transparent electrode. (b) Chemical structure of P3HT and $PC_{61}BM$. (c) J–V characteristics of PSCs with ITO and a H_2SO_4-treated Clevios™ PH1000 PEDOT:PSS film as the anodes under AM 1.5G illumination (100 mW cm^{-2}). (Reproduced from Y. J. Xia et al., *Adv. Mater.*, 24, 2436, 2012.)

PEDOT:PSS film has a thickness of 70 nm. The device with the H_2SO_4-treated PEDOT:PSS film anode exhibited high photovoltaic performance: short-circuit current (J_{sc}) = 9.29 mA cm^{-2}, open-circuit voltage (V_{oc}) = 0.59 V, fill factor (FF) = 0.65, and power conversion efficiency (PCE) = 3.56%. The photovoltaic performance is comparable to that of the control devices using ITO anode. Photovoltaic performances of OSCs with posttreated PEDOT:PSS films as the transparent electrode are summarized in Table 3.2.

Next-Generation Transparent Electrode Materials for Organic Solar Cells 55

TABLE 3.2
Photovoltaic Performance of OSCs with PEDOT:PSS as the Transparent Electrode

Transparent Electrode	Treatment Chemical	Thickness (nm)	R_s (Ω/sq)	T (%)	Active Layer	J_{sc} (mA/cm²)	V_{oc} (V)	FF	PCE (%)	Year	Ref.
Glass/PV4071	—	50	1.01×10^6	93	P3HT:PC$_{61}$BM	0.79	0.45	0.29	0.1	2008	[76]
Glass/PV4071	Ethanol	50	1.19×10^5	93	P3HT:PC$_{61}$BM	6.74	0.58	0.3	1.16	2008	[76]
Glass/PV4071	Methoxyethanol	50	7.3×10^3	93	P3HT:PC$_{61}$BM	9.5	0.58	0.57	3.13	2008	[76]
Glass/PV4071	1,2-dimethoxyethane	50	6.04×10^5	94	P3HT:PC$_{61}$BM	4.32	0.57	0.32	0.78	2008	[76]
Glass/PV4071	EG	50	5.1×10^3	93	P3HT:PC$_{61}$BM	8.99	0.59	0.64	3.39	2008	[76]
Glass/Baytron P	CuCl$_2$:H$_2$O	130	455	—	P3HT:PC$_{61}$BM	2.85	0.52	0.29	0.43	2010	[39]
Glass/Baytron P	CuBr$_2$:H$_2$O	130	409	—	P3HT:PC$_{61}$BM	6.62	0.52	0.31	1.08	2010	[39]
Glass/Baytron P	DMCSP:H$_2$O	130	950	—	P3HT:PC$_{61}$BM	7	0.6	0.3	1.24	2010	[39]
Glass/Baytron P	DDMAP:H$_2$O	130	1570	—	P3HT:PC$_{61}$BM	8.51	0.57	0.43	2.08	2010	[39]
Glass/Baytron P	DNSPN:H$_2$O	130	836	—	P3HT:PC$_{61}$BM	8.25	0.55	0.4	1.79	2010	[39]
Glass/Clevios P	EtOH:H$_2$O	130	1054	—	P3HT:PC$_{61}$BM	8.99	0.61	0.52	2.87	2011	[26]
Glass/Clevios P	Acetonitrile:H$_2$O	130	974	—	P3HT:PC$_{61}$BM	7.84	0.63	0.51	2.51	2011	[26]
Glass/PH1000:6%EG	EG 30 min	<140	>65	>80	ZnPc:C$_{60}$	7.98	0.55	0.58	2.54	2011	[40]
PET/PH1000:6%EG	EG 30 min	<140	>65	>80	ZnPc:C$_{60}$	6.94	0.54	0.52	1.95	2011	[40]
Glass/PH1000:6%EG	EG 15 min	—	—	—	ZnPc:C$_{60}$	8.3	0.57	0.57	2.7	2011	[77]
Glass/PH1000	HFA:H$_2$O	164	46	>80	P3HT:PC$_{61}$BM	9.44	0.59	0.64	3.57	2012	[41]
Glass/PH1000	H$_2$SO$_4$:H$_2$O	109	39	>80	P3HT:PC$_{61}$BM	9.29	0.59	0.65	3.56	2012	[57]
Glass/PH1000	Methanol	50	197	>90	P3HT:PC$_{61}$BM	9.32	0.58	0.66	3.57	2012	[44]
Glass/PH1000	Methanol	50	164	>90	P3HT:PC$_{61}$BM	9.51	0.58	0.67	3.71	2012	[44]
Glass/PH1000	Methanol	50	147	>90	P3HT:PC$_{61}$BM	9.41	0.58	0.66	3.62	2012	[44]
Glass/PH1000	DMSO	55	131	>80	P3HT:PC$_{61}$BM	7.29	0.58	0.65	2.74	2013	[78]
Glass/PH1000	H$_2$SO$_4$	<100	46	90	PTB7:PC$_{71}$BM	13.9	0.73	0.65	6.6	2013	[79]

Note: The conductivity of the PEDOT:PSS films was enhanced through a posttreatment.

3.3 CARBON-BASED NANOMATERIAL

Carbon-based nanomaterials, including carbon nanotubes (CNTs) and graphene, can have high conductivity and high transparency in the visible range, high mechanical flexibility and excellent chemical stability. They emerged as candidates for transparent electrode in the last decade.

3.3.1 Carbon Nanotubes

3.3.1.1 Structure and Properties of Carbon Nanotubes

Since the first report on carbon nanotubes (CNTs) by Sumio Iljima in 1991,[80] CNTs have been a hot topic in research due to their interesting structure and properties. A CNT can be considered as a roll-up of monolayer(s) of sp^2 hybridized carbon atoms that are covalently bonded. Both the rolling angle (chirality) and radius determine the electronic structure and properties of CNTs. CNTs can be classified into single-walled carbon nanotubes (SWNT) and multiwalled carbon nanotubes (MWNT) in terms of the number of "walls." A SWNT is usually a mixture of semiconducting and metallic CNTs owing to the different chiralities and tube radii, and MWNT is always metallic because of statistical probability and restrictions on the relative diameters of the individual tubes that form one MWNT. As shown in Figure 3.7, the chirality affects the energy band structure of SWNTs. The SWNT with (9,3) chirality is metallic, and the one with (8,4) chirality is semiconductive.[81]

CNTs are usually prepared by chemical vapor deposition (CVD). A substrate deposited with a catalyst is put in a furnace of >800°C. Carbon feedstock gas and hydrogen gas are supplied to the furnace. The CNTs grow directly on the substrate.

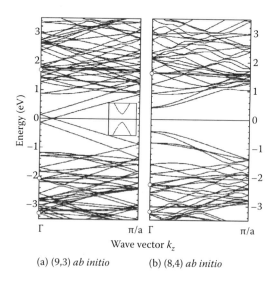

(a) (9,3) *ab initio*   (b) (8,4) *ab initio*

FIGURE 3.7 *Ab initio* band structure of two chiral SWNTs. (a) (9,3) quasimetallic nanotube. (b) (8,4) semiconducting nanotube. (Reproduced from S. Reich et al., *Phys. Rev. B*, 65, 155411, 2002.)

The diameter, length, and wall number of CNTs are affected by the experimental conditions. Arc discharge, laser ablation, and high-pressure carbon monoxide (HiPCO) are also used to fabricate CNTs. These CNTs are usually in powder form. They should be purified by removing amorphous carbon and catalysts. They can be dispersed in solution of a surfactant and then deposited to CNT films by filtration, coating, or printing. CNTs were also directly dispersed in liquid surfactants without any solvent.[82] The CNTs form gels with liquid surfactants. These gels can be processed into CNT films as well.

The resistance of CNT films is determined by the charge transport across the CNTs. The intra-CNT resistance is significantly lower than the contact resistance among CNTs. Both sheet resistance and transmittance decrease as the thickness increases. Hu et al. modeled the relationship between the transmittance (T) and sheet resistance (R_s) by the following equation[83]:

$$T = \left(1 + \frac{1}{2R_s}\sqrt{\frac{\mu_0 \sigma_{op}}{\varepsilon_0 \sigma_{dc}}}\right)^{-2} = \left(1 + \frac{188(\Omega)}{R_s}\frac{\sigma_{op}}{\sigma_{dc}}\right)^{-2}$$

where μ_0 and ε_0 are the free space permeability and permittivity, respectively, and σ_{op} and σ_{dc} are the optical and dc conductivities, respectively. The value of σ_{op} is related to the CNT network density rather than the variables, such as the CNT–CNT junction, CNT length, and doping. It is about 200 S/cm at 550 nm.

As shown in Figure 3.8, CNTs do not have strong absorption in the visible range,[82] but CNTs can scatter light. SWNTs are more transparent than MWNTs because SWNTs have fewer carbon layers/walls that absorb light.

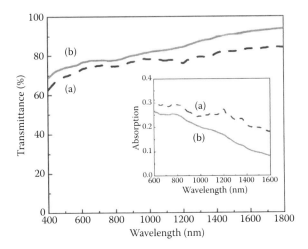

FIGURE 3.8 UV-Vis-IR transmittance spectra of a SWCNT film (a) before and (b) immediately after the HNO_3 treatment. The inset shows the corresponding absorption spectra. (Reproduced from X. Mei, J. Ouyang, *J. Mater. Chem.*, 21, 17842, 2011.)

3.3.1.2 Application of Carbon Nanotubes as the Transparent Electrode of OSCs

CNT films can have high transmittance and low sheet resistance. They were investigated as the transparent electrode of OSCs. The first attempt to use CNT films as the electrode of OSCs was done by Ago et al. in 1999.[84] They synthesized MWNTs via pyrolysis and then functionalized the MWNTs by oxidation in a solution of HNO_3 and H_2SO_4. These functionalized MWNTs could be dispersed in water without surfactant. The MWNT thin films were fabricated by spin coating the water solution dispersed with MWNTs onto glass or quartz substrate. The OSCs has architecture of a poly(*p*-phenylene vinylene) (PPV) layer sandwiched between MWNT and Al, substrate/MWNT/PPV/Al. Under a weak monochromatic light illumination at 485 nm of 37 µW/cm² on the semitransparent Al side, they recorded a V_{oc} of 0.90 V, J_{sc} of 0.56 µA/cm², FF of 0.23, and an overall PCE of 0.081%. The poor photovoltaic efficiency is related to both the photoactive layer and the high sheet resistance of functionalized MWNT films.

Wu et al. successfully demonstrated highly transparent and conductive SWNT films through filtration in 2004.[85] Their SWNT films exhibited a transparency greater than 70% in the visible light range and a sheet resistance of 30 Ω/sq at a film thickness of 50 nm. Since then, CNT films prepared by filtration were studied as the transparent electrode of OSCs by many laboratories. The photovoltaic performances of OSCs with a CNT transparent electrode after Wu's work are summarized in Table 3.3.

The conductivity of CNT films is determined by the charge transport across the CNTs. It strongly depends on the preparation method of the CNT films. Ulbricht et al. could produce a free-standing CNT film by drawing laterally from the side of a CNT forest (Figure 3.9).[86] This film was further densified by using the surface tension effect of an imbibed liquid, such as methanol. CNT films produced by drawing exhibited orientation-dependent conductivity behavior. The conductivity along the CNT orientation is much higher than that perpendicular to the CNT orientation. OSCs with such a CNT film as the anode had an optimal PCE of 1.32%. Alternatively, van de Lagemaat et al. pattern SWNTs on glass by spray coating.[87] The inks were formulated by dispersing acid-purified SWNTs in water and alcohol. They obtained SWNT films with a surface resistance of about 50 Ω/sq and transmittance of about 70% at 650 nm. The P3HT:PC$_{61}$BM OSCs showed a PCE of 1.50%, with a good J_{sc} (9.24 mA/cm²) and V_{oc} (0.56 V) but poor FF (0.29).

The rough surface of CNT films leads to high leakage current and poor photovoltaic efficiency for OSCs.[88] Smooth CNT films were obtained by a PDMS transfer method.[89] CNT films were first prepared by vacuum filtration and were then transferred by PDMS stamp to a PET substrate. Finally, a layer of PEDOT:PSS was coated on the CNT films. The rms roughness of the CNT/PEDOT:PSS film was reduced to less than 10 nm over a surface of 25 µm² (Figure 3.10a and b). They observed a PCE of 2.5% on P3HT:PCBM OSCs (Figure 3.10c). Furthermore, the OSCs showed an excellent stability during mechanical distortion.

In order to prepare smooth CNT films in a large area, Tenent et al. developed an ultrasonic spraying method to prepare CNT films in 2009.[90] The SWNT was

TABLE 3.3
Photovoltaic Performance of OSCs with CNTs as the Transparent Electrode

Cell Architecture	Thickness (nm)	R_s (Ω/sq)	T (%)	J_{sc} (mA/cm²)	V_{oc} (V)	FF	PCE (%)	Year	Ref.
Quartz/SWNT/PEDOT:PSS/P3HT:PC$_{61}$BM/Ga:In	300	282	45	6.65	0.50	0.30	0.99	2005	[91]
Glass/CNT/PEDOT:PSS/P3HT:PC$_{61}$BM/Al	—	717	>80	5.47	0.49	0.49	1.32	2006	[86]
Glass/SWNT/PEDOT:PSS/P3HT:PC$_{61}$BM/Al	—	50	70	9.24	0.56	0.29	1.50	2006	[87]
PET/SWNT/PEDOT:PSS/P3HT:PC$_{61}$BM/Al	30	200	85	7.8	0.61	0.52	2.5	2006	[88]
Glass/FWNT/P3HT:PC$_{61}$BM/Al	—	86	70	4.46	0.36	0.38	0.61	2009	[92]
GZO/P3HT:PC$_{61}$BM/PEDOT:PSS/MWNT/PEDOT:PSS	—	500–700	>85	3.4	0.51	0.26	0.45	2009	[93]
Glass/SWNT/PEDOT:PSS/P3HT:PC$_{61}$BM/Ca/Al	40	60	>70	11.50	0.58	0.48	3.10	2009	[90]
Glass/SWNT/PEDOT:PSS/P3HT:PC$_{61}$BM/Ca/Al	—	56	70	13.78	0.57	0.53	4.13	2010	[94]
Glass/SWNT/P3HT:PC$_{61}$BM/Ca/Al	—	24	50	11.39	0.54	0.55	3.37	2010	[94]
Glass/SWNT(DCE)/PEDOT:PSS/P3HT:PC$_{61}$BM/LiF/Al	24	128	90	9.90	0.55	0.43	2.30	2010	[95]
Glass/SWNT(H$_2$O:SDS)/PEDOT:PSS/P3HT:PC$_{61}$BM/LiF/Al	26	57	65	7.30	0.59	0.46	2.20	2010	[95]
Glass/SWNT(H$_2$O:SDBS)/PEDOT:PSS/P3HT:PC$_{61}$BM/LiF/Al	28	68	70	6.70	0.55	0.31	1.20	2010	[95]
Glass/SWNT/PEDOT:PSS/P3HT:PC$_{61}$BM/LiF/Al	—	411	92	10.52	0.56	0.35	2.05	2011	[96]
ITO/n-C60/C60/ZnPc:C$_{60}$/p-BF-DPB/MWNT	20	250	38	4.5	0.57	0.58	1.5	2012	[97]

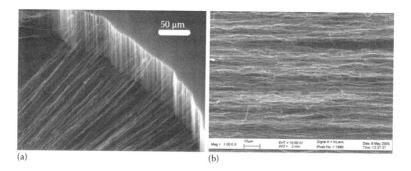

FIGURE 3.9 (a) SEM of CNT film being drawn from a CNT forest. (b) Undensified single layer sheet of free-standing CNT film. (Reproduced from R. Ulbricht et al., *Phys. Stat. Sol. B* 243, 3528, 2006.)

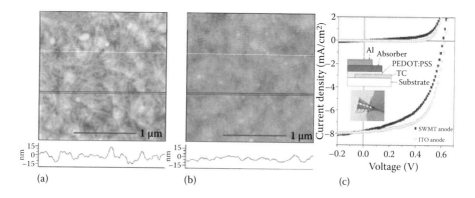

FIGURE 3.10 AFM images of SWNT films (a) before and (b) after PEDOT:PSS coating. The rms roughness is 7 and 3.5 nm, respectively. (c) J–V curves of P3HT:PC$_{61}$BM devices under AM 1.5G illumination. Inset: Schematic of device structure and photograph of a highly flexible OSC using a SWNT film on PET. (Reproduced from M. W. Rowell et al., *Appl. Phys. Lett.*, 88, 233506, 2006.)

first dispersed in an aqueous solution of sodium carboxymethyl cellulose (CMC). The SWNTs formed a film via ultrasonic spray coating. CMC was removed through the exposure of the SWNT films to nitric acid. The roughness of such a CNT film was ~3 nm. The overall PCE of the P3HT:PC$_{61}$BM OSCs was 3.1%. They further improved the efficiency to 4.13%.[94] They also showed that the role of PEDOT:PSS appeared to be maintaining a high shunt resistance by physically blocking parallel paths. Without it, the OSC suffered from reduced V_{oc} and FF.

Because the as-prepared SWNTs are a mixture of both metallic and semiconductive CNTs, removal of the semiconductive ones can improve the conductivity of CNT films. The SWNTs could be sorted via density gradient ultracentrifugation (DGU).[98] Tyler et al. found that metallic-enriched SWNT yielded greater PCE for OSCs than semiconductive SWNT (2.05% vs. 0.04%).[96]

3.3.2 Graphene

3.3.2.1 Preparation, Structure, and Properties of Graphene

Graphene is a single layer of sp^2 hybridized carbon atoms. Unlike CNT that forms a long tube and has the charge transport along one-dimension, graphene has a 2-D structure. Figure 3.11 presents the energy band structure of graphene.[99] The π band touches the π* band at some singularity points. Graphene is thus metallic. But the charge carrier density near the Fermi level is quite low.

There are three popular ways to prepare graphene films. The first one is the mechanical exfoliation of graphite.[100] Adhesive Scotch tapes can be used for the exfoliation. Although the graphene films by mechanical exfoliation can have good quality, it is very small and meaningful only for fundamental study. The second one is CVD.[101] Cu or Ni foils are put in a furnance of about 1000°C. Carbon feedstock gas and hydrogen are supplied to the furnace. Graphene of monolayer or a few layers grows on the metal foils. The layer number can be controlled. The graphene films on metal foils can be transferred to another substrate. The third one is the oxidation and exfoliation of graphite with strong acids and then reduction of graphene oxide at high temperature or with a reducing agent.[102]

Figure 3.12 presents the UV-Vis-NIR transmittance spectra of graphene films of one to four layers.[101] The transmittance in the visible range decreases with the increase of the layer number. The transmittance is more than 95% at 550 nm for the monolayer graphene film. It reduces to about 90% for the four-layer graphene film.

3.3.2.2 Application of Graphene as the Transparent Electrode of OSCs

Thin graphene films can have high transparency in the visible range and high conductivity. They were used as the transparent electrode of OSCs. Table 3.4 lists the photovoltaic performance of OSCs with a graphene transparent electrode. Before 2008, methods to fabricate single-layer graphene film in a large area were still lacking, but a few groups had started the attempt to use graphene film as the transparent electrode for OSCs. Wang et al. employed a bottom-up chemical approach to form graphene films.[103] At first, polycyclic aromatic hydrocarbon (PAH) was deposited on

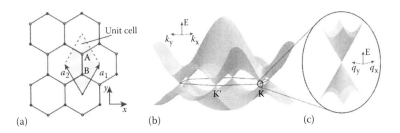

FIGURE 3.11 (a) Graphene lattice with two carbon atoms per unit cell denoted by A and B. (b) Electronic dispersion of graphene. The π-band (bottom) and the π*-band (top) touch each other at singularity points (K and K′ points). (c) Zoom of the dispersion at the singularity points. The dispersion has a conical form with an approximately linear slope. (Reproduced from J. Güttinger et al., *Rep. Prog. Phys.*, 75, 126502, 2012.)

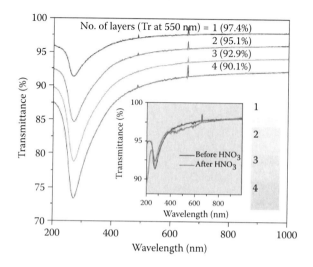

FIGURE 3.12 Absorption spectra of graphene films. (Reproduced from S. Bae et al., *Nat. Nanotechnol.*, 5, 574, 2010.)

quartz via spin coating. Then, the PAH molecules were cross-linked by high temperature annealing at 1100°C in Ar. This process could produce ultrasmooth, transparent, and conductive graphene films. The thickness of such a graphene film could be tuned by controlling the PAH layer thickness. The OSCs with such a graphene film as the transparent electrode can have a PCE of 1.7% PCE.[104,105]

It is difficult to process graphene by solution processing. An alternative way is to prepare graphene oxide (GO) by solution processing and subsequently reduce GO into graphene. In 2008, Wu et al. and Eda et al. separately reported the use of reduced graphene oxide (rGO) as the transparent electrode of OSCs.[106,107] Wu et al. deposited the GO film by spin coating, and Eda et al. obtained the film by filtering the GO solution and then dissolving the ester membrane by acetone.[108] Besides the reduction at high temperatures or with a chemical reducing agent, Kymakis et al. introduced a laser to reduce GO.[109] Because the conversion from GO to rGO cannot be 100%, there are defects in the rGO film, which diminish the film's conductivity. The transparent rGO films usually exhibit a high sheet resistance above 1000 Ω/sq, rendering a low PCE of ~1% for OSCs.[106,107,109–111] The optoelectronic properties of the rGO film could be improved by constructing a mesh-like structure. Zhang et al. showed that the rGO meshes could have a sheet resistance of 608 Ω/sq. Accordingly, the PCE of the OSCs was improved to 2.04%.[112]

The first demonstration of large-area nearly defect-free graphene films as the transparent electrode for OSCs was reported by Wang et al. in 2009.[113] Highly crystalline graphene films containing a few graphene layers were synthesized by CVD[114] and then transferred onto glass by a PDMS stamp. The graphene films exhibited sheet resistances between 210 and 1350 Ω/sq. However, the graphene films have a low work function of 4.2 eV and a hydrophobic surface, leading to a low PCE of only 0.21%. After improving the surface properties by UV-ozone treatment or

TABLE 3.4
Photovoltaic Performance of OSCs with Graphene as the Transparent Electrode

Cell Architecture	Thickness (nm)	R_s (Ω/sq)	T (%)	J_{sc} (mA/cm²)	V_{oc} (V)	FF	PCE (%)	Year	Ref.
Quartz/graphene/P3HT:PC$_{61}$BM/Ag	—	1.8×10^4	85	0.36	0.38	0.25	0.29	2008	[103]
Quartz/rGO/CuPc/C$_{60}$/BCP/Ag	4–7	1–5×10^5	85–95	2.1	0.48	0.34	0.4	2008	[106]
Glass/rGO/PEDOT:PSS/P3HT:PC$_{61}$BM/Al	14	4×10^4	65	~0.4	~0.58	—	~0.1	2008	[107]
Quartz/rGO/PEDOT:PSS/P3HT:PC$_{61}$BM/ZnO/Al	10	1.1×10^5	68	8.2	0.44	0.31	1.12	2009	[110]
Glass/graphene/PEDOT:PSS/P3HT:PC$_{61}$BM/LiF/Al	6–30	210–1350	72–91	6.05	0.55	0.51	1.71	2009	[113]
Quartz/graphene/PEDOT:PSS/P3HT:PC$_{61}$BM/Al	17	1.6×10^5	81	4.82	0.54	0.26	0.68	2010	[115]
Quartz/graphene/PEDOT:PSS/CuPc/C$_{60}$/BCP/Ag	—	300–500	91–97	9.15	0.43	0.42	1.63	2010	[116]
PET/graphene/PEDOT:PSS/CuPc/C$_{60}$/BCP/Al	—	500	75	4.73	0.48	0.52	1.18	2010	[117]
Glass/rGO/PEDOT:PSS/P3HT:PC$_{61}$BM/LiF/Al	25	1.79×10^4	69	1.18	0.46	0.25	0.13	2010	[111]
ITO/ZnO/P3HT:PC$_{61}$BM/GO/graphene	3.4	<200	>80	10.50	0.54	0.44	2.50	2011	[118]
Glass/graphene/MoO$_x$/PEDOT:PSS/P3HT:PC$_{61}$BM/LiF/Al	1.3	~100	85	8.5	0.59	0.51	2.5	2011	[119]
Glass/graphene/PEDOT:PSS/P3HT:PC$_{61}$BM/Ca/Al	1.3	374	84	6.91	0.52	0.33	1.17	2012	[120]
Glass/ITO/ZnO/P3HT:PC$_{61}$BM/PEDOT:PSS/graphene/Au	0.3	96	90	11.97	0.56	0.45	3.04	2012	[121]
Glass/ITO/ZnO/P3HT:PC$_{61}$BM/PEDOT:PSS/graphene/Au	0.3	96	90	10.39	0.59	0.38	2.30	2012	[121]
Glass/graphene/PEDOT:PSS/P3HT:PC$_{61}$BM/Ca/Al	<15	450	>90	8	0.58	0.59	2.60	2012	[122]
PET/graphene/PEDOT:PSS/P3HT:PC$_{61}$BM/Ca/Al	<15	450	>90	8.3	0.56	0.56	2.54	2012	[122]
Quartz/graphene/PEDOT/DBP/C$_{60}$/BCP/Al	—	450	92	5.89	0.89	0.48	2.49	2012	[121]
Glass/graphene/PEDOT:PSS/P3HT:PC$_{61}$BM/Al	1.3	450	~90	9.91	0.53	0.34	1.79	2012	[123]
Glass/rGO/PEDOT:PSS/P3HT:PC$_{61}$BM/Al	16.4	1.6×10^3	70	5.62	0.57	0.34	1.1	2013	[109]
Glass/ITO/ZnO/P3HT:PC$_{61}$BM/PEDOT:PSS/graphene(Au)	0.7	—	—	10.6	0.60	0.50	3.17	2013	[124]
Quartz/graphene/PEDOT:PSS/P3HT:PC$_{61}$BM/Ca/Al	~1	2.5×10^3	~60	6.33	0.58	0.48	1.76	2013	[104]
Quartz/graphene/PEDOT:PSS/P3HT:PC$_{61}$BM/Ca/Al	21	1.46×10^3	52	5.43	0.58	0.55	1.73	2013	[105]
Quartz/rGO grid/PEDOT:PSS/P3HT:PC$_{61}$BM/LiF/Al	100	608	55	6.35	0.54	0.58	2.04	2013	[112]

nonconvalent functionalization with pyrene buanoic acid succidymidyl ester, the PCEs could be enhanced to 0.74% and 1.71%, respectively. The wetting problem could also be circumvented by vapor deposition of PEDOT as reported by Park et al.[125] One year later, Wang's group developed a layer-by-layer (LBL) transferring method to prepare graphene films with better conductivity. They were able to obtain a sheet resistance of about 100 Ω/sq and a transmittance of around 85% at 550 nm for a four-layer graphene film. The OSCs exhibited a PCE of 2.5%.[119] A couple of laboratories also used CVD-grown graphene films as the transparent electrode in OSCs. The device performance is strongly dependent on the conductivity of the graphene film.[115–117,120]

Because a CVD-grown graphene film has to be transferred from its growing substrate (usually transition metal foil) to other substrates, it becomes relatively easy to use graphene films as the top electrode of OSCs. Lee et al. were the first to report semitransparent inverted OSCs with a graphene film as the top electrode.[118] A graphene film was transferred onto the polymer layer of an OSC by a thermal releasing tape. Efficiencies of 2.50% and 1.88% were recorded when the OSC was illuminated from the ITO side and the graphene side, respectively. Liu et al. optimized the experimental conditions for the semitransparent inverted OSCs with a graphene top electrode.[121] The maximum PCE obtained was 3% for an active area of 6 mm^2, and it decreased to 2.3% when the active area increased to 50 mm^2. Later this group demonstrated an even higher PCE of 3.17%. They also found that multilayer graphene films were impermeable to air, which thus greatly improved the device stability.[124]

Flexible OSCs with a graphene transparent electrode were also demonstrated. Lee et al. showed the flexible OSCs could retain a PCE of 2.5%–2.6%, regardless of the bending conditions, even up to a 5.2 mm bending radius.[122] Kim et al. even made a stretchable graphene transparent electrode.[126] Because graphene is more transparent than ITO in the ultraviolet (UV) region, Zhao et al. demonstrated enhanced UV response in OSCs with a graphene transparent electrode and proposed its application as an UV photodetector.[121]

3.4 METAL NANOSTRUCTURES

Metals are well known for their excellent electrical conductivity. But metals are generally opaque to light. A transparent electrode made from metal becomes possible only if (1) the metal film is thin enough, and/or (2) there are holes in the film. A couple of techniques have been developed to fabricate metal nanostructures as transparent electrodes. The metal nanostructures are classified into metal nanomeshes, metal nanowire grids, and ultrathin metal films in this article.

3.4.1 Metal Nanomeshes

The booming of the semiconductor industry and development of nanolithography technology has largely expanded the gearbox for the fabrication of well-defined nanostructures. By employing nanolithography, metal nanomeshes with feature size in submicrometers were fabricated. They were studied as the transparent electrode of OSCs. Some recent publications on this topic are listed in Table 3.5.

TABLE 3.5
Photovoltaic Performance of OSCs with Metal Nanomeshes as the Transparent Electrode

Nanowire	Width (μm)	Height (nm)	Spacing (μm)	R_s (Ω/sq)	T (%)	Active Layer	J_{sc} (mA/cm²)	V_{oc} (V)	FF	PCE (%)	Year	Ref.
Ag	20–40	~100	100–800	~0.5	>85	APFO-Green 5:PC$_{61}$BM	3.67	0.58	0.47	1	2007	[127]
Au	0.4	40	10 and 0.7	24	84	P3HT:PC$_{61}$BM	5.5	0.57	0.62	1.96	2008	[128]
Cu	0.4	40	10 and 0.7	28	83	P3HT:PC$_{61}$BM	5.71	0.57	0.63	2.06	2008	[128]
Ag	0.4	40	10 and 0.7	23	78	P3HT:PC$_{61}$BM	5.34	0.58	0.65	2	2008	[128]
Cu	0.7	40	10 and 0.7	22	78	P3HT:PC$_{61}$BM	5.7	0.6	62	2.1	2010	[129]
Ag	0.055	40	—	—	77	CuPc:C$_{60}$	5.21	0.5	0.5	1.32	2010	[130]
Au	—	18	0.43	—	<85	P3HT:PC$_{61}$BM	9.6	0.53	0.61	3.12	2012	[131]
Au	0.05	30	0.2	2.2	81	P3HT:PC$_{61}$BM	10.4	0.62	0.67	4.4	2013	[132]

The replacement of ITO by metallic meshes was first reported by Tvingstedt et al. in 2007.[114] A master template was first made on a negative photoresist SU8 by photolithography. The template was then replicated with a PDMS stamp. Subsequently, the PDMS stamp was placed onto a desired substrate, a silver ink was flowed in through the microchannels between the PDMS stamp and the substrate, driven by the capillary effect and a pressure difference set up by a vacuum. After drying, parallel silver lines with a width of 20–40 μm, a height of ~100 nm, and a spacing of 100–800 μm can be fabricated on glass with lengths of up to 5 cm. The silver grid showed an extremely low sheet resistance of ~0.5 Ω/sq at an average transmittance of more than 85% in the visible light range. OSCs with APFO-Green 5 and PC$_{61}$BM as the active materials exhibited a PCE of 1%.

Kang et al. designed a rectangular grid pattern with a small feature size to improve the conductance and the transmittance (Figure 3.13a).[128] The small feature size was achieved by wet chemical etching of the SiO$_2$ grating. The rectangular grid pattern of the mold was then created by performing nanoimprint lithography twice using two grating molds with an orientation orthogonal to each other. PDMS stamps were created using the mold, and then Au, Cu, or Ag could be deposited onto the PDMS stamp by electron-beam evaporation to form the metal grid. The grid was finally transferred to the PEDOT:PSS/glass substrate by the same PDMS stamp. Following this protocol, Kang et al. were able to make grids with a width and spacing as small as 400 nm and 700 nm, respectively.[128] The height of the meshes could be controlled precisely at 40 nm. The sheet resistance can be below 30 Ω/sq at a transmittance of around 80%. The PCEs of P3HT:PC$_{61}$BM OSCs with different metal meshes were all around 2%. This group later demonstrated the ability to fabricate such metal meshes on flexible PET substrate using the roll-to-roll transfer printing process.[129]

The metal nanomeshes can have a shape different from a rectangle. Zhu et al. reported gold nanomeshes with a hexagonal array of periodic circular holes that were fabricated by using nanosphere lithography (Figure 3.13b).[131] In this technique, polystyrene nanospheres with a diameter of 430 nm and a size deviation of less than 3%

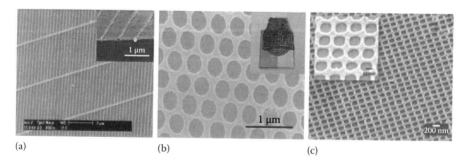

(a) (b) (c)

FIGURE 3.13 SEM images of (a) rectangular Cu meshes on PEDOT:PSS-coated glass (inset shows the cross section of the grid), (b) Au nanomesh with a hexagonal array of periodic circular holes on glass substrate (inset is a photograph of the 1.5 × 1.5 cm^2 glass substrate with Au nanomesh on the left half), (c) Au nanomesh with 175 nm diameter and 200 nm pitch hole arrays (inset is a magnified SEM image with scale bar of 200 nm). (Reproduced from M. G. Kang et al., *Adv. Mater.*, 20, 4408, 2008; J. F. Zhu et al., *Appl. Phys. Lett.*, 100, 143109, 2012; S. Y. Chou et al., *Opt. Exp.*, 21, A60, 2013.)

were self-assembled to form a monolayer of polystyrene nanospheres. This monolayer was then transferred onto a clean glass substrate. Oxygen reactive-ion etching was used to etch the edge of the polystyrene nanospheres and reduced their size. Subsequently, metals were deposited by electron beam evaporation. After removing the polystyrene nanospheres by ultrasonication in chloroform, a Au nanomesh was obtained. Replacing ITO with such a Au nanomesh yielded a J_{sc} of 9.6 mA/cm², a V_{oc} of 0.53 V, a FF of 0.61, and a PCE of 3.12% for P3HT:PC$_{61}$BM OSCs.

Recently, Chou and Ding fabricated Au nanomeshes with a grid spacing of only 200 nm by nanoimprint lithography (Figure 3.13c).[132] The 30-nm-thick Au nanomeshes exhibited a very low sheet resistance of 2.2 Ω/sq at a transmittance of 81%. The P3HT:PC$_{61}$BM OSCs using such Au nanomesh as the transparent anode showed a PCE of 4.4%. This efficiency exceeded that of the control device with an ITO anode (2.9%). The performance enhancement was mainly contributed by the gain in photocurrent as a result of the improved light absorption due to the plasmonic cavity effect.

3.4.2 METAL NANOWIRE GRIDS

The advancement of nanotechnology enables controlled synthesis of metal nanowires of high aspect ratio. Like CNTs, metal nanowires are solution-processable and highly conductive. The contact resistance among metal nanowires (NWs) is even smaller than that among CNTs. It could be further reduced by welding metal NWs at their junctions by the plasmonic effect.[133]

Typically metal NWs are synthesized by reducing a metal precursor in the solution in the presence of specific capping agents, such as poly(vinyl pyrrolidine) and ethylenediamine.[134] The role of the capping agent is to bind preferentially to some facets of the nucleus/seed of the metal NWs and lead to the preferential growth along the axial direction. Zhang et al. reported the synthesis of ultralong Cu NWs in a liquid-crystalline medium of hexadecylamine and cetyltriamonium bromide.[135] Guo et al. harvested well-dispersed Cu NWs with an uniform diameter through a catalytic scheme involving nickel ions.[136] Nowadays, metal NWs can be produced chemically on a large scale at low cost. They are now also commercially available.

Table 3.6 summarizes the works on exploiting metal NWs as the transparent electrode of OSCs. Lee et al. were the first to use solution-processed Ag NW networks as the transparent electrode of OSCs in 2008.[137] The Ag NWs had an average length of 8.7 ± 3.7 μm and an average diameter of 103 ± 17 nm. They were drop cast on a substrate to form a 2-D network. Because Ag NWs were dispersed by insulating PVP surfactants, the initial sheet resistance of the Ag NW network was more than 1 kΩ/sq. After partially decomposing PVP at 200°C, the sheet resistance could be reduced by one order of magnitude. By tuning the loading of Ag NWs, they demonstrated Ag nanowire grids with a sheet resistance of 10 Ω/sq at a transmittance of 85%. They also found that these electrodes could be bent to a radius of 4 mm without affecting the sheet resistance. The initial OSCs with Ag NWs as the transparent electrode exhibited a PCE of only 0.38%. But they improved the PCE to 2.5% by using P3HT:PC$_{61}$BM as the photoactive layer and the Ag NW network as the top electrode two years later.[138] Semitransparent OSCs were also demonstrated.[139]

TABLE 3.6
Photovoltaic Performance of OSCs with Metal Nanowire Grids as the Transparent Electrode

Nanowire	Length (μm)	Diameter (nm)	R_s (Ω/sq)	T (%)	Active Layer	J_{sc} (mA/cm²)	V_{oc} (V)	FF	PCE (%)	Year	Ref.
Ag	8.7 ± 3.7	103 ± 17	16	86	CuPc/PTCBI	1.83	—	—	0.38	2008	[137]
Ag	—	—	10	88	P3HT:PC$_{61}$BM	10.59	0.51	0.46	2.5	2010	[138]
Ag	—	—	15–25	75–85	CuPc/C$_{60}$	1.91	0.44	0.55	0.63	2010	[139]
Cu	>1000	100	50	90	P3HT:PC$_{61}$BM	10.4	0.55	0.53	3	2010	[140]
Ag	<100	<100	29	97	P3HT:PC$_{61}$BM	10.1	0.56	0.61	3.45	2011	[141]
Ag	—	—	15	—	P3HT:PC$_{61}$BM	9.51	0.6	0.68	3.85	2012	[142]
Ag	—	—	22	85	P3HT:PC$_{61}$BM	7.12	0.59	0.63	2.62	2012	[143]
Ag	—	—	38.7	85	P3HT:PC$_{61}$BM	8.93	0.59	0.62	3.23	2012	[144]
Ag	—	—	14	94	PBDTTPD:PC$_{71}$BM	8.83	0.95	0.6	5.03	2013	[145]
Ag	—	—	7	85	P3HT:PC$_{61}$BM	5.83	0.56	0.65	2.13	2013	[146]
Ag	7	90	25	81	P3HT:PC$_{61}$BM	7.1	0.56	0.59	2.31	2013	[147]
Ag	10	30	7	83	P3HT:Si-PCPDTBT:PC$_{61}$BM	9.5	0.59	0.62	3.5	2013	[148]

To fill the void space among the metal NWs can improve the charge collection of OSCs. Leem et al. put a TiO$_x$ layer of about 200 nm thick on Ag NWs for OSCs. The TiO$_x$ layer was used for electron collection. They obtained a PCE of 3.45%.[141] One year later, Ajuria et al. used ZnO nanoparticles with a diameter of 20–25 nm to replace TiO$_x$ as the electron transfer layer. They recorded a PCE of 3.85% for OSCs with the same organic donor and acceptor materials.[142] Recently, Beiley et al. pointed out the sheet resistance of the filling material in the voids of metal NW network must be less than 1 × 10^9 Ω/sq in order to overcome the resistive losses arising from the lateral transport of charges to the nearest metal wire.[145] They used a 75-nm-thick ZnO layer as the filling material for the Ag NWs. They observed a PCE of 5.0% on semitransparent OSCs with a low bandgap donor polymer, poly(di(2-ethylhexyloxy)benzol[1,2-*b*:4,5-*b'*]dithiohene-*co*-octylthieno[3,4-*c*]pyrrole-4,6-dione) (PBDTTPD), and ITO as the transparent top electrode.

Metal NWs could also be fabricated by electrospining. Wu et al. obtained Cu NWs with a length of over 1000 μm and diameter of around 100 nm.[140] The Cu NW network could have a sheet resistance of 50 Ω/sq at the transmittance of 90%. The OSCs showed a PCE of 3%.

Because the metal NWs could be well dispersed and stabilized in solution, large-scale fabrication of a metal NW transparent electrode via solution processing techniques is possible. In addition to laboratory-scale spin coating, other solution-based processing techniques have been reported, including brush painting,[144] spray coating,[146] dip coating,[147] etc.

3.4.3 METAL ULTRATHIN FILMS

When a metal film is only a few nanometers thick, it is transparent or semitransparent and can be used as the transparent electrode of OCSs. Stec et al. fabricated ultrathin gold films with the aid of a mixed monolayer of molecular adhesives.[149] They found that small molecules, such as 3-aminopropyltrimethoxysilane and 3-mercaptopropyltrimethoxysilane, were able to facilitate Au nucleation and promoting uniform film growth at a low thickness. The Au thin films with micrometer holes could have a sheet resistance of 27 Ω/sq at a transmittance of above 80%. They demonstrated that the ultrathin Au film could be used as a transparent electrode. A PCE of 3.36% was obtained for OSCs.[149] Ajuria et al. also achieved a PCE of 2.57% on P3HT:PC$_{61}$BM OSCs by replacing ITO with a 5-nm-thick gold layer.[150]

Ultrathin metal films with a low work function and good air stability were demonstrated by Hutter et al. in 2012.[151] A 0.8-nm-thick Al layer was deposited on a 7-nm-thick Cu film. Upon exposure to air, an ultrathin self-limiting ternary oxide layer forms at the surface. This oxide layer can protect the underlying Cu against oxidation. The Cu/AlO$_x$ films have a work function of 3.5 eV and a sheet resistance of 16.5 Ω/sq. The work function and sheer resistance only slightly change after 32 hr in air. Inverted OSCs, Cu:AlO$_x$/PCDTBT:PC$_{71}$BM/MoO$_3$/Al, could have a PCE of 4.0%.[151]

The metal ultrathin film can be readily deposited onto the photoactive layer to make a semitransparent OSC when it also has a transparent bottom electrode. Jen et al. demonstrated semitransparent OSCs with a PCE of ~6%.[152,153]

Metal nanostructures have one unique property, that is, plasmonic effect, which can potentially improve the light absorption of OSCs. In 2008, Reilly et al. fabricated a transparent electrode with surface-plasmon enhanced effect for OSCs.[154] The transparent electrode was a 30-nm-thick Ag film with random holes of 92 nm in diameter. UV-Vis spectra suggested the addition of subwavelength holes to the Ag film increased the magnitude of transmission and modulated the spectrum of light that passed through the film. This could thus improve J_{sc} of OSCs.

3.5 TRANSPARENT CONDUCTING OXIDE RATHER THAN ITO

Transparent conducting oxides (TCOs) are a unique type of materials with simultaneously high optical transparency and high electrical conductivity. Among all TCOs, ITO is widely used as the transparent electrode of optoelectronic devices. Owing to concern on the scarcity of indium, inexpensive alternatives for ITO are essential. The photovoltaic performances of OSCs with TCOs rather than ITO are summarized in Table 3.7.

One of the closest TCOs to ITO is fluorine-doped tin oxide ($F:SnO_2$ or FTO). It exhibits a similar sheet resistance to ITO but is slightly less transparent. The advantage of FTO lies in its thermal stability. It is able to retain its good optoelectronic properties at elevated temperatures of above 400°C, at which the properties of ITO degrade. So it is widely employed as the working electrodes in dye-sensitized solar cells (DSSCs) that need high-temperature sintering. However, when it comes to OSCs, the surface of FTO is much rougher than ITO and might cause shorting of the OSCs because OSCs are about 500 times thinner than DSSCs.[155] Yang and Forrest overcame this problem by depositing the CuPc and C_{60} organic layers via organic vapor phase deposition (OVPD).[156] Unlike growth using vacuum thermal evaporation in which the molecules follow radial trajectories from source to substrate, the molecules in OVPD diffuse through a boundary layer before reaching the substrate at random incident angles. Hence, the molecules can diffuse into the rough surfaces that are otherwise unreachable by vacuum thermal evaporation. They observed a PCE of 2.5% on the small molecule OSCs.

TCOs based on ZnO have also attracted much attention. ZnO is a wide bandgap (3.3 eV) semiconductor. Its conductivity can be greatly improved through doping with group III elements, such as Ga (GZO) or Al (AZO).[157–159] The films generally have a sheet resistance of less than 30 Ω/sq, a transmittance of above 80% in the visible range, and a work function (4.2 eV). OSCs with a GZO transparent electrode exhibited medium efficiencies due to low FF.[157–159] AZO can have conductivity and transparency as good as ITO. However, high OSC performance is still yet to be demonstrated.

Besides GZO and AZO, hydrogenated In-doped ZnO (In:ZnO:H)[160] and W-doped In_2O_3 ($W:In_2O_3$)[161] have also been proposed. But their properties are not as good as other TCOs.

Next-Generation Transparent Electrode Materials for Organic Solar Cells

TABLE 3.7
Photovoltaic Performance of OSCs with Transparent Conducting Oxides as the Transparent Electrode

TCO	Deposition Technique	Thickness (nm)	R_s (Ω/sq)	T (%)	Active Layer	J_{sc} (mA/cm²)	V_{oc} (V)	FF	PCE (%)	Year	Ref.
F:SnO₂	—	750	12	70–80	CuPc/C₆₀	9.1	0.61	—	2.5	2006	[156]
F:SnO₂	—	—	15	~80	P3HT:PC₆₁BM	11.27	0.59	0.63	4.31	2010	[162]
F:SnO₂	—	—	<15	>80	P3HT:PC₆₁BM	10.80	0.59	0.69	4.41	2011	[163]
Ga:ZnO	Pulsed laser deposition	—	—	>85	CuPc/C₆₀	9	0.50	0.35	1.25	2007	[157]
Ga:ZnO	Sputtering	400	14.3	90	P3HT:PC₆₁BM	8.52	0.52	0.32	1.4	2007	[158]
Ga:ZnO	Atomic layer deposition	300	24	84	P3HT:PC₆₁BM	7.5	0.60	0.56	2.5	2013	[159]
Al:ZnO	Sputtering	300	8	>80	DCV5T-Et/C₆₀	8.7	1.02	0.43	2.9	2007	[164]
Al:ZnO	Sputtering	541	26	92	CuPc/C₆₀	4.53	0.50	0.55	1.3	2009	[165]
Al:ZnO (O₃)	Sputtering	1000	4.59	87	P3HT:PC₆₁BM	8.94	0.50	0.45	2.01	2010	[166]
Al:ZnO	Atomic layer deposition	170	500	—	PHT/PTCDI	1.97	0.39	0.54	0.43	2010	[167]
Al:ZnO	Sputtering	257	20	85	P3HT:PC₆₁BM	9.05	0.61	0.56	3.06	2012	[168]
Al:ZnO	Dip coating	2000	25	—	Cu(In,Ga)Se₂	33.4	0.63	0.70	14.7	2013	[169]
Al:ZnO	Sputtering	–/0.5	10	91	CuPc/C₆₀	—	—	—	1.4	2008	[170]
Al:ZnO/Au	Sputtering	—	31	85	P3HT:PC₆₁BM	10.1	0.55	0.56	3.15	2010	[171]
Al:ZnO/NiO	Sputtering	362	20	82	PTB7:PC₇₁BM	12.1	0.73	0.72	6.3	2013	[172]
Al:ZnO/TiO₂	Sputtering	200	175	88	MEH-PPV:PC₆₁BM	1.68	0.8	0.39	0.53	2008	[160]
In:ZnO:H	Electron beam evaporation	120	40	78	P3HT:PC₆₁BM	8.73	0.64	0.63	3.52	2011	[161]
W:In₂O₃	Sputtering	—	18	—	P3HT:PC₆₁BM	11.2	0.58	0.55	3.60	2012	[173]
TiO₂:ITO	Sputtering	200	18.06	81	P3HT:PC₆₁BM	8.33	0.60	0.64	3.18	2012	[174]
TiO₂:ITO	Pulsed laser deposition	250	20	80	P3HT:PC₆₁BM	9.90	0.59	0.66	3.83	2013	[175]
Zn₀.₃In₁.₄Sn₀.₃O₃	Pulsed laser deposition	250	20	80	PTB7:PC₇₁BM	14.7	0.74	0.68	7.41	2013	[175]

3.6 MULTILAYER THIN FILMS

The original proposal of oxide/metal/oxide (O/M/O) multilayer thin film (MTF) was for tackling the technological difficulty of fabricating high-quality ITO on plastic substrate. Early in this century, ITO films of low resistivity were hardly achievable without high-temperature treatment. The high-temperature process is incompatible with a plastic substrate. To solve this problem, Fahland et al. inserted a thin Ag layer between two ITO layers to improve the overall electrical property.[176] Later, people found that it was unnecessary for the oxide layers to be transparent. In fact, any dielectric material with a high refractive index and moderate electrical conductivity was suitable for this application. In the O/M/O structure, the in-plane conductivity was mainly contributed by the metal layer, which must be thick enough to overcome its percolation threshold. The reflectivity rising from the metal layer could be significantly reduced due to the optical interference by the oxide layers with a high refractive index.

MTFs were studied as the transparent electrode of OSCs. Table 3.8 summarizes recent works. In 2009, Cattin et al. chose MoO_3 for the oxide layers because it was an efficient hot-collection material in OSCs. They found that the percolation threshold of the Ag layer was 9.5 nm and obtained the optimum MTF by sequentially depositing 45 nm of MoO_3, 10 nm of Ag, and another 37.5 nm of MoO_3 on glass. The MoO_3/Ag/MoO_3 films can have a sheet resistance of 57.7 Ω/sq and a transmittance of about 80% at 500 nm. Small molecule CuPc:C_{60} OSCs with MoO_3/Ag/MoO_3 as the transparent electrode had a PCE of 1.15%.[175]

Because the MTFs are fabricated by vacuum deposition, they can be readily used as the top electrode of OSCs. Tao et al. adopted MoO_3/Ag/MoO_3 and fabricated semitransparent P3HT:$PC_{61}BM$ OSCs on ITO glass. The optimal PCE was 0.96% when the cells were illuminated from the MTF side.[177]

Other materials were also exploited to construct the O/M/O MTFs, such as zinc tin oxide (ZTO) or V_2O_5 for the oxides and Au or Cu for the metal.[178–181] The two oxide layers in MTFs do not need to be the same material. Han et al. selected WO_3 as the inner oxide layer because its high work function and electrical conductivity made it a good anode buffer layer in OSCs. ZnS was also used as the outer oxide layer for the transmittance tuning purpose. After the optimization of the oxide layer thicknesses, they achieved a maximum PCE of 2.92% for P3HT:$PC_{71}BM$ OSCs. The OSCs also had good mechanical flexibility. They could be bent for 5000 cycles with little effect on the photovoltaic performance.[182]

The optical properties could be further improved by carefully designing the MTF architecture. Ham et al. chose WO_3 as the inner oxide layer because it had a high refractive index ($n = 1.99$) or a low extinction coefficient ($k = 0.024$). In comparison, V_2O_5 has an n of 2.24 and k of 0.047. The n and k values are 1.8, $k = 0.01$ for NiO, respectively. WO_3 is thus more transparent. Regarding the optical design of the outer oxide layer, their computer simulation showed enhanced transmittance over a broad range of wavelengths could be achieved from a dielectric with a refractive index of $n = 2.1$ and a low extinction coefficient ($k \sim 0$). Consequently, Ta_2O_5 ($n = 2.19$, $k = 0.0001$) was selected. The resultant WO_3 (40 nm)/Ag (10 nm)/Ta_2O_5 (35 nm) MTFs can have a sheet resistance of 11 Ω/sq and a transmittance of 90%, which was very close to the properties of ITO on glass.[183]

TABLE 3.8
Photovoltaic Performance of OSCs with Oxide/Metal/Oxide Multilayer Thin Films (MTFs) as the Transparent Electrode

MTF (Thickness in nm)	R_s (Ω/sq)	T (%)	Active Layer	J_{sc} (mA/cm^2)	V_{oc} (V)	FF	PCE (%)	Year	Ref.
MoO$_3$(37.5)/Ag(10)/MoO$_3$(45)	57.7	~80	CuPc/C$_{60}$	4.9	0.43	0.52	1.15	2009	[184]
MoO$_3$(1)/Ag(10)/MoO$_3$(20)	—	~90	P3HT:PC$_{61}$BM	2.72	0.57	0.62	0.96	2009	[177]
ITO(40)/Ag(16)/ITO(40)	4.4	86	P3HT:PC$_{61}$BM	9.22	0.54	0.65	3.25	2009	[185]
WO$_3$(30)/Ag(12)/ZnS(30)	—	—	P3HT:PC$_{71}$BM	6.83	0.61	0.7	2.92	2010	[182]
ZTO(35)/Ag(12)/ZTO(35)	3.96	86	P3HT:PC$_{61}$BM	7.95	0.55	0.59	2.55	2011	[178]
V$_2$O$_5$(10)/Ag(13)/V$_2$O$_5$(40)	—	90	P3HT:PC$_{61}$BM	3.79	0.57	0.61	1.3	2011	[179]
ZTO(20)/Ag(8)/ZTO(39)	8.8	82	P3HT:PC$_{61}$BM	6.4	0.55	0.62	2.2	2011	[180]
MoO$_3$(5)/Au(10)/MoO$_3$(40)	12	80	P3HT:PC$_{61}$BM	7.1	0.58	0.59	2.4	2012	[181]
WO$_3$(20)/Cu(12)/ZnS(40)	12.7	60	P3HT:PC$_{71}$BM	8.1	0.6	0.7	3.4	2012	[186]
MoO$_3$(1)/Ag(12)/MoO$_3$(10)/Alq$_3$(50)	~10	~50	CuPc/C$_{60}$	8.99	0.52	0.55	2.54	2012	[187]
MoO$_3$(2)/Ag(6)/MoO$_3$(10)	6.2	80	PCDTBT:PC$_{71}$BM	8.4	0.85	0.62	4.44	2012	[188]
WO$_3$(20)Ag(15)/WO$_3$(40)	—	—	P3HT:PC$_{61}$BM	5.85	0.59	0.58	2.00	2012	[189]
MoO$_3$(5)/Ag(13)/MoO$_3$(40)	5	~80	P3HT:PC$_{61}$BM	4.56	0.58	0.50	1.35	2012	[190]
MoO$_3$(5)/Ag(13)/MoO$_3$(40)	5	~80	PCDTBT:PC$_{71}$BM	7.36	0.90	0.47	3.08	2012	[190]
MoO$_3$(10)/Ag(9)/ZnO(20)	5	70	P3HT:PC$_{61}$BM	7.0	0.59	0.62	2.3	2013	[191]
WO$_3$(40)/Ag(10)/Ta$_2$O$_5$(35)	11	90	P3HT:PC$_{61}$BM	7.4	0.63	0.61	2.9	2013	[183]
TiO$_x$(5)/Ag(15)/ZnS(25)	—	—	PCDTBT:PC$_{71}$BM	9.8	0.90	0.58	5.1	2013	[192]
MoO$_3$(3)/Au(1)/Ag(7)/MoO$_3$(5)/Alq$_3$(45)	19	83	F4-ZnPc/C$_{60}$	11.7	0.67	0.60	4.7	2013	[193]
ZnO(30)/Ag(14)/ZnO(30)	7	86	P3HT:PC$_{61}$BM	9.5	0.54	0.50	2.58	2013	[194]
MoO$_x$(5)/Ag(10)/MoO$_x$(10)	9	70	SubPc/C$_{60}$	5.15	1.03	0.65	3.45	2013	[195]
ITO(30)/Ag(8)/ITO(30)	10	~60	PTB7-F20:PC$_{71}$BM	12.53	0.65	0.58	4.72	2013	[196]
ITO(30)/AgO$_x$(8)/ITO(30)	10	~70	PTB7-F20:PC$_{71}$BM	14.62	0.67	0.60	5.88	2013	[196]

Depending on the choice of the oxide layer, the MTF electrode can be used as a cathode for electron extraction. Han et al. used TiO_x as the inner oxide layer because TiO_x has a low work function and can form good Ohmic contact with the $PC_{61}BM$ acceptor.[192] Thus, efficient OSCs were demonstrated using MTF TEs.

Recently, Yun et al. reported a new MTF structure that showed better optoelectronic properties than the O/M/O structures mentioned above.[196] By the inclusion of two or three atomic percent of AgO_x into the Ag layer sandwiched between 30-nm-thick ITO layers, the refractive index of the Ag layer was sharply increased without decreasing the electrical conductivity of the MTF films. The AgO_x formation led to a decrease in the degree of light reflection by the original Ag film and a better morphology for the top ITO layer. Gaining from more transparent ITO/AgO_x/ITO MTF films, flexible OSCs with low-bandgap polymer PTB7 showed almost a PCE of 5.88%, higher than that of the OSCs with ITO/Ag/ITO.[196] Jin et al. showed O/M/O MTFs were suitable for large-area OSCs. They demonstrated 25.0 cm^2 monolithic submodules with MoO_3/Ag/MoO_3 as the top anode and observed PCEs of 1.35% and 3.08% for $P3HT:PC_{61}BM$ and $PCDTBT:PC_{71}BM$ OSCs, respectively.[190]

3.7 SUMMARY AND OUTLOOK

Although ITO is the most popular transparent electrode material for OSCs today, it is the most expensive component in OSCs.[197] Moreover, it has poor mechanical flexibility. In order to develop flexible OSCs at low cost, ITO should be replaced by other transparent conductive materials. A couple of materials, including conducting polymers, carbon nanotubes, graphene, metal nanostructures, transparent conducting oxides other than ITO, and multilayer thin films, have been reported as alternatives for ITO. Each material has its own merits but also some drawbacks. No single technology can satisfy all the requirements for the transparent electrode of OSCs at this moment. Nevertheless, their properties have been constantly improved after the understanding of these materials becomes deeper and deeper.

ACKNOWLEDGMENT

This work was supported by a research grant from the Ministry of Education, Singapore (R-284-000-113-112).

REFERENCES

1. Li, Y.F. Molecular design of photovoltaic materials for polymer solar cells: Toward suitable electronic energy levels and broad absorption. *Acc. Chem. Res.* 45, 723 (2012).
2. Krebs, F.C.; Espinosa, N.; Hösel, M.; Søndergaard, R.R.; Jørgensen, M. 25th anniversary article: Rise to power–OSC-based solar parks. *Adv. Mater.* 26, 29 (2013).
3. Lunt, R.R.; Bulovic, V. Transparent, near-infrared organic photovoltaic solar cells for window and energy-scavenging applications. *Appl. Phys. Lett.* 98, 113305 (2011).
4. Chen, C.C.; Dou, L.; Zhu, R.; Chung, C.H.; Song, T.B.; Zheng, Y.B.; Hawks, S.; Li, G.; Weiss, P.S.; Yang, Y. Visibly transparent polymer solar cells produced by solution processing. *ACS Nano* 6, 7185 (2012).

5. Chen, Y.H.; Chen, C.W.; Huang, Z.Y.; Lin, W.C.; Lin, L.Y.; Lin, F.; Wong, K.T.; Lin, H.W. Organic solar cells: Microcavity-embedded, colour-tuneable, transparent organic solar cells. *Adv. Mater.* 26, 1144 (2014).
6. Chen, C.C.; Dou, L.; Gao, J.; Chang, W.H.; Li, G.; Yang, Y. High-performance semi-transparent polymer solar cells possessing tandem structures. *Energy Environ. Sci.* 6, 2714 (2013).
7. Inganäs, O. Organic photovoltaics: Avoiding indium. *Nat. Photon.* 5, 201 (2011).
8. Chipman, A. A commodity no more. *Nature* 449, 131 (2007).
9. Skotheim, T.A. *Handbook of Conducting Polymers*, M. Dekker, New York, (1986).
10. Xia, Y.; Ouyang, J. Significant different conductivities of the two grades of poly(3,4-ethylenedioxythiophene):poly(styrenesulfonate), Clevios P and Clevios PH1000, arising from different molecular weights. *ACS Appl. Mater. Interfaces* 4, 4131 (2012).
11. Pei, Q.; Zuccarello, G.; Ahlskog, M.; Inganäs, O. Electrochromic and highly stable poly(3,4-ethylenedioxythiophene) switches between opaque blue-black and transparent sky blue. *Polymer* 35, 1347 (1994).
12. Bhandari, S.; Deepa, M.; Singh, S.; Gupta, G.; Kant, R. Redox behavior and optical response of nanostructured poly(3,4-ethylenedioxythiophene) films grown in a camphorsulfonic acid based micellar solution. *Electrochim. Acta* 53, 3189 (2008).
13. Chelawat, H.; Vaddiraju, S.; Gleason, K. Conformal, conducting poly(3,4-ethylenedioxythiophene) thin films deposited using bromine as the oxidant in a completely dry oxidative chemical vapor deposition process. *Chem. Mater.* 22, 2864 (2010).
14. Ha, Y.H.; Nikolov, N.; Pollack, S.K.; Mastrangelo, J.; Martin, B.D.; Shashidhar, R. Towards a transparent, highly conductive poly(3,4-ethylenedioxythiophene). *Adv. Funct. Mater.* 14, 615 (2004).
15. Lefebvre, M.; Qi, Z.; Rana, D.; Pickup, P.G. Chemical synthesis, characterization, and electrochemical studies of poly(3,4-ethylenedioxythiophene)/poly(styrene-4-sulfonate) composites. *Chem. Mater.* 11, 262 (1999).
16. Padmalekha, K.G.; Admassie, S. Electrochromic, magnetotransport and AC transport properties of vapor phase polymerized PEDOT (VPP PEDOT). *Synth. Met.* 159, 1885 (2009).
17. Fabretto, M.; Muller, M.; Zuber, K.; Murphy, P. Influence of PEG-ran-PPG surfactant on vapour phase polymerised PEDOT thin films. *Macromol. Rapid Commun.* 30, 1846 (2009).
18. Winther-Jensen, B.; West, K. Vapor-phase polymerization of 3,4-ethylenedioxythiophene: A route to highly conducting polymer surface layers. *Macromolecules* 37, 4538 (2004).
19. Winther-Jensen, B.; Breiby, D.W.; West, K. Base inhibited oxidative polymerization of 3,4-ethylenedixoythiophene with iron(III)tosylate. *Synth. Met.* 152, 1 (2005).
20. Fabretto, M.; Jariego-Moncunill, C.; Autere, J.P.; Michelmore, A.; Short, R.; Murphy, P. High conductivity PEDOT resulting from glycol/oxidant complex and glycol/polymer intercalation during vacuum vapour phase polymerization. *Polymer* 52, 1725 (2011).
21. Zuber, K.; Fabretto, M.; Hall, C.; Murphy, P. Improved PEDOT conductivity via suppression of crystallite formation in Fe(III) tosylate during vapor phase polymerization. *Macromol. Rapid Commun.* 29, 1503 (2008).
22. Evans, D.; Fabretto, M.; Mueller, M.; Zuber, K.; Short, R.; Murphy, P. Structure-directed growth of high conductivity PEDOT from liquid-like oxidant layers during vacuum vapor phase polymerization. *J. Mater. Chem.* 22, 14889 (2012).
23. Groenendaal, L.; Jonas, F.; Freitag, D.; Peilartzik, H.; Reynolds, J.R. Poly(3,4-ethylenedioxythiophene) and its derivatives: past, present, and future. *Adv. Mater.* 12, 481 (2000).
24. Cao, Y.; Yu, G.; Menon, R.; Heeger, A.J. Polymer light-emitting diodes with polyethylene dioxythiophene–polystyrene sulfonate as the transparent anode. *Synth. Met.* 87, 171 (1997).

25. Lang, U.; Müller, E.; Naujoks, N.; Dual, J. Microscopical investigations of PEDOT:PSS thin films. *Adv. Funct. Mater.* 19, 1215 (2009).
26. Xia, Y.; Ouyang, J. PEDOT:PSS films with significantly enhanced conductivities induced by preferential solvation with cosolvents and their application in polymer photovoltaic cells. *J. Mater. Chem.* 21, 4927 (2011).
27. Kim, J.Y.; Jung, J.H.; Lee, D.E.; Joo, J. Enhancement of electrical conductivity of poly(3,4-ethylenedioxythiophene)/poly(4-styrenesulfonate) by a change of solvents. *Synth. Met.* 126, 311 (2002).
28. Ouyang, J. "Secondary doping" methods to significantly enhance the conductivity of PEDOT:PSS for its application as transparent electrode of optoelectronic devices. *Displays* 34, 423 (2013).
29. Ouyang, J.; Xu, Q.; Chu, C.W.; Yang, Y.; Li, G.; Shinar, J. On the mechanism of conductivity enhancement in poly(3,4-ethylenedioxythiophene):poly(styrene sulfonate) film through solvent treatment. *Polymer* 45, 8443 (2004).
30. Crispin, X.; Jakobsson, F.L.E.; Crispin, A.; Grim, P.C.M.; Andersson, P.; Volodin, A.; van Haesendonck, C.; Van der Auweraer, M.; Salaneck, W.R.; Berggren, M. The origin of the high conductivity of poly(3,4-ethylenedioxythiophene)-poly(styrenesulfonate) (PEDOT-PSS) plastic electrodes. *Chem. Mater.* 18, 4354 (2006).
31. Nardes, A.M.; Janssen, A.J.R.; Kemerink, M.A. A morphological model for the solvent-enhanced conductivity of PEDOT:PSS thin films. *Adv. Funct. Mater.* 18, 865 (2008).
32. Döbbelin, M.; Marcilla, R.; Salsamendi, M.; Pozo-Gonzalo, C.; Carrasco, P.M.; Pomposo, J.A.; Mecerreyes, D. Influence of ionic liquids on the electrical conductivity and morphology of PEDOT:PSS films. *Chem. Mater.* 19, 2147 (2007).
33. Fan, B.; Mei, X.; Ouyang, J. Significant conductivity enhancement of conductive poly(3,4-ethylenedioxythiophene):poly(styrenesulfonate) films by adding anionic surfactants into polymer solution. *Macromolecules* 41, 5971 (2008).
34. Pettersson, L.A.A.; Ghosh, S.; Inganäs, O. Optical anisotropy in thin films of poly(3,4-ethylenedioxythiophene)–poly(4-styrenesulfonate). *Org. Electron.* 3, 143 (2002).
35. Jönsson, S.K.M.; Birgersonb, J.; Crispin, X.; Greczynski, G.; Osikowicz, W.; van der Gon, A.W.D.; Salaneck, J.R.; Fahlman, M. The effects of solvents on the morphology and sheet resistance in poly(3,4-ethylenedioxythiophene)–polystyrenesulfonic acid (PEDOT–PSS) films. *Synth. Met.* 139, 1 (2003).
36. Reyes-Reyes, M.; Cruz-Cruz, I.; Lopez-Sandoval, R. Enhancement of the electrical conductivity in PEDOT:PSS films by the addition of dimethyl sulfate. *J. Phys. Chem. C* 114, 20220 (2010).
37. Xia, Y.; Ouyang, J. Significant conductivity enhancement of conductive poly(3,4-ethyl enedioxythiophene):poly(styrenesulfonate) films through a treatment with organic carboxylic acids and inorganic acids. *ACS Appl. Mater. Interfaces* 2, 474 (2010).
38. Xia, Y.; Ouyang, J. Salt-induced charge screening and significant conductivity enhancement of conducting poly(3,4-ethylenedioxythiophene):poly(styrenesulfonate). *Macromolecules* 42, 4141 (2009).
39. Xia, Y.; Zhang, H.; Ouyang, J. Highly conductive PEDOT:PSS films prepared through a treatment with zwitterions and their application in polymer photovoltaic cells. *J. Mater. Chem.* 20, 9740 (2010).
40. Kim, Y.H.; Sachse, C.; Machala, M.L.; May, C.; Muller-Meskamp, L.; Leo, K. Highly conductive PEDOT:PSS electrode with optimized solvent and thermal post-treatment for ito-free organic solar cells. *Adv. Funct. Mater.* 21, 1076 (2011).
41. Xia, Y.J.; Sun, K.; Ouyang, J.Y. Highly conductive poly(3,4-ethylenedioxythiophene): poly(styrene sulfonate) films treated with an amphiphilic fluoro compound as the transparent electrode of polymer solar cells. *Energy Environ. Sci.* 5, 5325 (2012).

42. Sun, K.; Xia, Y.; Ouyang, J. Improvement in the photovoltaic efficiency of polymer solar cells by treating the poly(3,4-ethylenedioxythiophene):poly(styrenesulfonate) buffer layer with co-solvents of hydrophilic organic solvents and hydrophobic 1,2-dichlorobenzene. *Sol. Energy Mater. Sol. Cells* 97, 89 (2012).
43. Xia, Y.; Ouyang, J. Highly conductive PEDOT:PSS films prepared through a treatment with geminal diols or amphiphilic fluoro compounds. *Org. Electron.* 13, 1785 (2012).
44. Alemu, D.; Wei, H.Y.; Ho, K.C.; Chu, C.W. Highly conductive PEDOT:PSS electrode by simple film treatment with methanol for ITO-free polymer solar cells. *Energy Environ. Sci.* 5, 9662 (2012).
45. Cruz-Cruz, I.; Reyes-Reyes, M.; Aguilar-Frutis, M.A.; Rodriguez, A.G.; López-Sandoval, R. Study of the effect of DMSO concentration on the thickness of the PSS insulating barrier in PEDOT:PSS thin films. *Synth. Met.* 160, 1501 (2010).
46. Nardes, A.M.; Kemerink, M.; de Kok, M.M.; Vinken, E.; Maturova, K.; Janssen, R.A.J. Conductivity, work function, and environmental stability of PEDOT:PSS thin films treated with sorbitol. *Org. Electron.* 9, 727 (2008).
47. Peng, B.; Guo, X.; Cui, C.; Zou, Y.; Pan, C.; Li, Y. Performance improvement of polymer solar cells by using a solvent-treated poly(3,4-ethylenedioxythiophene):poly(styrenesulfonate) buffer layer. *Appl. Phys. Lett.* 98, 243308 (2011).
48. Ouyang, J.; Chu, C.-W.; Chen, F.-C.; Xu, Q.; Yang, Y. Polymer optoelectronic devices with highly-conductivity poly(3,4-ethylenedioxythiophene) anodes. *J. Macromol. Sci.; Pure Appl. Chem.* 41, 1497–1511 (2004).
49. Ouyang, J.; Chu, C.W.; Chen, F.C.; Xu, Q.; Yang, Y. High-conductivity poly(3,4-ethylenedioxythiophene):poly(styrene sulfonate) film and its application in polymer optoelectronic devices. *Adv. Funct. Mater.* 15, 203–208 (2005).
50. Xia, Y.; Ouyang, J. Anion effect on salt-induced conductivity enhancement of poly(3,4-ethylenedioxythiophene):poly(styrenesulfonate) films. *Org. Electron.* 11, 1129–1135 (2010).
51. Badre, C.; Marquant, L.; Alsayed, A.M.; Hough, L.A. Highly conductive poly(3,4-ethylenedioxythiophene):poly (styrenesulfonate) films using 1-ethyl-3-methylimidazolium tetracyanoborate ionic liquid. *Adv. Funct. Mater.* 22, 2723 (2012).
52. Ouyang, J.; Guo, T.F.; Yang, Y.; Higuchi, H.; Yoshioka, M.; Nagatsuka, T. High-performance, flexible polymer light-emitting diodes fabricated by a continuous polymer coating process. *Adv. Mater.* 14, 915 (2002).
53. Lipomi, D.J.; Lee, J.A.; Vosgueritchian, M.; Tee, B.C.K.; Bolander, J.A.; Bao, Z. Electronic properties of transparent conductive films of PEDOT:PSS on stretchable substrates. *Chem. Mater.* 24, 373 (2012).
54. Vosgueritchian, M.; Lipomi, D.J.; Bao, Z. Highly conductive and transparent PEDOT:PSS films with a fluorosurfactant for stretchable and flexible transparent electrodes. *Adv. Funct. Mater.* 22, 421 (2012).
55. Lim, F.J.; Ananthanarayanan, K.; Luther, J.; Ho, G.W. Influence of a novel fluorosurfactant modified PEDOT:PSS hole transport layer on the performance of inverted organic solar cells. *J. Mater. Chem.* 22, 25057 (2012).
56. Howden, R.M.; McVay, E.D.; Gleason, K.K. CVD poly(3,4-ethylenedioxythiophene) conductivity and lifetime enhancement via acid rinse dopant exchange. *J. Mater. Chem. A* 1, 1334 (2013).
57. Xia, Y.J.; Sun, K.; Ouyang, J.Y. Solution-processed metallic conducting polymer films as transparent electrode of optoelectronic devices. *Adv. Mater.* 24, 2436 (2012).
58. Joo, J.; Long, S.M.; Pouget, J.P.; Oh, E.J.; MacDiarmid, A.G.; Epstein, A.J. Charge transport of the mesoscopic metallic state in partially crystalline polyanilines. *Phys. Rev. B* 57, 9567 (1998).
59. Ouyang, J. Solution-processed PEDOT:PSS films with conductivities as indium tin oxide through a treatment with mild and weak organic acids. *ACS Appl. Mater. Interfaces* 5, 13082 (2013).

60. Zhang, F.; Johansson, M.; Andersson, M.R.; Hummelen, J.C.; Inganäs, O. Polymer photovoltaic cells with conducting polymer anodes. *Adv. Mater.* 14, 662 (2002).
61. Na, S.I.; Kim, S.S.; Jo, J.; Kim, D.Y. Efficient and flexible ITO-free organic solar cells using highly conductive polymer anodes. *Adv. Mater.* 20, 4061 (2008).
62. Ahlswede, E.; Mühleisen, W.; bin Moh Wahi, M.W.; Hanisch, J.; Powalla, M. Highly efficient organic solar cells with printable low-cost transparent contacts. *Appl. Phys. Lett.* 92, 143307 (2008).
63. Zhou, Y.H.; Zhang, F.L.; Tvingstedt, K.; Barrau, S.; Li, F.H.; Tian, W.J.; Inganäs, O. Investigation on polymer anode design for flexible polymer solar cells. *Appl. Phys. Lett.* 92, 233308 (2008).
64. Lim, Y.F.; Lee, S.; Herman, D.J.; Lloyd, M.T.; Anthony, J.E.; Malliaras, G.G. Spray-deposited poly(3,4-ethylenedioxythiophene):poly(styrenesulfonate) top electrode for organic solar cells. *Appl. Phys. Lett.* 93, 193301 (2008).
65. Hau, S.K.; Yip, H.-L.; Zou, J.; Jen, A.K.Y. Indium tin oxide-free semi-transparent inverted polymer solar cells using conducting polymer as both bottom and top electrodes. *Org. Electron.* 10, 1401 (2009).
66. Do, H.; Reinhard, M.; Vogeler, H.; Puetz, A.; Klein, M.F.G.; Schabel, W.; Colsmann, A.; Lemmer, U. Polymeric anodes from poly(3,4-ethylenedioxythiophene): poly(styrenesulfonate) for 3.5% efficient organic solar cells. *Thin Solid Films* 517, 5900 (2009).
67. Na, S.I.; Wang, G.; Kim, S.S.; Kim, T.W.; Oh, S.H.; Yu, B.K.; Lee, T.; Kim, D.Y. Evolution of nanomorphology and anisotropic conductivity in solvent-modified PEDOT:PSS films for polymeric anodes of polymer solar cells. *J. Mater. Chem.* 19, 9045 (2009).
68. Nickel, F.; Puetz, A.; Reinhard, M.; Do, H.; Kayser, C.; Colsmann, A.; Lemmer, U. Cathodes comprising highly conductive poly(3,4-ethylenedioxythiophene):poly (styrenesulfonate) for semi-transparent polymer solar cells. *Org. Electron.* 11, 535 (2010).
69. Dong, Q.; Zhou, Y.; Pei, J.; Liu, Z.; Li, Y.; Yao, S.; Zhang, J.; Tian, W. All-spin-coating vacuum-free processed semi-transparent inverted polymer solar cells with PEDOT:PSS anode and PAH-D interfacial layer. *Org. Electron.* 11, 1327 (2010).
70. Na, S.I.; Yu, B.K.; Kim, S.S.; Vak, D.; Kim, T.S.; Yeo, J.S.; Kim, D.Y. Fully spray-coated ITO-free organic solar cells for low-cost power generation. *Sol. Energy Mater. Sol. Cells* 94, 1333 (2010).
71. Zhou, Y.; Cheun, H.; Choi, S.; Potscavage, W.J.; Fuentes-Hernandez, C.; Kippelen, B. Indium tin oxide-free and metal-free semitransparent organic solar cells. *Appl. Phys. Lett.* 97, 153304 (2010).
72. Zhou, Y.H.; Cheun, H.; Choi, S.; Fuentes-Hernandez, C.; Kippelen, B. Optimization of a polymer top electrode for inverted semitransparent organic solar cells. *Org. Electron.* 12, 827 (2011).
73. Lipomi, D.J.; Tee, B.C.K.; Vosgueritchian, M.; Bao, Z. Stretchable organic solar cells. *Adv. Mater.* 23, 1771 (2011).
74. Colsmann, A.; Reinhard, M.; Kwon, T.-H.; Kayser, C.; Nickel, F.; Czolk, J.; Lemmer, U.; Clark, N.; Jasieniak, J.; Holmes, A.B.; Jones, D. Inverted semi-transparent organic solar cells with spray coated, surfactant free polymer top-electrodes. *Sol. Energy Mater. Sol. Cells* 98, 118 (2012).
75. Tait, J.G.; Worfolk, B.J.; Maloney, S.A.; Hauger, T.C.; Elias, A.L.; Buriak, J.M.; Harris, K.D. Spray coated high-conductivity PEDOT:PSS transparent electrodes for stretchable and mechanically-robust organic solar cells. *Sol. Energy Mater. Sol. Cells* 110, 98 (2013).
76. Hsiao, Y.S.; Whang, W.T.; Chen, C.P.; Chen, Y.C. High-conductivity poly(3,4-ethylen edioxythiophene):poly(styrene sulfonate) film for use in ITO-free polymer solar cells. *J. Mater. Chem.* 18, 5948 (2008).

77. Kim, Y.H.; Sachse, C.; Hermenau, M.; Fehse, K.; Riede, M.; Muller-Meskamp, L.; Leo, K. Improved efficiency and lifetime in small molecule organic solar cells with optimized conductive polymer electrodes. *Appl. Phys. Lett.* 99, 113305 (2011).
78. Yeo, J.S.; Yun, J.M.; Kim, D.Y.; Kim, S.S.; Na, S.I. Successive solvent-treated PEDOT:PSS electrodes for flexible ITO-free organic photovoltaics. *Sol. Energy Mater. Sol. Cells* 114, 104 (2013).
79. Kim, N.; Kee, S.; Lee, S.H.; Lee, B.H.; Kahng, Y.H.; Jo, Y.R.; Kim, B.J.; Lee, K. Highly conductive PEDOT:PSS nanofibrils induced by solution-processed crystallization. *Adv. Mater.* 26, 2268 (2014).
80. Iijima, S. Helical microtubules of graphitic carbon. *Nature* 354, 56 (1991).
81. Reich, S.; Thomsen, C.; Ordejón, P. Electronic band structure of isolated and bundled carbon nanotubes. *Phys. Rev. B* 65, 155411 (2002).
82. Mei, X.; Ouyang, J. Highly conductive and transparent single-walled carbon nanotube thin films fabricated by gel coating. *J. Mater. Chem.* 21, 17842 (2011).
83. Hu, L.; Hecht, D.S.; Grüner, G. Percolation in transparent and conducting carbon nanotube networks. *Nano Lett.* 4, 2513 (2004).
84. Ago, H.; Petritsch, K.; Shaffer, M.S.P.; Windle, A.H.; Friend, R.H. Composites of carbon nanotubes and conjugated polymers for photovoltaic devices. *Adv. Mater.* 11, 1281 (1999).
85. Wu, Z.C.; Chen, Z.H.; Du, X.; Logan, J.M.; Sippel, J.; Nikolou, M.; Kamaras, K.; Reynolds, J.R.; Tanner, D.B.; Hebard, A.F.; Rinzler, A.G. Transparent, conductive carbon nanotube films. *Science* 305, 1273 (2004).
86. Ulbricht, R.; Jiang, X.; Lee, S.; Inoue, K.; Zhang, M.; Fang, S.; Baughman, R.; Zakhidov, A. Polymeric solar cells with oriented and strong transparent carbon nanotube anode. *Phys. Stat. Sol. B* 243, 3528 (2006).
87. van de Lagemaat, J.; Barnes, T.M.; Rumbles, G.; Shaheen, S.E.; Coutts, T.J.; Weeks, C.; Levitsky, I.; Peltola, J.; Glatkowski, P. Organic solar cells with carbon nanotubes replacing In_2O_3: Sn as the transparent electrode. *Appl. Phys. Lett.* 88, 233503 (2006).
88. Rowell, M.W.; Topinka, M.A.; McGehee, M.D.; Prall, H.J.; Dennler, G.; Sariciftci, N.S.; Hu, L.B.; Gruner, G. Organic solar cells with carbon nanotube network electrodes. *Appl. Phys. Lett.* 88, 233506 (2006).
89. Zhou, Y.; Hu, L.; Grüner, G. A method of printing carbon nanotube thin films. *Appl. Phys. Lett.* 88, 123109 (2006).
90. Tenent, R.C.; Barnes, T.M.; Bergeson, J.D.; Ferguson, A.J.; To, B.; Gedvilas, L.M.; Heben, M.J.; Blackburn, J.L. Ultrasmooth, large-area, high-uniformity, conductive transparent single-walled-carbon-nanotube films for photovoltaics produced by ultrasonic spraying. *Adv. Mater.* 21, 3210 (2009).
91. Du Pasquier, A.; Unalan, H.E.; Kanwal, A.; Miller, S.; Chhowalla, M. Conducting and transparent single-wall carbon nanotube electrodes for polymer-fullerene solar cells. *Appl. Phys. Lett.* 87, 203511 (2005).
92. Feng, Y.; Ju, X.; Feng, W.; Zhang, H.; Cheng, Y.; Liu, J.; Fujii, A.; Ozaki, M.; Yoshino, K. Organic solar cells using few-walled carbon nanotubes electrode controlled by the balance between sheet resistance and the transparency. *Appl. Phys. Lett.* 94, 123302 (2009).
93. Tanaka, S.; Zakhidov, A.A.; Ovalle-Robles, R.; Yoshida, Y.; Hiromitsu, I.; Fujita, Y.; Yoshino, K. Semitransparent organic photovoltaic cell with carbon nanotube-sheet anodes and Ga-doped ZnO cathodes. *Synth. Met.* 159, 2326 (2009).
94. Barnes, T.M.; Bergeson, J.D.; Tenent, R.C.; Larsen, B.A.; Teeter, G.; Jones, K.M.; Blackburn, J.L.; van de Lagemaat, J. Carbon nanotube network electrodes enabling efficient organic solar cells without a hole transport layer. *Appl. Phys. Lett.* 96, 243309 (2010).

95. Kim, S.; Yim, J.; Wang, X.; Bradley, D.D.C.; Lee, S.; Demello, J.C. Spin- and spray-deposited single-walled carbon-nanotube electrodes for organic solar cells. *Adv. Funct. Mater.* 20, 2310 (2010).
96. Tyler, T.P.; Brock, R.E.; Karmel, H.J.; Marks, T.J.; Hersam, M.C. Electronically monodisperse single-walled carbon nanotube thin films as transparent conducting anodes in organic photovoltaic devices. *Adv. Energy Mater.* 1, 785 (2011).
97. Kim, Y.H.; Muller-Meskamp, L.; Zakhidov, A.A.; Sachse, C.; Meiss, J.; Bikova, J.; Cook, A.; Leo, K. Semi-transparent small molecule organic solar cells with laminated free-standing carbon nanotube top electrodes. *Sol. Energy Mater. Sol. Cells* 96, 244 (2012).
98. Arnold, M.S.; Green, A.A.; Hulvat, J.F.; Stupp, S.I.; Hersam, M.C. Sorting carbon nanotubes by electronic structure using density differentiation. *Nat. Nanotechnol.* 1, 60 (2006).
99. Güttinger, J.; Molitor, F.; Stampfer, C.; Schnez, S.; Jacobsen, A.; Dröscher, S.; Ihn, T.; Ensslin, K. Transport through graphene quantum dots. *Rep. Prog. Phys.* 75, 126502 (2012).
100. Geim, A.K.; Novoselov, K.S. The rise of graphene. *Nat. Mater.* 6, 183 (2007).
101. Bae, S.; Kim, H.; Lee, Y.; Xu, X.; Park, J.S.; Zheng, Y.; Balakrishnan, J.; Lei, T.; Kim, H.R.; Song, Y.I.; Kim, K.S.; Ozyilmaz, B.; Ahn, J.H.; Hong, B.H.; Iijima, S. Roll-to-roll production of 30-inch graphene films for transparent electrodes. *Nat. Nanotechnol.* 5, 574 (2010).
102. Mei, X.; Zheng, H.; Ouyang, J. Ultrafast reduction of graphene oxide with Zn powder in neutral and alkaline solutions at room temperature promoted by the formation of metal complexes. *J. Mater. Chem.* 22, 9109 (2012).
103. Wang, X.; Zhi, L.J.; Tsao, N.; Tomovic, Z.; Li, J.L.; Mullen, K. Transparent carbon films as electrodes in organic solar cells. *Angew. Chem. Int. Ed.* 47, 2990 (2008).
104. Na, S.I.; Noh, Y.J.; Son, S.Y.; Kim, T.W.; Kim, S.S.; Lee, S.; Joh, H.I. Efficient organic solar cells with solution-processed carbon nanosheets as transparent electrodes. *Appl. Phys. Lett.* 102, 043304 (2013).
105. Na, S.I.; Lee, J.S.; Noh, Y.J.; Kim, T.W.; Kim, S.S.; Joh, H.I.; Lee, S. Efficient ITO-free polymer solar cells with pitch-converted carbon nanosheets as novel solution-processable transparent electrodes. *Sol. Energy Mater. Sol. Cells* 115, 1 (2013).
106. Wu, J.B.; Becerril, H.A.; Bao, Z.N.; Liu, Z.F.; Chen, Y.S.; Peumans, P. Organic solar cells with solution-processed graphene transparent electrodes. *Appl. Phys. Lett.* 92, 263302 (2008).
107. Eda, G.; Lin, Y.Y.; Miller, S.; Chen, C.W.; Su, W.F.; Chhowalla, M. Transparent and conducting electrodes for organic electronics from reduced graphene oxide. *Appl. Phys. Lett.* 92, 233305 (2008).
108. Eda, G.; Fanchini, G.; Chhowalla, M. Large-area ultrathin films of reduced graphene oxide as a transparent and flexible electronic material. *Nat. Nanotechnol.* 3, 270 (2008).
109. Kymakis, E.; Savva, K.; Stylianakis, M.M.; Fotakis, C.; Stratakis, E. Flexible organic photovoltaic cells with in situ nonthermal photoreduction of spin-coated graphene oxide electrodes. *Adv. Funct. Mater.* 23, 2742 (2013).
110. Su, Q.; Pang, S.; Alijani, V.; Li, C.; Feng, X.; Müllen, K. Composites of graphene with large aromatic molecules. *Adv. Mater.* 21, 3191 (2009).
111. Xu, Y.; Long, G.; Huang, L.; Huang, Y.; Wan, X.; Ma, Y.; Chen, Y. Polymer photovoltaic devices with transparent graphene electrodes produced by spin-casting. *Carbon* 48, 3308 (2010).
112. Zhang, Q.; Wan, X.J.; Xing, F.; Huang, L.; Long, G.K.; Yi, N.B.; Ni, W.; Liu, Z.B.; Tian, J.G.; Chen, Y.S. Solution-processable graphene mesh transparent electrodes for organic solar cells. *Nano Res.* 6, 478 (2013).
113. Wang, Y.; Chen, X.; Zhong, Y.; Zhu, F.; Loh, K.P. Large area, continuous, few-layered graphene as anodes in organic photovoltaic devices. *Appl. Phys. Lett.* 95, 063302 (2009).

114. Li, X.S.; Cai, W.W.; An, J.H.; Kim, S.; Nah, J.; Yang, D.X.; Piner, R.; Velamakanni, A.; Jung, I.; Tutuc, E.; Banerjee, S.K.; Colombo, L.; Ruoff, R.S. Large-area synthesis of high-quality and uniform graphene films on copper foils. *Science* 324, 1312 (2009).
115. Kalita, G.; Matsushima, M.; Uchida, H.; Wakita, K.; Umeno, M. Graphene constructed carbon thin films as transparent electrodes for solar cell applications. *J. Mater. Chem.* 20, 9713 (2010).
116. Park, H.; Rowehl, J.A.; Kim, K.K.; Bulovic, V.; Kong, J. Doped graphene electrodes for organic solar cells. *Nanotechnology* 21, 505204 (2010).
117. Arco, L.G.D.; Zhang, Y.; Schlenker, C.W.; Ryu, K.; Thompson, M.E.; Zhou, C. Continuous, highly flexible, and transparent graphene films by chemical vapor deposition for organic photovoltaics. *ACS Nano* 4, 2865 (2010).
118. Lee, Y.Y.; Tu, K.H.; Yu, C.C.; Li, S.-S.; Hwang, J.Y.; Lin, C.C.; Chen, K.H.; Chen, L.C.; Chen, H.L.; Chen, C.W. Top laminated graphene electrode in a semitransparent polymer solar cell by simultaneous thermal annealing/releasing method. *ACS Nano* 5, 6564 (2011).
119. Wang, Y.; Tong, S.W.; Xu, X.F.; Özyilmaz, B.; Loh, K.P. Interface engineering of layer-by-layer stacked graphene anodes for high-performance organic solar cells. *Adv. Mater.* 23, 1514 (2011).
120. Choi, Y.Y.; Kang, S.J.; Kim, H.K.; Choi, W.M.; Na, S.I. Multilayer graphene films as transparent electrodes for organic photovoltaic devices. *Sol. Energy Mater. Sol. Cells* 96, 281 (2012).
121. Liu, Z.K.; Li, J.H.; Sun, Z.H.; Tai, G.A.; Lau, S.P.; Yan, F. The application of highly doped single-layer graphene as the top electrodes of semitransparent organic solar cells. *ACS Nano* 6, 810 (2012).
122. Lee, S.; Yeo, J.S.; Ji, Y.; Cho, C.; Kim, D.Y.; Na, S.I.; Lee, B.H.; Lee, T. Flexible organic solar cells composed of P3HT:PCBM using chemically doped graphene electrodes. *Nanotechnology* 23, 344013 (2012).
123. Zhao, Z.Y.; Fite, J.D.; Haldar, P.; Lee, J.U. Enhanced ultraviolet response using graphene electrodes in organic solar cells. *Appl. Phys. Lett.* 101, 063305 (2012).
124. Liu, Z.; Li, J.; Yan, F. Package-free flexible organic solar cells with graphene top electrodes. *Adv. Mater.* 25, 4296 (2013).
125. Park, H.; Howden, R.M.; Barr, M.C.; Bulovic, V.; Gleason, K.; Kong, J. Organic solar cells with graphene electrodes and vapor printed poly(3,4-ethylenedioxythiophene) as the hole transporting layers. *ACS Nano* 6, 6370 (2012).
126. Kim, K.S.; Zhao, Y.; Jang, H.; Lee, S.Y.; Kim, J.M.; Kim, K.S.; Ahn, J.H.; Kim, P.; Choi, J.Y.; Hong, B.H. Large-scale pattern growth of graphene films for stretchable transparent electrodes. *Nature* 457, 706 (2009).
127. Tvingstedt, K.; Inganas, O. Electrode grids for ITO-free organic photovoltaic devices. *Adv. Mater.* 19, 2893 (2007).
128. Kang, M.G.; Kim, M.S.; Kim, J.S.; Guo, L.J. Organic solar cells using nanoimprinted transparent metal electrodes. *Adv. Mater.* 20, 4408 (2008).
129. Kang, M.G.; Park, H.J.; Ahn, S.H.; Guo, L.J. Transparent Cu nanowire mesh electrode on flexible substrates fabricated by transfer printing and its application in organic solar cells. *Sol. Energy Mater. Sol. Cells* 94, 1179 (2010).
130. Kang, M.G.; Xu, T.; Park, H.J.; Luo, X.G.; Guo, L.J. Efficiency enhancement of organic solar cells using transparent plasmonic Ag nanowire electrodes. *Adv. Mater.* 22, 4378 (2010).
131. Zhu, J.F.; Zhu, X.D.; Hoekstra, R.; Li, L.; Xiu, F.X.; Xue, M.; Zeng, B.Q.; Wang, K.L. Metallic nanomesh electrodes with controllable optical properties for organic solar cells. *Appl. Phys. Lett.* 100, 143109 (2012).
132. Chou, S.Y.; Ding, W. Ultrathin, high-efficiency, broad-band, omni-acceptance, organic solar cells enhanced by plasmonic cavity with subwavelength hole array. *Opt. Exp.* 21, A60 (2013).

133. Garnett, E.C.; Cai, W.; Cha, J.J.; Mahmood, F.; Connor, S.T.; Greyson Christoforo, M.; Cui, Y.; McGehee, M.D.; Brongersma, M.L. Self-limited plasmonic welding of silver nanowire junctions. *Nat. Mater.* 11, 241 (2012).

134. Rathmell, A.R.; Bergin, S.M.; Hua, Y.L.; Li, Z.Y.; Wiley, B.J. The growth mechanism of copper nanowires and their properties in flexible, transparent conducting films. *Adv. Mater.* 22, 3558 (2010).

135. Zhang, D.; Wang, R.; Wen, M.; Weng, D.; Cui, X.; Sun, J.; Li, H.; Lu, Y. Synthesis of ultralong copper nanowires for high-performance transparent electrodes. *J. Am. Chem. Soc.* 134, 14283 (2012).

136. Guo, H.; Lin, N.; Chen, Y.; Wang, Z.; Xie, Q.; Zheng, T.; Gao, N.; Li, S.; Kang, J.; Cai, D.; Peng, D.L. Copper nanowires as fully transparent conductive electrodes. *Sci. Rep.* 3, 2323 (2013).

137. Lee, J.Y.; Connor, S.T.; Cui, Y.; Peumans, P. Solution-processed metal nanowire mesh transparent electrodes. *Nano Lett.* 8, 689 (2008).

138. Gaynor, W.; Lee, J.Y.; Peumans, P. Fully solution-processed inverted polymer solar cells with laminated nanowire electrodes. *ACS Nano* 4, 30 (2010).

139. Lee, J.Y.; Connor, S.T.; Cui, Y.; Peumans, P. Semitransparent organic photovoltaic cells with laminated top electrode. *Nano Lett.* 10, 1276 (2010).

140. Wu, H.; Hu, L.; Rowell, M.W.; Kong, D.; Cha, J.J.; McDonough, J.R.; Zhu, J.; Yang, Y.; McGehee, M.D.; Cui, Y. Electrospun metal nanofiber webs as high-performance transparent electrode. *Nano Lett.* 10, 4242 (2010).

141. Leem, D.S.; Edwards, A.; Faist, M.; Nelson, J.; Bradley, D.D.C.; de Mello, J.C. Efficient organic solar cells with solution-processed silver nanowire electrodes. *Adv. Mater.* 23, 4371 (2011).

142. Ajuria, J.; Ugarte, I.; Cambarau, W.; Etxebarria, I.; Tena-Zaera, R.; Pacios, R. Insights on the working principles of flexible and efficient ITO-free organic solar cells based on solution processed Ag nanowire electrodes. *Sol. Energy Mater. Sol. Cells* 102, 148 (2012).

143. Lim, J.W.; Cho, D.Y.; Eun, K.T.; Choa, S.H.; Na, S.I.; Kim, J.H.; Kim, H.K. Mechanical integrity of flexible Ag nanowire network electrodes coated on colorless PI substrates for flexible organic solar cells. *Sol. Energy Mater. Sol. Cells* 105, 69 (2012).

144. Lim, J.W.; Cho, D.Y.; Jihoon, K.; Na, S.I.; Kim, H.K. Simple brush-painting of flexible and transparent Ag nanowire network electrodes as an alternative ITO anode for cost-efficient flexible organic solar cells. *Sol. Energy Mater. Sol. Cells* 107, 348 (2012).

145. Beiley, Z.M.; Christoforo, M.G.; Gratia, P.; Bowring, A.R.; Eberspacher, P.; Margulis, G.Y.; Cabanetos, C.; Beaujuge, P.M.; Salleo, A.; McGehee, M.D. Semi-transparent polymer solar cells with excellent sub-bandgap transmission for third generation photovoltaics. *Adv. Mater.* 25, 7020 (2013).

146. Krantz, J.; Stubhan, T.; Richter, M.; Spallek, S.; Litzov, I.; Matt, G.J.; Spiecker, E.; Brabec, C.J. Spray-coated silver nanowires as top electrode layer in semitransparent P3HT:PCBM-based organic solar cell devices. *Adv. Funct. Mater.* 23, 1711 (2013).

147. Sachse, C.; Muller-Meskamp, L.; Bormann, L.; Kim, Y.H.; Lehnert, F.; Philipp, A.; Beyer, B.; Leo, K. Transparent, dip-coated silver nanowire electrodes for small molecule organic solar cells. *Org. Electron.* 14, 143 (2013).

148. Guo, F.; Zhu, X.; Forberich, K.; Krantz, J.; Stubhan, T.; Salinas, M.; Halik, M.; Spallek, S.; Butz, B.; Spiecker, E.; Ameri, T.; Li, N.; Kubis, P.; Guldi, D.M.; Matt, G.J.; Brabec, C.J. ITO-free and fully solution-processed semitransparent organic solar cells with high fill factors. *Adv. Energy Mater.* 3, 1062 (2013).

149. Stec, H.M.; Williams, R.J.; Jones, T.S.; Hatton, R.A. Ultrathin transparent Au electrodes for organic photovoltaics fabricated using a mixed mono-molecular nucleation layer. *Adv. Funct. Mater.* 21, 1709 (2011).

150. Ajuria, J.; Etxebarria, I.; Cambarau, W.; Munecas, U.; Tena-Zaera, R.; Jimeno, J.C.; Pacios, R. Inverted ITO-free organic solar cells based on p and n semiconducting oxides. New designs for integration in tandem cells, top or bottom detecting devices, and photovoltaic windows. *Energy Environ. Sci.* 4, 453 (2011).
151. Hutter, O.S.; Stec, H.M.; Hatton, R.A. An indium-free low work function window electrode for organic photovoltaics which improves with in-situ oxidation. *Adv. Mater.* 25, 284 (2012).
152. Chen, K.S.; Salinas, J.F.; Yip, H.L.; Huo, L.; Hou, J.; Jen, A.K.Y. Semi-transparent polymer solar cells with 6% PCE, 25% average visible transmittance and a color rendering index close to 100 for power generating window applications. *Energy Environ. Sci.* 5, 9551 (2012).
153. Chueh, C.C.; Chien, S.C.; Yip, H.L.; Salinas, J.F.; Li, C.Z.; Chen, K.S.; Chen, F.C.; Chen, W.C.; Jen, A.K.Y. Toward high-performance semi-transparent polymer solar cells: Optimization of ultra-thin light absorbing layer and transparent cathode architecture. *Adv. Energy Mater.* 3, 417 (2013).
154. Reilly, T.H.; van de Lagemaat, J.; Tenent, R.C.; Morfa, A.J.; Rowlen, K.L. Surface-plasmon enhanced transparent electrodes in organic photovoltaics. *Appl. Phys. Lett.* 92, 243304 (2008).
155. Zheng, H.; Neo, C.Y.; Ouyang, J. Highly efficient iodide/triiodide dye-sensitized solar cells with gel-coated reduce graphene oxide/single-walled carbon nanotube composites as the counter electrode exhibiting an open-circuit voltage of 0.90 V. *ACS Appl. Mater. Interfaces* 5, 6657 (2013).
156. Yang, F.; Forrest, S.R. Organic solar cells using transparent SnO_2–F anodes. *Adv. Mater.* 18, 2018 (2006).
157. Bhosle, V.; Prater, J.T.; Yang, F.; Burk, D.; Forrest, S.R.; Narayan, J. Gallium-doped zinc oxide films as transparent electrodes for organic solar cell applications. *J. Appl. Phys.* 102, 023501 (2007).
158. Owen, J.; Son, M.S.; Yoo, K.-H.; Ahn, B.D.; Lee, S.Y. Organic photovoltaic devices with Ga-doped ZnO electrode. *Appl. Phys. Lett.* 90, 033512 (2007).
159. Song, Y.S.; Seong, N.J.; Choi, K.J.; Ryu, S.O. Optical and electrical properties of transparent conducting gallium-doped ZnO electrodes prepared by atomic layer deposition for application in organic solar cells. *Thin Solid Films* 546, 271 (2013).
160. Park, Y.R.; Jung, D.; Kim, Y.S. Organic solar cells with hydrogenated In-doped ZnO replacing Sn-doped In_2O_3 as transparent electrode. *Jpn. J. Appl. Phys.* 47, 516 (2008).
161. Hu, Z.; Zhang, J.; Chen, X.; Ren, S.; Hao, Z.; Geng, X.; Zhao, Y. Performance of electron beam deposited tungsten doped indium oxide films as anodes in organic solar cells. *Sol. Energy Mater. Sol. Cells* 95, 2173 (2011).
162. Baek, W.H.; Choi, M.; Yoon, T.S.; Lee, H.H.; Kim, Y.S. Use of fluorine-doped tin oxide instead of indium tin oxide in highly efficient air-fabricated inverted polymer solar cells. *Appl. Phys. Lett.* 96, 133506 (2010).
163. Hu, Z.; Zhang, J.; Hao, Z.; Hao, Q.; Geng, X.; Zhao, Y. Highly efficient organic photovoltaic devices using F-doped SnO2 anodes. *Appl. Phys. Lett.* 98, 123302 (2011).
164. Schulze, K.; Maennig, B.; Leo, K.; Tomita, Y.; May, C.; Hüpkes, J.; Brier, E.; Reinold, E.; Bäuerle, P. Organic solar cells on indium tin oxide and aluminum doped zinc oxide anodes. *Appl. Phys. Lett.* 91, 073521 (2007).
165. Murdoch, G.B.; Hinds, S.; Sargent, E.H.; Tsang, S.W.; Mordoukhovski, L.; Lu, Z.H. Aluminum doped zinc oxide for organic photovoltaics. *Appl. Phys. Lett.* 94, 213301 (2009).
166. Park, J.H.; Ahn, K.J.; Park, K.I.; Na, S.I.; Kim, H.K. An Al-doped ZnO electrode grown by highly efficient cylindrical rotating magnetron sputtering for low cost organic photovoltaics. *J. Phys. D: Appl. Phys.* 43, 115101 (2010).
167. Saarenpää, H.; Niemi, T.; Tukiainen, A.; Lemmetyinen, H.; Tkachenko, N. Aluminum doped zinc oxide films grown by atomic layer deposition for organic photovoltaic devices. *Sol. Energy Mater. Sol. Cells* 94, 1379 (2010).

168. De Sio, A.; Chakanga, K.; Sergeev, O.; von Maydell, K.; Parisi, J.; von Hauff, E. ITO-free inverted polymer solar cells with ZnO:Al cathodes and stable top anodes. *Sol. Energy Mater. Sol. Cells* 98, 52 (2012).
169. Hagendorfer, H.; Lienau, K.; Nishiwaki, S.; Fella, C.M.; Kranz, L.; Uhl, A.R.; Jaeger, D.; Luo, L.; Gretener, C.; Buecheler, S.; Romanyuk, Y.E.; Tiwari, A.N. Highly transparent and conductive ZnO:Al thin films from a low temperature aqueous solution approach. *Adv. Mater.* 26, 632 (2014).
170. Bernède, J.C.; Berredjem, Y.; Cattin, L.; Morsli, M. Improvement of organic solar cell performances using a zinc oxide anode coated by an ultrathin metallic layer. *Appl. Phys. Lett.* 92, 083304 (2008).
171. Sun, N.; Fang, G.; Qin, P.; Zheng, Q.; Wang, M.; Fan, X.; Cheng, F.; Wan, J.; Zhao, X. Bulk heterojunction solar cells with NiO hole transporting layer based on AZO anode. *Sol. Energy Mater. Sol. Cells* 94, 2328 (2010).
172. Formica, N.; Ghosh, D.S.; Martinez-Otero, A.; Chen, T.L.; Martorell, J.; Pruneri, V. Ultrathin oxidized Ti to increase stability and smoothness of Al doped ZnO transparent conductors for high efficiency indium-free polymer solar cells. *Appl. Phys. Lett.* 103, 183304 (2013).
173. Lee, K.S.; Lim, J.W.; Kim, H.K.; Alford, T.L.; Jabbour, G.E. Transparent conductive electrodes of mixed TiO_2-indium tin oxide for organic photovoltaics. *Appl. Phys. Lett.* 100, 213302 (2012).
174. Lim, J.W.; Na, S.I.; Kim, H.K. Anatase TiO_2 and ITO co-sputtered films for an indium-saving multicomponent electrode in organic solar cells. *Sol. Energy Mater. Sol. Cells* 98, 409 (2012).
175. Zhou, N.; Buchholz, D.B.; Zhu, G.; Yu, X.; Lin, H.; Facchetti, A.; Marks, T.J.; Chang, R.P.H. Ultraflexible polymer solar cells using amorphous zinc–indium–tin oxide transparent electrodes. *Adv. Mater.* 26, 1098 (2014).
176. Fahland, M.; Karlsson, P.; Charton, C. Low resisitivity transparent electrodes for displays on polymer substrates. *Thin Solid Films* 392, 334 (2001).
177. Tao, C.; Xie, G.H.; Liu, C.X.; Zhang, X.D.; Dong, W.; Meng, F.X.; Kong, X.Z.; Shen, L.; Ruan, S.P.; Chen, W.Y. Semitransparent inverted polymer solar cells with MoO_3/Ag/MoO_3 as transparent electrode. *Appl. Phys. Lett.* 95, 053303 (2009).
178. Choi, Y.Y.; Choi, K.H.; Lee, H.; Kang, J.W.; Kim, H.K. Nano-sized Ag-inserted amorphous $ZnSnO_3$ multilayer electrodes for cost-efficient inverted organic solar cells. *Sol. Energy Mater. Sol. Cells* 95, 1615 (2011).
179. Shen, L.; Xu, Y.; Meng, F.X.; Li, F.M.; Ruan, S.P.; Chen, W.Y. Semitransparent polymer solar cells using V_2O_5/Ag/V_2O_5 as transparent anodes. *Org. Electron.* 12, 1223 (2011).
180. Winkler, T.; Schmidt, H.; Flugge, H.; Nikolayzik, F.; Baumann, I.; Schmale, S.; Weimann, T.; Hinze, P.; Johannes, H.H.; Rabe, T.; Hamwi, S.; Riedl, T.; Kowalsky, W. Efficient large area semitransparent organic solar cells based on highly transparent and conductive ZTO/Ag/ZTO multilayer top electrodes. *Org. Electron.* 12, 1612 (2011).
181. Cao, W.R.; Zheng, Y.; Li, Z.F.; Wrzesniewski, E.; Hammond, W.T.; Xue, J.G. Flexible organic solar cells using an oxide/metal/oxide trilayer as transparent electrode. *Org. Electron.* 13, 2221 (2012).
182. Han, S.; Lim, S.; Kim, H.; Cho, H.; Yoo, S. Versatile multilayer transparent electrodes for ito-free and flexible organic solar cells. *IEEE J. Sel. Topics Quantum Electron.* 16, 1656 (2010).
183. Choi, H.W.; Theodore, N.D.; Alford, T.L. ZnO-Ag-MoO_3 transparent composite electrode for ITO-free, PEDOT: PSS-free bulk-heterojunction organic solar cells. *Sol. Energy Mater. Sol. Cells* 117, 446 (2013).
184. Cattin, L.; Morsli, M.; Dahou, F.; Abe, S.Y.; Khelil, A.; Bernede, J.C. Investigation of low resistance transparent MoO_3/Ag/MoO_3 multilayer and application as anode in organic solar cells. *Thin Solid Films* 518, 4560 (2009).

185. Jeong, J.A.; Kim, H.K. Low resistance and highly transparent ITO-Ag-ITO multilayer electrode using surface plasmon resonance of Ag layer for bulk-heterojunction organic solar cells. *Sol. Energy Mater. Sol. Cells* 93, 1801 (2009).
186. Lim, S.; Han, D.; Kim, H.; Lee, S.; Yoo, S. Cu-based multilayer transparent electrodes: A low-cost alternative to ITO electrodes in organic solar cells. *Sol. Energy Mater. Sol. Cells* 101, 170 (2012).
187. Schubert, S.; Hermenau, M.; Meiss, J.; Muller-Meskamp, L.; Leo, K. Oxide sandwiched metal thin-film electrodes for long-term stable organic solar cells. *Adv. Funct. Mater.* 22, 4993 (2012).
188. Sergeant, N.P.; Hadipour, A.; Niesen, B.; Cheyns, D.; Heremans, P.; Peumans, P.; Rand, B.P. Design of transparent anodes for resonant cavity enhanced light harvesting in organic solar cells. *Adv. Mater.* 24, 728 (2012).
189. Yu, W.J.; Shen, L.; Meng, F.X.; Long, Y.B.; Ruan, S.P.; Chen, W.Y. Effects of the optical microcavity on the performance of ITO-free polymer solar cells with $WO_3/Ag/WO_3$ transparent electrode. *Sol. Energy Mater. Sol. Cells* 100, 226 (2012).
190. Jin, H.; Tao, C.; Velusamy, M.; Aljada, M.; Zhang, Y.; Hambsch, M.; Burn, P.L.; Meredith, P. Efficient, large area ITO-and-PEDOT-free organic solar cell sub-modules. *Adv. Mater.* 24, 2572 (2012).
191. Ham, J.; Kim, S.; Jung, G.H.; Dong, W.J.; Lee, J.L. Design of broadband transparent electrodes for flexible organic solar cells. *J. Mater. Chem. A* 1, 3076 (2013).
192. Han, D.; Lee, S.; Kim, H.; Jeong, S.; Yoo, S. Cathodic multilayer transparent electrodes for ITO-free inverted organic solar cells. *Org. Electron.* 14, 1477 (2013).
193. Schubert, S.; Meiss, J.; Müller-Meskamp, L.; Leo, K. Improvement of transparent metal top electrodes for organic solar cells by introducing a high surface energy seed layer. *Adv. Energy Mater.* 3, 438 (2013).
194. Vedraine, S.; El Hajj, A.; Torchio, P.; Lucas, B. Optimized ITO-free tri-layer electrode for organic solar cells. *Org. Electron.* 14, 1122 (2013).
195. Xu, W.F.; Chin, C.C.; Hung, D.W.; Wei, P.K. Transparent electrode for organic solar cells using multilayer structures with nanoporous silver film. *Sol. Energy Mater. Sol. Cells* 118, 81 (2013).
196. Yun, J.; Wang, W.; Bae, T.S.; Park, Y.H.; Kang, Y.C.; Kim, D.H.; Lee, S.; Lee, G.H.; Song, M.; Kang, J.W. Preparation of flexible organic solar cells with highly conductive and transparent metal-oxide multilayer electrodes based on silver oxide. *ACS Appl. Mater. Interfaces* 5, 9933 (2013).
197. Emmott, C.J.M.; Urbina, A.; Nelson, J. Environmental and economic assessment of ITO-free electrodes for organic solar cells. *Sol. Energy Mater. Sol. Cells* 97, 14 (2012).

Section II

Modeling

4 Relating Synthesis Parameters to the Morphology of the Photoactive Layer in Organic Photovoltaic Solar Cells Using Molecular Dynamics Simulations

S. M. Mortuza and Soumik Banerjee

CONTENTS

4.1	Introduction	90
4.2	Theoretical Background of MD Simulations	92
4.3	Molecular Simulations of Systems Relevant to OPVs	95
	4.3.1 Relating Synthesis Parameters to Morphology Using MD	95
	4.3.1.1 Relative Concentration of Polymers and Acceptor Species in the Blend	96
	4.3.1.2 Choice of Solvent	98
	4.3.1.3 Solute Concentration in Solvent	102
	4.3.1.4 Chemical Structure of Polymers and Acceptor Species	103
	4.3.2 Other Studies Related to the Morphology of Photoactive Layers Using MD	104
4.4	Outlook and Future Directions	107
References		108

4.1 INTRODUCTION

Photovoltaic solar cells, which utilize incidental solar radiation to generate electrical energy, are widely considered as potential renewable energy devices capable of meeting the increasing energy demand.[1] Photovoltaic solar cells are often classified based on the materials employed to achieve energy conversion, which include both inorganic and organic carbon-based materials. Despite the dominant use of inorganic materials in solar cell technology, organic photovoltaic (OPV) solar cells have generated significant scientific interest, owing to the inexpensive fabrication process, light weight, ease of manufacturing, flexible structure, and low operating temperature.[2,3] Although state-of-the-art OPV solar cells lack the efficiency for large-scale applications and have stability issues,[4] their low production costs[5] have resulted in significant research efforts to improve their efficiency. Such efforts have led to a substantial increase in the power conversion efficiency (PCE) over the past decade with a maximum efficiency of 12% reported recently.[6]

The fundamental steps involved in conversion of solar energy to electrical energy include[7] (1) absorption of the incident photon in photoactive layers comprising electron donor and acceptor materials that lead to the generation of an exciton (electron-hole pair); (2) the diffusion of the exciton to the electron donor–acceptor interface; (3) the separation of the exciton into an electron and a hole at the interface; (4) the hopping of separated electrons and holes from an individual acceptor or donor to another; and (5) the transport of these electrons and holes to their corresponding electrodes. Each step has an efficiency associated with it, and their product determines the external quantum efficiency (EQE),[8] the ratio of the number of photogenerated electrons collected to the number of incident photons.

$$EQE = \eta_A \eta_{diff} \eta_{diss} \eta_{tr} \eta_{cc} \quad (4.1)$$

In Equation 4.1, η_A is the photon absorption efficiency and is determined by the optical absorption coefficient of the photoactive layer and its thickness.[9] The parameter η_{diff} is the exciton diffusion yield, which represents the ability of excitons to diffuse through the polymer without recombination.[9] The length scale for exciton diffusion in order to preclude recombination is limited to ~10 nm.[10] Recombination losses majorly impede attainment of high efficiencies in OPV. It is therefore extremely important to minimize exciton recombination losses by reducing the effective diffusion length scale, which can be achieved by maximizing the donor–acceptor interface within a photoactive layer.[11] The term η_{diss} is the exciton dissociation yield, which is the probability that the hole and electron will be separated at the interface of the electron donor and acceptor materials. This is determined by the energy difference between the lowest unoccupied molecular orbital (LUMO) of the donor and acceptor.[9] The charge carrier transport yield η_{tr} involves hopping of the separated electron (hole) from one acceptor (donor) to another. The charge collection yield η_{cc} represents the ability of the charges to be transported from the interface of the electron donor and acceptor materials to the electrodes. In order to increase the charge collection efficiency, adequate networks of electron donor and acceptor materials are required to facilitate the transportation of hole and electron, respectively, to the electrodes.

Relating Synthesis Parameters to the Photoactive Layer

Of the abovementioned steps, efficient execution of steps (2) and (5) depends on the morphology of the photoactive layers of the OPV cells. Organic bulk heterojunctions (BHJ), which comprise blends of conjugated polymers and functionalized fullerenes as electron donors and acceptors, respectively, were designed for efficient exciton dissociation and thus achieve the abovementioned second step efficiently.[12,13] Dispersing acceptor species within the polymer in BHJs leads to an increase in the interfacial area between donor and acceptor materials, thus facilitating exciton separation. However, maximization of the interfacial area between the electron and donor materials, and formation of suitable networks of these materials for transport of the charge from the interface to the electrodes is still a challenge. The requirements of maximizing such interface area and creating such networks indicate the importance of morphology of photoactive layers for better device efficiency. Based on these requirements, conjugated polymers, such as poly(3-hexylthiophene) (P3HT), and fullerene derivatives, such as [6,6]-phenyl-C61-butyric acid methyl ester (PCBM), are the most popular choices for electron donor and acceptor species, respectively, in BHJs.[11,14–17] Figure 4.1 presents an ideal morphology of photoactive layers of OPV with sufficient interfacial area for exciton dissociation and presence of adequate donor and acceptor networks for transporting holes and electrons to the respective electrodes.

The photoactive layer in a BHJ is commonly synthesized through spin coating[18] of a solvent containing the acceptor nanoparticles and a polymer.[9] The solution, comprising polymer–nanoparticle blends, is prepared by mixing both species and dissolving it in solvents. The solution-based synthesis of the photoactive layer and subsequent annealing control its morphology and result in specific distributions of donor–acceptor interfaces. Specifically, the following factors govern the morphology of the polymer–nanoparticle blends: (i) the relative concentration of the polymer and acceptor species, (ii) the choice of solvent, (iii) the solute concentration in the solvent, and (iv) the chemical structure of the donor and acceptor species.[19] Despite being studied extensively,[20] a consensus on the morphology of polymer–nanoparticle blends and suitable synthesis parameters has not yet emerged.[7]

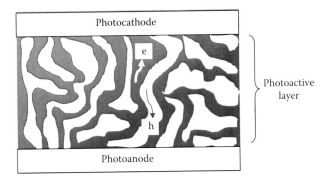

FIGURE 4.1 A schematic diagram illustrating a portion of BHJ, which consists of a bi-continuous donor and acceptor phase, is presented. The dark region corresponds to the electron acceptor material and the lighter domain corresponds to the electron donor material. The notations e and h correspond to electron and hole, respectively.

The ultimate choice of synthesis parameters is determined by the structure of agglomerates of acceptor molecules and polymers and their assembly and disassembly in photoactive layers. The molecular interactions among solvent molecules, polymeric units, and acceptor species during photoactive layer formation, in turn, influence the formation of such agglomerate structures. Therefore atomistic simulations, which can directly account for intermolecular interactions, are ideal tools for studying these systems. In particular, molecular dynamics (MD) simulations have been successfully used to study interactions among different molecular components in nanocomposites and are able to connect atomic-level details to the bulk properties of such systems.[7,21–34] The goal of this chapter is to provide an overview of the scope of molecular modeling in predicting optimal synthesis parameters for OPVs, based on some of the recent research efforts reported in the scientific literature. To that effect, this chapter is structured to include a brief background and discussion on the fundamental theory for MD simulations (Section 4.2) and discussions of results from recent molecular simulations based studies that relate the synthesis parameters to the morphology of photoactive layers in OPVs (Section 4.3). Finally, a broader outlook and possible future directions for applying molecular simulations to the field of OPVs are laid out (Section 4.4).

4.2 THEORETICAL BACKGROUND OF MD SIMULATIONS

Molecular dynamics simulations utilize a computer or a network of computers to simulate the physical motion of atoms and molecules in a system. In MD simulations, the macroscopic properties of a system are explored by studying microscopic systems numerically. The connection between these numerical simulations and macroscopic properties is made via the principles of statistical mechanics. For instance, the velocity distribution of the interacting particles in the system within an ensemble (collection of all possible systems, which have different microscopic states but identical macroscopic or thermodynamic states), which follows the Maxwell-Boltzmann distribution, can be directly related to the temperature of the system. The principle objective of the simulations is to attain energetically favorable states of the system of atoms and analyze them to obtain valuable structural and thermodynamic information.

Classical MD simulation consists of the numerical, step-by-step solution of the classical equations of motion, which for an atom i within a system comprising a collection of N atoms may be written:

$$m_i \ddot{r}_i = F_i, \qquad (4.2)$$

where m_i is the mass of the atom, $\ddot{r}_i$ is the position vector of that atom, and F_i is the force acting on that atom. Integration of Newton's laws of motion, using various algorithms, such as Verlet,[35] Velocity Verlet,[36] and LeapFrog,[37] generates molecular trajectories in space and time. The force on an atom is defined as the negative gradient of the potential energy function, which is a function of atomic positions of N atoms in a system.

$$F_i = -\frac{\partial U(r_1, \ldots, r_N)}{\partial r_i} \qquad (4.3)$$

Relating Synthesis Parameters to the Photoactive Layer

This potential function involves summing over a large number of bonded and nonbonded terms.

$$U = U_{bond} + U_{angle} + U_{dihedral} + U_{nonbonded} \tag{4.4}$$

$$U_{nonbonded} = U_{vdw} + U_{electrostatic} \tag{4.5}$$

The bond potential between a pair of covalently bonded atoms is expressed in various forms, such as Morse potential and harmonic potential. The total bond potential, U_{bond}, modeled as a harmonic function, is defined as

$$U_{bond} = \sum_{bonds} \frac{1}{2} k_b (r - b)^2, \tag{4.6}$$

where k_b is the force constant, and r and b are the bond length and the equilibrium bond length, respectively.

A harmonic function is commonly used to describe the potential for the angle between three covalently bonded atoms. The total potential in the system for such angular interactions, U_{angle}, is defined as

$$U_{angle} = \sum_{angles} \frac{1}{2} k_\theta (\theta - \theta_0)^2, \tag{4.7}$$

where k_θ is the bond bending force constant, and θ and θ_0 are the bond angle and the equilibrium bond angle, respectively. The bond angles θ are between successive bond vectors, and therefore involve three atoms.

The total dihedral potential in the system $U_{dihedral}$, is often defined as

$$U_{dihedral} = \sum_{dihedrals} k_\varphi (1 + \cos[n\varphi + \varphi_0]), \tag{4.8}$$

where k_φ is a rotation constant, n is the periodicity of the potential (number of minima/maxima) in the interval $[0, 2\pi]$, φ is torsion angle, and φ_0 is the corresponding equilibrium angle. The torsion angles φ are defined in terms of three connected bonds, hence involving four atomic coordinates.

The total configurational potential energy due to van der Waals interactions in the system U_{vdw}, described typically by Lennard-Jones (LJ) pairwise potential, can be expressed as

$$U_{vdw} = \sum_{i=1}^{N-1} \sum_{j=i+1}^{N} 4\varepsilon \left(\frac{\sigma^{12}}{r_{ij}} - \frac{\sigma^6}{r_{ij}} \right), \tag{4.9}$$

where ε is the depth of the potential well, σ is the finite distance at which the interparticle potential is zero, and r_{ij} is the distance between the particles i and j.

The net pairwise electrostatic interactions in the system, $U_{electrostatic}$, can be expressed as:

$$U_{electrostatic} = \sum_{i=1}^{N-1} \sum_{j=i+1}^{N} \frac{q_i q_j}{4\pi\varepsilon_0 r_{ij}}, \quad (4.10)$$

where r_{ij} represents the distance between the particles i and j with charges and q_i and q_j, respectively, and ε_0 is the permittivity of free space.

The set of parameters, such as the equilibrium bond lengths, bond angles, force constants, partial charges, and LJ parameters, that are employed in the above equations and pertain to specific atoms or coarse-grained (CG) sites are collectively known as a force field. Such parameters and constants are derived from either experimental work or high-level quantum mechanical calculations. The force fields can be classified as "all atom" and "united atom" force fields. The parameters for every type of atom in a system are provided by all-atom force fields, and a united-atom force field combines hydrogen and carbon atoms in each terminal methyl and each methylene bridge and treats them as a single interaction center. Some of the widely used force fields for modeling organic systems in OPVs are AMBER,[38] CHARMM,[39] and OPLS.[40] Popular MD simulation packages, which utilize such force fields for molecular modeling, are LAMMPS,[41] Gromacs,[42] DL_Poly,[43] and Tinker.[44]

The length and time scales accessible to all-atom MD simulations are restricted to approximately 1000 Å and 10^{-7} s, respectively. Although the thickness of the photoactive layer is approximately 100 nm (1000 Å), the size in other dimensions (length and width) is much greater than 100 nm. Molecular modeling of such systems requires simulation of billions of atoms over time scales up to several seconds, which is not possible without substantial reduction of the number of equations of motion. Therefore, CG models are frequently used for relatively long simulations of such systems. The CG method reduces the number of interacting sites in molecules to some beads instead of all the atoms while maintaining the accuracy in the dynamics and structural properties desired from a simulation. Coarse graining can also refer to the removal of certain degrees of freedom (e.g., vibrational modes between two atoms) by freezing the bonds, angles, or torsional degrees of freedom. Although the CG method provides crude representations of molecules, it allows bridging the gap between simulation and experimental time and length scales. Several research groups have used the CG technique to relate the processing parameters with the morphology of photoactive layers. As a representative case, Figure 4.2 shows coarse graining schemes, whereby Lee et al.[28] considered whole P3HT and PCBM as two separate beads, and Marsh et al.[30] and Jankowski et al.[33] considered three beads (two for side chains and one for backbone) for polymers for their CG studies.

Experimental characterization of the morphology of BHJ layers have mostly relied on transmission electron microscopy (TEM)[45] and atomic force microscopy (AFM).[18] However, experimental characterization of the three-dimensional microstructure of the BHJ layer is not trivial with these techniques. Therefore, the detailed structure within the BHJ layer is not revealed. In contrast, all-atom and coarse-grained molecular dynamics (CGMD) can provide valuable information regarding the three-dimensional structure across the entire thickness of the BHJ. The inherent

Relating Synthesis Parameters to the Photoactive Layer

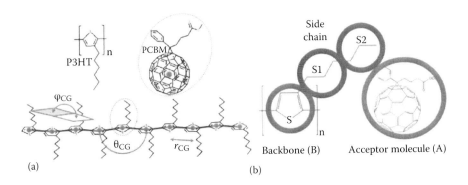

FIGURE 4.2 (See color insert.) (a) Chemical structures of P3HT and PCBM are shown. Pink beads designate the CG representation of P3HT monomers, and a blue bead designates the CG representation of a PCBM molecule. The intramolecular degrees of freedom between the CG particles of P3HT monomers are highlighted, where r_{CG} represents bond length, θ_{CG} represents bond angle, and φ_{CG} represents planar angle. (Reproduced from C.-K. Lee et al., *Energy and Environmental Science* 2011, 4, 4124–4132, 2011, with permission from the Royal Society of Chemistry.) (b) Coarse-grained model of conjugated polymer with backbone beads (B) and side-chain beads (S1 and S2) and acceptor molecule (A) are shown. (Marsh, H. S.; Jayaraman, A., Morphological studies of blends of conjugated polymers and acceptor molecules using langevin dynamics simulations. *Journal of Polymer Science Part B-Polymer Physics* 51, 64–77, 2013. Copyright Wiley-VCH Verlag GmbH & Co. KGaA. Reproduced with permission.)

molecular-scale resolution combined with the predictive capability has made MD a very suitable and popular numerical approach to study organic heterojunctions.

4.3 MOLECULAR SIMULATIONS OF SYSTEMS RELEVANT TO OPVs

The performance of BHJ cells is determined by morphological characteristics, such as the following:

1. The interface-to-volume ratio of the blend, which should be large enough to allow for efficient exciton dissociation without recombination.
2. The percolation ratio, defined by the volume fraction of electron donor and acceptor phases that percolates through the blend, should be high enough for efficient charge transport and collection at both electrodes to generate photocurrents.

These quantities, in turn, depend on several synthesis parameters, which are mentioned in Section 4.1. Predicting these synthesis parameters by repetitive experiments is time-consuming and expensive. Therefore, MD simulations have been performed for systems relevant to OPVs in order to predict these synthesis parameters and help experimentalists with the selection of favorable parameters.

4.3.1 Relating Synthesis Parameters to Morphology Using MD

It is essential to characterize the shapes of the nanoparticles and polymer agglomerates at various conditions because the nanoparticle–polymer interfacial area is critical to the performance of photovoltaic cells. Sections 4.3.1.1–4.3.1.4 are structured to address molecular modeling studies of distinct synthesis parameters that affect the agglomeration of acceptor materials and polymers, their crystalline or amorphous structures, and the domain sizes and shapes. Collectively, these factors define the morphology of the photoactive layer.

4.3.1.1 Relative Concentration of Polymers and Acceptor Species in the Blend

The concentrations of the acceptor species and polymers play a crucial role both during synthesis as well as during the annealing process and determine the PCE.[46,47] Although polymer concentration determines the photogeneration efficiency, acceptor concentration controls charge transport and recombination losses.[45] If acceptor molecules and polymers are miscible (acceptor molecules are completely dispersed in polymers) in the photoactive layer, charge transport is strongly suppressed because no percolation pathway is formed. On the other hand, if these components are immiscible, the interfacial area between nanoparticles and polymers is relatively low. Therefore, partial miscibility between the conjugated polymer and acceptor molecules is required to create percolation pathways for electrons and holes to travel to the electrodes[48–50] and to simultaneously increase the interfacial area between these components. The relative concentration between the components in the active layer has a significant effect on the partial miscibility of these components.[48–50] It has been shown that acceptor species, such as PCBMs, form clusters within the blend, and the size of clusters increases with the increase of PCBM concentration.[26,47,51,52] Increase in the size of PCBM clusters leads to the formation of percolation paths and thus improves the photocurrent.[46] However, beyond a certain concentration of PCBM, the PCBM clusters cause mechanical stress on the metal electrode, therefore, possibly damaging the interface.[46] An experimental study of Chirvase et al.[46] suggested an optimal PCBM concentration of 50% by weight fraction in P3HT:PCBM blends. At a fundamental level, the intermolecular interactions between acceptor molecules and polymers at various concentrations determine the agglomeration and relative arrangement of these components. Therefore, analyzing these interactions among the components of active layers can help predict the morphology.

In an effort to understand the effect of the relative concentration of acceptor species and polymers on the agglomeration of acceptor species, Mortuza et al.[27] performed MD simulations of systems comprising P3HT oligomers with four monomeric units and PCBM in weight ratios 1:1 and 1:0.5. These systems correspond to PCBM mass fractions of 50% and 33.33%, respectively. The study showed that PCBMs form a cluster in both systems due to strong π–π interaction among their fullerene moieties. The extent of agglomeration of PCBMs increases with their increasing concentration, leading to the formation of percolation pathways for electron transfer. However, the study also demonstrated the strong π–π interaction of fullerene and a phenyl ring of individual PCBMs with the thiophene rings of P3HT. Additionally, the alkyl side chains of P3HT also interact strongly with the ester moiety of PCBM and contribute toward increasing its overall interactions with P3HT.[53] As a result of such interactions, individual PCBMs dissolve easily in P3HT, resulting in substantial intercalation of PCBM in P3HT. Figure 4.3a and b show a magnified snapshot of a

Relating Synthesis Parameters to the Photoactive Layer

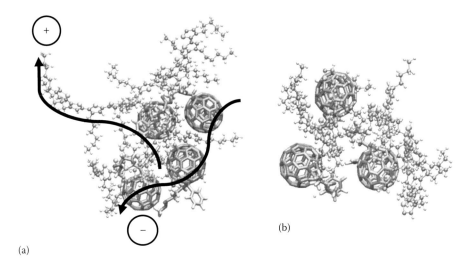

FIGURE 4.3 (See color insert.) The molecular arrangement of P3HT and PCBM in systems of (a) 1:1 P3HT:PCBM and (b) 1:0.5 P3HT:PCBM at 310 K are shown. Although at a higher concentration of PCBM clear percolation networks are formed, it is relatively sparse at low concentrations.

subsection of the simulation domain that illustrates the molecular arrangement of P3HT and PCBM in systems comprising P3HT and PCBM in weight ratios 1:1 and 1:0.5, respectively, at 310 K. The aggregation pattern of P3HT and PCBM in Figure 4.3a qualitatively demonstrates a greater contact area of P3HT and PCBM compared to that in Figure 4.3b. Figure 4.3a and b also illustrates that a relatively low concentration of PCBM in P3HT leads to sparse networks for charge transfer. In contrast, a greater concentration of PCBM in P3HT leads to formation of suitable networks for transporting holes to the photoanode and electrons to the photocathode, respectively. The simulations conducted by Mortuza et al.[27] demonstrated that the interfacial area as well as the size of individual clusters of PCBMs could be tuned based on the relative concentration of PCBM to reduce recombination losses and, at the same time, facilitate percolation pathways.

Lee et al.[28] performed a detailed study to characterize the nanoscale morphologies of various P3HT:PCBM blends to determine optimal weight ratios. They employed CGMD to simulate systems comprising P3HT:PCBM in weight ratios 2:1, 1:1, 1:2, and 1:3. The goal was to understand the effect of the relative concentration of P3HT and PCBM on the resultant morphology of the BHJ. The study employed a spatial-discretization scheme to quantify several morphological quantities, such as domain sizes, the interface-to-volume ratio, and the percolation ratio of both donor and acceptor phases. The study illustrated that the interface-to-volume ratio is maximum at a P3HT:PCBM weight ratio of 1:1. The study of Lee et al.[28] also demonstrated that, at a weight ratio of 1:1, the percolation ratios of both P3HT and PCBM phases are approximately equal. The greater weight ratio of P3HT (PCBM) increases the percolation ratio of P3HT (PCBM). Although greater percolation ratios of P3HT compared to that of PCBM enhance hole conductivity, the bulk of the electrons generated at the interfaces would be trapped within isolated PCBM domains and will never be transported to the anode. The opposite phenomenon would be observed if the percolation ratio of PCBM is greater than that of P3HT. Hence, from the perspective of the nanoscale morphology of P3HT:PCBM blend, the highest interface-to-volume ratio and balanced percolation ratios between donor and acceptor phases result in the optimal P3HT:PCBM weight ratio 1:1.[7,28] Lee et al.[29] extended their study by investigating the morphology of photoactive layers comprising poly-2,5-bis(3-tetradecylthiophene-2-yl)thieno[3,2-b]-thiophene (PBTTT) and PCBM at various concentrations. The study showed that, in contrast to that of P3HT:PCBM blends, the optimum weight ratio of a PBTTT:PCBM blend at which the highest interface-to-volume ratio and balanced percolation ratios between donor and acceptor phases are achieved is 1:3. The greater number of thiophene rings in a monomeric unit of PBTTT compared to that in P3HT results in more favorable interactions between PBTTT and PCBM compared to that between P3HT and PCBM. Hence, PCBM has a greater solubility in PBTTT than in P3HT. Therefore, in PBTTT:PCBM blends, a greater number of PCBM molecules needs to be added to ensure precipitation of the pure PCBM phase to create an adequate network for electron transfer to the cathode. The interface-to-volume ratio in the PBTTT:PCBM blend is found to be lower than that of the P3HT:PCBM blend at similar weight ratios to these blends. Therefore, BHJ cells comprising P3HT and PCBM are expected to perform better than those containing a PBTTT:PCBM blend. Figure 4.4 shows the detailed comparison of morphology between the two distinct systems,[54–57] P3HT:PCBM and PBTTT:PCBM, studied by Lee et al.[29]

4.3.1.2 Choice of Solvent

Solvents play an important role in controlling the degree of crystallinity of polymers and nanoparticles that, in turn, improve the transport of charge carriers by forming percolation pathways for both carriers.[58] In order to form interspersed crystalline domains of polymers and acceptors, a solvent is expected to dissolve these components completely, produce a highly uniform mixture, and allow for generation of an active layer that is uniformly composed of small domains of polymer or acceptors after drying.[58] Good solubility of acceptor molecules and polymers in solvents, a slow evaporation rate of the solvent after spin coating,[59] and subsequent annealing induce a high degree of crystallinity of polymer domains in the photoactive layers. Some of the common organic solvents used to prepare polymer-PCBM photovoltaic cells are toluene, xylene, chlorobenzene (CB), mono-chlorobenzene (MCB), o-dichlorobenzene (ODCB), chloroform, anisole, and indane.[9,60–62] The size of clusters of acceptors and polymers within active layer changes enormously with the choice of solvent.[47] This phenomenon is due to the varying solubility of acceptors and polymers in the solvents.

The solubility of a material in a solvent is characterized by the solubility parameter (δ), which is the square root of the cohesive energy density and describes the attractive strength between the molecules of the materials.[63] In order for a material to dissolve, this cohesive energy needs to be overcome as the molecules are separated from each other and surrounded by the solvent. The solubility of a material in a solvent increases as the difference between their corresponding δ values decreases. Materials with values of δ similar to those of solvents are likely to be soluble in

Relating Synthesis Parameters to the Photoactive Layer

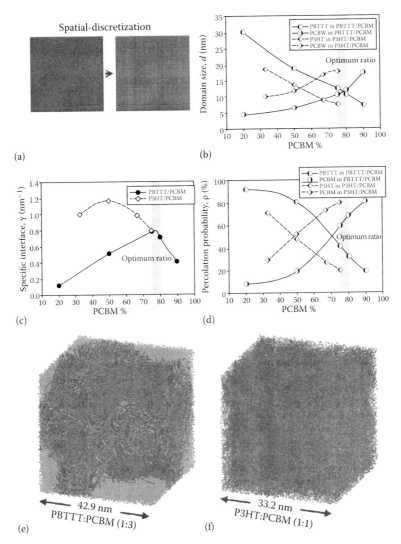

FIGURE 4.4 (See color insert.) (a) A spatial-discretization scheme to evaluate the morphological properties of photoactive layers. Polymer and PCBM domains are colored in red and blue, respectively. (b) Domain size, (c) specific interfacial area, and (d) percolation probabilities are plotted with respect to the PCBM weight percentage of the PBTTT:PCBM blends (solid lines and symbols). The morphological properties of P3HT:PCBM blends are also plotted (dashed lines) for comparison purposes. (From Lee, C.-K. et al., *Energy and Environmental Science*, 4, 4124–4132, 2011.) The gray-shadowed areas highlight the optimal blending ratio for the PBTTT:PCBM blend from experiments. (e) Snapshots of PBTTT:PCBM (blending ratio 1:3) and (f) P3HT:PCBM (blending ratio 1:1) are shown, and PCBM molecules are transparent for clarity. (Reprinted with permission from Lee, C.-K. and Pao, C.-W., *Journal of Physical Chemistry C*, 116, 12455–12461. Copyright 2012, American Chemical Society; Hwang, I.-W. et al., *Journal of Physical Chemistry C*, 112, 7853–7857, 2008; Mayer, A. C. et al., *Advanced Functional Materials*, 19, 1173–1179, 2009; Cates, N. C. et al., *Nano Letters*, 9, 4153–4157, 2009; Rance, W. L. et al., *ACS Nano*, 5, 5635–5646, 2011.)

these solvents. For instance, the δ values of P3HT and MEH-PPV are 18.6 MP$^{1/2}$ and 18.0 MP$^{1/2}$, respectively, which are close to that of CB (19.6 MP$^{1/2}$), toluene (18.2 MP$^{1/2}$), and chloroform (19.0 MP$^{1/2}$).[64,65] Therefore, CB, toluene, and chloroform are expected to be effective solvents for dissolving these polymers.[58] On the other hand, PCBM has very high solubility in polar solvents, such as CB, MCB, and ODCB, compared to nonpolar solvents, such as toluene and chloroform.[66] The above discussions clearly indicate that molecular interactions of solvent with polymers and acceptor molecules during photoactive layer formation in OPV cells influence the solubility of these components in solvents and, hence, the morphology of photoactive layers. Therefore, analyzing the molecular interaction of solvents with donor and acceptor materials is one of the key factors for the improvement of the morphology.

The molecular interactions between acceptor molecules and polymers with solvents with relatively high boiling points, such as CB (405 K) and DCB (452 K), have been investigated in many studies.[23,34,67] The solvents with high boiling points allow more time for crystallization of nanoparticles and polymers within the photoactive layer.[68] A previous study by Rispens et al.[68] suggested that MCB-PCBM blends exhibit a crystal packing in which electrons can hop easily in three dimensions. Therefore, they concluded that MCB is a good solvent for synthesizing the photoactive layer. Tummala et al.[34] simulated triclinic and monoclinic PCBM crystals with or without ODCB, CB, and chloroform to investigate the influence of the solvent on packing, orientation, and thermodynamic transitions. The study showed that the packing of PCBM increases after removing solvents. When the solvent is removed from the triclinic cocrystal, considerable structural rearrangement occurs, resulting in a deviation of 15.4% in the unit-cell parameters vis-à-vis the reference case of CB cocrystal and a 6.6% decrease in cell volume. The study also indicated that the remnant solvent molecules in the cocrystal hinder the formation of close contacts between PCBM, and therefore, the presence of solvents as an impurity has an adverse effect on the performance of OPVs.[34] The study of Tummala et al.[34] also analyzed the solubility of PCBM in these solvents by determining the solubility parameter (δ) of PCBM. The δ values of triclinic and monoclinic PCBMs are 22.7 MP$^{1/2}$ and 21.5 MP$^{1/2}$, respectively, which are similar to that of ODCB (22.3 MP$^{1/2}$). Therefore, PCBMs solvated in ODCB are expected to result in uniform dispersion into the P3HT polymer matrix.[69] The δ values of CB (19.6 MP$^{1/2}$), toluene (18.2 MP$^{1/2}$), and chloroform (19.0 MP$^{1/2}$) clearly indicate that PCBMs have a greater solubility in ODCB than in these solvents. The solubility of PCBM has its origins in the molecular interactions with the solvent. For instance, chloroform molecules are attracted strongly to the ester moiety of PCBM compared to the fullerene moiety. In contrast, toluene and fullerene display strong π–π interactions. However, toluene shows relatively weak interactions with the ester moiety of PCBM. Therefore, PCBMs are partially soluble in toluene and chloroform. On the other hand, the Coulombic interactions between chlorine atoms of ODCB and ester moieties of PCBMs, and π–π interaction among aromatic carbon atoms of ODCB and fullerene moieties of PCBMs cause greater solubility of PCBMs in ODCB compared to that in other solvents. However, due to the inherently toxic properties of halogenated solvents, Schmidt-Hansberg et al.[62] suggested indane as a nonhalogenated replacement. Moreover, indane is expected to dissolve both

Relating Synthesis Parameters to the Photoactive Layer

PCBM and polymer due to the strong π–π interaction of aromatic carbon atoms of indane with fullerene moieties of PCBM and conjugated polymer rings.

Mortuza et al.[26] performed MD simulations to simulate PCBM in three solvents: (i) toluene, (ii) indane, and (iii) toluene-indane mixtures to correlate solvent–acceptor interactions with the size of the agglomerate structure of PCBM. The study showed that PCBMs form large clusters in toluene and they are relatively dispersed in indane (Figure 4.5a), indicating a greater solubility of PCBM in indane compared to that in toluene. The observed greater solubility of PCBM in indane compared to that in toluene is due to the greater number of carbon atoms in the former solvent than the latter,[66] which leads to an enhanced interaction between PCBM and solvent molecules in the former case. To quantitatively determine the binding affinity between fullerene moieties of PCBMs in solvents, Mortuza et al.[26] calculated the potential of mean force (PMF) between a pair of fullerene molecules in toluene and indane at 310 K, as shown in Figure 4.5b. Figure 4.5b shows that the depth of the PMF of fullerenes in indane is approximately 0.4 kcal/mol, which is shallower than that in toluene (0.85 kcal/mol).[26] The greater depth of well in toluene indicates that the binding energy between fullerenes is greater in toluene than in indane. Therefore, it is easier for fullerenes to escape the potential well in indane and form a dispersed solution than in toluene. The PMFs also indicate a smaller induced repulsion in toluene compared to that in indane. Due to this phenomenon, PCBMs form larger clusters in toluene than in indane. Therefore, based on the solubility of PCBM, indane is a better choice of solvent than toluene and the toluene-indane mixture for fabricating the photoactive layer of OPV solar cells. Varying the relative concentration of toluene and indane in mixed solvent systems might provide a way to tune the extent of

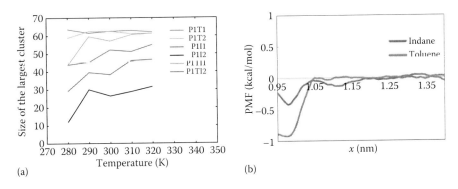

FIGURE 4.5 (See color insert.) (a) Cluster size of PCBMs has been characterized by evaluating the size of the largest cluster with temperature at various concentrations. The notations P1T1, P1T2, P1I1, P1I2, P1TI1, and P1TI2 correspond to the simulated systems in the study of Mortuza et al.[26] in which systems with toluene, indane, and toluene-indane mixtures have the letters T, I, and TI in the names, respectively, and 1 and 2 refer to different concentrations. (b) Potential of mean force (PMF) for fullerene molecules in toluene and indane at 310 K are presented. (Reprinted with permission from S. M. Mortuza, S. Banerjee, *Journal of Chemical Physics*, 137. Copyright 2012, American Institute of Physics.)

agglomeration during the solvent-based processing of photoactive layers. Mortuza et al.[26] also analyzed the effects of concentration of PCBM on the molecular rearrangement and agglomeration of PCBM in various solvents that will be discussed later.

Lee et al.[70] performed CGMD simulations of systems of MEH-PPV in chloroform and toluene to understand the agglomeration of polymer in solvents. The study showed that toluene has a strong affinity toward the MEH-backbone, but it is not attracted to the side chains of the polymer. On the other hand, chloroform is relatively attracted to the side chains, but it does not have any affinity toward the backbone of MEH-PPV. Hence, the study of Lee et al.[70] concluded that the solubility of MEH-PPV is low in chloroform and toluene, which conflicts with the interpretation based on the solubility parameter of MEH-PPV, chloroform, and toluene discussed above. However, because PCBMs have low solubility in toluene and chloroform, perhaps researchers should not use these solvents to synthesize photoactive layers. MD simulations are helpful tools that provide such information *a priori* for guiding experiments. None of the previous studies, based on MD, considered combined systems comprising polymers, acceptor species, and solvent. The lack of such studies is rooted in the fact that the photoactive layer consists of only acceptor species and polymers because the solvent has already evaporated after spin coating. However, because both acceptors and polymers are mixed and solvated together to generate BHJ, it is important to understand the ternary interactions of acceptor molecules and polymers in solvents. In order to fill this gap, a ternary system comprising solvent, acceptors, and polymers should be modeled as part of future studies to select the most favorable solvent for synthesis of photoactive layers of OPV.

4.3.1.3 Solute Concentration in Solvent

The relative concentrations of solute and solvent play an important role in the agglomeration of solutes during spin coating. To determine the effect of concentration of PCBM on their molecular arrangement and agglomeration, Mortuza et al.[26] simulated systems comprising PCBM and the three aromatic solvents mentioned in Section 4.3.1.2 in weight ratios 1:1 and 1:2, which correspond to PCBM mass fractions of 50% and 33.33%, respectively. As expected, the study illustrated that the size of the largest cluster of PCBM is greater in systems with a high concentration of PCBM (Figure 4.5a). Agglomeration of PCBMs increases with an increase in concentration of PCBM in solvents.[47,51,52] Frigerio et al.[23] simulated systems with PCBM in ODCB and MCB in order to understand the effect of temperature and concentration of PCBM on the agglomeration and crystallinity of PCBM clusters in these solvents. The study showed that the association of PCBM with each other increases with the increase of concentration of PCBM in the simulated solvents. At high concentrations of PCBM, the number of molecules in a solvation shell around PCBM clusters is not sufficient to solvate the clusters. Under such conditions, strong $\pi-\pi$ interaction among fullerene moieties of PCBMs leads to highly coordinated clusters. During synthesis of photoactive layers, the solute concentration should be chosen such that all solute molecules dissolve in the solvent. Otherwise, uniform blend composition is not achieved, and hence, the morphology of the photoactive layer becomes less favorable, and the device performance deteriorates. Although the above studies[23,26] have focused on systems with a high concentration of PCBM, they provide fundamental

Relating Synthesis Parameters to the Photoactive Layer

insight that can guide the selection of optimal concentration of acceptor molecules and solvents that lead to desired morphologies of the photoactive layer.

4.3.1.4 Chemical Structure of Polymers and Acceptor Species

The choice of the electron donor (polymer) and acceptor species, which is tied to their respective molecular structures, has significant influence on the resulting morphology of the photoactive layer. Several studies have employed MD simulations to evaluate the effect of specific structural factors, such as regioregularity of the polymer, presence of specific chemical moieties, chain length, degree of polymerization, and numbers of side chains on the morphology of photoactive layers.[7,30,33] Carrillo et al.[7] performed CGMD simulations of systems comprising PCBM and P3HT, in which the degree of polymerization (N) varied from 10 to 150 in order to investigate the effect of chemical structure of P3HT on the extent of phase segregation in the P3HT:PCBM blends. For a weight ratio of 1:1 and short chain length ($N = 10$), the authors observed a smectic/layered phase-segregated structure with alternate layers of P3HT and PCBM (see Figure 4.6). In contrast, such morphology is not observed for higher values of N, suggesting that chain length of P3HT plays an important role in the morphology. Figure 4.6 shows that increase of P3HT chain length enhances the number of pure domains of P3HT and PCBM. The study illustrated that the average size of such pure P3HT domains in a phase-segregated state decreases with an increase in N. For instance, for a P3HT:PCBM weight ratio of 1:1 and $N = 50$, the average P3HT domain size D is approximately 18 nm. In contrast, for a same weight ratio, the value of D is approximately 10 nm and 8 nm when $N = 100$ and $N = 150$, respectively. The suggested maximum value of D, to reduce recombination losses, is 10 nm for diffusion of exciton to the interface of P3HT and PCBM. Hence, longer chain length, which leads to low average domain size of P3HT, is desired to limit exciton recombination. The average domain size of P3HT is related to the interfacial area A by the relation $DA \sim L^3 \varnothing_v$, where $\varnothing_v$ is the volume fraction of P3HT domains, and L^3 is the total volume of the system.[7] The study showed that the interfacial area-to-volume ratio, which has an inverse relation to D, increases with the increase of chain length of P3HT. For example, for a weight ratio of 1:1 and $N = 50$, the interfacial area-to-volume ratio is approximately 0.24 nm^{-1}, and it is approximately 0.36 nm^{-1} when $N = 150$. Overall, for P3HT:PCBM blends with 1:1 weight

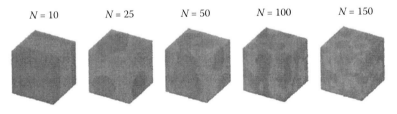

FIGURE 4.6 (See color insert.) Effects of the P3HT degree of polymerization, N, on the morphology of P3HT:PCBM blends are presented at a weight ratio of 1:1. P3HT monomers and PCBM beads are colored in red and blue, respectively. (From J.-M. Y. Carrillo, R. Kumar, M. Goswami, B. G. Sumpter, W. M. Brown, *Physical Chemistry Chemical Physics*, 15, 17873–17882, 2013. Reproduced by permission of The Royal Society of Chemistry.)

ratio, it is advantageous to have longer chain lengths ($N > 100$) because an increase in the number of monomers in a polymer unit leads to more pure domains, low average domain size of P3HT, and an increase in the interfacial area-to-volume ratio.[7]

Scientists have already identified numerous polymers for use in OPVs based on their bandgaps, hole mobilities, and conjugation length. However, ongoing research in various scientific groups is continuing to explore new polymers that are amenable to better processibility and lead to optimized morphologies.[30,33] Molecular modeling can eliminate the trial-and-error process involved in synthesis and characterization of polymers for OPV applications and help identify novel polymers with great potential to improve device performance. In order to understand the effect of physical (size and flexibility) and chemical (pair-wise interaction strengths) features of donor and acceptor molecules on the morphology of photoactive layers, Marsh et al.[30] performed CGMD simulations of systems comprising a generic polymer and PCBM. The study showed that generic conjugated polymers with rigid backbones and constrained side chains cause a layer-by-layer morphology of polymers and amorphous acceptors whereas flexible backbones and unconstrained side chains facilitate a highly ordered acceptor arrangement. Thus, polymers with flexible backbones and unconstrained side chains are great choices in order to achieve enhanced performance of OPV devices. The increase in length of polymers and number of side chains can result in such flexible backbones in polymers. The study of Marsh et al.[30] also demonstrated that increasing the diameter of acceptor species decreases ordering of the acceptor domain. Therefore, although there are other candidates with larger diameters, such as $PC_{71}BM$, $PC_{61}BM$ has been the most popular acceptor. Although sizes of donor and acceptor molecules influence the morphology of photoactive layers, the orientation of side chains of polymers facilitates the ordering of polymers and molecular arrangement of acceptors in photoactive layers. Jankowski et al.[33] investigated the effect of orientation of side chains on the morphology of photoactive layers by performing CGMD simulations of PCBMs in various polymers, such as P3HT, poly-(2,2′:5′,2″-3,3″-dihexylterthiophene) (PTTT) and poly(3,4-dihexyl-2,2′-bithiophene) (PDHBT). Jankowski et al.[33] showed that the orientation of side chains of polymers determines the molecular arrangement of PCBM in polymers and hence the overall interfacial area in the polymer: PCBM blends. If the gap between side chains of contiguous monomeric units is too small to accommodate an acceptor molecule without significantly bending the side chains or backbone, the intercalation of PCBM is restricted. Because the van der Waals diameter of PCBM is approximately 1 nm, the distance between two side chains in polymers should be at least 1.3 nm in order to intercalate a PCBM.[33] The results presented in the abovementioned studies provide great insight and demonstrate that MD-based studies have excellent scope to predict specific configurations that favor the formation of interfaces between polymers and nanoparticles. In the near future, MD simulations can therefore be used to identify polymers that lead to such enhanced morphologies.

4.3.2 OTHER STUDIES RELATED TO THE MORPHOLOGY OF PHOTOACTIVE LAYERS USING MD

From the studies presented and discussed in Sections 4.3.1.1–4.3.1.4, we can conclude that processing parameters play an important role in determining the extent

Relating Synthesis Parameters to the Photoactive Layer

of crystallinity of polymer and nanoparticle domains and therefore have profound impact on the performance of OPVs. It is expected that the crystalline structure of these components will be hindered at the interface of nanoparticles and polymers due to their strong molecular interactions.[29] On the other hand, bulk polymers or nanoparticles need to be in crystalline form so that electrons and holes can hop through them to reach the electrodes. MD simulations are able to estimate the crystallinity of polymers by evaluating the angles between main chains of polymers. For instance, Reddy et al.[25] performed MD simulations of a system comprising thiophene oligomers and fullerenes. They determined the orientation of polymers at the interface and at the bulk. The study demonstrates that the formation of ordered configurations of polymers, which promote the transfer of charges, is strongly disrupted at the proximity of nanoparticles compared to that in bulk oligothiophene systems (Figure 4.7). The strong π–π interactions among fullerene and thiophene rings cause such disruption. In contrast, thiophene units that are at a distance from the interface do not interact significantly with PCBM domains due to π–π stacking of thiophene rings and are therefore in a more ordered state. Therefore, in the photoactive layer, PCBMs and polymers occur in an amorphous phase near the interfaces while bulk polymer and PCBM are in crystalline phases. This is tied to the river-and-stream model for photoactive layers that was recently proposed based on neutron scattering experimental diagnostics.[71]

In order to evaluate the ordering of polymers in bulk, Takizawa et al.[72] simulated P3HT molecules in anisole. Their results indicated that the initial nucleation

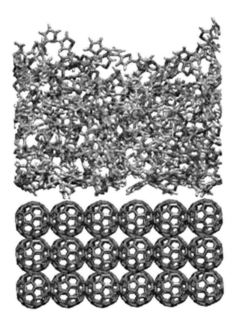

FIGURE 4.7 (See color insert.) Molecular snapshot of the orientation of thiophene oligomers at the fullerene-oligothiophene interface is shown. Oligothiophene molecules are placed on top of fullerene molecules. (Reprinted with permission from Reddy, S. Y. and Kuppa, V. K., *Journal of Physical Chemistry C*, 116, 14873–14882. Copyright 2012, American Chemical Society.)

of P3HT in solution begins with ring ordering, followed by main chain ordering and side chain ordering. Initially, the neighboring thiophene rings align very quickly due to π–π interaction and the neighboring main chains align their directions with respect to each other, which takes more than five times longer than the ring ordering process. During these quick ordering processes, the planar structure of thiophene rings and the strong interaction between them play an important role. The ordering of side chains occurs through the packing of main chains. The ordering of side chains is a relatively slow process and takes almost five times longer than the ordering of the main chain. The slow ordering dynamics of side chains is one of the factors that cause anisotropic growth of nuclei, during which the crystalline domain grows faster in the direction of the normal vectors of rings than in the direction of side chain vectors. A schematic adopted from this study,[72] which shows ordered conjugated P3HT domains in solutions, is presented in Figure 4.8. These fundamental studies[25,72] provide insight into some of the basic criteria to choose conjugated polymers for OPVs.

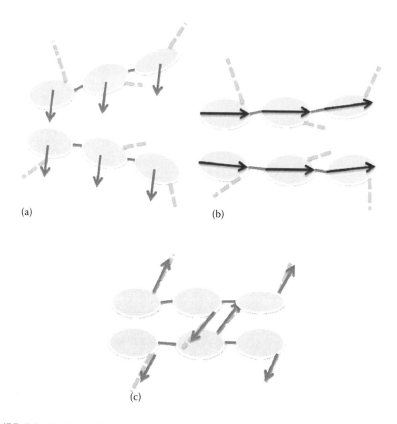

FIGURE 4.8 A schematic representation of the ordering mechanism of P3HT in solutions is shown. (a) Initially, the neighboring thiophene rings align. (b) Next, the neighboring main chains arrange their directions. At this stage, side chains are in various directions. (c) Finally, formation alignment of flexible side chains occurs. (Reprinted with permission from Takizawa, Y. et al., *Journal of Physical Chemistry B*, 117, 6282–6289. Copyright 2013, American Chemical Society.)

A recent study by Pan et al.[73] utilized MD simulations to simulate PCBM and P3HT blends. The goal of this study was to build up highly ordered columnar structures within the active layer by introducing a second polymer, which is immiscible with the first, possibly causing lateral phase segregation into columnar structures. The study found that the introduction of polystyrene organizes the P3HT into columnar phases decorated by PCBM at the interface. The columnar structure of P3HT with abundant PCBM particles at the interface generates high interfacial area, which enhances the exciton dissociation and charge separation.[74]

Another very recent study[75] introduced single-walled carbon nanotubes (SWCNTs) as alternate electron acceptor molecules in the photoactive layer. The study evaluated the interaction energy between SWCNTs and polymers, including P3HT, MDMO-PPV, and MEH-PPV, by MD simulation. The interaction energy was calculated based on the difference between the potential energy of the system and the potential energy for the polymer and SWCNTs. The study illustrated that the SWCNTs interact more strongly with P3HT, compared to that with other polymers. The study also evaluated the radii of gyration of polymers that were absorbed on the surface of SWCNTs in order to understand the morphology. The greater radius of gyration indicates expanded polymer on the surface of SWCNTs and, hence, a greater interfacial area. The study showed that the radius of gyration is the greatest for P3HTs on SWCNTs surfaces. This outcome signifies that SWCNTs can be possible candidates for acceptor materials.

As illustrated through numerous examples in this chapter, the critical advantages of molecular modeling tools is that they enable a systematic choice of electron donor and acceptor materials, solvents, and synthesis parameters before fabricating and testing the device experimentally.

4.4 OUTLOOK AND FUTURE DIRECTIONS

OPV solar cells have become popular due to their light weight, flexibility, and ease of manufacturing. Additionally, the cells have no long-term environmental and health issues, and they have the potential to generate cost-effective and efficient energy. However, in order to realize OPV devices as energy harvesters for our long-term energy needs, significant improvements in performance are required. The design and fabrication of a high-efficiency OPV device requires precise control over the nanoscale morphology, molecular ordering, and interfacial properties of all components in the device. Moreover, it is also important to select favorable synthesis parameters for fabrication of OPVs since these parameters control the morphology of the photoactive layer of OPVs and, hence, the device performance. Instead of using traditional trial-and-error approaches for selecting these parameters, a rational predictive approach is highly essential. Molecular dynamics simulations are favorable tools that can guide the selection of favorable synthesis conditions for OPVs.

The results presented in this chapter, based on classical MD and CGMD simulations performed by various research groups, demonstrate recent attempts toward systematic studies of the relationships of various synthesis parameters with the morphology. We have addressed the effect of parameters, such as relative concentrations of polymers and acceptor molecules, choice of solvent, and chemical structure of the donor and acceptor molecules on the agglomerate structures of acceptors and polymers. There are several other studies,[76–79] which employed ab-initio MD or density functional theory (DFT) to understand the electronic structure, bandgaps, and electron and hole mobilities of photoactive materials. Combining classical all-atom and course-grained MD simulations with ab-initio MD and DFT calculations will help to bridge the gaps between various scales involved in energy conversion processes in OPVs, such as exciton generation, separation into an electron and a hole, and transportation of separated charges from the interface to electrodes. Such combined modeling efforts will be helpful in relating the underlying physics of these energy conversion processes with the morphology of photoactive layers. This predictive modeling approach will help in significantly improving the performance of OPVs.

REFERENCES

1. Lunt, R. R.; Osedach, T. P.; Brown, P. R.; Rowehl, J. A.; Bulovic, V., Practical roadmap and limits to nanostructured photovoltaics. *Advanced Materials* 2011, 23, 5712–5727.
2. Dang, M. T.; Wantz, G.; Bejbouji, H.; Urien, M.; Dautel, O. J.; Vignau, L.; Hirsch, L., Polymeric solar cells based on P3HT:PCBM: Role of the casting solvent. *Solar Energy Materials and Solar Cells* 2011, 95, 3408–3418.
3. Facchetti, A., Polymer donor-polymer acceptor (all-polymer) solar cells. *Materials Today* 2013, 16, 123–132.
4. Jorgensen, M.; Norrman, K.; Krebs, F. C., Stability/degradation of polymer solar cells. *Solar Energy Materials and Solar Cells* 2008, 92, 686–714.
5. Po, R.; Carbonera, C.; Bernardi, A.; Tinti, F.; Camaioni, N., Polymer- and carbon-based electrodes for polymer solar cells: Toward low-cost, continuous fabrication over large area. *Solar Energy Materials and Solar Cells* 2012, 100, 97–114.
6. Heliatek. 12.0% efficient small molecule tandem device. Available at http://www.heliatek.com/newscenter/latest_news/neuer-weltrekord-fur-organische-solarzellen-heliatek-behauptet-sich-mit-12-zelleffizienz-als-technologiefuhrer/?lang=en (accessed October 7, 2014).
7. Carrillo, J.-M. Y.; Kumar, R.; Goswami, M.; Sumpter, B. G.; Brown, W. M., New insights into the dynamics and morphology of P3HT:PCBM active layers in bulk heterojunctions. *Physical Chemistry Chemical Physics* 2013, 15, 17873–17882.
8. Moliton, A.; Nunzi, J. M., How to model the behaviour of organic photovoltaic cells. *Polymer International* 2006, 55, 583–600.
9. Saunders, B. R.; Turner, M. L., Nanoparticle-polymer photovoltaic cells. *Advances in Colloid and Interface Science* 2008, 138, 1–23.
10. Shaw, P. E.; Ruseckas, A.; Samuel, I. D. W., Exciton diffusion measurements in poly(3-hexylthiophene). *Advanced Materials* 2008, 20, 3516–3520.
11. van Bavel, S.; Sourty, E.; de With, G.; Frolic, K.; Loos, J., Relation between photoactive layer thickness, 3D morphology, and device performance in P3HT/PCBM bulk-heterojunction solar cells. *Macromolecules* 2009, 42, 7396–7403.
12. Yu, G.; Gao, J.; Hummelen, J. C.; Wudl, F.; Heeger, A. J., Polymer photovoltaic cells—Enhanced efficiencies via a network of internal donor-acceptor heterojunctions. *Science* 1995, 270, 1789–1791.
13. Halls, J. J. M.; Walsh, C. A.; Greenham, N. C.; Marseglia, E. A.; Friend, R. H.; Moratti, S. C.; Holmes, A. B., Efficient photodiodes from interpenetrating polymer networks. *Nature* 1995, 376, 498–500.

14. Treat, N. D.; Brady, M. A.; Smith, G.; Toney, M. F.; Kramer, E. J.; Hawker, C. J.; Chabinyc, M. L., Interdiffusion of PCBM and P3HT reveals miscibility in a photovoltaically active blend. *Advanced Energy Materials* 2011, 1, 82–89.
15. Hauch, J. A.; Schilinsky, P.; Choulis, S. A.; Childers, R.; Biele, M.; Brabec, C. J., Flexible organic P3HT:PCBM bulk-heterojunction modules with more than 1 year outdoor lifetime. *Solar Energy Materials and Solar Cells* 2008, 92, 727–731.
16. Hwang, I.-W.; Moses, D.; Heeger, A. J., Photoinduced carrier generation in P3HT/PCBM bulk heterojunction materials. *Journal of Physical Chemistry C* 2008, 112, 4350–4354.
17. Vanlaeke, P.; Swinnen, A.; Haeldermans, I.; Vanhoyland, G.; Aernouts, T.; Cheyns, D.; Deibel, C.; D'Haen, J.; Heremans, P.; Poortmans, J.; Manca, J. V., P3HT/PCBM bulk heterojunction solar cells: Relation between morphology and electro-optical characteristics. *Solar Energy Materials and Solar Cells* 2006, 90, 2150–2158.
18. Ma, W. L.; Yang, C. Y.; Gong, X.; Lee, K.; Heeger, A. J., Thermally stable, efficient polymer solar cells with nanoscale control of the interpenetrating network morphology. *Advanced Functional Materials* 2005, 15, 1617–1622.
19. Günes, S.; Neugebauer, H.; Sariciftci, N. S., Conjugated polymer-based organic solar cells. *Chemical Reviews* 2007, 107, 1324–1338.
20. Collins, B. A.; Tumbleston, J. R.; Ade, H., Miscibility, crystallinity, and phase development in P3HT/PCBM solar cells: Toward an enlightened understanding of device morphology and stability. *Journal of Physical Chemistry Letters* 2011, 2, 3135–3145.
21. Cheung, D. L.; Troisi, A., Theoretical study of the organic photovoltaic electron acceptor PCBM: Morphology, electronic structure, and charge localization. *Journal of Physical Chemistry C* 2010, 114, 20479–20488.
22. MacKenzie, R. C. I.; Frost, J. M.; Nelson, J., A numerical study of mobility in thin films of fullerene derivatives. *Journal of Chemical Physics* 2010, 132, 064904-1–064904–6.
23. Frigerio, F.; Casalegno, M.; Carbonera, C.; Nicolini, T.; Meille, S. V.; Raos, G., Molecular dynamics simulations of the solvent- and thermal history-dependent structure of the PCBM fullerene derivative. *Journal of Materials Chemistry* 2012, 22, 5434–5443.
24. Reddy, S. Y.; Kuppa, V. K., Molecular dynamics simulations of organic photovoltaic materials: Investigating the formation of pi-stacked thiophene clusters in oligothiophene/fullerene blends. *Synthetic Metals* 2012, 162, 2117–2124.
25. Reddy, S. Y.; Kuppa, V. K., Molecular dynamics simulations of organic photovoltaic materials: Structure and dynamics of oligothiophene. *Journal of Physical Chemistry C* 2012, 116, 14873–14882.
26. Mortuza, S. M.; Banerjee, S., Molecular modeling study of agglomeration of 6,6-phenyl-C61-butyric acid methyl ester in solvents. *Journal of Chemical Physics* 2012, 137, 244308-1–244308-12.
27. Mortuza, S. M.; Banerjee, S., Molecular modeling of nanoparticles and conjugated polymers during synthesis of photoactive layers of organic photovoltaic solar cells. In *AIChE Annual Meeting*, San Francisco, 2013.
28. Lee, C.-K.; Pao, C.-W.; Chu, C.-W., Multiscale molecular simulations of the nanoscale morphologies of P3HT:PCBM blends for bulk heterojunction organic photovoltaic cells. *Energy and Environmental Science* 2011, 4, 4124–4132.
29. Lee, C.-K.; Pao, C.-W., Solubility of 6,6-phenyl-C-61-butyric acid methyl ester and optimal blending ratio of bulk heterojunction polymer solar cells. *Journal of Physical Chemistry C* 2012, 116, 12455–12461.
30. Marsh, H. S.; Jayaraman, A., Morphological studies of blends of conjugated polymers and acceptor molecules using langevin dynamics simulations. *Journal of Polymer Science Part B-Polymer Physics* 2013, 51, 64–77.
31. Huang, D. M.; Faller, R.; Do, K.; Moule, A. J., Coarse-grained computer simulations of polymer/fullerene bulk heterojunctions for organic photovoltaic applications. *Journal of Chemical Theory and Computation* 2010, 6, 526–537.

32. Huang, D. M.; Moule, A. J.; Faller, R., Characterization of polymer-fullerene mixtures for organic photovoltaics by systematically coarse-grained molecular simulations. *Fluid Phase Equilibria* 2011, 302, 21–25.
33. Jankowski, E.; Marsh, H. S.; Jayaraman, A., Computationally linking molecular features of conjugated polymers and fullerene derivatives to bulk heterojunction morphology. *Macromolecules* 2013, 46, 5775–5785.
34. Tummala, N. R.; Mehraeen, S.; Fu, Y.-T.; Risko, C.; Bredas, J.-L., Materials-scale implications of solvent and temperature on 6,6-Phenyl-C61-butyric acid methyl ester (PCBM): A theoretical perspective. *Advanced Functional Materials* 2013, 23, 5800–5813.
35. Verlet, L., Computer "experiments" on classical fluids. I. Thermodynamical properties of Lennard-Jones molecules. *Physical Review* 1967, 159, 98–103.
36. Andersen, H. C., Rattle—A velocity version of the shake algorithm for molecular-dynamics calculations. *Journal of Computational Physics* 1983, 52, 24–34.
37. Hockney, R. W.; Eastwood, J. W., *Computer Simulation Using Particles*. McGraw-Hill, New York, 1981.
38. Cornell, W. D.; Cieplak, P.; Bayly, C. I.; Gould, I. R.; Merz, K. M.; Ferguson, D. M.; Spellmeyer, D. C.; Fox, T.; Caldwell, J. W.; Kollman, P. A., A second generation force field for the simulation of proteins, nucleic acids, and organic molecules. *Journal of the American Chemical Society* 1995, 117, 5179–5197.
39. Brooks, B. R.; Bruccoleri, R. E.; Olafson, B. D.; States, D. J.; Swaminathan, S.; Karplus, M., CHARMM—A program for macromolecular energy, minimization, and dynamics calculations. *Journal of Computational Chemistry* 1983, 4, 187–217.
40. Jorgensen, W. L.; Maxwell, D. S.; Tirado Rives, J., Development and testing of the OPLS all-atom force field on conformational energetics and properties of organic liquids. *Journal of the American Chemical Society* 1996, 118, 11225–11236.
41. Plimpton, S., Fast parallel algorithms for short-range molecular-dynamics. *Journal of Computational Physics* 1995, 117, 1–19.
42. Hess, B.; Kutzner, C.; van der Spoel, D.; Lindahl, E., GROMACS 4: Algorithms for highly efficient, load-balanced, and scalable molecular simulation. *Journal of Chemical Theory and Computation* 2008, 4, 435–447.
43. Smith, W.; Todorov, I. T.; Leslie, M., The DL_POLY molecular dynamics package. *Zeitschrift Fur Kristallographie* 2005, 220, 563–566.
44. Ponder, J. W.; Richards, F. M., An efficient Newton-like method for molecular mechanics energy minimization of large molecules. *Journal of Computational Chemistry* 1987, 8, 1016–1024.
45. Yang, X. N.; Loos, J.; Veenstra, S. C.; Verhees, W. J. H.; Wienk, M. M.; Kroon, J. M.; Michels, M. A. J.; Janssen, R. A. J., Nanoscale morphology of high-performance polymer solar cells. *Nano Letters* 2005, 5, 579–583.
46. Chirvase, D.; Parisi, J.; Hummelen, J. C.; Dyakonov, V., Influence of nanomorphology on the photovoltaic action of polymer-fullerene composites. *Nanotechnology* 2004, 15, 1317–1323.
47. Hoppe, H.; Niggemann, M.; Winder, C.; Kraut, J.; Hiesgen, R.; Hinsch, A.; Meissner, D.; Sariciftci, N. S., Nanoscale morphology of conjugated polymer/fullerene-based bulk-heterojunction solar cells. *Advanced Functional Materials* 2004, 14, 1005–1011.
48. Vakhshouri, K.; Kozub, D. R.; Wang, C.; Salleo, A.; Gomez, E. D., Effect of miscibility and percolation on electron transport in amorphous poly(3-Hexylthiophene)/phenyl-C-61-butyric acid methyl ester blends. *Physical Review Letters* 2012, 108, 026601-1–026601-5.
49. Kim, J. Y.; Frisbie, D., Correlation of phase behavior and charge transport in conjugated polymer/fullerene blends. *Journal of Physical Chemistry C* 2008, 112, 17726–17736.
50. Morana, M.; Koers, P.; Waldauf, C.; Koppe, M.; Muehlbacher, D.; Denk, P.; Scharber, M.; Waller, D.; Brabec, C., Organic field-effect devices as tool to characterize the bipolar transport in polymer/fullerene blends: The case of P3HT-PCBM. *Advanced Functional Materials* 2007, 17, 3274–3283.

Relating Synthesis Parameters to the Photoactive Layer

51. Hoppe, H.; Sariciftci, N. S., Morphology of polymer/fullerene bulk heterojunction solar cells. *Journal of Materials Chemistry* 2006, 16, 45–61.
52. Laiho, A.; Majumdar, H. S.; Baral, J. K.; Jansson, F.; Osterbacka, R.; Ikkala, O., Tuning the electrical switching of polymer/fullerene nanocomposite thin film devices by control of morphology. *Applied Physics Letters* 2008, 93, 203309-1–203309-3.
53. Yao, Y.; Hou, J.; Xu, Z.; Li, G.; Yang, Y., Effect of solvent mixture on the nanoscale phase separation in polymer solar cells. *Advanced Functional Materials* 2008, 18, 1783–1789.
54. Hwang, I.-W.; Kim, J. Y.; Cho, S.; Yuen, J.; Coates, N.; Lee, K.; Heeney, M.; McCulloch, I.; Moses, D.; Heeger, A. J., Bulk heterojunction materials composed of poly(2,5-bis(3-tetradecylthiophen-y1-2)thieno 3,2-b thiophene): Ultrafast electron transfer and carrier recombination. *Journal of Physical Chemistry C* 2008, 112, 7853–7857.
55. Mayer, A. C.; Toney, M. F.; Scully, S. R.; Rivnay, J.; Brabec, C. J.; Scharber, M.; Koppe, M.; Heeney, M.; McCulloch, I.; McGehee, M. D., Bimolecular crystals of fullerenes in conjugated polymers and the implications of molecular mixing for solar cells. *Advanced Functional Materials* 2009, 19, 1173–1179.
56. Cates, N. C.; Gysel, R.; Beiley, Z.; Miller, C. E.; Toney, M. F.; Heeney, M.; McCulloch, I.; McGehee, M. D., Tuning the properties of polymer bulk heterojunction solar cells by adjusting fullerene size to control intercalation. *Nano Letters* 2009, 9, 4153–4157.
57. Rance, W. L.; Ferguson, A. J.; McCarthy-Ward, T.; Heeney, M.; Ginley, D. S.; Olson, D. C.; Rumbles, G.; Kopidakis, N., Photainduced carrier generation and decay dynamics in intercalated and non-intercalated polymer: Fullerene bulk heterojunctions. *ACS Nano* 2011, 5, 5635–5646.
58. Reisdorffer, F.; Haas, O.; Le Rendu, P.; Nguyen, T. P., Co-solvent effects on the morphology of P3HT:PCBM thin films. *Synthetic Metals* 2012, 161, 2544–2548.
59. Nilsson, S.; Bernasik, A.; Budkowski, A.; Moons, E., Morphology and phase segregation of spin-casted films of polyfluorene/PCBM blends. *Macromolecules* 2007, 40, 8291–8301.
60. Martens, T.; D'Haen, J.; Munters, T.; Beelen, Z.; Goris, L.; Manca, J.; D'Olieslaeger, M.; Vanderzande, D.; De Schepper, L.; Andriessen, R., Disclosure of the nanostructure of MDMO-PPV:PCBM bulk hetero-junction organic solar cells by a combination of SPM and TEM. *Synthetic Metals* 2003, 138, 243–247.
61. Liu, J.; Shi, Y. J.; Yang, Y., Solvation-induced morphology effects on the performance of polymer-based photovoltaic devices. *Advanced Functional Materials* 2001, 11, 420–424.
62. Schmidt-Hansberg, B.; Sanyal, M.; Grossiord, N.; Galagan, Y.; Baunach, M.; Klein, M. F. G.; Colsmann, A.; Scharfer, P.; Lemmer, U.; Dosch, H.; Michels, J.; Barrena, E.; Schabel, W., Investigation of non-halogenated solvent mixtures for high throughput fabrication of polymer/fullerene solar cells. *Solar Energy Materials and Solar Cells* 2012, 96, 195–201.
63. Brandrup, J.; Immergut, E. H.; Grulke, E. A., *Polymer Handbook*, 4th ed. John Wiley and Sons, New York, 1999.
64. Cossiello, R. F.; Akcelrud, L.; Atvars, D. Z., Solvent and molecular weight effects on fluorescence emission of MEH-PPV. *Journal of the Brazilian Chemical Society* 2005, 16, 74–86.
65. Ananthakrishnan, N.; Padmanaban, G.; Ramakrishnan, S.; Reynolds, J. R., Tuning polymer light-emitting device emission colors in ternary blends composed of conjugated and nonconjugated polymers. *Macromolecules* 2005, 38, 7660–7669.
66. Ruoff, R. S.; Tse, D. S.; Malhotra, R.; Lorents, D. C., Solubility of c-60 in a variety of solvents. *Journal of Physical Chemistry* 1993, 97, 3379–3383.
67. Casalegno, M.; Zanardi, S.; Frigerio, F.; Po, R.; Carbonera, C.; Marra, G.; Nicolini, T.; Raos, G.; Meille, S. V., Solvent-free phenyl-C61-butyric acid methyl ester (PCBM) from clathrates: Insights for organic photovoltaics from crystal structures and molecular dynamics. *Chemical Communications* 2013, 49, 4525–4527.

68. Rispens, M. T.; Meetsma, A.; Rittberger, R.; Brabec, C. J.; Sariciftci, N. S.; Hummelen, J. C., Influence of the solvent on the crystal structure of PCBM and the efficiency of MDMO-PPV:PCBM "plastic" solar cells. *Chemical Communications* 2003, 17, 2116–2118.
69. Kawano, K.; Sakai, J.; Yahiro, M.; Adachi, C., Effect of solvent on fabrication of active layers in organic solar cells based on poly(3-hexylthiophene) and fullerene derivatives. *Solar Energy Materials and Solar Cells* 2009, 93, 514–518.
70. Lee, C. K.; Hua, C. C.; Chen, S. A., Single-chain and aggregation properties of semiconducting polymer solutions investigated by coarse-grained Langevin dynamics simulation. *Journal of Physical Chemistry B* 2008, 112, 11479–11489.
71. Yin, W.; Dadmun, M., A new model for the morphology of P3HT/PCBM organic photovoltaics from small-angle neutron scattering: Rivers and streams. *ACS Nano* 2011, 5, 4756–4768.
72. Takizawa, Y.; Shimomura, T.; Miura, T., Simulation study of the initial crystallization processes of poly(3-hexylthiophene) in solution: Ordering dynamics of main chains and side chains. *Journal of Physical Chemistry B* 2013, 117, 6282–6289.
73. Pan, C.; Li, H.; Akgun, B.; Satijia, S. K.; Zhu, Y.; Xu, D.; Ortiz, J.; Gersappe, D.; Rafailovich, M. H., Enhancing the efficiency of bulk heterojunction solar cells via templated self-assembly. *Macromolecules* 2013, 46, 1812–1819.
74. Ferenczi, T. A. M.; Mueller, C.; Bradley, D. D. C.; Smith, P.; Nelson, J.; Stingelin, N., Organic semiconductor: Insulator polymer ternary blends for photovoltaics. *Advanced Materials* 2011, 23, 4093–4097.
75. Zaminpayma, E.; Mirabbaszadeh, K., Investigation of molecular interaction between single-walled carbon nanotubes and conjugated polymers. *Polymer Composites* 2012, 33, 548–554.
76. Mizuseki, H.; Igarashi, N.; Belosludov, R. V.; Farajian, A. A.; Kawazoe, Y., Theoretical study of phthalocyanine-fullerene complex for a high efficiency photovoltaic device using ab initio electronic structure calculation. *Synthetic Metals* 2003, 138, 281–283.
77. Blase, X.; Attaccalite, C.; Olevano, V., First-principles GW calculations for fullerenes, porphyrins, phtalocyanine, and other molecules of interest for organic photovoltaic applications. *Physical Review B* 2011, 83, 115103-1–115103-9.
78. Berube, N.; Gosselin, V.; Gaudreau, J.; Cote, M., Designing polymers for photovoltaic applications using ab initio calculations. *Journal of Physical Chemistry C* 2013, 117, 7964–7972.
79. Napoles-Duarte, J. M.; Reyes-Reyes, M.; Ricardo-Chavez, J. L.; Garibay-Alonso, R.; Lopez-Sandoval, R., Effect of packing on the cohesive and electronic properties of methanofullerene crystals. *Physical Review B* 2008, 78, 035425-1–035425-7.

5 Insights Obtained from Modeling of Organic Photovoltaics
Morphology, Interfaces, and Coupling with Charge Transport

*Rajeev Kumar, Jan-Michael Carrillo,
Monojoy Goswami, and Bobby G. Sumpter*

CONTENTS

5.1	Introduction	114
5.2	Issues with Which Modeling Can Provide Insights	116
	5.2.1 Exciton Generation, Dissociation, and Recombination	117
	5.2.2 Effects of Macromolecular Characteristics of Conjugated Polymers on Morphology	118
	5.2.2.1 Equilibrium Aspects	118
	5.2.2.2 Nonequilibrium Effects of Additives and Processing Conditions	118
	5.2.3 Interfacial Properties of Electrode–Donor–Acceptor Interfaces	119
	5.2.4 Effects of Morphology on the Charge Generation and Recombination	119
5.3	Insights Obtained from Simulation Approaches	120
	5.3.1 Electronic Structure Calculations	120
	5.3.2 Classical Mechanics-Based Approaches	122
	5.3.2.1 Molecular Dynamics	123
	5.3.2.2 Monte Carlo	129
	5.3.3 Continuum-Based Approaches	130
	5.3.4 Multiscale Methods	131
5.4	Discussion and Outlook	134
References		135

5.1 INTRODUCTION

Societal need for renewable energy has led to the advent of research areas such as organic photovoltaics (OPVs).[1,2] Enormous interest in the OPVs is premised on the facts that sunlight is the most abundant renewable energy source and conjugated polymers can be used for solar energy conversion. Usage of polymers in OPV devices promises low cost, flexibility, and fabrication from earth-abundant elements. Despite these advantages, the power conversion efficiencies (PCEs) of the OPVs are not on a par with inorganic semiconductors. Therefore a large amount of the fundamental and applied research in the field of OPVs is focused on providing useful insights into improving the PCEs of OPVs.

The PCE is defined as the ratio of power generated to the incident solar power impinging on the photovoltaic device. The PCE of an OPV device depends on a number of factors. In order to obtain some insight into how the PCE depends on various parameters, we consider a simpler case of a single p-n junction-based solar cell. Shockley and Queisser[3] studied such a system in 1961 and computed an upper theoretical limit assuming that all avoidable losses are absent. As per definition of the PCE, the maximum efficiency corresponds to maximum power generated per unit energy impinged on the solar cell. The power generated is material-specific and depends on three main quantities: rates of electron-hole pair (known as an exciton[4]) generation, their dissociation into "free charges," and recombination by various mechanisms. Physically, a "hole" refers to a vacant electronic site in a molecular orbital resulting from excitation of the electron from the orbital and has a unit positive charge.

The exciton generation rate depends on geometrical factors, such as the angle of incidence of solar radiation, the absorption coefficient of the material, and the bandgap, E_g, between the conduction and valence bands—also known as the lowest unoccupied molecular orbital (LUMO) and the highest occupied molecular orbital (HOMO), respectively. Modeling the sun as a black body at temperature T_s, ignoring radiative losses from the solar cell by treating it as a black body at temperature $T_c = 0$ and neglecting geometrical factors related to incidence angle and assuming that all photons with energy $E > E_g = h\nu_g$ are absorbed and lead to generation of an electronic charge e at voltage V_g (so that $eV_g = h\nu_g$); "ultimate efficiency" is found to be a function of $\frac{E_g}{k_B T_s}$. The ultimate efficiency is found to be approximately 44% for $\frac{E_g}{k_B T_s} = 2.2$, which corresponds to $E_g = 1.1$ eV for $T_s = 6000$ K.

In an actual device, not every photon with energy $E > E_g = h\nu_g$ leads to an absorption and generation of an electronic charge. The ratio of the charge generation rate and the absorption rate of solar photons defines the internal quantum efficiency (IQE). In the calculation of the ultimate efficiency, the IQE is assumed to be unity, and in experiments, it is found to be less than unity. A somewhat related quantity for the charge generation is the exciton dissociation efficiency, which depends on electrostatic binding energy and the presence of contacts between donor and acceptor domains. A schematic showing the formation and dissociation of an exciton is shown in Figure 5.1.

Generation of charges and exciton pairs leads to a nonequilibrium state with higher energy. Electrons and holes tend to reestablish equilibrium by recombining

Insights Obtained from Modeling of Organic Photovoltaics

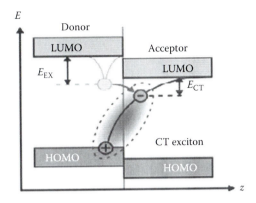

FIGURE 5.1 Schematic energy (E) band diagram near an electron donor–acceptor interface for a single junction. Presence of lower energy bands on the acceptor side leads to dissociation of a singlet exciton to form a charge-transfer exciton/exciplex (dashed oval). E_{EX} is the single exciton binding energy, and E_{CT} is the charge-transfer exciton binding energy. (Adapted from Muntwiler, M. et al., *Phys. Rev. Lett.*, 101, 196403, 2008. With permission.)

in a number of ways and lowering the energy. Different ways of recombining[4] are broadly classified into radiative and nonradiative mechanisms. Radiative mechanisms involve direct recombination of electrons and holes so that the electron moves back to HOMO with the emission of a photon/radiation. Other types of radiative mechanisms involve recombination through traps and surface recombination, in which an electron recombines with a hole to form an exciton and the exciton decays by radiation. Nonradiative mechanisms involve transfer of energy of an electron to another electron in the conduction band creating a highly energetic carrier (known as Auger recombination) or to lattice vibrations (phonons). Note that nonradiative processes also lead to generation of charge carriers. The recombination rate affects the PCE of a photovoltaic cell in a significant manner due to its direct effect on the number of charge carriers, electrons, and holes.

Current, I, generated by any solar cell is the rate at which charges (electrons and holes) are being collected from the solar cell. At equilibrium, the number of charges generated and collected must be equal and is given by the steady state condition[3]: $I = e\,(J_{gr} - J_{cr} + J_{gnr} - J_{cnr})$. Here, J_{gr}, J_{cr}, J_{gnr}, and J_{cnr} are the rates of charge generation by radiative, recombination by radiative, generation by nonradiative, and recombination by nonradiative mechanisms, respectively. Invoking dependences of different rates on the voltage for a p-n junction, the current-voltage relationship was constructed by Shockley and Queisser, and the PCE was computed as a function of $\dfrac{E_g}{k_B T_s}, \dfrac{T_c}{T_s}$, geometrical factors and IQE. Maximum efficiency of around 30% was found by varying various parameters.

These calculations reveal the importance of different factors that affect the PCE. In order to maximize the PCE, exciton generation and its dissociation into charge carriers needs to be maximized. At the same time, recombination needs to be minimized along with an optimization of the current collection. The requirement for the optimization of current collection arises from the fact that contacts between electrodes, electron donors, and acceptors affect charge transport. Issues such as the work function of electrodes and the resulting potential barrier to the electronic current are topics of extensive research. Unlike inorganic semiconductors, which have an almost pure crystalline structure, OPVs have crystalline and amorphous domains due to the use of conjugated polymers,[2] which complicate generation of electron-hole pairs and charge conduction. Thiophene-based conjugated polymers are used as electron donors along with fullerene-based acceptors. Also, instead of a p-n junction (which is created by doping in a single crystal), junctions between donors and acceptors (known as heterojunctions) are widely used. Experimentally, it has been shown that the simplest kind of heterojunction, which corresponds to a bilayer device containing a donor and acceptor in a thin film, is not the best OPV device in terms of the PCE. Instead, devices based on a mixture of donors and acceptors, which can phase segregate to form various heterojunctions in bulk—known as bulk heterojunctions[1] (BHJs)—have higher PCEs. Note that the distribution of donors and acceptors in a BHJ is a result of phase segregation and depends on the processing conditions along with specific nature of donors and acceptors.

Due to the interplay of various parameters, such as polymer characteristics, processing conditions, etc., it is very hard to understand the roles played by different intrinsic processes in OPV devices. With recent advances in computational power and modeling techniques, it is not surprising that simulations of OPVs have become more popular and are beginning to provide very useful insights. In this chapter, we present a number of areas where simulations can provide fundamental insights. Discussion and future directions are presented in Section 5.4.

5.2 ISSUES WITH WHICH MODELING CAN PROVIDE INSIGHTS

It is widely accepted that material structure and its mechanical as well as transport properties are intimately connected. Simulations provide very useful information about this particular connection by exploring chemical and physical mechanisms for the self-assembly and dynamics of molecular and hybrid materials. For example, in the context of OPVs, the PCE and underlying current–voltage (I–V) curves resulting from exposure to sunlight are strongly dependent on the morphology, device architecture, and molecular interactions. Analogous to inorganic semiconductors, such as silicon, photocurrent in conjugated polymers, used in OPVs as an electron donor, can be tweaked in a number of ways, for example, by applying an electric field or due to impurities. At the fundamental level, photocurrent changes due to a change in the structure of energy bands and a change in bandgap between the conduction and the valence band.

Theoretical and computational chemical physics and materials sciences have reached a status that is ideally positioned to offer considerable aid in the quest to understand materials structure–property–transport relationships. Additionally, the rapid progression of computational hardware (1000× performance in the last 5 years) with the power to numerically solve the key underlying differential equations for large system size, often means that realistic system sizes in relevant environments can now be simulated. Using such simulations, insights into a number of issues relevant for OPVs can be obtained. We categorize these issues into four categories:

a. Exciton generation, dissociation, and recombination
b. Effects of macromolecular characteristics of conjugated polymers on morphology
c. Interfacial properties of electrode–donor–acceptor interfaces
d. Effects of morphology on the charge generation and recombination

Details about the kind of insights that can be obtained by applying various modeling techniques are presented in the Section 5.2. Specific applications of modeling techniques are reviewed in Section 5.3.

5.2.1 Exciton Generation, Dissociation, and Recombination

From the discussion on p-n junction, it is clear that the generation, dissociation, and decay of excitonic states plays a critical role in both optoelectronic and photovoltaic systems. These quasiparticles, generated primarily by photoexcitation are held together by Coulombic interactions.[5,6] At the molecular level, excitonic states can correspond to localized or delocalized electron-hole densities that extend over the underlying material. Appropriate electronic coupling to charge-transfer states[5] can give rise to efficient exciton dissociation,[7] which is one key to enhancing OPV efficiency (see the schematic in Figure 5.1 showing the formation of a charge-transfer exciton/exciplex). For heterojunctions, exciton decay toward a charge-separated state is strongly influenced/controlled by molecular-level electronic interactions at the interface. In addition to the important energetic criteria of the band-offset (Figure 5.1), fast charge transfer is another critical component to the overall PV efficiency. Reducing the probability of exciton recombination in the donor phase by strong wave function overlap to excited states can open channels for fast adiabatic electron transfer.[8–10] By engineering the molecular chemistry at the surface, the electronic structure can be modified to promote charge transfer bridging states[11] that can facilitate transfer[12] as well as significantly modulate the open circuit voltage (V_{oc}).[13] Similarly, chemical modification of a semiconducting polymer backbone can potentially be exploited to manipulate the underlying electronic structure.[14]

Experimentally measureable quantities that directly relate to device performance metrics, such as the PCE include[15] short-circuit current (J_{sc}), open-circuit voltage (V_{oc}), and fill factor (FF). J_{sc}, V_{oc}, and FF are defined as the current at zero voltage, voltage at zero current, and ratio of maximum power that can be obtained from the solar cell to the actual power obtained, respectively. The product of these three quantities divided by the total solar input power yields the PCE. J_{sc} depends on the external quantum efficiency (EQE), defined as the number of charge carriers generated per photon impinged (in contrast to absorbed, which defines IQE) on the solar cell. However, V_{oc} is still not well understood. Many authors equate it to the energetic difference between the ionization potential (HOMO) of the donor and the electron affinity (LUMO) of the acceptor.[16] Because such information is a direct outcome of the underlying electronic structure, quantum chemistry and electronic structure calculations can provide a direct means for accessing them and making appropriate correlations.[17-20]

Advent of semilocal and especially nonlocal descriptions of exchange and correlation effects within the density functional theory (DFT) now allows describing bonding in molecules and solids with sufficient accuracy to compare quantitatively to experiments for many classes of systems. It is therefore becoming possible to develop a semiquantitative description of a large number of systems and processes relevant to OPV development. We note that considerable progress is still needed in designing efficient electronic structure algorithms that can accurately treat excited state processes and that can speed up the required evaluation of energy and forces (see Section 5.3.1).

5.2.2 Effects of Macromolecular Characteristics of Conjugated Polymers on Morphology

5.2.2.1 Equilibrium Aspects

Analysis of the exciton generation, dissociation, and recombination as presented here reveals that optoelectronic properties in a heterojunction are strongly dependent on morphology and interfacial structure. Also, it is widely appreciated that the morphology of the active layer in an OPV device is a result of phase segregation[21,22] between the conjugated polymer and fullerene-based electron acceptors. From extensive work in the area of polymer physics,[23,24] it is known that the phase segregation must depend on polymer chain characteristics, such as its molecular weight, architecture (effects of side chains, regioregularity, etc.), and interaction energy with electron acceptors and substrates. Simulations can provide useful insights into the effects of macromolecular characteristics on the morphology in OPVs by simulating self-assembly in thin films containing conjugated polymers and electron acceptors. For such purposes, molecular dynamics[25] (MD), Monte-Carlo[25] (MC), and field theoretical simulations[23] (FTS) are ideal tools. Different techniques have certain advantages and disadvantages, which we discuss in Section 5.3. Nevertheless, these three simulation techniques have been used extensively to obtain insights into morphologies in OPVs.

5.2.2.2 Nonequilibrium Effects of Additives and Processing Conditions

In order to optimize the PCE of OPVs, the mixture of electron donor and acceptor is dissolved in a common solvent containing a small amount of additives, which are also volatile. The choice of a solvent additive is based on the criterion[26] that it should be less volatile than the parent solvent so that it can help in overcoming kinetic barriers by imparting prolonged mobility to reach a true equilibrium state. Experimentally,[26] it has been shown that the use of additives tends to increase the PCEs of OPVs. The origin of such an increase in PCE resulting from the use of solvent additives is a topic of extensive research. In particular, it is not clear how the solvent and additive evaporation rates and any residual additive in the thin film containing the donor–acceptor blend affects the internal structure. Modeling solvent evaporation along with the self-assembly of donor–acceptor blends requires a multiscale approach coupling molecular details of phase transitions with the self-assembly processes inside thin films. However, there have been a few attempts to model the effects of solvent evaporation[27,28] on the morphology in OPVs in the absence of any additive using a continuum-based approach, such as the Cahn-Hilliard-Cook formalism. These simulations provide qualitative insights into the effects of solvent annealing on OPV thin film structures, which has been a topic of extensive research in recent years.

5.2.3 Interfacial Properties of Electrode–Donor–Acceptor Interfaces

The surface of a semiconductor tends to have energy levels[4] lying in the forbidden gap between the HOMO and LUMO. Furthermore, the metal–semiconductor interface tends to have an energy barrier (known as the Schottky barrier[4] for the electronic flow from the semiconductor to the metal). The origin of the barrier lies in the additional bands near the surfaces, which can trap holes or electrons and space-charge distribution near the metal–semiconductor interfaces, which results in an electrostatic potential. In addition, if one of the surfaces is charged, for example, in the case of an electrode, then the energy barrier can be reduced or increased based on the sign of the charge on the surface. Space charge distribution turns out to be an important factor in determining the energy barrier. Research has been conducted to modify cathode and anode electrodes so as to enhance current collection, reduce the barrier, and enhance the PCE. For example, a thin film of poly(3,4-ethylenedioxythiophene) (PEDOT) and poly(styrene sulphonate) (PSS) on indium tin oxide (ITO) is used as an anode buffer layer in a number of studies[29,30] (a conventional configuration). However, devices using ITO covered with cesium carbonate as a cathode (known as an inverted configuration) have been shown to outperform[31] the conventional OPV devices in terms of PCEs. The differences in the PCEs between the conventional and an inverted configuration have been shown[31] to be a result of vertical phase segregation of donors and acceptors due to different interfacial interaction energies. For example, in the inverted configuration, the electron acceptors are populated next to the cathode instead of the anode in the case of the conventional configuration. Modeling based on DFT can provide insights into the effects of polymer chain orientation, interface dipoles on the energy barrier at interfaces involving the electrode, donor, or acceptor.

5.2.4 Effects of Morphology on the Charge Generation and Recombination

Our understanding of how morphology affects charge transport and other electronic properties in an OPV device is at the infancy stage. Conjectures about the ideal morphology for optimum PCE has been put forward in the literature.[2] However, there is a need for systematic studies which can shed light on how morphology affects electronic properties. Preliminary work has been done on coupling drift-diffusion models relating electron, hole, and exciton transport to local structure using continuum modeling.[32,33] In principle, both the morphology and charge transport need to be determined self-consistently. However, we are not aware of any such simulation approach so far. Such investigations will provide guidelines about how global current-voltage relationships can be improved by manipulating local structure.

In Section 5.3, we give a brief overview of some of the primary computational modeling tools that have been deployed to study physicochemical properties/processes in OPVs.

5.3 INSIGHTS OBTAINED FROM SIMULATION APPROACHES

5.3.1 Electronic Structure Calculations

Quantum mechanical electronic structure theory offers a suitable framework to explore many properties and structures of materials and interfaces that are fundamentally important to optoelectronics and photovoltaics. In particular, considerable work has been accomplished in terms of predicting electronic energy levels (HOMO, LUMO), bandgap, dipole moment, electron and hole mobilities, and electronic band offsets at interfaces (for band engineering) to highlight systems that are good candidates for low-bandgap organics and polymers[34] (see Figure 5.2). The most popular computational approach in this regard has been DFT. DFT proposes to solve electronic structure problems using the electron charge density, $\rho(r)$, as the fundamental variable,[35] which is formally based on the Hohenberg and Kohn theorems.[36] In practice, DFT is applied using the Kohn-Sham ansatz (KS), using a mean-field approach.[37] The KS method represents the density as a linear combination of the inner products of independent spin-orbital functions and the energy as a functional of $\rho(r)$ as

$$E^{DFT}[\rho] = T_o[\rho] + E[\rho] + E_{xc}[\rho] + \int \rho(\vec{r}) v_{ext}(\vec{r}) \, d\vec{r} + V_{NN} \quad (5.1)$$

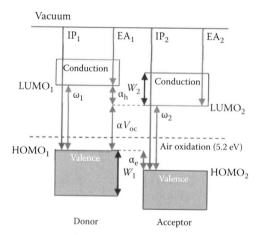

FIGURE 5.2 Schematic diagram for energy levels of a model heterojunction and various quantities that can be computed from electronic structure calculations. (Adapted from Rajendran, A. et al., *The Journal of Physical Chemistry A*, 116, 12153–12162, 2012. With permission.)

where the first and second terms are the kinetic energy of independent particles, $T_o[\rho]$, and the Coulomb interaction energy ($E[\rho] = \frac{1}{2} \int \int dr' dr \rho(r')\rho(r)/|r'-r|$). The term $v_{ext}(r)$ is the potential generated by the nuclei and experienced by the electrons, and V_{NN} is the nuclear repulsion energy for a fixed nuclear configuration. In Equation 5.1, the contribution $E_{xc}[\rho]$ is the exchange-correlation energy, which includes the electron exchange interaction as well as the many body contributions to the kinetic and electron–electron repulsion potentials (that is, $V_{ee}[\rho]$) that are not included in $T_o[\rho]$ or $E[\rho]$, that is, $E_{xc}[\rho] = V_{ee}[\rho] + T[\rho] - E[\rho] - T_o[\rho]$. The explicit expression of E_{xc} remains unknown, but there are many approaches that have shown satisfactory results using standard approximations, such as general gradient approximations and hybrid functionals. Such approaches are grouped according to their treatment of the density into "generations" or "ladder rungs."[38] The most common are based on the local density approximation (LDA) or generalized gradient approximation (GGA).[39] Although some functionals have shown impressive results, those are not always transferable for every problem and can often fail for the description of long-range interactions and excited states. The origins of these difficulties are attributed to the incorrect cancellation of electron self-interaction[40] and incorrect treatment of dynamic correlation, among others. There are many strategies to avoid these problems, some of which involve the inclusion of explicit terms from wave function theories (hybrid functionals), treatments with optimized effective potentials, an adjustment to the asymptotic correction of exchange correlation potentials, and the addition of empirical energy terms. For more details on DFT and post-DFT methods for the study of OPVs, we refer the reader to recent reviews.[41–43]

The pursuit of conjugated organic molecular and polymeric compounds as potential electronic materials in new-generation optoelectronic and photovoltaic devices has recently compounded the long interest of many scientists in the basic photophysics of such species.[18] Currently, there is considerable activity aimed at synthesizing new compounds and formulating structure property relationships, especially in terms of absorption spectroscopy and electrochemistry, which determine bandgap behavior and light-harvesting bandwidth, plus the ability to donate or accept electrons, respectively.[44] Effective computational prediction of the electronic properties of suitable molecules can provide guidance to target the most promising variants of any synthetic family of compounds, depending on the specific properties desired. Recent progress is highlighted by the Harvard Clean Energy Project[45–47] as well as many other efforts from various groups.[48,49]

However, it should be noted that conventional DFT in both its static and time-dependent (TDDFT) formulations suffers from limitations in destabilizing occupied states and overstabilizing unoccupied states in conjugated molecules. This tends to pose some difficulties in directly using information to describe charge transfer, and absolute energies are sometimes suspect. However, in general, useful trends can often be obtained (as mentioned above). Efforts to put DFT on more firm ground for predicting charge transfer and donor–acceptor levels have included, among many others, using orbital-dependent DFT based on Koopman's condition. This approach shows promise for computing energies accurate to a few tenths of an electron volt.[50]

The carrier mobility for photovoltaic materials (ability of a charge to move in the bulk material) is an important parameter that has been examined and estimated using DFT-based approaches. A recent review gives an overview of different first principle models to address scattering regimes in which (1) the molecular internal vibration severely induces charge self-trapping and, thus, the hopping mechanism dominates; (2) both intermolecular and intramolecular scatterings play roles, so the Holstein-Peierls polaron model is applied; and (3) charge is well delocalized with a coherence length comparable to an acoustic phonon wavelength so that a deformation potential approach is more appropriate.[51] Numerous studies have shown results that provide new insight into the microscopic origins of high-performing active layer materials.[52–55] Recent work has also successfully demonstrated DFT-based approaches to determine experimentally validated results for electron mobilities in fullerene networks.[56,57]

In photovoltaic cells, ultrathin interlayers placed between the active layer and a conducting electrode can often improve device performance.[58–62] Such interlayers, when properly designed, improve numerous parameters simultaneously (V_{oc}, J_{sc}, and FF), leading to increased PCE in conventional device architectures and the fabrication of more efficient and air-stable inverted and tandem devices.[63–65] Optimizing device efficiency requires a fundamental understanding of charge carrier transport within the BHJ and across interfaces within the BHJ and between the BHJ and the electrodes. A detailed account of how surface electrostatics are affected by the presence of surface monolayers and are dependent on their molecular orientation was discussed in recent publications.[66,67] Computational studies show the frontier electronic states of poly (3-hexylthiophene) (P3HT) monolayers to be impacted by the orientation angle between the polymer chain and the confining surface. Moreover, surface dipoles resulting from molecular assemblies of thiophene oligomers lying coincident with or perpendicular to the plane of the surface can have a dramatic impact on electronic structure and device-relevant properties. In general, self-assembled monolayers of oriented dipolar molecules physisorbed on metal surfaces generate ample collective electric fields, and subsequently, electrostatics dictates key factors for energy-level alignment and molecular electronic structure. The fundamental problem requires addressing the electronic structure for a coupled system that is composed of a thin film of π-conjugated molecules possessing weak intermolecular interactions on a metal (electrode) surface. A rigorous *ab initio* approach termed "self-consistent periodic image-charges embedding" (SPICE) was recently developed to investigate the evolution of the work function in self-assembled monolayers on conductive substrates.[68] This approach allows using a high level of theory for extended molecular arrays while retaining collective effects observed in a molecular layer. Similarly, DFT slab-model calculations have shown considerable promise for understanding the evolution of interfacial electronic structure.[69] A recent perspective on interfacial electronic structure details additional insight.[70]

5.3.2 Classical Mechanics-Based Approaches

Although many of the intrinsic processes governing OPV operation are quantum mechanical in nature (for example, exciton generation/recombination, charge carrier mobility, charge transfer), structural and dynamical processes of long chain molecules at the nanometer length scales typically used in OPVs are better described within classical mechanics. Molecular dynamics (MD) and Monte Carlo (MC) methods are well-tested classical mechanics techniques, which have been successfully used in addressing polymer physics problems.[71,72]

For molecular-based materials, these methods require the specification of potential energy functions (usually known as force fields) to describe the many various relevant body interactions. Because one of the main interests here is to understand the structure and morphology in solution-cast thin films, the use of potential functions with a proven record of accurate prediction of structure and vibrational spectra for conjugated organic molecules are required. Such potentials are divided into two parts: (i) bonded interactions, typically composed of harmonic or Morse oscillators for the bond-stretching and angle-bending terms (both in- and out-of-plane), truncated Fourier series for the torsion interactions (regular dihedral and improper), and (ii) nonbonded interactions typically composed of van der Waals type, such as Lennard-Jones 6–12, and electrostatics, such as Coulomb. Enormous research is being dedicated to improve the force fields to generate a more accurate representation of the "close to real" force fields for a particular system. Several standard force fields fall into this category, such as the MM2, MM3, MM4, Dreiding, UFF, MMFF, CHARMM, AMBER, GROMACS, TRIPOS, and OPLS.[73]

Most of these simulations can be performed using fully atomistic models in which interactions between each atomic site are relevant. The advantage of fully atomistic simulation lies in the fact that the underlying chemistry and physics can be more "accurately" addressed by the model system. On the other hand, as the number of particles and degrees of freedom increases, the computational cost and time increases enormously. In order to approach "realistic" length and time scales, coarse-grained approaches are often used.[74–76] The coarse graining consists of replacing an atomistic description of a molecule with a lower-resolution model that averages away some of the finer details of the interactions while still preserving the underlying physical and chemical characteristics. Numerous coarse-grained models (differing in the metric used to obtain the coarse-grained potential, such as the fitting to the forces or to structural features but also in the definition of the coarse grain) have been developed that are critical to many long-chain molecular processes, such as polymer, lipid membranes, and proteins. Coarse graining can also refer to the removal of certain degrees of freedom (for example, vibrational modes between two atoms) by freezing the bonds, bends, or torsional degrees of freedom, but more typically it implies that two or more atoms are collapsed into a single particle representation (the so-called united atom model was one of the first popular coarse-grained models). Fundamentally, the level to which a system may be coarse grained is bound by the accuracy in the dynamics and structural properties desired from a simulation. We refer to several review articles that provide explicit details.[77–79]

5.3.2.1 Molecular Dynamics

MD simulations[80–82] essentially consist of numerically integrating Hamilton's, Newton's, or Lagrange's equations of motion using small integration time steps. Although these equations are valid for any set of conjugate positions and momenta, the use of Cartesian coordinates greatly simplifies the kinetic energy term (Newton's equations of motion). The resulting coupled first-order, ordinary differential equations can be solved using numerical integrators.[83] Despite the simplicity of MD methods, the simulation of millions of atoms over long time scales, such as seconds, is not possible without a substantial reduction in the number of equations of motion, a feat typically achieved by coarse graining. This approach not only reduces the number of equations, but if proper separation of the time scales for motion can be achieved, it also allows the use of considerably larger integration time steps and thus helps bridge the large gap between simulation and experimental time scales. In order to show the nature of insights obtained from coarse-grained MD simulations, the authors describe one such recent application employing a supercomputer.[74]

One of the systems that has been studied extensively is based on an alkyl thiophene known as poly (3-hexylthiophene) (P3HT) (as an electron donor) and fullerene-based [6,6]-phenyl-C61-butyric acid methyl ester (PCBM) (an electron acceptor). Coarse graining for the P3HT:PCBM blend has been done by a number of workers to capture different levels of detail in the coarse-grained models. One such coarse-graining procedure has been outlined by Lee, Pao, and Chu[75] (LPC), which requires running atomistic and coarse-grained simulations. The atomistic simulations used the Dreiding force field,[84] and force field parameters for the coarse-grained model were determined by matching structural information, such as radial distribution function, bond angles, etc., between the coarse-grained model and atomistic simulations. In the coarse-grained model developed by Lee, Pao, and Chu,[75] the P3HT monomer and the PCBM were coarse grained as single Lennard-Jones (LJ) beads so that effects of side chains and regioregularity cannot be captured using the model. The P3HT polymer consists of the LJ beads connected through a series of harmonic bonds.

We have used such a coarse-grained model to understand the kinetics of phase segregation in active layers of OPVs using MD simulations. These simulations were performed using LAMMPS[85] MD code with GPU acceleration.[86] The initial configuration of the simulation box consisted of m randomly distributed P3HT chains with N degree of polymerization in a random walk configuration mixed with randomly distributed PCBM beads at a low number density in a simulation box that is periodic in x, y, and z directions. For details of the simulations, we refer readers to the work of Carrillo and coworkers.[74] In Figure 5.3, the coarse-graining procedure for P3HT and PCBM are described together with a snapshot of the simulation box after a 400 ns stimulation in of the isobaric-isothermal (NPT) ensemble.

The MD simulations relate microscopic details about conformational characteristics of the polymer chain, such as the persistent length, radius of gyration, and alignment of chain segments to macroscopic phase segregation/morphology. Coarse-grained model parameters are set by the constraints placed on bond angles, dihedrals, etc., which are determined using atomistic simulations and chemical details of the macromolecules. For example, for the LPC force field parameters, the persistence length of the P3HT polymer, l_p, is set at 37.8 Å, which is close to the experimental value of 33 Å. Relating the radius of gyration to the length of a persistent segment and bond angle, it turns out one persistent segment consists of approximately 10 beads for the LPC force field.

Simulations of P3HT:PCBM blends can be used to map out the phase diagram by a number of MD runs focusing on different parameters. For example, in order to understand

Insights Obtained from Modeling of Organic Photovoltaics 125

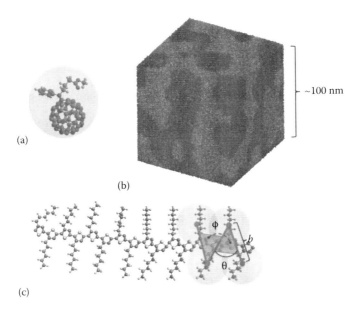

FIGURE 5.3 PCBM coarse grained as a single LJ bead (a). P3HT monomer coarse grained as a single LJ bead and the coarse-grained polymer bond-length, bond-angle, and dihedral angle representations (b). Snapshot of the simulation box for 1:1 weight ratio of P3HT (gray):PCBM (black) after 400 ns of isobaric-isothermal annealing at 423 K and 1 atm (c).

the effects of thermal annealing and mixing ratio of P3HT:PCBM, the authors run a set of calculations for a given chain length, and results for the morphology are presented in Figure 5.4. From the snapshots, it appears that there is a particular mixing ratio beyond which the P3HT and PCBM are uniformly mixed. For example, in Figure 5.4, the uniform mixing takes place at P3HT volume fractions greater than 0.86.

Similarly, effects of the polymer chain length on the morphologies can be studied by varying the degree of polymerization, N, keeping the mixing ratio fixed. Results of such MD runs are presented in Figure 5.5. It is found that the polymer chain length plays a key role in determining the morphology. For example, for $N = 10$ and weight fraction $\phi_m = 0.5$, the morphology is found to be smectic/lamellar-like with alternating domains of P3HT and PCBM. For higher values of N, bicontinuous morphologies expected from spinodal decomposition, such as phase segregation, are found. The smectic morphology found for $N = 10$ is similar to the one stabilized by entropy in mixtures of rods and spheres. Note that in our estimation, a persistent segment consists of 10 beads and reveals that the chains with $N = 10$ can be approximated by short rods. Also, the coarse-grained model used approximates the PCBM by a single spherical bead. We are not aware of any experimental result corroborating such morphology in OPVs containing short thiophenes. Nevertheless, the MD simulations can be used to map out the full morphology diagram.

Dissociation of an exciton into an electron and hole takes place at the interface between the donor and acceptor (cf. Figure 5.1). This, in turn, means that in an inhomogeneous medium, such as those shown in Figure 5.4, optimization of the

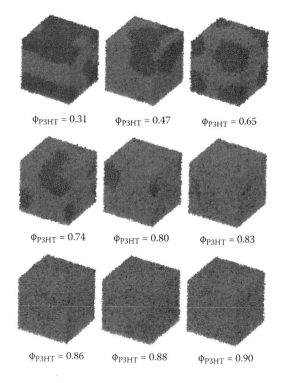

FIGURE 5.4 Snapshots of P3HT (gray):PCBM (black spheres) MD simulations with $N = 100$, $m = 1200$ at different values of P3HT volume fraction.

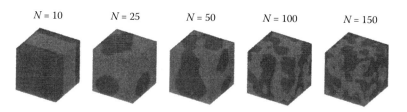

FIGURE 5.5 Simulation snapshots after 400 ns for systems with $m = 32{,}400$, weight fraction $\phi_m = 0.5$, and different degrees of polymerization N.

interfacial area between the P3HT and PCBM is of vital importance for exciton dissociation. Computation of the interfacial area from three-dimensional structures involving different kinds of particles can be readily done using the Laguerre or radical Voronoi tessellation schemes, which have been implemented in the Voro++ library.[87] Details about the Voronoi tessellation schemes can be found in Rycroft.[87] In particular, the Laguerre tessellation method takes into account the different bead size in calculating the Voronoi volumes. Furthermore, the tessellation scheme provides the surface area of each face of the Voronoi polyhedron. This information is used to calculate the interfacial area in the morphologies obtained from MD runs by summing the surface areas that belong to two Voronoi polyhedrons with different

Insights Obtained from Modeling of Organic Photovoltaics

type of seed (that is, PCBM or P3HT). Furthermore, the tessellation scheme provides the nearest neighbor information, which we can be used to quantify the connectivity of domains to their respective electrode for runs involving substrates.

We have calculated the interface area-to-volume ratio, γ, at different ϕ_{P3HT} and N. In Figure 5.6a, we observe a steady increase in γ as a function of ϕ_{P3HT} followed by a plateau value at the range of ϕ_{P3HT} = 0.6 to 0.8. When the system changes from a two-phase to a homogenous system (ϕ_{P3HT} ~ 0.86), we observe a steep rise in the value of γ followed by a steep fall when there are low to zero amounts of PCBM. For the system with different N and ϕ_m = 0.5 (see Figure 5.6b) γ is a nonmonotonic function in which N = 10 has the highest value. From N = 25 to 150, for the large system (m = 32,400), N = 150 has a large value for γ highlighting the importance of the chain degree of polymerization in the morphology of the BHJ when longer chains have larger γ.

Recombination of electrons and holes depends on the lifetime and a length scale determined by the exciton diffusion constant. The length scale for the exciton diffusion is ~10 nm. As the excitons are generated in a domain-containing electron donor, the size of the domains turns out to be a very important factor in affecting the recombination rate. Based on these arguments, the ideal domain size of the electron donor domain should be around 10 nm. The domain size, d, for a phase-segregating mixture depends on the kinetics of phase segregation ($d \sim t^\alpha$, t being the time), presence of interfaces, etc. For phase segregation dominated by diffusion-limited processes, α = 1/3, and the presence of interfaces modify this exponent. For the MD snapshots, the domain size can be computed by constructing the scattering function $S(\vec{q})$ defined as

$$S(\vec{q}) = \frac{1}{N \times m} \sum_{k,j}^{N \times m} \exp\left(i\left(\vec{q} \cdot \left(\vec{r}_k - \vec{r}_j\right)\right)\right) \qquad (5.2)$$

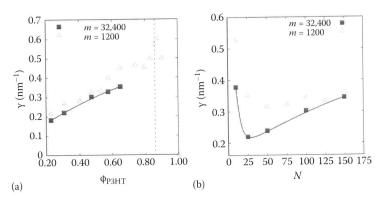

FIGURE 5.6 Interface area-to-volume ratio, γ as a function of ϕ_{P3HT} with N = 100 after t = 400 ns (a); as a function of N at P3HT weight fraction of ϕ_m = 0.5, after t = 400 ns for m = 32,400 and t = 360 ns for m = 1200 (b). The dotted line in (a) is ϕ_{P3HT} = 0.86 when the morphology changes from two-phase to a homogenous phase system.

where the summation is over all of the P3HT monomers, and $\vec{r}_k$ is the radius vector of the kth monomer, and $\vec{q}$ is the wave vector. A typical $S(\vec{q})$ is presented in Figure 5.7 and shows a peak at $q^* = \frac{2\pi}{d}$. The peak at nonzero wave vector characterizes the average domain size and corresponds to a P3HT domain size of ~20 nm in Figure 5.7. For systems undergoing phase separation, Fratzl and Lebowitz have proposed a heuristic "universal" formula,[88] which can be used to extract the relevant correlation length and domain sizes. In Figure 5.7, we have used the formula to extract correlation length. The same information can be extracted in real space using the radial distribution function. The relationship between the first peak and the domain size can be seen by plotting the radial distribution function of P3HT $(g(r)_{P3HT-P3HT})$ and the Debye correlation function,[89] $\Gamma(r) = \frac{\rho_i \rho_j}{\rho^2}$ where ρ_i and ρ_j are the local fluctuations in the number density from its average value and $r = |r_i - r_j|$ is the distance between two points i and j. ρ^2 is the average value of the square of the number density fluctuations. Note that $\Gamma(r)$ can be fitted with an equation of the form[88]

$$\Gamma(r) = \left(\frac{d}{2\pi r}\right)\exp(-\lambda r)\sin\left(\frac{2\pi r}{d}\right) \quad (5.3)$$

where λ^{-1} is the correlation length. Figure 5.7 illustrates the relationship of q^*, d, $S(q)$, $g(r)_{P3HT-P3HT}$ and $\Gamma(r)$.

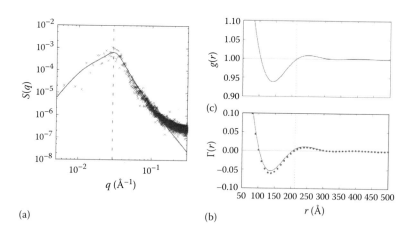

FIGURE 5.7 (a) Scattering function, $S(q)$ (black crosses) for the 50% by weight P3HT:PCBM with $m = 32{,}000$ and $N = 100$ at $t = 100$ ns. The solid line is $q^* = 0.0295$ Å^{-1}, and the solid black line corresponds to Fratzl and Lebowitz fit. The corresponding radial distribution function $g(r)_{P3HT-P3HT}$ (b) and Debye correlation function $\Gamma(r)$ (filled circles) (c) with $d = \frac{2\pi}{q^*} = 213$ Å (broken black line). The blue line in (c) is Equation 5.3 where $d = 205 \pm 10$ Å.

5.3.2.2 Monte Carlo

MC methods[90] used in macromolecular science begin by constructing a Markov chain generated by random changes moderated by an acceptance criterion, which is used to generate new chain configurations commensurate with the prescribed thermodynamic ensemble. Using an algorithm for generating a sequence of changes (for example, the Metropolis algorithm), a move is only accepted if the new configuration space leads to a lower energy configuration with the probability of the Boltzmann factor. As the sequence length increases, it quickly becomes necessary to introduce a series of biased moves in which additional information about the system is incorporated into the MC selection process to maintain a detailed balance. The most commonly used biased sampling techniques are the continuum configuration-bias and concerted rotation moves.[91] These modern algorithms or slightly modified versions can efficiently generate dense fluid polymer systems for chain lengths of 30–100 monomers. Longer polymer chain lengths pose additional convergence problems and often require the use of other types of biased moves, in particular the double-bridging moves.[92,93] Addition of solvent via continuum models (implicit solvation models)[94] or explicit atoms can also be implemented. A primary interest in using the MC method is to ensure adequate equilibration of longer chain polymers, that is, those with hundreds of monomers. The advantage of the MC over the MD method is that MC can reach a much longer time and length scale with larger time steps; however, MC is not suitable for understanding relaxation on longer time scales. MD methods can also be used but generally require considerably longer computational times to equilibrate as they often spend long times in local minima (the typical force calculation requires evaluation of $O(N^2)$ energy terms for every step). The force field used in MC is similar to the MD simulation: nonbonded interactions are modeled as van der Waals and electrostatics, and bonded interactions are modeled using bond, angle, and dihedral potentials.

A useful extension of the MC method is to introduce a stochastic time variable into the simulation to allow modeling of rates of specified processes. This approach is referred to as kinetic Monte Carlo (KMC) and is often used to examine aspects of crystal-growth dynamics[95] and to estimate charge transport.[96] For example, Liu et al.[97] used KMC to examine the growth of a C_{60} phase on graphite. Extensions of KMC to treat nucleation and growth in solutions can also be achieved.[98] KMC relies on the charge transport mechanism in organic materials, which are the focus of this discussion. Typically, organic materials have weak electronic coupling between molecules, thereby giving rise to a series of hopping events. The rate of hopping can be described by the Marcus equation.[99] KMC treats the prefactor in the Marcus equation as a fitting parameter that may or may not be isotropic. As KMC uses the time-averaged electronic coupling of the electron and the hole, computational overhead of simulating an entire OPV device gets reduced. The KMC simulation results using this approach match fairly well with experimental data for polymer,[100] polymer-fullerene[101] OPV devices, and organic LEDs.[102] An example of such a comparison is shown in Figure 5.8.[103]

KMC can be used to study charge separation, density of states, and charge delocalization. Furthermore, the KMC models provide unique opportunities to couple

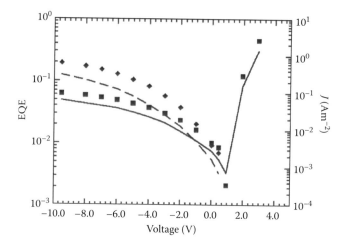

FIGURE 5.8 Experimental external quantum efficiency, EQE (diamonds), and current density, J (squares), for poly[9,9-dioctylfluorene-co-bis[N,N''-(4-butylphenyl)]bis(N,N''-phenyl-1,4-phenylene)diamine] (PFB) and poly(9,9-dioctylfluorene-co-benzothiadiazole) (FBBT) bilayer OPV. Dashed and solid lines are predictions from the KMC simulations. (From Kimber, R.G.E. et al., *Physical Review B*, 86, 235206, 2012.)

structural and dynamical properties at the nanoscale (derived from MD) with the mesoscopic continuum models.

5.3.3 Continuum-Based Approaches

These approaches include modeling of structure and transport properties using methods based on continuum mechanics. *Field theory* (FT) and phase field modeling are the two main computational schemes, which can access longer time and length scales. FT representations can be effectively obtained from coarse-grained models and offer an advantage of significantly reducing the number of differential equations for the particles to a diffusion equation that can be numerically solved using a self-consistent algorithm.[23,104,105] Another advantage of the FT approach is that it ensures thermodynamic equilibrium of the final morphology and thus provides access to modeling much larger length scales. Such improvement is difficult to obtain with the particle-based approaches (atomistic or coarse-grained) due to the existence of a very high-dimensional potential energy landscape plagued with multiple local minima. The FT approach can also incorporate realistic polymer models that are often difficult for macroscopic, continuum simulations. The coarse-grained Hamiltonian of a dense melt of polymer chains contains chain stretching or bending penalties along with other interactions. The theory can be easily generalized to a larger number of blocks to include architectural asymmetry, such as branched copolymer architectures and surface/substrate corrugation, and to hybrids with nanoparticles. Transition from a particle description to field configuration is established using formally exact FT transformations.

The self-consistent field theory (SCFT) corresponds to the saddle-point configuration of the fields. Due to the fact that the Hamiltonian is complex in the FT approach, evaluation of the functional integrals over the fields can be done using the Complex-Langevin (CL) sampling scheme.[23,104,105] The latter goes beyond the saddle point approximation and is a formally exact evaluation of ensemble average properties. In both cases, a system of equations needs to be solved self-consistently using an iterative algorithm to solve for the number density and field configurations. To study time-dependent properties, a dynamic mean field theory[23,104,105] (DSCFT) can be constructed by using the SCFT as the equilibrium state.

The SCFT has been proven to be highly successful in describing the complex morphologies in multiblock copolymer melts and even for nanoparticles in polymers.[104–107] SCFT simulations for a dense melt of copolymers employ chemical specificity of monomers through excluded volume parameters, which controls the strength of monomeric interactions. The SCFT, coupled with a drift-diffusion model for electrons, holes, and excitons, has been used to explore the effects of morphology on the charge transport in rod-coil block copolymer melts,[33] where the rod-like block represents a conjugated polymer (that is, a donor), and the coil-like block represents electron acceptor moieties. Using such an approach, current-voltage relationships and the PCE of OPVs for different morphologies can be predicted. Mobilities of electrons, holes, and excitons, the dielectric constant of the medium, and temperature are the input parameters for such a model. In addition, the generation, decay, and recombination rates of excitons are assumed to be of a certain functional form on the basis of semiconductor physics. Using such a model, the effects of anisotropic hole mobilities along and perpendicular to the electron donor polymer backbones have been studied. Such simulations reveal the importance of morphology in affecting the device properties such as short-circuit current, open-circuit voltage, etc. For example, it was shown that a lamellar morphology between electrodes with alternating domains in contact with electrodes (perpendicular lamellar) has better PCE than a lamellar morphology, in which only one of the domains is in contact with the electrodes (parallel lamellar). The difference arises from the presence of direct conducting pathways in a perpendicular lamellar in contrast to the parallel lamellar. Also, the effects of parallel and perpendicular orientation of the rod-like block can be studied using the model.

The DSCFT can be used to study the effects of processing conditions, such as the solvent and additive evaporation. DSCFT has been used to study such effects in the case of block copolymer assembly in thin films.[108] However, we are not aware of any work in relation to the OPVs. With the recent advances in the modeling of semiflexible polymers[109] using the SCFT and the fact that conjugated polymers are semiflexible, we envision that the DSCFT will provide very useful insights into the effects of additives on OPV device properties. Preliminary work based on the Cahn-Hilliard-Cook formalism focusing on the effects of solvent evaporation has been reported.[27] Noting that the Cahn-Hilliard-Cook formalism can be derived as a limiting form of the DSCFT for weak density gradients, we can imagine an application of the DSCFT for improving our understanding of the effects of solvent additives in the near future.

5.3.4 Multiscale Methods

Multiscale methods generally involve employing two or more of the above computational approaches in order to treat a broader set of processes that may span multiple length and time scales.[110–112] Typically, the so-called multiscale modeling is actually more of a multistep approach (a loose integration), whereby a needed parameter/information at one length or time scale is obtained using an appropriate computational approach that is subsequently used to start another type of simulation for a different scale. For example, the issue of miscibility of alkyl thiophenes in different solvents and fullerene derivatives has been addressed using a multiscale approach. Here, we describe such a protocol in the analysis of miscibility between P3HT and PCBM employing MD simulations and the continuum-based Flory-Huggins theory.[24]

It has been reported that PCBM is miscible with P3HT for PCBM volume fractions, ϕ_{PCBM} less than 15%–22%.[113,114] In our simulations of different P3HT weight fractions, we found that at ϕ_{PCBM} greater than 14% (ϕ_{P3HT} < 86%), the PCBM beads begin to aggregate, forming PCBM domains as shown in Figure 5.4. This is further corroborated by the radial distribution function of PCBM (see Figure 5.9) where at ϕ_{P3HT} = 86% a second peak located at ~2 nm begins to develop, and higher-order peaks are evident, such as those seen in ϕ_{P3HT} ≤ 83%. To further investigate the miscibility of the coarse-grained P3HT-PCBM system, we have calculated the Flory-Huggins χ parameter and analyzed the miscibility of the PCBM in P3HT in the framework of the Flory-Huggins mean field approach to polymer solutions.

Within the Flory-Huggins[115] theory of binary mixtures, the energy change on mixing per lattice site is

$$\Delta U_{mix} = \chi \phi_v (1 - \phi_v) k_B T \qquad (5.4)$$

where χ is the Flory interaction parameter, and ϕ_v is the polymer volume fraction. The energy change on mixing per lattice site, ΔU_{mix} can be estimated by running three separate simulations: (1) simulation of pure PCBM, (2) simulation of pure P3HT polymers, and (3) simulation of the mixture. In each of the three simulations, the pair energy per bead is[25]

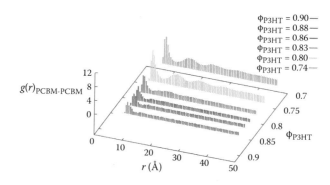

FIGURE 5.9 Radial distribution function of PCBM $g(r)_{PCBM-PCBM}$ as a function of ϕ_{P3HT} from of MD simulations of P3HT:PCBM with $N = 100$, $m = 1200$.

Insights Obtained from Modeling of Organic Photovoltaics

$$\frac{U}{n_{TOTAL}} = 2\pi\rho \int_0^\infty r^2 u(r) g(r) \, dr \qquad (5.5)$$

where n_{TOTAL} is the total number of beads, ρ is the number density, $u(r)$ is the LJ pair potential, and $g(r)$ is the radial distribution function. Fortunately, in MD, the pair energy is evaluated in order to evaluate the forces acting on each of the particles. In this regard, the pair energy per bead is already available, and Equation 5.5 needs not be evaluated separately. The energy change of mixing per lattice site is therefore

$$\Delta U_{mix} = \frac{(n_{P3HT} + n_{PCBM})_3 (U/n_{TOTAL})_3}{(V/v_0)_3} - \frac{(n_{P3HT})_3 (U/n_{TOTAL})_2}{(V/v_0)_3} - \frac{(n_{PCBM})_3 (U/n_{TOTAL})_1}{(V/v_0)_3} \qquad (5.6)$$

where the subscript in the parentheses denotes the type of simulation, that is, $(\ldots)_1$, $(\ldots)_2$, and $(\ldots)_3$ denotes the values derived from the pure PCBM, pure P3HT polymer, and the mixture of P3HT and PCBM simulations, respectively. V is volume of the simulation box, and v_0 is the reference volume, which is taken to be the volume of a PCBM bead.

The volume fraction of P3HT, ϕ_v, can be estimated by taking the ratio of the Voronoi volumes belonging to the P3HT monomer divided by the volume of the simulation box. The radical Voronoi tessellation provides the weighted Voronoi cell volumes. In a similar fashion, the smallest unit volume, v_0, can be estimated by averaging the volume of the PCBM Voronoi cells that do not touch a P3HT monomer. For the system with $N = 100$, $m = 1200$, and P3HT weight fraction of 0.5, the value of the P3HT volume fraction is $\phi_v = 0.475$ and the lattice volume, $v_0 = 891.5$ Å^3.

For the aforementioned system, ΔU_{mix} equals 0.203 kcal/mol, and using Equation 5.6 results in a Flory interaction parameter, χ that is equal to 0.97 ± 0.10. The experimentally determined value for the P3HT:PCBM blend is 0.86. The origin of such a small discrepancy in the value of χ can be attributed to experimental factors, such as the polydisperse nature of polymers and lack of π–π interactions in the MD simulations. Nevertheless, the estimated value of the χ parameter can be used to get a qualitative picture as described here.

According to the Flory-Huggins equation of polymer solutions, the critical point is given by $\chi_c = \frac{1}{2} + \frac{1}{\sqrt{N_1}}$ where N_1 is the number of lattice sites occupied by a single polymer chain. N_1 can be determined by dividing the total Voronoi volume occupied by a polymer chain by the lattice volume v_0. For $N = 100$, $N_1 = 15.47$, and $\chi_c = 0.754$. $\chi > \chi_c$ suggests that the mixture opposes mixing (cf. Figure 5.10), which is in qualitative agreement with the MD simulations in which P3HT chains blended with PCBM are found to be phase segregated.

However, note that the mean field approach does not agree with the MD simulations for $\phi_{P3HT} = 0.8$–0.86, where the simulations show phase segregation, and the

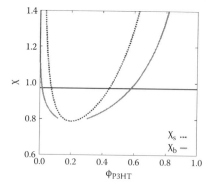

FIGURE 5.10 The phase diagram based on the Flory–Huggins theory for P3HT:PCBM blend with $\chi = 0.97 \pm 0.1$ determined using the MD simulations. The dashed line represents the spinodal curve, and the gray represents the coexistence curves.

theory predicts uniform mixing as per Figure 5.10. The source of such a discrepancy lies in the mean field nature of the theory and neglect of chain stretching, the semiflexible nature of the P3HT chain and the alignment of the P3HT segments in the blend. A more developed theory is needed to take into account all of these effects. Nonetheless, the mean field approach provides a phase diagram of a binary system that is in qualitative agreement with the simulations, and such a comparison reveals that the PCBM acts like a poor solvent for the P3HT chain. For example, $\phi_{P3HT} = 0.31$ is in the two-phase region, and $\phi_{P3HT} = 0.47$ and 0.65 have bicontinuous morphologies and are in the metastable region, and $\phi_{P3HT} = 0.9$ is in the uniformly mixed region. These morphologies qualitatively agree with the snapshots in Figure 5.4. Similar treatment can be done for atomistic simulations,[116] which can take into account π–π interactions, and a more accurate value of the Flory-Huggins parameter can be computed. Such multiscale treatments can provide useful information regarding the miscibility of conjugated polymers in different solvents and with electron acceptors.

5.4 DISCUSSION AND OUTLOOK

The field of optoelectronics and OPVs requires scientific studies via an interdisplinary research environment involving fundamentals of semiconductors along with polymer physics. It is clear that the prediction of optoelectronic properties of an OPV device, such as the current-voltage relationship, is a multiscale problem. As an outlook, we envision that multiscale methods will provide useful insights into intricate coupling between optoelectronic properties, morphology, and chemical details of electron donors and acceptors. These methods will include details about the regioregularity, polymer crystallization based on π-stacking in thiophenes, processing conditions, and surface effects due to buffer layers at electrodes.

Also, it is very important to realize that progress towards unraveling the underlying physicochemical processes that control the structure and macroscopic physical, mechanical, electrical, and transport properties of materials requires tight integration of theory, modeling, and simulation with precision synthesis, advanced experimental characterization, and device measurements. In this synergistic approach, theory and simulation can greatly accelerate the process of materials discovery by providing atomic level understanding of physicochemical phenomena and for making predictions of trends. In particular, this approach can provide understanding, prediction, and exploration of new materials and conditions before they are realized in the laboratory to illuminate connections between experimental observations and to help identify new materials for targeted synthesis.

REFERENCES

1. Hoppe, H. and Sariciftci, N. S. Organic solar cells: An overview. *J. Mater. Res.* 19, 1924–1945 (2004).
2. Coakley, K. M. and McGehee, M. D. Conjugated polymer photovoltaic cells. *Chem. Mater.* 16, 4533–4542 (2004).
3. Shockley, W. and Queisser, H. J. Detailed balance limit of efficiency of p–n junction solar cells. *J. Appl. Phys.* 32, 510–519 (1961).
4. Smith, R. A. *Semiconductors*, 2nd edn. (Cambridge University Press, New York, 1978).
5. Muntwiler, M., Yang, Q., Tisdale, W. A. and Zhu, X.-Y. Coulomb barrier for charge separation at an organic semiconductor interface. *Phys. Rev. Lett.* 101, 196403 (2008).
6. Zhu, X.-Y., Yang, Q. and Muntwiler, M. Charge-transfer excitons at organic semiconductor surfaces and interfaces. *Acc. Chem. Res.* 42, 1779–1787 (2009).
7. Kawatsu, T., Coropceanu, V., Ye, A. and Bredas, J. L. Quantum-chemical approach to electronic coupling: Application to charge separation and charge recombinaiton pathways in a model molecular donor-acceptor system for organic solar cells. *J. Phys. Chem. C.* 112, 3429–3433 (2008).
8. Kanai, Y. and Grossman, J. C. Insights on interfacial charge transfer across P3HT/fullerene photovoltaic heterojunction from ab initio calculations. *Nano Lett.* 1967–1972 (2007).
9. Marchiori, C. F. N. and Koehler, M. Dipole assisted exciton dissociation at conjugated polymer/fullerene photovoltaic interfaces: A molecular study using density functional theory calculations. *Syn. Met.* 160, 643–650 (2010).
10. McMahon, D. P., Cheung, D. L. and Troisi, A. J. Why holes and electrons seperate so well in polymer/fullerence photovoltaic cells. *J. Phys. Chem. Lett.* 2, 2737–2741 (2011).
11. Kanai, Y., Wu, Z. and Grossman, J. C. Charge separation in nanoscale photovoltaic materials: Recent insights from first-principles electronic structure theory. *J. Mater. Chem.* 20, 1053–1061 (2010).
12. Adams, D. M. et al. Charge transfer on the nanoscale: Current status. *J. Phys. Chem. B* 107, 6668–6697 (2003).
13. Tada, Y., Geng, Q., Wei, K., Hashimoto, K. and Tajima K. Tailoring organic heterojunction interfaces in bilayer polymer photovoltaic devices. *Nat. Mat.* 10, 450–455 (2011).
14. Trukhanov, V. A., Bruevich, V. V. and Parachuk, D. Y. Effect of doping on performance of organic solar cells. *Phys. Rev. B.* 84, 205318 (2011).
15. Benanti, T. L. and Venkataraman, D. Organic solar cells: An overview focusing on active layer morphology. *Photosynth. Res.* 87, 73–81 (2006).
16. Scharber, M. C, Wuhlbacher, D., Koppe, M., Denk, P., Waldauf, C., Heeger, A.J. and Brabec, C.L. Design rules for donors in bulk-heterojunction solar cells—Towards 10% energy-conversion efficiency. *Adv. Mater.* 18, 789 (2006).
17. Coropceanu, V., Cornil, J., da Silva, D. A., Olivier, Y., Silbey, R. and Bredas, J. L. Charge transport in organic semiconductors. *Chem. Rev.* 107, 926–952 (2007).
18. Bredas, J. L., Norton, J. E., Cornil, J. and Coropceanu, V. Molecular understanding of organic solar cells: The challenges. *Acc. Chem. Res.* 42, 1691–1699 (2009).

19. Beljonne, D., Cornil, J., Muccioli, L., Zannoni, C., Bredas, J.L. and Castet, F. Electronic processes at organic-organic interfaces: Insight from modeling and implications for opto-electronic devices. *Chem. Mater.* 23, 591–609 (2011).
20. Grozema, F. C. and Siebbeles, L. D. A. Mechanism of charge transport in self-organizing organic materials. *Int. Rev. Phys. Chem.* 27, 87–138 (2008).
21. Hoppe, H. and Sariciftci, N. S. Morphology of polymer/fullerene bulk heterojunction solar cells. *J. Mater. Chem.* 16, 45–61 (2006).
22. Brabec, C. J. et al. Polymer–fullerene bulk-heterojunction solar cells. *Adv. Mater.* 22, 3839–3856 (2010).
23. Fredrickson, G. H. *The Equilibrium Theory of Inhomogeneous Polymers*. (Oxford University Press, Oxford, UK, 2006).
24. de Gennes, P. G. *Scaling Concepts in Polymer Physics*. (Cornell University Press, New York, 1979).
25. Hansen, J. P. and McDonald, I. R. *Theory of Simple Liquids*. (Academic Press, New York, 1976).
26. Lee, J. K. et al. Processing additives for improved efficiency from bulk heterojunction solar cells. *J. Am. Chem. Soc.* 130, 3619–3623 (2008).
27. Kouijzer, S. et al. Predicting morphologies of solution processed polymer:fullerene blends. *J. Am. Chem. Soc.* 135, 12057–12067 (2013).
28. Michels, J. J. and Moons, E. Simulation of surface-directed phase separation in a solution-processed polymer/PCBM blend. *Macromolecules* 46, 8693–8701 (2013).
29. Wang, H., Shah, M., Ganesan, V., Chabinyc, M. L. and Loo, Y.-L. Tail state-assisted charge injection and recombination at the electron-collecting interface of P3HT:PCBM bulk-heterojunction polymer solar cells. *Adv. Energy Mater.* 2, 1447–1455 (2012).
30. Alonzo, J. et al. Assembly and organization of poly(3-hexylthiophene) brushes and their potential use as novel anode buffer layers for organic photovoltaics. *Nanoscale* 5, 9357–9364 (2013).
31. Xu, Z. et al. Vertical phase separation in poly(3-hexylthiophene): Fullerene derivative blends and its advantage for inverted structure solar cells. *Adv. Funct. Mater.* 19, 1227–1234 (2009).
32. Buxton, G. A. and Clarke, N. Computer simulation of polymer solar cells. *Modell. Simul. Mater. Sci. Eng.* 15, 13 (2007).
33. Shah, M. and Ganesan, V. Correlations between morphologies and photovoltaic properties of rod–coil block copolymers. *Macromolecules* 43, 543–552 (2009).
34. Rajendran, A., Tsuchiya, T., Hirata, S. and Iordanov, T. D. Predicting properties of organic optoelectronic materials: Asymptotically corrected density functional study. *J. Phys. Chem. A* 116, 12153–12162 (2012).
35. Parr, R. and Yang, W. *Density Functional Theory of Atoms and Molecules*. (Oxford University Press Oxford, UK, 1989).
36. Hohenberg, P. and Kohn, W. Inhomogeneous electron gas. *Phys. Rev.* 136, B864–B871 (1964).
37. Kohn, W. and Sham, L. J. Self-consistent equations including exchange and correlation effects. *Phys. Rev.* 140, A1133–A1138 (1965).
38. Perdew, J. P. and Schmidt, K. *Jacob's Ladder of Density Functional Approximations for the Exchange-Correlation Energy*. AIP Conf. Proc. 577, 1 (2001).
39. Perdew, J. P. and Kurth, S. A. *Primer in Density Functional Theory*. (Springer Verlag, Berlin, 2003).
40. Perdew, J. P. and Zunger, A. Self-interaction correction to density-functional approximations for many electron systems. *Phys. Rev. B* 23, 5048–5079 (1981).
41. Sumpter, B. G. and Meunier, V. Can computational approaches aid in untangling the inherent complexity of practical organic photovoltaic systems? *J. Polym. Sci. B: Polym. Phys.* 50, 1071–1089 (2012).

42. Risko, C., McGehee, M. D. and Bredas, J.-L. A quantum-chemical perspective into low optical-gap polymers for highly efficient organic solar cells. *Chem. Sci.* 2, 1200–1218 (2011).
43. Kanai, Y., Neaton, J. B. and Grossman, J. C. Theory and simulation of nanostructured materials for photovoltaic applications. *Comput. Sci. Eng.* 12, 18–27 (2010).
44. De Mitri, N., Monti, S., Prampolini, G. and Barone, V. Absorption and emission spectra of a flexible dye in solution: A computational time-dependent approach. *J. Chem. Theory Comput.* 9, 4507–4516 (2013).
45. Hachmann, J. et al. Lead candidates for high-performance organic photovoltaics from high-throughput quantum chemistry—The Harvard Clean Energy Project. *Energy Environ. Sci.* 7, 698–704 (2014).
46. Hachmann, J. et al. The Harvard Clean Energy Project: Large-scale computational screening and design of organic photovoltaics on the world community grid. *J. Phys. Chem. Lett.* 2, 2241–2251 (2011).
47. Olivares-Amaya, R. et al. Accelerated computational discovery of high-performance materials for organic photovoltaics by means of cheminformatics. *Energy Environ. Sci.* 4, 4849–4861 (2011).
48. Jackson, N. E. et al. Structural and conformational dispersion in the rational design of conjugated polymers. *Macromolecules* 47, 987–992 (2014).
49. Wykes, M., Milián-Medina, B. and Gierschner, J. Computational engineering of low bandgap copolymers. *Front Chem* 1, 1–35 (2013).
50. Dabo, I. et al. Donor and acceptor levels of organic photovoltaic compounds from first principles. *Phys. Chem. Chem. Phys.* 15, 685–695 (2013).
51. Shuai, Z., Wang, L. and Li, Q. Evaluation of charge mobility in organic materials: From localized to delocalized descriptions at a first-principles level. *Adv. Mater.* 23, 1145–1153 (2011).
52. Liu, T. and Troisi, A. Understanding the microscopic origin of the very high charge mobility in PBTTT: Tolerance of thermal disorder. *Adv. Funct. Mater.* 24, 925–933 (2014).
53. Zade, S. S. and Bendikov, M. Study of hopping transport in long oligothiophenes and oligoselenophenes: Dependence of reorganization energy on chain length. *Chemistry – A European Journal* 14, 6734–6741 (2008).
54. McMahon, D. P. et al. Relation between microstructure and charge transport in polymers of different regioregularity. *J. Phys. Chem. C* 115, 19386–19393 (2011).
55. Cheung, D. L., McMahon, D. P. and Troisi, A. A realistic description of the charge carrier wave function in microcrystalline polymer semiconductors. *J. Am. Chem. Soc.* 131, 11179–11186 (2009).
56. Aguirre, J. C. et al. Understanding local and macroscopic electron mobilities in the fullerene network of conjugated polymer-based solar cells: Time-resolved microwave conductivity and theory. *Adv. Funct. Mater.* 24, 784–792 (2014).
57. Tamura, H. and Tsukada, M. Role of intermolecular charge delocalization on electron transport in fullerene aggregates. *Phys. Rev. B* 85, 054301 (2012).
58. Peumans, P., Yakimov, A. and Forrest, S. R. Small molecular weight organic thin-film photodetectors and solar cells. *J. Appl. Phys.* 93, 3693–3723 (2003).
59. Schlaf, R. et al. Photoemission spectroscopy of LiF coated Al and Pt electrodes. *J. Appl. Phys.* 84, 6729–6736 (1998).
60. White, M. S., Olson, D. C., Shaheen, S. E., Kopidakis, N. and Ginley, D. S. Inverted bulk-heterojunction organic photovoltaic device using a solution-derived ZnO underlayer. *Appl. Phys. Lett.* 89, (2006).
61. Dou, L. T. et al. Tandem polymer solar cells featuring a spectrally matched low-bandgap polymer. *Nat. Photon.* 6, 180–185 (2012).
62. Kim, J. Y. et al. New architecture for high-efficiency polymer photovoltaic cells using solution-based titanium oxide as an optical spacer. *Adv. Mater.* 18, 572–576 (2006).

63. Ameri, T., Li, N. and Brabec, C. J. Highly efficient organic tandem solar cells: A follow up review. *Energy Environ. Sci.* 6, 2390–2413 (2013).
64. Yip, H. L. and Jen, A. K. Y. Recent advances in solution-processed interfacial materials for efficient and stable polymer solar cells. *Energy Environ. Sci.* 5, 5994–6011 (2012).
65. Dou, L. et al. A decade of organic/polymeric photovoltaic research. *Adv. Mater.* 25, 29 (2012).
66. Heimel, G., Salzmann, I., Duhm, S. and Koch, N. Design of organic semiconductors from molecular electrostatics. *Chem. Mater.* 23, 359–377 (2011).
67. Heimel, G., Salzmann, I., Duhm, S., Rabe, J. P. and Koch, N. Intrinsic surface dipoles control the energy levels of conjugated polymers. *Adv. Funct. Mater.* 19, 3874–3879 (2009).
68. Romaner, L., Heimel, G., Ambrosch-Draxl, C. and Zojer, E. The dielectric constant of self-assembled monolayers. *Adv. Funct. Mater.* 18, 3999–4006 (2008).
69. Natan, A., Zidon, Y., Shapira, Y. and Kronik, L. Cooperative effects and dipole formation at semiconductor and self-assembled-monolayer interfaces. *Phys. Rev. B* 73, 193310 (2006).
70. Monti, O. L. A. Understanding interfacial electronic structure and charge transfer: An electrostatic perspective. *J. Phys. Chem. Lett.* 3, 2342–2351 (2012).
71. Frenkel, D. and Smit, B. *Understanding Molecular Simulation-From Algorithms to Applications*. (Academic Press, San Diego, 2001).
72. Leach, A. R. *Molecular Modeling: Principles and Applications*. (Prentice Hall, New York, 2001).
73. Schlick, T. *Molecular Modeling and Simulation: An Interdisciplinary Guide Interdisciplinary Applied Mathematics: Mathematical Biology*. (Springer-Verlag, New York, 2000).
74. Carrillo, J.-M. Y., Kumar, R., Goswami, M., Sumpter, B. G. and Brown, W. M. New insights into the dynamics and morphology of P3HT:PCBM active layers in bulk heterojunctions. *Phys. Chem. Chem. Phys.* 15, 17873–17882 (2013).
75. Lee, C.-K., Pao, C.-W. and Chu, C.-W. Multiscale molecular simulations of the nanoscale morphologies of P3HT:PCBM blends for bulk heterojunction organic photovoltaic cells. *Energy Environ. Sci.* 4, 4124–4132 (2011).
76. To, T. T. and Adams, S. Modelling of P3HT:PCBM interface using coarse-grained force field derived from accurate atomistic force field. *Phys. Chem. Chem. Phys.* 16, 4653–4663 (2014).
77. Rudd, R. E. Coarse-grained molecular dynamics for computer modeling of nanomechanical systems. *Int. J. Multiscale Comput. Eng.* 2, 1–157 (2004).
78. Nielsen, S. O., Lopez, C. F., Srinivas, G. and Klein, M. L. Coarse grain models and the computer simulation of soft materials. *J. Phys.: Condens. Mater.* 16, R481 (2004).
79. Lipkowitz, K. B., Cundari, T. R. and Faller, R. *Coarse-Grain Modeling of Polymers: Reviews in Computational Chemistry*, Vol. 23. (John Wiley and Sons, Inc., 2007).
80. Allen, M. P. and Tildesley, D. J. *Computer Simulation of Liquids*. (Oxford University Press, Oxford, 1989).
81. McCammon, J. A. and Harvey, S. C. *Dynamics of Proteins and Nucleic Acids*. (Cambridge University Press, New York, 1987).
82. Rapaport, D. C. *The Art of Molecular Dynamics Simulation*, 2nd edn. (Cambridge University Press, New York, 2004).
83. Gray, S. K., Noid, D. W. and Sumpter, B. G. Symplectic integrators for large-scale molecular-dynamics simulations—A comparison of several explicit methods. *J. Chem. Phys.* 101, 4062–4072 (1994).
84. Mayo, S. L., Olafson, B. D. and Goddard, W. A. DREIDING: A generic force field for molecular simulations. *J. Phys. Chem.* 94, 8897–8909 (1990).
85. Plimpton, S. Fast parallel algorithms for short-range molecular dynamics. *J. Comput. Phys.* 117, 1–19 (1995).

86. Brown, W. M., Wang, P., Plimpton, S. J. and Tharrington, A. N. Implementing molecular dynamics on hybrid high performance computers—Short range forces. *Comput. Phys. Commun.* 182, 898–911 (2011).
87. Rycroft, C. H. VORO++: A three-dimensional Voronoi cell library in C++. *Chaos: An Interdisciplinary Journal of Nonlinear Science* 19, 041111 (2009).
88. Fratzl, P. and Lebowitz, J. L. Universality of scaled structure functions in quenched systems undergoing phase separation. *Acta Metall.* 37, 3245–3248 (1989).
89. Debye, P., Anderson, H. R. and Brumberger, H. Scattering by an inhomogeneous solid. II. The correlation function and its application. *J. Appl. Phys.* 28, 679–683 (1957).
90. Binder, K. *The Monte Carlo Method in Condensed Matter Physics.* (Springer, Berlin, 1995).
91. Ferguson, D. M., Siepmann, J. I. and Truhlar, D. G. *Monte Carlo Methods in Chemical Physics in Advances in Chemical Physics*, Vol. 105 (eds. S. A. Rice, I. Prigogine) (John Wiley and Sons, Inc., New York, 1999).
92. Karayiannis, N. C., Giannousaki, A. E., Mavrantzas, V. G. and Theodorou, D. N. Atomistic Monte Carlo simulation of strictly monodisperse long polyethylene melts through a generalized chain bridging algorithm. *J. Chem. Phys.* 17, 5465–5479 (2002).
93. Karayiannis, N. C., Giannousaki, A. E. and Mavrantzas, V. G. An advanced Monte Carlo method for the equilibration of model long-chain branched polymers with a well-defined molecular architecture: Detailed atomistic simulation of an H-shaped polyethylene melt. *J. Chem. Phys.* 118, 2451–2454 (2003).
94. Mennucci, B. Continuum solvation models: What else can we learn from them? *J. Phys. Chem. Lett.* 1, 1666–1674 (2010).
95. Kratzer, P. *Monte Carlo and Kinetic Monte Carlo Methods*—A Tutorial. (Forschungzentrum Julich, 2009).
96. Nelson, J., Kwiatkowski, J. J., Kirkpatric, J. and Frost, J. M. Modeling charge transport in organic photovolatic materials. *Acc. Chem. Res.* 42, 1768–1778 (2009).
97. Liu, H., Lin, Z. and Zhigilei, L. V. Fracture structures in fullerene layers: Simulation of the growth process. *J. Phys. Chem. C.* 112, 4687–4695 (2008).
98. Piana, S. and Gale, J. D. Three-dimensional kinetic Monte Carlo simulation of crystal growth from solution. *J. Cryst. Growth* 294, 46–52 (2006).
99. Marcus, R. A. Electron transfer reactions in chemistry: Theory and experiment (Nobel Lecture). *Ang. Chem. Int. Ed. Engl.* 32, 1111–1121 (1993).
100. Yan, H. et al. Influence of annealing and interfacial roughness on the performance of bilayer donor/acceptor polymer photovoltaic devices. *Adv. Funct. Mater.* 20, 4329–4337 (2010).
101. van Eersel, H., Janssen, R. A. J. and Kemerink, M. Mechanism for efficient photoinduced charge separation at disordered organic heterointerfaces. *Adv. Funct. Mater.* 22, 2700–2708 (2012).
102. Murat, M. et al. Molecular-scale simulation of electroluminescence in a multilayer white organic light-emitting diode. *Nat. Mater.* 12, 652–658 (2013).
103. Kimber, R. G. E., Wright, E. N., O'Kane, S. E. J., Walker, A. B. and Blakesley, J. C. Mesoscopic kinetic Monte Carlo modeling of organic photovoltaic device characteristics. *Phys. Rev. B* 86, 235206 (2012).
104. Fredrickson, G. H., Ganesan, V. and Drolet, F. Field-theoretic computer simulation methods for polymers and complex fluids. *Macromolecules* 35, 16–39 (2002).
105. Fredrickson, G. H. Computational field theory of polymers: Opportunities and challenges. *Soft Matter* 2, 1329 (2007).
106. Hammond, M. R., Sides, S. W., Fredrickson, G. H., Kramer, E. J., Ruokolainen, J. and Hahn, S.F. Adjustment of block copolymer nanodomain sizes at lattice defect sites. *Macromolecules* 36, 8712-8716 (2003).

107. Bosse, A. W., Sides, S. W., Katsov, K., Garcia-Cervera, C. J. and Fredrickson, G. H. Defects and their removal in block copolymer thin film simulations. *J. Poly. Sci. B: Polymer Physics* 44, 2495–2511 (2006).
108. Paradiso, S. P., Delaney, K. T., García-Cervera, C. J., Ceniceros, H. D. and Fredrickson, G. H. Block copolymer self assembly during rapid solvent evaporation: Insights into cylinder growth and stability. *ACS Macro Lett.* 3, 16–20 (2013).
109. Jiang, Y. and Chen, J. Z. Y. Self-consistent field theory and numerical scheme for calculating the phase diagram of wormlike diblock copolymers. *Phys. Rev. E* 88, 042603 (2013).
110. Karsten, B. P., Bijleveld, J. C., Viani, L., Cornil, J., Gierschner, J. and Janssen, R. A. J. *J. Mater. Chem.* 19, 5343–5350 (2009).
111. Muller, C., Wang, E. G., Andersson, L. M., Tvingstedt, K., Zhou, Y., Andersson, M. R. and Inganas, O. Influence of molecular weight on the performance of organic solar cells based on fluorene derivative. *Adv. Funct. Mater.* 20, 2124–2131 (2010).
112. Karsten, B. P., Viani, L., Gierschner, J., Cornil, J. and Janssen, R. A. J. An oligomer study on small band gap polymers. *J. Phys. Chem. A.* 112, 10764–10773 (2008).
113. Collins, B. A. et al. Molecular miscibility of polymer–fullerene blends. *J. Phys. Chem. Lett.* 1, 3160–3166 (2010).
114. Chen, H., Hegde, R., Browning, J. and Dadmun, M. D. The miscibility and depth profile of PCBM in P3HT: Thermodynamic information to improve organic photovoltaics. *Phys. Chem. Chem. Phys.* 14, 5635–5641 (2012).
115. Flory, P. J. *Principles of Polymer Chemistry*. (Cornell University Press, New York, 1953).
116. Caddeo, C. and Mattoni, A. Atomistic investigation of the solubility of 3-alkylthiophene polymers in tetrahydrofuran solvent. *Macromolecules* 46, 8003–8008 (2013).

Section III

Morphology, Interface, Charge Transport, and Defect States

6 Photoexcited Carrier Dynamics in Organic Solar Cells

Sou Ryuzaki and Jun Onoe

CONTENTS

6.1　Overview 143
6.2　Photocurrent Conversion Processes 144
6.3　Photogenerated Carrier Dynamics in the Vicinity of Donor/Acceptor
　　　Interfaces 147
　　　6.3.1　Introduction 147
　　　6.3.2　Impedance Spectroscopic Study of Heterojunction OPV Cells 148
6.4　Structural Effects of Organic Films on Exciton and Carrier Dynamics 153
　　　6.4.1　Introduction 153
　　　6.4.2　Effects of Intermolecular Charge-Transfer Excitons on the
　　　　　　　External Quantum Efficiency for Heterojunction OPV Cells 154
6.5　The Correlation between Open-Circuit Voltage and Photogenerated
　　　Carrier Dynamics 156
　　　6.5.1　Introduction 156
　　　6.5.2　Influence of Charge Accumulation of Photogenerated Carriers
　　　　　　　in the Vicinity of Donor–Acceptor Interface on the Open-Circuit
　　　　　　　Voltage of Heterojunction OPV Cells 157
6.6　Concluding Remarks and Perspectives 163
References 164

6.1　OVERVIEW

Understanding the photoexcited carrier dynamics in organic photovoltaic (OPV) cells helps us to improve their power conversion efficiency (η) for practical use because some breakthroughs in the recent decade have been based on bulk-heterojunctions (BHJs) between donor (D) and acceptor (A) materials. The BHJ plays a role of increasing the D-A interface area involving photocarrier generation regions, and thus the number of photocarriers increases at the interfaces [1,2]. Although the BHJ can improve the η of OPV cells by up to ca. 8%, the improving rate of η does not significantly increase [3–5]. This is presumably because there are still major unclear points for BHJ as follows: (1) the nanostructure of the BHJ (although its

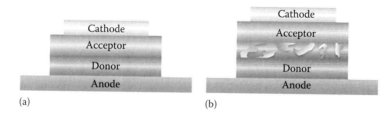

FIGURE 6.1 Schematic illustration of typical structures used for organic photovoltaic cells: (a) heterojunction type and (b) BHJ type.

schematic illustration has been shown) and (2) the reproducibility of BHJ fabrication. Figure 6.1 schematically illustrates typical structures of (a) a D-A hetero double-layered OPV cell and (b) a BHJ-OPV cell. As shown in Figure 6.1b, because the BHJ is fabricated by coevaporation of donor and acceptor materials or by spin-coating conductive polymers, the D-A interface thus formed is too complex to analyze and reproduce the nanostructure compared to that of a double-layered heterojunction [2,6]. These facts make it difficult to discuss the photoexcited carrier dynamics at the BHJ D-A interface. Accordingly, the reason why the BHJ improves the η has been unclear so far.

In addition, both the electronic states (corresponding to a built-in potential, etc.) at the D-A interface under illumination conditions and the origins of open circuit voltage (V_{oc}), which are related to the carrier dynamics, have also remained unclear [7–11]. For these reasons, we have not obtained any decisive breakthroughs to develop the OPV cells commercially available.

In the present chapter, we will discuss photogenerated carrier dynamics, electronic states at the D-A interface under illumination conditions, and the origins of V_{oc}, using an OPV cell [ITO/Zn(OEP)/C_{60}/Al] with a D-A hetero double-layered structure. Here, ITO, Zn(OEP), and Al express indium-tin-oxide, zinc octaethylporphyrin, and aluminum, respectively.

Contrary to the BHJ-OPV cells with a complex D-A interface, an OPV cell with a simple D-A hetero double-layered structure is more preferable to discuss the photoexcited carrier dynamics and to clarify what factors significantly contribute to the η because experimental results thus obtained are easy to be analyzed on the basis of the simple structured OPV cells.

6.2 PHOTOCURRENT CONVERSION PROCESSES

As shown in Figure 6.1, the device structure of the present OPV cell has only one D-A interface. The Fermi level (E_F), the vacuum level (VL), and the energy levels of the highest occupied molecular orbital (HOMO) and the lowest unoccupied molecular orbital (LUMO) for both donor and acceptor films play an important role for the generation rate of excitons, carriers (electrons and holes), and current in OPV cells under photoirradiation [12].

Figure 6.2 illustrates an energy-level alignment based on the Mott-Schottky model before and after the anode, donor, acceptor, and cathode films contact each

Photoexcited Carrier Dynamics in Organic Solar Cells

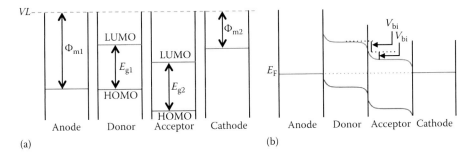

FIGURE 6.2 Schematic representation of the energy-level alignment based on the Mott-Schottky model: (a) before and (b) after the contact of individual films.

other [12–14]. Here, Φ_m and E_g denote a work function of individual electrodes and an energy gap of the D and A materials, respectively.

As shown in Figure 6.2, when neutral organic and metal films are separated from each other, the energy diagram of each film is aligned using the common VL. After the contact of individual films, their E_F levels equilibrate to be the same due to charge transfer (either electrons or holes) at each interface in association with a band bending [15]. This band bending induces an electrical field (in other words, generates a built-in potential: V_{bi}) in the vicinity of the individual interfaces for the OPV cell. However, these interfacial electronic structures after the contact have not yet been understood precisely. Indeed, another alignment model considering the VL shift at a metal/organic interface has been reported recently [16].

As shown in Figure 6.3, the fundamental mechanism of the photogenerated current in OPV cells has been considered as the following four elementary processes:

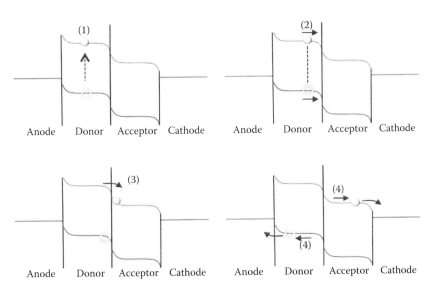

FIGURE 6.3 Four elementary processes of photogenerated current in OPV cells.

(1) excitons are generated by photoabsorption of a donor film, (2) excitons thus formed diffuse to the D-A interface, (3) excitons are dissociated to electrons and holes at the D-A interface, and (4) electrons and holes are, respectively, moved to and collected at each electrode.

For the first process (1), excitons are generated by photoabsorption of donor films. Here, the exciton is defined to be an electron-hole pair via Coulombic attractive force, which can be classified into the Mott-Wannier and Frenkel types. For the former type, the electron-hole interaction is weak, which results in the exciton radius being much larger than the lattice constant [17–19]. In this case, the binding energy of this type of exciton is typically in the order of 0.1 eV. On the other hand, for the latter type, the Coulombic interaction is very strong, and the exciton tends to have a radius on the same order as the unit cell. Thus, the electron and hole are localized in the same unit cell. In this case, the binding energy of the Frenkel-type exciton is in the order of 1.0 eV [20]. In addition, if a donor film consists of small organic molecules, excitons are further categorized into inter- and intramolecular type excitons [21].

For the second process (2), the excitons thus generated diffuse in the organic donor film to the D-A interface. The exciton diffusion length corresponding to the moving distance for its lifetime has been reported to be 5–50 nm [20] for organic donor films, which depends on the crystal structure, molecular orientation of organic films, and the exciton types [6]. Accordingly, the exciton diffusion length plays an important role of determining a thickness of organic donor film because excitons generated in a region with a distance to the D-A interface longer than the exciton diffusion length disappear by electron-hole recombination prior to reaching to the interface. Because the exciton diffusion length is typically shorter than the film thickness of 50–100 nm for OPV cells, this physical quantity significantly affects the photocurrent conversion efficiency [12].

For the third process (3), excitons are separated to electrons and holes at the D-A interface by a difference in the electron affinity between the donor and acceptor films. In case of a large electric field induced by a band bending in the vicinity of the D-A interface, excitons are more efficiently separated to the carriers. When the band-bending region is thin, carrier injection is enhanced by quantum tunneling between the D and A layers [12,16].

For the forth process (4), the electrons and holes thus formed are transported to and collected at the cathode and anode, respectively. In particular, a high resistivity of organic films (for example, 10^8–10^9 Ωcm for C_{60} single crystal [22,23]) prevents, however, their movement to each electrode. Therefore, the process (4) also reduces the η significantly [12].

In addition to the four elementary processes described above, if an acceptor film absorbs a visible light as well as a donor film, it is noticed that excitons generated in the acceptor film also contribute to the photocurrent of OPV cells.

The external quantum efficiency (EQE) is defined to be the number of electrons working as a current divided by the number of incident photons and expressed in terms of the following four kinds of the efficiencies [12]:

$$EQE = \eta_A \eta_{ED} \eta_{ES} \eta_{CT}. \tag{6.1}$$

Here, the η_A is the absorption efficiency of incident photons for organic films, the η_{ED} is the efficiency of exciton diffusion to the D-A interface, the η_{ES} is the generation efficiency of carriers from exciton separation at the D-A interface, the η_{CT} is the efficiency both for carriers transporting to each electrode and for collection at the electrode. The extinction coefficient and thickness of organic films determine the η_A whereas the exciton diffusion length determines the η_{ED}. Although the details of the η_{ES} and η_{CT} have not been well understood, it has been recognized that the D-A interface conditions, the film resistivity, and the morphology of organic–metal interfaces affect the above efficiencies significantly.

The EQE depends on light wavelength and then is compared to the photon flux of incident light. Thus, the EQE can be given by the following equation:

$$\text{EQE}(\lambda) = \frac{n_e}{n_p} = \frac{hc}{\lambda q} \frac{I_{SC}(\lambda)}{P_{0m}(\lambda)}. \tag{6.2}$$

Here, n_e and n_p are the number of photogenerated electrons and incident photons, respectively; h is the Planck's constant; c is the speed of light; λ is the wavelength of incident light; q is the elementary charge; and P_{0m} is the incident optical power of monochromatic light. Equation 6.2 generally follows the absorption spectra of the materials used for OPV cells.

The photocurrent (I_{ph}) generated under the illumination of AM1.5 at a given applied voltage V can be calculated by integrating the product of the EQE and the solar spectra of AM1.5 over the incident light wavelength λ,

$$I_{ph}(V) = \int \frac{\lambda q}{hc} \text{EQE}(\lambda, V) S(\lambda) \, d\lambda. \tag{6.3}$$

Here, $S(\lambda)$ is the incident optical power with spectral shape of AM1.5 [8]. Accordingly, the short-circuit current I_{SC} can be evaluated from the following formula:

$$I_{SC} = I_{ph}(0) \int \frac{\lambda q}{hc} \text{EQE}(\lambda, 0) S(\lambda) \, d\lambda. \tag{6.4}$$

6.3 PHOTOGENERATED CARRIER DYNAMICS IN THE VICINITY OF DONOR/ACCEPTOR INTERFACES

6.3.1 Introduction

As described in Section 6.2, understanding the photoexcited carrier dynamics in OPV cells is important to improve the η for their practical use. In general, in order to discuss the carrier dynamics in organic devices, the device structure has been often replaced with an equivalent circuit model consisting of contact resistance, interfacial resistance, film resistance, and film capacitance as shown in Figure 6.4 [24].

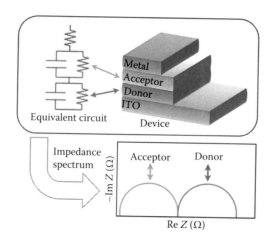

FIGURE 6.4 Schematic illustration of a typical heterojunction OPV cell, a corresponding equivalent circuit model, and impedance spectra attributed to each layer.

Because both the resistance and capacitance comprise information not only on the dynamics of excitons and carriers (mobility, recombination, accumulation, etc.), but also on the electronic states (energy gap and bending, etc.) for the OPV cells, a suitable equivalent circuit model helps us to elucidate the photoexcited carrier dynamics in the OPV cells [25]. The equivalent circuit corresponding to OPV cells can be determined using impedance spectroscopy. This method is a powerful tool to measure in situ and nondestructively the resistance and capacitance of both individual films and their interfaces for OPV cells under working conditions [26].

In the present section, we address the impedance spectroscopic studies of an OPV cell with a D-A double-layered heterojunction under dark and illumination (100 mW/cm², AM1.5) conditions and discuss the photoexcited carrier dynamics by analyzing the impedance spectra obtained under the both conditions with the corresponding equivalent circuits for each condition [27]. We fabricated an OPV cell [ITO/Zn(OEP)/ C_{60}/Al] consisting of 20-nm-thick Zn(OEP) and 30-nm-thick C_{60} films, respectively, used as donor and acceptor layers because Zn(OEP) has a high photoabsorption coefficient in the visible light region [28], C_{60} has a high electron affinity [29], and furthermore, a strong interaction between Zn(OEP) and C_{60} is expected to generate carriers efficiently from excitons at the D-A interface [30,31].

6.3.2 Impedance Spectroscopic Study of Heterojunction OPV Cells

Impedance spectroscopy yields nondestructively the impedance (Z) of each organic film and/or interface in OPV cells by using the following equation [26]:

$$Z = R_S + \frac{R_P}{1+\omega^2 R_P^2 C^2} - j\frac{\omega R_P^2 C}{1+\omega^2 R_P^2 C^2}. \tag{6.5}$$

Photoexcited Carrier Dynamics in Organic Solar Cells

Here, the R_S and R_P denote the series and parallel resistances, respectively, and the C is the capacitance and the ω is the angular frequency ($\omega = 2\pi f$). Accordingly, both the R and C are evaluated by fitting the impedance spectra with a modified Equation 6.5 obtained from a suitable equivalent circuit model corresponding to the OPV cells.

Figure 6.5 shows (a) comparison between the experimental impedance spectra (symbols) of the OPV cell [ITO/20-nm-thick Zn(OEP)/30-nm-thick C_{60}/Al] under dark conditions and calculated impedance spectra (solid line) from the equivalent circuit shown in Figure 6.5d, (b) the cross-sectional scanning electron microscopic (SEM) image of the Zn(OEP)/C_{60} heterojunction films formed on the ITO substrate, (c) the dependence of film thickness on impedance spectra, and (d) an equivalent circuit model corresponding to the present OPV cell.

Figure 6.5a shows the impedance spectrum for the present OPV cell, which seems to be only one semicircle under the dark condition. This is different from an expectation that the double-layered Zn(OEP)/C_{60} film structure (Figure 6.5b) would show two different semicircles corresponding to each layer as shown in Figure 6.4. When the time constant (the product of R and C) of the Zn(OEP) and C_{60} films was almost the same as for each other, namely $R_{OEP} \times C_{OEP} \cong R_{C60} \times C_{C60}$ [26], such as the one

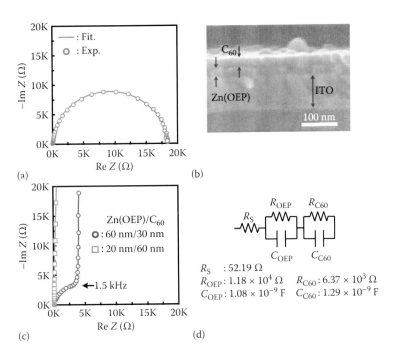

FIGURE 6.5 (a) Experimental (symbols) and simulated (solid line) impedance spectra for the OPV cell [ITO/Zn(OEP)/C_{60}/Al] under dark condition, (b) a cross-sectional SEM image of Zn(OEP)/C_{60} alternative deposited film on an ITO substrate, (c) impedance spectra for the OPV cells with different film thickness pairs of Zn(OEP)/C_{60}: 60 nm/30 nm (circle) and 20 nm/60 nm (square), and (d) the equivalent circuit of the present OPV cell under dark conditions.

semicircle shown in Figure 6.5a was obtained. Indeed, a change in the time constant by increasing the Zn(OEP) film thickness in the OPV cells gives rise to a drastic increase in the impedance in a lower (less than ca. 1.5 kHz) frequency region shown in Figure 6.5c. On the other hand, when the C_{60} film thickness increases, the impedance also significantly increases in the whole frequency region. These results indicate that the semicircle shown in Figure 6.5a consists of two different components attributed to both C_{60} and Zn(OEP) films.

Accordingly, the model shown in Figure 6.5d consisting of one series resistance (R_S) and two RC circuits of the individual films are considered to be suitable for the present OPV cells. Indeed, this model well reproduces the impedance spectra of Figure 6.5a with a capacitance of 1.29×10^{-9} F for the 30-nm-thick C_{60} film used as a fixed parameter. This value was calculated from the reported relative permittivity value of 4.4 [32]. In addition, the values of $R_{OEP} = 1.18 \times 10^4 \, \Omega$, $C_{OEP} = 1.08 \times 10^{-9}$ F, and $R_{C60} = 6.37 \times 10^3 \, \Omega$, obtained by fitting the model with the impedance spectra, are comparable to reported values, respectively. Given that the time constant was different by more than 10 times between both films, the impedance spectrum would clearly consist of two semicircles.

Figure 6.6a shows the experimental (diamonds) and calculated (solid line) impedance spectra of the OPV cell under the illumination condition along with those obtained under dark (circles and solid line) conditions.

The impedance spectrum obtained under illumination conditions seems to consist of two different semicircles. However, the equivalent circuit used for the dark conditions cannot reproduce the spectrum shown in Figure 6.6a. This indicates that an equivalent circuit model needs three or more RC circuits in order to reproduce

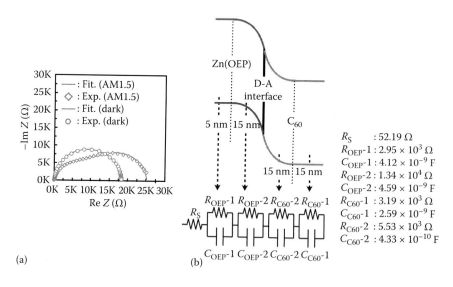

FIGURE 6.6 (a) Experimental (symbols) and simulated (solid line) impedance spectra of the OPV cell [ITO/Zn(OEP)/C_{60}/Al] under illumination of AM1.5 (diamond) and under dark (circle) conditions, and (b) the equivalent circuit for the OPV cells under illumination conditions along with schematic illustration of the energy diagram for the Zn(OEP)/C_{60} layered OPV cells.

the impedance spectrum obtained under illumination. It is reasonable to consider that the additional *RC* circuits are attributed to the vicinity of the D-A interface because the carriers generated by the dissociation of photogenerated excitons at the D-A interface tend to be accumulated in the vicinity of the interface due to the high resistivity of D-A organic films used in the OPV cells (for example, 10^8–10^{14} Ωcm for pristine C_{60} films [22]). Because the accumulation of photogenerated carriers in the vicinity of the D-A interface changes the electronic states, such as band-bending and built-in potential, etc., the resistance and capacitance near the D-A interface are reasonably considered to be different from those in the region far from the interface. Thus, we added two other *RC* circuits in the vicinity of the D-A interface to the equivalent circuit used for the dark conditions as shown in Figure 6.6b.

Figure 6.6b schematically illustrates the band-bending model applied to the present OPV cells along with its corresponding equivalent circuit consisting of a series resistance (R_S) and four different resistances (R_{OEP}-1 and R_{OEP}-2, R_{C60}-2 and R_{C60}-1) and capacitances (C_{OEP}-1 and C_{OEP}-2, C_{C60}-2 and C_{C60}-1). The present equivalent circuit model well reproduces the experimental impedance spectrum, using the fitting values shown in Figure 6.6b. In the present fitting procedure, R_{OEP}-1, C_{OEP}-1, R_{C60}-1, and C_{C60}-1 are evaluated by substituting the *R* and *C* values obtained under the dark condition into the following Equations 6.6 and 6.7 because the electronic states of each film far from the D-A interface can be considered to be constant before and after photoirradiation,

$$R = \rho \frac{d}{S}, \tag{6.6}$$

and

$$C = \varepsilon \frac{S}{d}. \tag{6.7}$$

Here, ρ, ε, d, and S denote resistivity, permittivity, film thickness, and cross-sectional area, respectively. Because $R = 2.95 \times 10^3$ Ω and $C = 4.12 \times 10^{-9}$ F obtained from the fitting are respectively smaller and larger by almost four times than $R_{OEP} = 1.18 \times 10^4$ Ω and $C_{OEP} = 1.08 \times 10^{-9}$ F, the *RC* circuit with the *R* (2.95×10^3 Ω) and *C* (4.12×10^{-9} F) values is suitable for the part of the Zn(OEP) film far from the D-A interface, and the film thickness *d* corresponding to the region can be estimated to be 5 nm from Equations 6.6 and 6.7 with a constant of ρ, ε, and *S*. In a similar manner, the region far from the D-A interface in the C_{60} film corresponds to the *RC* circuit with $R = 3.19 \times 10^3$ Ω and $C = 2.59 \times 10^{-9}$ F because the *R* (3.19×10^3 Ω) and *C* (2.59×10^{-9} F) are, respectively, smaller and larger by ca. two times than $R_{C60} = 6.37 \times 10^3$ Ω and $C_{C60} = 1.29 \times 10^{-9}$ F obtained under the dark condition. Then, the thickness of the region can be estimated to be 15 nm by using Equations 6.6 and 6.7. Accordingly, the individual D-A interfacial regions are both estimated to be 15 nm for each, which is a typical thickness of band bending for organic films [33–35].

In the region of the C_{60} film in the vicinity of the D-A interface, photogenerated electrons are accumulated. Thus, the space charge limited current (SCLC) model is reasonably considered [36]. The accumulation of photogenerated electrons caused by a high resistance of organic films induces the space charges, limiting the charge injection from the outside. According to the SCLC, the current density (J) can be given as the following equation:

$$J = \frac{9}{8}\mu\varepsilon\theta\frac{V^2}{d^3} \quad \left[\text{where, } \theta = \frac{\rho_f}{\rho_f + \rho_t}\right]. \tag{6.8}$$

Here, μ, V, ρ_f, and ρ_t are, respectively, carrier mobility, applied voltage, free carrier density, and trapped charge density. Because Equation 6.8 shows a decrease in J with increasing ρ_t, the accumulation of photogenerated electrons in the vicinity of the D-A interface reduces the J, which indicates an increment in resistance. In addition, the limitation of charge injection directly makes capacitance decrease. Consequently, both R and C in the 15-nm-thick interfacial region of the C_{60} film should be respectively larger and smaller than the values of $R = 3.19 \times 10^3$ Ω and $C = 2.58 \times 10^{-9}$ F obtained from Equations 6.6 and 6.7 by substituting the R and C values for the dark condition. In this reasoning, the RC circuit consisting of R_{C60}-2 = 5.33 × 10^3 Ω and C_{C60}-2 = 4.33 × 10^{-10} F is appropriate to be used for the interfacial region.

The RC circuit consisting of R_{OEP}-2 = 1.34 × 10^4 Ω and C_{OEP}-2 = 4.59 × 10^{-9} F thus corresponds well to the 15-nm-thick interfacial region of the Zn(OEP) film. From Equations 6.6 and 6.7, the R_{OEP}-2 and C_{OEP}-2 are larger than the estimated values, $R = 8.5 \times 10^3$ Ω and $C = 1.44 \times 10^{-9}$ F, for the 15-nm-thick Zn(OEP) film. In a part of the Zn(OEP) film in the vicinity of the D-A interface, injected electrons are not limited, unlike the SCLC model, because the photogenerated holes are accumulated in the Zn(OEP) film near the D-A interface. When the space-charge effect is negligible, the J of organic films can be generally expressed as follows [37,38]:

$$J = q\mu EN \exp\left[-\frac{q\phi_B}{kT}\right]\exp\left[\frac{q}{kT}\sqrt{\frac{q}{4\pi\varepsilon}}\sqrt{E}\right]. \tag{6.9}$$

Here, q, E, N, ϕ_B, k, and T are, respectively, the elemental charge, electric field, density of states, an injection barrier height, the Boltzmann's constant, and the absolute temperature. According to Equation 6.9, the J decreases with increasing the permittivity. This indicates that R increases with C (see Equation 6.7). The correlation between R and C is comparable to the present results that R_{OEP}-2 and C_{OEP}-2 are both larger than estimated values as described above. The increase in the capacitance of Zn(OEP) film at the D-A interface under illumination may be due to the injected electrons being recombined with the accumulated holes near the D-A interface, which results in an increase in the capacity of electron store, that is, capacitance.

The nondestructive impedance spectroscopy provides not only the equivalent circuits corresponding to OPV cells, but also qualitative information on the carrier dynamics and the electronic states in the vicinity of the D-A interface before and after photoirradiation of OPV. The equivalent circuit models used here can be applied to the other heterojunction OPV cells because porphyrin and fullerene, which are typical donor and acceptor materials, respectively, are used in the present study and because the impedance spectra were analyzed on the basis of band-bending models.

In summary, impedance spectroscopy showed that the equivalent circuit model consisting of two *RC* circuits is suitable for the OPV cells under the dark condition and the equivalent circuit one consisting of the four *RC* circuits for the OPV cells under the illumination condition (100 mW/cm^2, AM1.5). The two additional *RC* circuits well express the accumulated carriers in the vicinity of the D-A interface because of a high resistivity of organic donor–acceptor films. Although the electronic states of the OPV cells upon photoirradiation is important to reveal the unknown device mechanisms and the photoexcited carrier dynamics, the characteristics of OPV cells have been discussed almost in terms of their static electronic states under the dark condition. Because impedance spectroscopy can investigate OPV cells under working nondestructively, it is a powerful tool to reveal the photoexcited carrier dynamics in the OPV cells before and after photoirradiation, especially in the vicinity of D-A interfaces and thus play a key role of optimizing the performance of OPV cells.

6.4 STRUCTURAL EFFECTS OF ORGANIC FILMS ON EXCITON AND CARRIER DYNAMICS

6.4.1 INTRODUCTION

Understanding the influences of film structure on the photoexcited carrier dynamics is important for improvement of η because the exciton-diffusion length and carrier mobility, which directly affect η, depend on the crystal structure and/or molecular orientation of D-A organic films [39]. However, there have been hitherto few studies on the improvement of the η and EQE from this point of view [40]. This is because the control of organic film and interfacial structures is not so easy. One of the most typical methods to control the organic film structure is annealing that results in the reorientation or recrystallization of organic molecules [40]. However, the reorientation changes the surface morphology, grain boundaries, etc., of organic films [40]. Accordingly, it is difficult to clarify the correlation between the film structure and the photoexcited carrier dynamics for OPV cells. To fabricate OPV cells with a high η, it is important to examine how the structural properties (crystal structure, grain size, and molecular orientation, etc.) of the organic films affect the photoexcited carrier dynamics relevant to the improvement of η and/or EQE.

In this section, the correlation between EQE and film structure is addressed by comparing two kinds of OPV cells [ITO/Zn(OEP)/C$_{60}$/Al], consisting of amorphous or crystalline 20-nm-thick Zn(OEP) films [41]. The former Zn(OEP) film was formed on a bare ITO substrate at room temperature by thermal deposition in an ultrahigh vacuum (UHV) chamber with a base pressure of 10^{-7} Pa [42–44] whereas the latter Zn(OEP) film was formed by postannealing (473 K for 1 min) of the 20-nm-thick amorphous Zn(OEP) film in UHV. The postannealing made the amorphous film crystallized with 20-nm-diameter grains comparable to the film thickness without a change in the surface roughness as shown in Figure 6.7.

The two kinds of Zn(OEP) films allow us to examine the correlation between film structure and photoexcited carrier dynamics for the heterojunction OPV cells. We will discuss what exactly improves the EQE on the basis of the photoexcited carrier dynamics in this section.

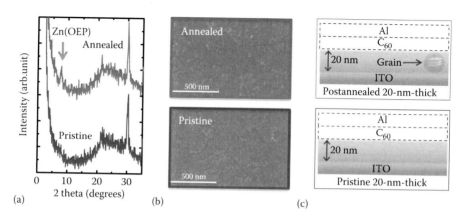

FIGURE 6.7 (a) XRD patterns of pristine (bottom) and annealed (top) 20-nm-thick Zn(OEP) films formed on an ITO. (b) SEM images of pristine (bottom) and annealed (top) 20-nm-thick Zn(OEP) film formed on an ITO. (c) Schematic illustrations of the film structures for pristine and crystalline 20-nm-thick Zn(OEP) films.

6.4.2 Effects of Intermolecular Charge-Transfer Excitons on the External Quantum Efficiency for Heterojunction OPV Cells

Figure 6.8 shows (a) the EQE of the [ITO/Zn(OEP)/C_{60}/Al] OPV cells with respect to the wavelength of irradiation, using amorphous (circle) and crystalline 20-nm-(triangle) thick Zn(OEP) films, and (b) the photoabsorption efficiency of Zn(OEP) molecules in dichloromethane (broken), amorphous (solid), and crystalline 20-nm-thick Zn(OEP) films (dot) along with that of a 30-nm-thick C_{60} film (dot-broken).

The maximum EQE at 400 nm in light wavelength was obtained to be 36% for the amorphous film and 42% for the annealed film, which are, respectively, comparable to and larger than that (35%) of OPV cells with buffer materials [45].

Both the C_{60}/Zn(OEP) interfacial states and the role of excitons generated in the C_{60} film are the same as for the OPV cells, respectively, because the surface morphology of each Zn(OEP) film used here remains almost unchanged by SEM as shown in Figure 6.7. Thus, this allows us to discuss the structural effects of the Zn(OEP) films on EQE. The increase in EQE may be due to the exciton-diffusion length becoming longer and/or the carrier mobility becoming larger in the crystalline Zn(OEP) film when compared to those in the amorphous one because the grain size is comparable to the film thickness (there are almost no grain boundaries in the crystalline film) as shown in Figure 6.7c. In fact, EQE increases to some extent when the Zn(OEP) film is crystallized in the range of 600–650 nm (see Figure 6.8a) although Zn(OEP) films exhibit no absorption band in this range. This indicates that the carriers (holes) generated from excitons in the C_{60} film move in the crystalline Zn(OEP) film more effectively than in the amorphous one. Thus, the mobility of holes in the crystalline Zn(OEP) film becomes larger to some extent.

It is interesting to note that the improvement of EQE depends on the wavelength of the incident light. The EQE increases by 37% on an average in the range of 420–520 nm whereas the average increment in the other range is obtained to be 16%.

Photoexcited Carrier Dynamics in Organic Solar Cells 155

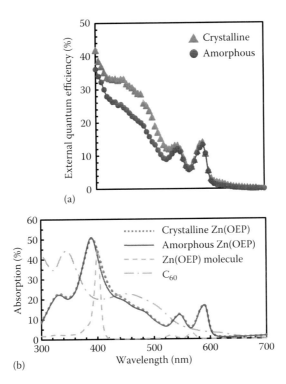

FIGURE 6.8 (a) EQE of the OPV cells using pristine (circle) and crystalline (triangle) 20-nm-thick Zn(OEP) films. (b) The photoabsorption efficiency of pristine (solid line), crystalline (dot) 20-nm-thick Zn(OEP), and Zn(OEP) molecules in dichloromethane (broken line) along with that of a 30-nm-thick C_{60} film (dot-broken line).

This may be due to the difference in exciton types: inter- or intramolecular excitons. Because the intramolecular excitons are localized within a Zn(OEP) molecule, the exciton diffusion length is not affected significantly by the structural properties (amorphous or crystalline) of the Zn(OEP) film. Accordingly, EQE is not remarkably improved after postannealing in the range of 520–630 nm, corresponding to the Q-band (transition from the ground singlet state to the first excited singlet state) of Zn(OEP) films, which causes intramolecular excitons. On the other hand, the photoabsorption in the range of 420–520 nm, leading to intermolecular excitons in Zn(OEP) film [41] results in the significant increment in EQE as shown in Figure 6.8a. In addition, the photoabsorption efficiency in the range becomes increased (by 1.06 times) slightly when the amorphous Zn(OEP) film changes to a crystalline one as shown in Figure 6.8b. This gives rise to an increase in the number of photogenerated intermolecular excitons in the crystalline Zn(OEP) film because a higher photoabsorption efficiency corresponds to a higher generation rate of excitons.

EQE for OPV cells can be expressed by the product of the four efficiencies, EQE = $\eta_A \eta_{ED} \eta_{ES} \eta_{CT}$ as described in Section 6.2. The present results find that the η_A and η_{CT} (hole) in the crystalline Zn(OEP) films slightly increase in the range of 420–520 nm and in the whole range, respectively, as described above. On the other hand, the η_{ES}

is considered to remain unchanged because of the similar D-A interfacial conditions for both OPV cells. These facts thus indicate that the η_{ED} enhances EQE considerably in the range of 420–520 nm in addition to the η_A and η_{CT} for the OPV cells consisting of the crystalline Zn(OEP) film. The improvement of η_{ED} implies that the diffusion length of the intermolecular excitons increases in crystalline Zn(OEP) films, which play a key role of increasing EQE for the OPV cells.

In summary, by using crystalline Zn(OEP) films with no grain boundaries, the maximum EQE obtained at 400 nm increased from 36% to 42%. Crystalline Zn(OEP) films play roles of increasing the number of intermolecular excitons by 1.06 times and of improving the mobility of carriers and intermolecular excitons in the Zn(OEP) films. Consequently, EQE shows the most remarkable improvement of 1.37 times in the range of 420–520 nm, which corresponds to the Zn(OEP) photoabsorption band resulting in intermolecular excitons. These results indicate that the film structure is more effective to excitons than to photogenerated carriers, and the improvement of the intermolecular exciton mobility play a key role of increasing EQE of the OPV cells. In addition, the improvement of EQE results in the increase in I_{SC} because I_{SC} can be calculated by integrating EQE with the incident light wavelength λ over the solar spectra of AM1.5 as expressed in Equation 6.4.

6.5 THE CORRELATION BETWEEN OPEN-CIRCUIT VOLTAGE AND PHOTOGENERATED CARRIER DYNAMICS

6.5.1 Introduction

As noted in Section 6.1, the determining factors of V_{oc} are still not yet understood. Thus, it is important to elucidate the origins of V_{oc} for enlarging the V_{oc} for practical use of OPV cells. Much attention has hitherto been paid to understanding the origins by considering the electronic states at D-A interfaces, such as the energy difference (ΔE_{HL}) between the HOMO of the donor material and the LUMO of acceptor. For some OPV cells, V_{oc} was reported to be proportional to ΔE_{HL} and/or to be less than ΔE_{HL} [7–11]. However, some other OPV cells did not exhibit such a relationship [46–48]. In the previous reports, the origins of V_{oc} have been often discussed on the basis of the electronic states at the D-A interface under dark conditions in spite of the fact that photogenerated carriers were reported to cause the change in the electronic states in the vicinity of their interfaces, such as the built-in potential (V_{bi}) [49,50]. For example, the surface potential of a dimethylquinquethiophene (DM5T) film on a platinum (Pt) substrate decreases from 190 mV to 20 mV upon photoirradiation because photogenerated holes are moved to the Pt substrate by a local electrical field at the interface. Thus, the correlation between the electronic states at the D-A interface upon photoirradiation and V_{oc} is important to clarify the origins of V_{oc}.

In this section, the correlations among V_{oc}, V_{bi}, and the photoexcited carrier dynamics in the vicinity of the D-A interface upon photoirradiation will be discussed using in situ nondestructive impedance spectroscopy [51]. The capacitance–voltage (C–V) characteristics obtained from impedance spectra can estimate the V_{bi} of individual films in the vicinity of the D-A interface [24,51]. On the other hand, to obtain the V_{oc}, we have examined the current–density versus voltage (J–V) characteristics of OPV cells consisting of 20-nm-thick Zn(OEP) and 30-nm-thick C_{60} layered films [ITO/Zn(OEP)/C_{60}/Al] under monochromatic light (400, 440, 460, 485, 500, 545, and 590 nm) irradiation and subsequently examined the dependence of the short-circuit current–density (J_{sc}), V_{oc}, and the fill factor (FF) on the wavelength of the monochromatic light. The FF is related to the mobility of photogenerated carriers in both D and A films. The difference in V_{oc} among the irradiation-light wavelengths in terms of V_{bi}, FF, and photoexcited carrier dynamics are finally discussed.

6.5.2 Influence of Charge Accumulation of Photogenerated Carriers in the Vicinity of Donor–Acceptor Interface on the Open-Circuit Voltage of Heterojunction OPV Cells

The band energy model and equivalent circuit described in Section 6.3 are applied to the present OPV cells with the Zn(OEP)/C_{60} heterojunction structure. As shown in Figure 6.9, the resistance and capacitance of each film are separated into two different ones in the region far from (R_{OEP}-1, R_{C60}-1 and C_{OEP}-1, C_{C60}-1) and near (R_{OEP}-2, R_{C60}-2 and C_{OEP}-2, C_{C60}-2) the D-A interface, respectively.

V_{oc} is predicted to be affected more significantly by the change in the electronic states in the vicinity of the D-A interface upon photoirradiation rather than by the organic/metal interface because excitons generated in organic films are basically dissociated only at the D-A interface (namely, carriers generated only at the D-A interface). Consequently, the built-in potentials V_{bi} of the Zn(OEP) and C_{60} films near the D-A interface under dark and photoirradiation conditions are discussed.

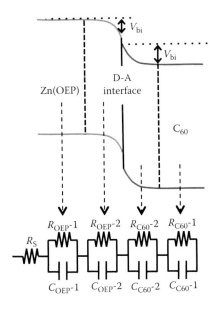

FIGURE 6.9 Equivalent circuit used for fitting the present impedance spectra along with schematic illustration of the Zn(OEP)/C_{60} layered organic photovoltaic cell.

The procedure of evaluating V_{bi} from the impedance spectra is briefly described as follows. For the 30-nm-thick C_{60} film, the capacitance of 1.23×10^{-9} F calculated from the reported relative permittivity value of 4.4 was used as a fixed parameter for fitting the impedance spectra obtained under dark conditions [32], that is, $(1/C_{C60}\text{-}1) + (1/C_{C60}\text{-}2) = 1/(1.23 \times 10^{-9} \text{ F})$. On the other hand, under photoirradiation, the $C_{OEP}\text{-}1$ and $C_{C60}\text{-}1$ values obtained under dark conditions are used as fixed parameters because the carriers are generated at only the D-A interface as described above. To estimate the V_{bi} of each organic film in the vicinity of the D-A interface (OEP-2 and C60-2), the C–V characteristics near the D-A interface are examined using the following Mott-Schottky equation often used for organic films [24,51],

$$C^{-2} = \frac{2(V_{bi} - V)}{S^2 q \varepsilon \varepsilon_0 N}. \tag{6.10}$$

Here, V, q, ε, ε_0, N, S, and C, respectively, denote the external bias voltage in the range of $-1.0 \sim +1.0$ V, the elementary charge, the relative dielectric constant, the permittivity in vacuum, charge density, device active area, and capacitance obtained using the above methods for each bias voltage. As seen by Equation 6.10, when C^{-2} is plotted as a function of V, V_{bi} can be evaluated by extrapolation of $C^{-2} = 0$.

Figure 6.10 shows the J–V characteristics of the OPV cells under monochromatic photoirradiation with a wavelength of 400, 440, 460, 485, 500, 545, and 590 nm. The inset shows the photoabsorption efficiency (AE) of 20-nm-thick Zn(OEP) and 30-nm-thick C_{60} films. Table 6.1 summarizes the results of J_{sc}, V_{oc}, FF, and EQE obtained for each wavelength.

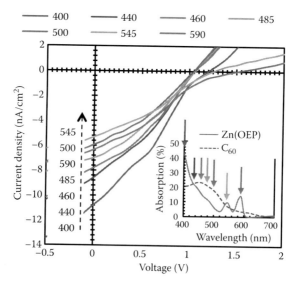

FIGURE 6.10 (See color insert.) J–V characteristics of [ITO/20-nm-thick Zn(OEP)/30-nm-thick C_{60}/Al] cells under irradiation with monochromatic light of 400, 440, 460, 485, 500, 545, and 590 nm wavelengths. The inset shows the absorption spectra of 20-nm-thick Zn(OEP) (solid) and 30-nm-thick C_{60} films.

Photoexcited Carrier Dynamics in Organic Solar Cells

TABLE 6.1
Summary of J_{sc}, V_{oc}, FF, and EQE of the [ITO/20-nm-thick Zn(OEP)/30-nm-thick C_{60}/Al] Cells under Each Irradiation Condition

Wavelength (nm)	J_{sc} (nA/cm²)	V_{oc} (V)	FF	EQE (%)
400	10.8	1.20	0.29	36.1
440	8.60	1.05	0.32	25.3
460	7.81	1.05	0.32	22.9
485	7.01	1.06	0.31	19.1
500	5.93	1.05	0.32	14.2
545	5.32	1.26	0.30	12
590	6.35	1.51	0.25	13.2

The EQE shown in Table 6.1 is quoted from Section 6.4. The V_{oc} is almost constant to be 1.05–1.06 V for 440, 460, 485, and 500 nm photoirradiations in which photons are absorbed remarkably by the C_{60} film, whereas it becomes larger to be 1.20, 1.26, and 1.51 V for 400, 545, and 590 nm photoirradiations in which photons are absorbed remarkably by the Zn(OEP) film, respectively. Because the dependence of V_{oc} on the light intensity for 400 and 590 nm photoirradiations indicates that the V_{oc} is independent of the light intensity for both cases, the V_{oc} is found to depend upon the wavelength of light. In addition, V_{oc} in the range of 1.05–1.51 V is much larger than the ΔE_{HL} of 0.22 V estimated for the Zn(OEP)/C_{60} interface [33,52]. Accordingly, the determining factors of V_{oc} are strongly related to the electronic states of the OPV cells upon photoirradiation.

Figure 6.11 shows the impedance spectra of the OPV cells under dark and photoirradiation conditions: dark (square), 400 nm (triangle), 500 nm (cross), and 590 nm (circle). Here, the individual solid lines show the simulated results corresponding to the experimental results. As shown in Figure 6.11, the impedance spectra change

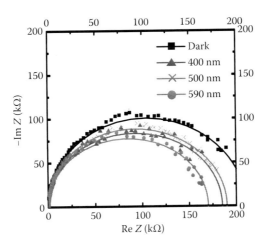

FIGURE 6.11 Impedance spectra of the OPV cell [ITO/Zn(OEP)/C_{60}/Al] under dark (square), 400 (triangle), 500 (cross), and 590 (circle) nm irradiation conditions.

before and after photoirradiation, and furthermore, their changes depend on the wavelength of irradiation light. However, two semicircles shown in Figure 6.6 are not observed. This may be due to a difference in wavelength. The AM1.5 light consisting of various wavelengths causes drastic change in the impedance spectra. The dependence of impedance spectra on the wavelength of incident light thus suggests that the capacitance of the organic films in the OPV cell also depends on the wavelength.

To estimate V_{bi} for each condition, the dependence of bias voltage (V) in the range of −1.0 V ∼ +1.0 V is examined. Figure 6.12 shows the experimental (dot) and simulated (solid line) impedance spectra for each condition: $V = 0$ V (circle), −0.4 V (square), and −1.0 V (triangle) before (dark) and under 400 nm, 500 nm, and 590 nm photoirradiations. It is found that the shape of each impedance spectrum is changed slightly by the applied voltage, and their radius decreases with increasing the applied voltage. Such results have been observed for BHJ OPV cells [24]. We next estimate the V_{bi} of each film near the D-A interface from the results of Figure 6.12.

Figure 6.13 shows the C^{-2}–V characteristics of Zn(OEP) [OEP-2] and C_{60} [C_{60}-2] films in the vicinity of the D-A interface under (a) dark, (b) 400 nm, (c) 500 nm, and (d) 590 nm photoirradiation conditions.

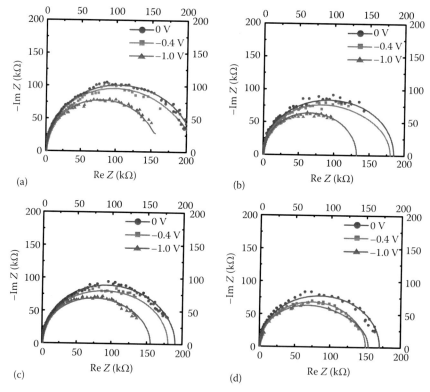

FIGURE 6.12 Dependence of impedance spectra of the OPV cell [ITO/Zn(OEP)/C_{60}/Al] on applied voltage for dark (a), 400 nm (b), 500 nm (c), and 590 nm (d) irradiation conditions. We chose typical impedance spectra obtained for 0 V (circle), −0.4 V (square), and −1.0 V (triangle) although impedance spectra were obtained by varying the applied voltage from −1.0 V to +1.0 V.

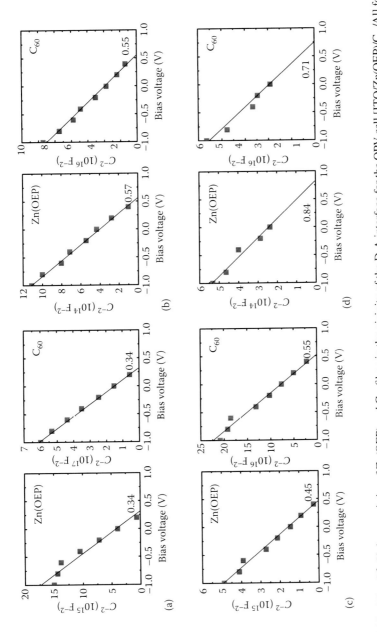

FIGURE 6.13 The $C^{-2}-V$ characteristics of Zn(OEP) and C_{60} films in the vicinity of the D-A interface for the OPV cell [ITO/Zn(OEP)/C_{60}/Al] for dark (a), 400 nm (b), 500 nm (c), and 590 nm (d) irradiation conditions.

For the dark condition, using Equation 6.10, V_{bi} of the C_{60} film is obtained to be 0.34 V that is comparable to that of 0.30–0.45 V for a C_{60} film on metal, semiconductor, and organic films on an ITO substrate [29,36,53]. This supports that the band-bending model shown in Figure 6.10 is reasonable to be used for the present OPV cells. The V_{bi} of the Zn(OEP) film at the D-A interface is the same value as for the C_{60} film.

On the other hand, when the OPV cells are irradiated with monochromatic light with wavelengths of 400, 500, and 590 nm, V_{bi} increases upon photoirradiation in a similar manner to V_{oc} as shown in Figure 6.10 and shows the highest values of 0.84 and 0.71 V for Zn(OEP) and C_{60} films, respectively, under 590 nm photoirradiation. The increase in V_{bi} upon photoirradiation may be due to the accumulation of photogenerated electrons and holes in the vicinity of the D-A interface because a high resistivity of both organic films probably prevents carriers (electrons and holes) from diffusing in the acceptor and donor films, respectively, as discussed in Section 6.3. Thus, V_{bi} should be strongly affected by carrier mobility because the degree of accumulation of photogenerated carriers increases with decreasing carrier mobility. On the other hand, because FF increases with carrier mobility, the wavelength-dependence of V_{bi} is predicted to exhibit an opposite dependence to that of FF [54].

Figure 6.14 shows a plot of the sum of V_{bi} (ΣV_{bi}) (square) obtained for both the Zn(OEP) and C_{60} films, FF (triangle), and V_{oc} (circle) as a function of monochromatic light wavelength. It is interesting to note that ΣV_{bi} is in good agreement with V_{oc}.

This implies that V_{bi} upon photoirradiation plays a dominant role in determining V_{oc}. Correspondingly, the wavelength-dependence of V_{oc} should also exhibit the opposite behavior to that of FF as well as V_{bi} because the dependence of V_{bi} is predicted to exhibit an opposite dependence to that of FF as described above. Indeed, the wavelength-dependence of both ΣV_{bi} and V_{oc} exhibits an opposite behavior to that of FF as seen in Figure 6.11. This indicates that the accumulation of photogenerated

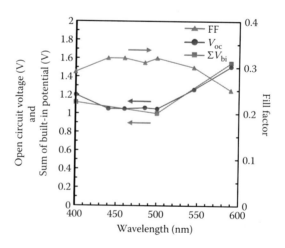

FIGURE 6.14 Plot of the open-circuit voltage V_{oc} (circle), the sum of V_{bi} (ΣV_{bi}) (square), and the fill factor FF (triangle) as a function of irradiation-light wavelength.

carriers in the interfacial region increases V_{bi} and finally enlarges V_{oc}. Such the behavior of FF and V_{oc} with respect to monochromatic light resembles that of J_{sc}, V_{oc}, and FF with respect to temperature [49]. FF (and J_{sc}) and V_{oc} were, respectively, reported to increase and decrease with increasing carrier mobility. The correlations among ΣV_{bi}, FF, and V_{oc} strongly suggest that accumulation of photogenerated carriers in the vicinity of the D-A interface plays a main role of determining V_{oc} for the OPV cells.

According to the nonequilibrium thermodynamics theory [54], the potential energy (E) driving the diffusion of photogenerated electrons and holes is given as the sum of the electrical and chemical potential energies, that is,

$$E_e = U_e + \mu_e$$
$$E_h = U_h + \mu_h. \tag{6.11}$$

Here, the U and μ denote the electrical and chemical potential energies, respectively, and the subscripts e and h mean electron and hole, respectively. The driving force for diffusion of photogenerated carriers can be expressed as ∇E, which is zero when the applied voltage V reaches V_{oc}. Thus, ∇U corresponds to $\nabla(V_{bi} - V)$ used in the present study as follows:

$$\nabla E = \nabla(V_{bi} - V) + \nabla \mu$$
$$0 = \nabla(V_{bi} - V_{oc}) + \nabla \mu. \tag{6.12}$$

Because $\nabla \mu$ has been recognized as a primary factor of the photocarrier diffusion to each electrode rather than $\nabla U = [\nabla(V_{bi} - V)]$ for the OPV cells, the photocurrent can be measured in spite of $V = V_{bi}$ [54]. However, as shown in Figure 6.14, because V_{bi} is almost equal to V_{oc}, the electrical potential (V_{bi}) in the D-A interfacial region upon photoirradiation plays a primary role of determining V_{oc} rather than μ and/or ΔE_{HL} for the OPV cells.

In summary, V_{bi} of each film in the vicinity of the D-A interface is found to increase upon photoirradiation because a high resistivity of each organic film accumulates photogenerated carriers in the vicinity of the D-A interfacial region. It is interesting to note that ΣV_{bi} is in good agreement with V_{oc}, which suggests that the electrical field caused by charge accumulation of photogenerated carriers in the vicinity of the D-A interface plays a main role of determining V_{oc} of OPV cells.

6.6 CONCLUDING REMARKS AND PERSPECTIVES

In the present chapter, we discussed the photoexcited carrier dynamics for OPV cells with a heterojunction double-layered structure. It was found that the excitons and photogenerated carriers affect the I_{SC}, V_{oc}, EQE, and η directly. The diffusion length of intermolecular excitons, which depends on the organic film structures, plays a key role of increasing I_{SC} and EQE. On the other hand, the V_{bi} in the vicinity of a D-A interface caused by the accumulation of photogenerated carriers was found to be the determining factor of V_{oc}. Accordingly, (i) an improvement of the diffusion length of intermolecular excitons by controlling film structures and (ii) a combination of donor and acceptor materials with both a low resistivity and a high V_{bi} are regarded as an essential approach to improve the power conversion efficiency η of OPV cells.

REFERENCES

1. Yu, G., Gao, J., Hummelen, J. C., Wudl, F., Heeger, A. J. "Polymer photovoltaic cells: Enhanced efficiencies via a network of internal donor-acceptor heterojunctions." *Science* 270 (1995): 1789–1791.
2. Peumans, P., Uchida, S., Forrest, S. R. "Efficient bulk heterojunction photovoltaic cells using small-molecular-weight irganic thin films." *Nature* 425 (2003): 158–162.
3. Chen, H. Y., Hou, J. H., Zhang, S. Q., Liang, Y. Y., Yang, G. W., Yang, Y., Yu, L. P., Wu, Y., Li, G. "Polymer solar cells with enhanced open-circuit voltage and efficiency." *Nat. Photon.* 3 (2009): 649–653.
4. Liang, Y. Y., Xu, Z., Xia, J. B., Tsai, S. T., Wu, Y., Li, G., Ray, C., Yu, L. P. "For the bright future—Bulk heterojunction polymer solar cells with power conversion efficiency of 7.4%." *Adv. Mater.* 22 (2010): E135–E138.
5. Green, M. A., Emery, K., Hishikawa, Y., Warta, W. "Solar cell efficiency tables." *Prog. Photovolt.: Res. Appl.* 19 (2011): 84–92.
6. Schmidt-Mende, L., Fechtenkotter, A., Mullen, K., Moons, E., Friend, R. H., Mackenzie, J. D. "Self-organized discotic liquid crystals for high-efficiency organic photovoltaics." *Science* 10 (2001): 1119–1122.
7. Brabec, C. J., Cravino, A., Meissner, D., Sariciftci, N. S., Fromherz, T., Rispens, M. T., Sanchez, L., Hummelen, J. C. "Origin of the open circuit voltage of plastic solar cells." *Adv. Funct. Mater.* 11 (2001): 374–380.
8. Rand, B. P., Burk, D. P., Forrest, S. R. "Offset energies at organic semiconductor heterojunctions and their influence on the open-circuit voltage of thin-film solar cells." *Phys. Rev. B* 75 (2007): 115327.
9. Giebink, N. C., Wiederrecht, G. P., Wasielewski, M. R., Forrest, S. R. "Ideal diode equation for organic heterojunctions. I. Derivation and application." *Phys. Rev. B* 82 (2010): 155305.
10. Giebink, N. C., Lassiter, B. E., Wiederrecht, G. P., Wasielewski, M. R., Forrest, S. R. "Ideal diode equation for organic heterojunctions. II. The role of polaron pair recombination." *Phys. Rev. B* 82 (2010): 155306.
11. Ryuzaki, S., Onoe, J. "Basic aspects for improving the energy conversion efficiency of hetero-junction organic photovoltaic cells." *Nano Rev.* 4 (2013): 21055.
12. Sun, S-S., Sariciftic, N. S. *Organic Photovoltaics: Mechanisms, Materials, and Devices.* (Florida, CRC Press, 2005).
13. Mott, N. F. "Note on the contact between a metal and an insulator or semi-conductor." *Proc. Camb. Philos. Soc.* 34 (1938): 568.
14. Schottky, W. "Deviations from Ohm's law in semiconductors." *Phys. Z* 41 (1940): 570.
15. Wang, H., Wang, J., Huang, H., Yan, X., Yan, D. "Organic heterojunction with reverse rectifying characteristics and its application in field-effect transistors." *Org. Electron.* 7 (2006): 369–374.
16. Ishii, H., Hayashi, N., Ito, E., Washizu, Y., Sugi, K., Kimura, Y., Niwano, M., Ouchi, Y., Seki, K. "Kelvin probe study of band bending at organic semiconductor/metal interfaces: Examination of Fermi level alignment." *Phys. Stat. Sol. (A)* 201 (2004): 1075–1094.
17. Frenkel, J. "On pre-breakdown phenomena in insulators and electronic semi-conductors." *Phys. Rev.* 54 (1928): 647–648.
18. Mott, N. "Electrons in glass." *Nobel Lect.* (1977): 403–413.

19. Wannier, G. H. "The structure of electronic excitation levels in insulating crystals." *Phys. Rev.* 52 (1937): 191–197.
20. Liang, W. Y. "Exitons." *Phys. Educ.* 5 (1970): 226–228.
21. Minami, N., Kazaoui, S., Ross, R. "Properties of electronic excited states of C_{60} thin films as seen from photoconductivity and luminescence behavior: Dependence on excitation wavelength and temperature." *Mol. Cryst. Liq. Cryst.* 256 (1994): 233–240.
22. Wen, C., Li, J., Kitazawa, K., Aida, T., Honma, I., Komiyama, H., Yamada, K. "Electrical conductivity of a pure C_{60} single crystal." *Appl. Phys. Lett.* 61 (1992): 2162–2163.
23. Mort, J., Ziolo, R., Machonkin, M., Huffmann, D. R., Ferguson, M. I. "Electrical conductivity studies of undoped solid films of $C_{60/70}$." *Chem. Phys. Lett.* 186 (1991): 284–286.
24. Belmonte, G., Munar, A., Barea, E. M., Bisquert, J., Ugarte, I., Pacios, R. "Charge carrier mobility and lifetime of organic bulk heterojunctions analyzed by impedance spectroscopy." *Org. Electron.* 9 (2008): 847–851.
25. Okachi, T., Nagase, T., Kobayashi, T., Naito, H. "Determination of charge-carrier mobility in organic light-emitting diodes by impedance spectroscopy in presence of localized states." *Jpn. J. Appl. Phys.* 47 (2008): 8965–8972.
26. Barsoukov, E., Macdonald, J. R. *Impedance Spectroscopy: Theory, Experiment, and Applications.* (New Jersey, Wiley-Interscience, 2005).
27. Ryuzaki, S., Onoe, J. "In situ nondestructive impedance spectroscopic study of nanostructured heterojunction organic photovoltaic cells." *Jpn. J. Appl. Phys.* 52 (2013): 06GD03.
28. Rispens, M. T., Hummelen, J. C. *Fullerenes: From Synthesis to Optoelectronic Properties.* (Heidelberg, Springer, 2002), 387.
29. Ohno, T. R., Chen, Y., Harvey, S. E., Kroll, G. H., Weaver, J. H., Haufler, R. E., Smalley, R. E. "C_{60} bonding and energy-level alignment on metal and semiconductor surfaces." *Phys. Rev. B* 44 (1991): 13747–13755.
30. Ryuzaki, S., Ishii, T., Onoe, J. "In situ x-ray photoelectron spectroscopic study of metalloporphyrin–fullerene alternative-deposited thin films." *Jpn. J. Appl. Phys.* 46 (2007): 5363–5366.
31. Ishii, T., Aizawa, N., Kanehama, R., Yamashita, M., Sugiura, K., Miyasaka, H. "Cocrystallites consisting of metal macrocycles with fullerenes." *Coord. Chem. Rev.* 226 (2002): 113–124.
32. Hebard, A. F., Haddon, R. C., Fleming, R. M., Kortan, A. R. "Deposition and characterization of fullerene films." *Appl. Phys. Lett.* 59 (1991): 2109–2111.
33. Peumans, P., Yakimov, A., Forrest, S. R. "Small molecular weigh organic thin-film photodetectors and solar cells." *J. Appl. Phys.* 93 (2003): 3693–3723.
34. Itoh, E., Kokubo, H., Shouriki, S., Iwamoto, M. "Surface potential of phthalocyanine Langmuir-Blodgett films on metal electrodes." *J. Appl. Phys.* 83 (1998): 372–376.
35. Tang, J. X., Zhou, Y. C., Liu, Z. T., Lee, C. S., Lee, S. T. "Interfacial electronic structures in an organic double-heterostructure photovoltaic cell." *Appl. Phys. Lett.* 93 (2008): 043512.
36. Payne, M. M., Parkin, S. R., Anthony, J. E., Kuo, C.-C., Jackson, T. N. "Organic field-effect transistors from solution-deposited functionalized acenes with mobilities as high as 1 cm^2/Vs." *J. Am. Chem. Soc.* 127 (2005): 4986–4987.
37. Sez, S. M. *Physics of Semiconductor Device.* (New York, Wiley-New York, 2nd edition, 1981), 254–270.
38. Lee, S., Hattori, R. "Derivation of current-voltage equation for OLED using device simulation." *J. Inform. Display* 10 (2009): 1212–1215.
39. Kampas, F. J., Gouterman, M. "Octaethylporphin films. II absorption and emission of amorphous films." *J. Lumin.* 14 (1976): 121–129.
40. Ma, W., Yang, C., Gong, X., Lee, K., Heeger, A. L. "Thermally stable, efficient polymer solar cells with nanoscale control of the interpenetrating network morphology." *Adv. Funct. Mater.* 15 (2005): 1617–1622.

41. Ryuzaki, S., Kai, T., Toda, Y., Adachi, S., Onoe, J. "Effects of inter-molecular charge-transfer excitons on the external quantum efficiency of zinc-porphyrin/C_{60} heterojunction photovoltaic cells." *J. Phys. D: Appl. Phys.* 44 (2011): 145103.
42. Ryuzaki, S., Onoe, J. "The crystallinity and surface morphology of zinc octaethylporphyrin thin films on an indium-tin-oxide substrate." *J. Appl. Phys.* 103 (2008): 033516.
43. Ryuzaki, S., Hasegawa, T., Onoe, J. "X-ray diffraction and infrared multiple-angle incidence resolution spectroscopic studies on the crystal structure and molecular orientation of zinc-porphyrin thin films on a SiO_2/Si substrate." *J. Appl. Phys.* 105 (2009): 113529.
44. Ryuzaki, S., Onoe, J. "X-ray diffraction and scanning electron microscopic studies on the crystal structure and surface/interface morphology of zinc-octaethylporphyrin films on an indium tin oxide substrate spin coated with 3,4-polyethylenedioxythiophene:poly styrenesulfonate." *J. Appl. Phys.* 106 (2009): 023526.
45. Peumans, P., Forrest, S. R. "Very-high-efficiency double-heterostructure copper phthalocyanine/C_{60} photovoltaic cells." *Appl. Phys. Lett.* 79 (2001): 126–128.
46. Yamanari, T., Taima, T., Sakai, J., Saito, K. "Origin of the open-circuit voltage of organic thin-film solar cells based on conjugated polymers." *Sol. Energy Mater. Sol. Cells* 93 (2009): 759–761.
47. Terao, Y., Sasabe, H., Adachi, C. "Correlation of hole mobility, exciton diffusion length, and solar cell characteristics in phthalocyanine/fullerene organic solar cells." *Appl. Phys. Lett.* 90 (2007): 103515.
48. Kumar, H., Kumar, P., Chaudhary, N., Bhardwaj, R., Chand, S., Jain, S. C., Kumar, V. "Effect of temperature on the performance of CuPc/C_{60} photovoltaic device." *J. Phys. D: Appl. Phys.* 42 (2009): 015102.
49. Yamada, H., Fukuma, T., Umeda, K., Kobayashi, K., Matsushige, K. "Local structures and electrical properties of organic molecular films investigated by non-contact atomic force microscopy." *Appl. Surf. Sci.* 188 (2002): 391–398.
50. Teich, S., Grafstrom, S., Eng, L. M. "Surface photovoltage of thin organic films studied by modulated photoelectron emission." *Surf. Sci.* 552 (2004): 77–84.
51. Ryuzaki, S., Onoe, J. "Influence of charge accumulation of photogenerated carriers in the vicinity of donor/acceptor interface on the open-circuit voltage of zinc-porphyrin/C_{60} heterojunction organic photovoltaic cells." *J. Phys. D: Appl. Phys.* 44 (2011): 265102.
52. Liu, X., Yeow, E. K. L., Velate, S., Steer, R. P. "Photophysics and spectroscopy of the higher electronic states of zinc metalloporphyrins: A theoretical and experimental study." *Phys. Chem. Chem. Phys.* 8 (2006): 1298–1309.
53. Hayashi, N., Ishii, H., Ouchi, Y., Seki, K. "Examination of band bending at buckminsterfullerene (C_{60})/metal interfaces by the Kelvin probe method." *J. Appl. Phys.* 92 (2002): 3784–3793.
54. Gregg, B. A. "The photoconversion mechanism of excitonic solar cells." *MRS Bull.* 305 (2005): 20–22.

7 Defect States in Organic Photovoltaic Materials, Thin Films, and Devices

John A. Carr and Sumit Chaudhary

CONTENTS

7.1 Introduction 168
7.2 Defect States in Organic Semiconductors: Background 168
7.3 Typical Measurement Techniques 169
    7.3.1 Thermally Stimulated Current (TSC) 169
    7.3.2 Fractional Thermally Stimulated Current (FTSC) 170
    7.3.3 Space Charge Limited (SCL) Current Modeling 171
    7.3.4 Capacitance versus Voltage (CV) 172
    7.3.5 Capacitance versus Frequency (CF) 173
    7.3.6 Drive-Level Transient Spectroscopy (DLTS) 174
    7.3.7 Open-Circuit Impedance Spectroscopy (IS) 174
7.4 Identification and Characterization of Defects in Neat Material Systems ... 175
    7.4.1 General Depiction 175
    7.4.2 Defect States in Donor Materials 177
        7.4.2.1 Defects with Relatively Shallow Activation Energies 177
        7.4.2.2 Defects with Relatively Deep Activation Energies 180
        7.4.2.3 Electron Traps 182
    7.4.3 Defect States in Acceptor Materials 182
    7.4.4 Origins of Defect States: Oxygen, Moisture, Structural, and Synthesis Residuals 184
7.5 Identification and Characterization of Defect States in Donor–Acceptor Blends 187
    7.5.1 General Depiction 187
    7.5.2 Defect States in the Donor and Acceptor Phases and New States in Blends 188
    7.5.3 Effects of Defects on the Performance of Organic Photovoltaic Devices 191
    7.5.4 Defect States at Organic–Electrode Interfaces 192
    7.5.5 Origins of Defect States: Oxygen, Structural, and Synthesis Residuals 193
7.6 Conclusions 195
References 196

7.1 INTRODUCTION

Organic photovoltaic (OPV) cells have been improving in performance over the last few years. But bottlenecks still exist, and the efficiencies are yet to reach the Schockley-Queisser limit (~23%).[1] Key bottlenecks are misalignment of energy levels, insufficient optical absorption, small exciton diffusion lengths, carrier recombination, and low charge carrier mobilities. Recent improvements in efficiencies are mainly due to the development of new materials and improved processing conditions. New materials that have been reported either show improved bandgaps that absorb more solar radiation or energy level offsets that improve cell voltage, charge dissociation, etc., or a combination of these properties. Nevertheless, irrespective of what the material is, the understanding of basic physical parameters and processes and their correlation to performance remain important to further improve the applicability and promise of this technology.

7.2 DEFECT STATES IN ORGANIC SEMICONDUCTORS: BACKGROUND

Defects, or states within the forbidden bandgap of materials, are one of the important physical parameters that directly impact the performance of OPVs. These defects can be shallow or deep within the bandgap as depicted in Figure 7.1. The origin of these defects can be extrinsic or intrinsic in nature.[2] Extrinsic defects are caused by chemical impurities, either introduced in materials during synthesis or in devices due to ambient exposure,[2–12] external bias-stress, or an electrochemical reaction involving electrical bias and oxygen.[13,14] Intrinsic defects are due to morphological disorder inherent in organic films; the morphology of these films is typically between the polycrystalline and amorphous regimes.[15–19] Morphological defects can be described as either noncovalent or covalent,[15] noncovalent referring to slight energetic perturbations leading to defects shallow in the bandgap[20] and covalent referring to higher energy perturbations that disrupt covalent bonds and cause defects deeper in the bandgap.[21]

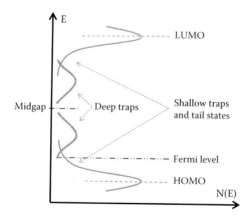

FIGURE 7.1 Exemplary distribution of density of defect states within the bandgap of an organic material.

Defect States in Organic Photovoltaic Materials, Thin Films, and Devices 169

Defects can act as charge carrier recombination centers, annihilating photogenerated charge, especially if located at donor–acceptor interfaces.[22–24] If charged, defects can also (1) reduce carrier mobility and transport by affecting the electrostatic potential in the device,[2] (2) lead to quenching of photogenerated excitons,[21,25] and (3) pin the Fermi-level within the bandgap, thus suppressing the built-in electric field.[26] In an OPV device, all these effects can lead to a reduction in the short circuit current density (J_{sc}), open circuit voltage (V_{oc}), and/or fill factor (FF); thus reducing the overall power conversion efficiency. In addition to efficiency, defects are also likely to affect stability and contribute to long-term degradation of these devices.[27,28]

On the other hand, some aspects of these unintentional defects may also be beneficial to OPVs. Wang et al. asked the question, "Do the defects make it work?"[20] The typical organic semiconductor with a bandgap of 2 eV is expected to have an intrinsic carrier concentration of about 1×10^3 to 1×10^7 cm^{-3}.[15,21] However, the actual carrier density reaches concentrations of about 1×10^{15} to 1×10^{19} cm^{-3}.[20,29–32] These extra carriers originate because some defects can produce electronic states near a band edge.[15] For example, oxygen can act as a p-type dopant, especially in the presence of light.[3,4,7,33–39] These excess carrier concentrations can alter carrier conductivities as well as the electric field at the donor–acceptor interface,[3,15,20,21,40–42] improving charge transport and exciton dissociation. Although, as the carrier from unintentional doping becomes a delocalized state, a localized site of opposite charge is left behind, this becomes a charge trap. Thus, there is a tradeoff involved.[3,41]

Nevertheless, good or bad, defect states in OPVs need to be well understood, and their identification, characterization, and mitigation are extremely important areas of research. *Identification* represents the need to profile the complete trap density of states (tDOS). *Characterization* represents the determination of the physical or chemical origins of these traps, their predominate characteristics, and their influence of the device performance. Some characteristics of interest are energetic locations, concentrations, disorder spreading, spatial location, type of carrier trapped, and capture/emission coefficients. *Mitigation or enhancement* represents the last step: reducing defects that have a negative effect on the devices while augmenting states that affect them positively.

This chapter presents a review and discussion of selected studies on the topic of defects in OPVs. First, we discuss typical measurement techniques used for defect investigation and their introductory concepts, equations, and relevant citations. Second, defect profiling in neat materials will be examined, followed by that of blended systems. Effects of traps on device performance and thoughts on trap origins will be discussed throughout. One should note that in our discussions, acceptor density (N_A) refers to shallow defect levels, which give an apparent p-doping in conjugated polymers/oligomers, whereas density of deep trap levels is marked N_T.

7.3 TYPICAL MEASUREMENT TECHNIQUES

7.3.1 THERMALLY STIMULATED CURRENT (TSC)

In the TSC method, the sample is first cooled to cryogenic temperatures, and excess charge carriers are generated (for example, optically or electrically) to induce filled trap states. Then, the sample is heated at a linear rate to emit stored charges, and the stimulated current is monitored, generating a current versus temperature spectrum. The concentration and mean energy of the dominant trap can then be calculated. Concentration is quantified by

$$\int_{\text{total}} I_{\text{TSC}}\, dt \leq qN_T \qquad (7.1)$$

where q is the elementary charge.[43,44] Note, this technique only gives a lower limit of the total trap density because recombination between thermally released electrons and holes can reduce the stimulated current. Incomplete trap filling or limited detrapping or both can also lower the measurement. The mean trap energy is quantified by

$$E_T = k_B T_{\max} \ln\left(\frac{T_{\max}^4}{\beta}\right) \qquad (7.2)$$

where k_B is the Boltzmann constant, $T_{\max}$ the temperature at the current peak, and β the heating rate.[45]

7.3.2 Fractional Thermally Stimulated Current (FTSC)

An extension of the conventional TSC method is FTSC, in which fractional heating cycles are used to further resolve the temperature/energy spectrum (Figure 7.2).[46] The sample is first cooled to a minimum temperature (T_{start}) and then trap filled. The sample is next heated to an intermediate end temperature (T_{stop}), which is less than the final temperature of interest. Finally, the sample is again cooled to T_{start} and then heated to the final temperature. This fractional cycle is repeated for increasing T_{stop} temperatures. Assuming the initial rise of the FTSC interval (Figure 7.2) is described by

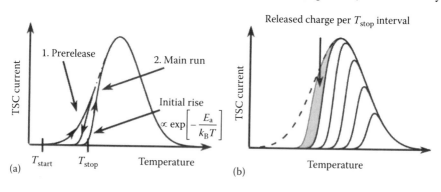

FIGURE 7.2 Schematics of the FTSC method: (a) The basic cycle consists of two individual TSC scans, and (b) the whole measurement is a replication of basic cycles with different T_{stop} temperatures. The activation energy of the initial rise describes the energetic trap depth, and the shaded area resembles the released charge. (Reprinted from *Progress in Photovoltaics: Research and Applications*, 12/2–3, S. S. Hegedus and W. N. Shafarman, 155–176, Copyright 2004, with permission from Elsevier.)

Defect States in Organic Photovoltaic Materials, Thin Films, and Devices

a Boltzmann-activated process, the activation energy for each T_{stop} can be found via the "initial rise method" as quantified by

$$I_{\text{iTSC}} \propto \exp\left(\frac{-E_T}{k_B T}\right) \quad (7.3)$$

where I_{iTSC} is the current of the initial rise.[46,47] The activation energy can then be coupled with a concentration, calculated from Equation 7.1, to build an energetic tDOS.

7.3.3 SPACE CHARGE LIMITED (SCL) CURRENT MODELING

Dark current–voltage measurements can also assist in determining trapping parameters. Neglecting diffusion and assuming a constant mobility, the current density in a single carrier, *p*-type material is given by

$$J = q\mu_p F(x) p_f(x) \quad (7.4)$$

where μ_p is the hole mobility, $F(x)$ the spatial distribution of the electric field, and $p_f(x)$ the spatial density of free holes.[48–50] The spatial distribution of the electric field is given by the Poisson equation

$$\frac{dF(x)}{dx} = \frac{q}{\varepsilon_s}[p_f(x) + p_T(x)] \quad (7.5)$$

where ε_s is the semiconductor permittivity and $p_T(x)$ the spatial density of trapped holes, defined by

$$p_T(x) = \int_0^\infty g(E) f(E, E_F, T) \, dE \quad (7.6)$$

where $g(E)$ is the DOS and $f(E, E_F, T)$ the Fermi function for occupation.[48–50] Thus, with the definition of $E_F(x)$,[49] numerical integration yields a current density, which can be coupled to a voltage via

$$V = \int_0^d F(x) \, dx \quad (7.7)$$

where d is the film thickness.[48] Extensive modeling yields information on the energetic distribution. If an exponential or Gaussian distribution of traps is assumed such that $p_f < p_T$, further simplifications can be made. For an exponential distribution,

$$g(E) = \frac{N_T}{k_B T_T} \exp\left(\frac{-E}{k_B T_T}\right) \quad (7.8)$$

where T_T is the characteristic temperature; the drift-only current density is approximated by

$$J = q^{1-l}\mu_p N_V \left(\frac{2l+1}{l+1}\right)^{l+1} \left(\frac{l}{l+1}\frac{\varepsilon_s}{N_T}\right)^l \frac{V^{l+1}}{d^{2l+1}} \tag{7.9}$$

where V is the voltage, and $l = T_T/T$.[51] For a Gaussian distribution,

$$g(E) = \frac{N_T}{\sqrt{2\pi}\sigma_T}\exp\left(-\frac{(E-E_T)^2}{2\sigma_T^2}\right) \tag{7.10}$$

two considerations are needed. For shallow trap centers, the current density is approximated by a modified Mott-Gurney law,

$$J = \frac{9}{8}\varepsilon_s\mu_p\theta\frac{V^2}{d^3} \tag{7.11}$$

where θ is a scaling factor,[52] and $\mu_p\theta$ is the effective mobility depending on the free-to-trapped charge ratio. For deep traps, the current is approximated by Equation 7.9 with a modified exponent (l') and trap concentration (N_T').[52]

7.3.4 Capacitance versus Voltage (CV)

Capacitance measurements have also been utilized to probe trap states. The CV method, for example, exploits the existence of a depletion region to measure both N_A and N_T. For instance, in an ideal p-type semiconductor–metal Schottky junction, the depletion capacitance is given by

$$C_d = \varepsilon_s \frac{A}{W}, \tag{7.12}$$

which is linearized to give the Mott-Schottky (MS) relationship,

$$\frac{1}{C^2} = \frac{2}{A^2 q\varepsilon_s N_A}(V_{bi} - V_{app}) \tag{7.13}$$

where W is the depletion width, N_A is the acceptor density, V_{bi} is the built-in voltage, and V_{app} is the applied bias.[53] For more accuracy, V_{bi} is replaced with V_D, where V_D is the diffusion potential: $qV_{bi} = E_F + qV_D$.[54] A plot of $1/C^2$ versus the applied bias yields a straight line with the slope related to N_A and the voltage-related intercept to V_{bi}. In the case of inhomogeneous doping, a spatial distribution of acceptor states is determined using the profiler equation

Defect States in Organic Photovoltaic Materials, Thin Films, and Devices

$$N_A(x) = \frac{C^3}{q\varepsilon_s A^2} \frac{dV}{dC} \tag{7.14}$$

where x is the distance from the junction.

In nonideal materials, however, careful consideration must be made to accurately interpret the results of these equations. When deep defects are present, they perturb the capacitance as the applied bias and/or AC frequency changes, and N_A, N_T, or some combination thereof may be measured by Equations 7.13 and 7.14. The effects of deep traps as a function of the applied bias was first discussed in 1973.[55] In the commonly encountered case when the trap emission rate ($e_{n,p}$) is slower than the AC frequency (ν_{AC}), it is typically assumed that the trap is frozen and has no contribution to the capacitance; it cannot emit during the course of the experiment. However, Kimmerling explains that if the trap emission is faster than the change in DC voltage (ΔV_{DC}), its charge can be uncovered as a function of the DC bias and contribute to the capacitance.[55] In such a case, $\nu_{AC} > e_{p,n} > \Delta V_{DC}$, and the "$N_A$" measured by Equation 7.13 or 7.14 represents

$$N(x) = N_T(x_T)\left[1 - \frac{W - x_T}{W}\right] + N_A(W) \tag{7.15}$$

where x_T is the spatial distance from the junction where E_T intersects E_F, and $W - x_T$ is assumed to be constant.[53,55] Thus, when x_T is small, $N_A(x)$ indeed represents N_A, and when it is large, $N_A(x)$ is a closer representation of $N_A + N_T$. Thereby, careful measurements must be made to accurately determine N_A, and, further, with careful measurement, N_T can also be ascertained. The influence of defect contribution to capacitance measurements as the AC frequency changes is discussed in Section 7.3.5.

7.3.5 Capacitance versus Frequency (CF)

Deep traps also contribute to the capacitance through a dynamic response to the AC small-signal frequency. This forms the basis for CF measurements, discussed in the work of Walter et al.[56] The CF method, also termed as admittance spectroscopy, relies on a frequency differential in which the DC bias is kept constant, and the frequency of the AC small signal is varied to include or exclude traps. In a p-type semiconductor, the thermal emission rate of a trap state is given by

$$\frac{1}{\tau_p} = e_p = N_V v_{th} \sigma_p \exp\left(\frac{-E_A}{k_B T}\right) \tag{7.16}$$

where N_V is the density of states in the valence band, v_{th} is the thermal velocity, σ_p is the capture cross-section, and E_A is the activation energy of the trap.[53,57] In inorganic devices, assuming that $v_{th} \propto T^{1/2}$, $N_V \propto T^{3/2}$, and σ_p independent of T, the prefactor in Equation 7.16 is typically written in terms of T^2,[53,57] or as a single, temperature-independent parameter, ω_0, called the attempt-to-escape frequency.[58,59] With the latter, the frequency of the AC small signal inherently defines an energy demarcation

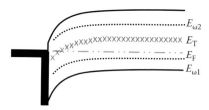

FIGURE 7.3 P-type/metal Schottky junction band diagram showcasing the CF experiment: E_F is the Fermi level, E_T the trap level, $E_{\omega 1}$ a high-frequency demarcation in which no states can respond, and $E_{\omega 2}$ a low-frequency demarcation in which all states respond.

that separates the defects with an emission rate slower than the applied frequency that can contribute to the capacitance from those that cannot.[53,56,58]

$$E_\omega = k_B T \ln\left(\frac{\omega_0}{\omega}\right) \quad (7.17)$$

where ω is the applied (angular) frequency. As ω varies from high to low frequencies, the demarcation energy is moved from below the Fermi level, where no states respond, to above the trap levels, where all states respond (Figure 7.3).[53,56] Note, only states that cross the Fermi level can contribute, and in the low frequency limit, it is only the states between the midgap and Fermi level that can be profiled.[56] An energetic profile of the tDOS can then be obtained using[56]

$$N_T(E_\omega) = -\frac{V_{bi}}{qW}\frac{dC}{d\omega}\frac{\omega}{k_B T}. \quad (7.18)$$

This profile is then typically fit with a Gaussian or exponential model, and trap parameters are extracted.

7.3.6 Drive-Level Transient Spectroscopy (DLTS)

DLTS also utilizes probing of the depletion capacitance. This technique studies trap characteristics by monitoring transient capacitance changes induced by a voltage or optical pulse.[53] The technique was introduced for crystalline materials and later extended to amorphous ones.[60,61] The approach is rigorous and yields several parameters, including trap band magnitudes, capture cross-section, activation energies, emission rates, and trap nature (majority or minority).[53] However, Sharma et al. have cautioned that due to low charge carrier mobility of some organic materials, DLTS may not be a suitable technique for such materials.[62]

7.3.7 Open-Circuit Impedance Spectroscopy (IS)

The IS method is an adaptation of the CF method at (typically) forward voltage potentials. The measurement results are expressed as Cole–Cole plots and can be used to extract the density of photogenerated carriers, mobilities, carrier lifetimes, and the

Defect States in Organic Photovoltaic Materials, Thin Films, and Devices 175

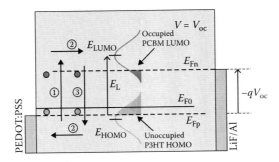

FIGURE 7.4 Band diagram of the P3HT:PCBM bulk heterojunction in steady-state illumination at open circuit ($V = V_{oc}$): Key physical processes depicted are excess holes and electrons are photogenerated (1) into the P3HT HOMO and PCBM LUMO manifolds, respectively. Charge carriers diffuse in the diode bulk (2) and finally recombine (3). Molecular orbitals spread in energy following Gaussian shapes. The occupancy level of LUMO and HOMO states depends on competing photogeneration and recombination rates. This governs the achievable V_{oc}, which is equal to the split between quasi Fermi levels. The DOS centers are at E_{LUMO} and E_{HOMO}, respectively. (Reprinted from *Solar Energy Materials and Solar Cells*, 94/2, G. Garcia-Belmonte, P. P. Boix, J. Bisquert, M. Sessolo and H. J. Bolink, 366–375, Copyright 2010, with permission from Elsevier.)

electron density of states.[26,63,64] For the latter, the impedance is measured in the open circuit condition, in which recombination balances generation (Figure 7.4).[64] The photovoltaic device is illuminated at different intensities, and a bias is applied to offset the photovoltage. The impedance examined is that of a chemical capacitance,[26,63] which is governed by changes in the electron quasi Fermi level (E_{Fn}) and defined by

$$\frac{C_\mu}{A} = Lq^2 \frac{dn}{dE_{Fn}} \qquad (7.19)$$

where L is the thickness of the photovoltaic layer.[64] Assuming a zero-temperature Fermi distribution,[64]

$$\frac{C_\mu}{A} = Lq^2 g(E_{Fn}). \qquad (7.20)$$

Thereby, with small movement of the polymer E_{Fp}, the fullerene DOS in a typical polymer/fullerene bulk heterjunction blend is obtained as E_{Fn} moves through the distribution at different illumination intensities (Figure 7.4).

7.4 IDENTIFICATION AND CHARACTERIZATION OF DEFECTS IN NEAT MATERIAL SYSTEMS

7.4.1 General Depiction

First, we visit the photothermal deflection spectra (PDS) of Goris et al. to understand wavelength-dependent absorption, which generally depicts the midgap tDOS in common OPV-relevant materials.[65] The spectra of poly(3-hexylthiophene) (P3HT), poly(p-phenylene vinylene) (PPV), and [6,6]-phenyl-C61-butyric acid methyl ester (PC$_{60}$BM) (neat films and blends) are shown in Figure 7.5. For now, we will focus on neat materials; blend spectra will be discussed later. The PDS spectra in Figure 7.5 establish the general picture of the energetic DOS. Weak absorption can be seen at lower energies, followed by a rapid increase at medium energies and strongest absorption at higher energies. This indicates the general depiction of molecular orbitals that are Gaussian in nature with a broadening that introduces a shallow defect distribution that trails into more localized, deep trap states.

For acceptor materials, this depiction is substantiated by optical experiments on C$_{60}$ and C$_{70}$ fullerene derivatives.[66,67] In donor materials, a thorough substantiating view is obtained through TSC measurements.[48] An indium tin oxide (ITO)/P3HT/Aluminum (Al) device structure was investigated.[48] The obtained TSC profile identified two peaks in the energetic tDOS (Figure 7.6). The low temperature peak at 93 K represents shallow

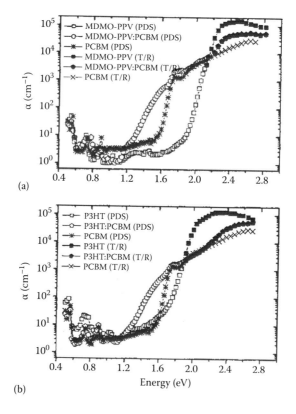

FIGURE 7.5 (a) PDS of MDMO-PPV, PCBM, and MDMO-PPV:PCBM. (b) PDS of P3HT, PCBM, and P3HT:PCBM. The full symbols (PCBM = star) correspond to data obtained from transmittance and reflectance measurements. (Reprinted with permission from L. Goris, A. Poruba, L. Hodakova, M. Vanecek, K. Haenen, M. Nesladek, P. Wagner, D. Vanderzande, L. De Schepper and J. Manca, *Applied Physics Letters*, 88, 052113, 2006, Copyright 2006, American Institute of Physics.)

Defect States in Organic Photovoltaic Materials, Thin Films, and Devices 177

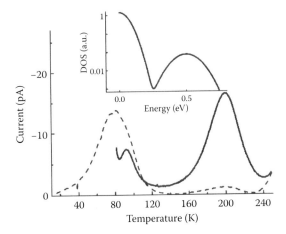

FIGURE 7.6 TSC signal from a thin P3HT film sandwiched between ITO and an aluminum electrode for the case of temperature of trap filling 77 K (solid line) and 4 K (dashed line). Inset shows the energy dependencies of the DOS. (Reprinted with permission from V. Nikitenko, H. Heil and H. Von Seggern, *Journal of Applied Physics*, 94, 2480, 2003, Copyright 2003, American Institute of Physics.)

levels, likely an Urbach tail from the Gaussian-shaped orbital. The high temperature peak at 200 K represents deeper states that are likely to be more localized trap levels.

7.4.2 Defect States in Donor Materials

7.4.2.1 Defects with Relatively Shallow Activation Energies

The low-temperature TSC peak in Figure 7.6 was confirmed and further resolved in another study focusing on neat P3HT films.[68] A low-temperature peak around 85 K with a shoulder around 50 K was observed (Figure 7.7a). The 85 K peak is in reasonable agreement with the previous TSC work (93 K). But the 200 K peak mentioned above was not reproduced in this study.[68] In another study, DLTS measurements revealed a similar finding with a single trap level at an activation energy of about 87 meV; again, no deeper band was found.[69] Interestingly, Schafferhans' shallower "shoulder band" also was not revealed in the DLTS study. Nonetheless, Schafferhans et al. estimated the total trap density in the studied energy regime to be approximately 10^{16} cm^{-3}.[68] For further quantification of the tDOS (Figure 7.7b), the authors applied the fractional $T_{start} - T_{stop}$ TSC technique (Figure 7.7a). Fractional TSC data clearly shows the overlap of two Gaussian-like distributions, indicating that there are two bands in a continuous distribution of shallow trap states.

Shallow activation levels in neat donor materials have also been observed in CV measurement studies.[7] One should note that the CV measurement records the response of only mobile charges. Thus, when complete ionization is not assumed, the CV measured quantity is the "ionized acceptor density" (N_A^-). This value should be considered a lower limit of the total shallow impurity concentration. As mentioned in Section 7.2, defects tend to *p*-dope OPV materials. With moderate doping, Fermi

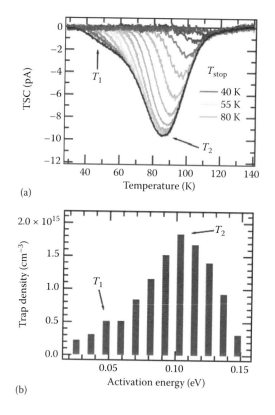

FIGURE 7.7 (a) Main runs of the different $T_{start} - T_{stop}$ cycles as well as the conventional TSC spectrum (black curve), revealing two trap states T_1 and T_2 and (b) the resulting DOOS distribution. (Reprinted with permission from J. Schafferhans, A. Baumann, C. Deibel, and V. Dyakonov, *Applied Physics Letters*, 93, 093303, 2008, Copyright 2008, American Institute of Physics.)

levels are around a few hundred meV above the highest occupied molecular orbital (HOMO). This, coupled with cathode work functions many hundred meVs lower, leads to the formation of a Schottky junction at the organic cathode interface.[70] The existence of this junction in P3HT diodes was shown by Dennler et al.[7] The authors leveraged depletion capacitance techniques to study defect states in the bandgap. Using CV measurements, they monitored charge carrier density as a function of air exposure (Figure 7.8). They found that the ionized acceptor density increased from 5.0×10^{16} to 1.0×10^{17} cm^{-3} as the exposure to air increased.[7] The charge density before exposure was in the same range as predicted by the TSC measurements,[68] again showing the presence of shallow states. These observations also support the claim that oxygen and humidity cause the *p*-type doping; this topic will be revisited in Section 7.4.4.

There are also other reports on probing defects using the CV method. Li. et al.[71] performed CV measurements at different frequencies on a ITO/PEDOT:PSS/P3HT/Al structure. They found ionized impurity concentrations between 6 and 10×10^{14} cm^{-3},

Defect States in Organic Photovoltaic Materials, Thin Films, and Devices

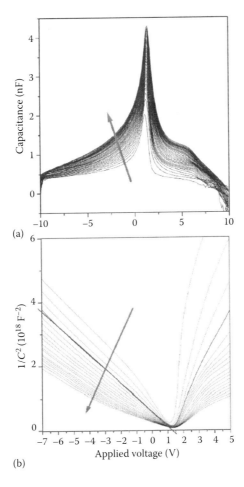

FIGURE 7.8 (a) Capacitance C and (b) $1/C^2$ of an ITO/P3HT/Al diode recorded at 1 kHz every 2 min in air versus the applied voltage V. The arrows indicate increasing time. The solid line in (b) is a linear fit of one intermediate curve. (Reprinted with permission from G. Dennler, C. Lungenschmied, N. S. Sariciftci, R. Schwödiauer, S. Bauer, and H. Reiss, *Applied Physics Letters*, 87, 163501, 2005, Copyright 2005, American Institute of Physics.)

much lower than that reported by Dennler et al.[7] This difference could be due to different levels of ambient exposure. Another strong possibility is processing condition–related; the polymer film thickness was ca. 1 μm for Dennler and ca. 5 μm for Li.[7,71] Thicker films dry more slowly, leading to better structural order and a lower apparent p-type doping.[72]

The CV method has also been applied to investigate shallow states in ITO/PPV/Al structures (Figure 7.9).[73] These samples were measured between −5 and 3 V_{DC} at an unmentioned frequency.[73] N_A^- was not quantified, but assuming a nominal device area, it can be estimated to be around 5×10^{16} cm^{-3}. This is similar to typically reported values. The above reports and discussion readily show that shallow states

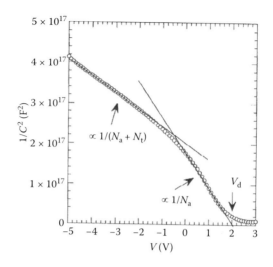

FIGURE 7.9 Inverse capacitance squared, $1/C^2$, against applied bias, V_{appl}, of an ITO/PPV/Al device. (Reprinted from *Synthetic Metals*, 111, A. Campbell, D. Bradley, E. Werner and W. Brütting, 273–276, Copyright 2000, with permission from Elsevier.)

are abundant in neat donor materials. These states are clearly distributed in energy and are at least partially ionized, giving rise to mobile charges.

Interestingly, Figure 7.9 yields further information. One can clearly see a break in the MS slope around $-0.75\ V_{DC}$. Both Dennler et al. and Li et al. also observed this change. In the former (Figure 7.8), the authors ascribed this break to inhomogeneous doping.[7] However, Li et al. argued that this may be due to nonuniform spatial doping *or* energetically deep trap band(s).[71] The formalization of the latter interpretation was given in a seminal work by Kimmerling and is discussed briefly in Section 7.3.4.[55] Campbell et al. took this interpretation and asserted the break is due to contributions from deeper traps. This indicates that deeper defects are present, which, because of the decrease in the MS slope, are acceptor-like states such that $N_T > N_A^-$.[73] Section 7.4.2.2 focuses on the discussion of these deep states in organic donor materials.

7.4.2.2 Defects with Relatively Deep Activation Energies

The presence of deeper defect levels in PPV was substantiated and quantified by DLTS measurements in the aforementioned work of Campbell et al.[73] Through temperature-dependent DLTS measurements and modeling, the authors reported a trap level ~750 meV above the HOMO level and with a concentration of $5 \times 10^{17}\ cm^{-3}$. This corroborates the interpretation of the CV data shown in Figure 7.9.

One recalls a similar defect band in P3HT was predicted by the above-discussed TSC measurements (Section 7.4.1), which can be further supported by ultraviolet and inverse photoemission spectroscopy measurements.[74] The photoemission spectra (Figure 7.10) show a feature (labeled "def") that cannot be ascribed to the molecular orbitals.[74] This distribution lies within the bandgap and thereby likely represents deep trap states. CF measurements can provide further insights on this suspected band.[30] These measurements revealed a Gaussian-shaped band in neat P3HT diodes

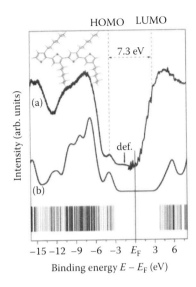

FIGURE 7.10 Occupied (left) and unoccupied (right) molecular orbital contributions of regioregular P3HT to the (a) photoemission and inverse photoemission spectra, respectively, are compared with (b) theory. The experimental HOMO–LUMO gap of P3HT (7.3 eV) is indicated as is a defect state (def.) not directly attributable to the molecular orbitals indicated. The inset shows a schematic of a single chain of regioregular P3HT. (D. Q. Feng, A. Caruso, Y. B. Losovyj, D. Shulz, and P. Dowben: *Polymer Engineering and Science*, 2007, 47, 9, 1359–1364. Copyright Wiley-VCH Verlag GmbH & Co. KGaA. Reproduced with permission.)

at ~390 meV above the HOMO with a concentration $N_T = 2.1 \times 10^{16}$ cm^{-3}.[30] These numbers are slightly different from PPV-based structures but are qualitatively similar. CV measurements in the same work yielded an ionized acceptor density of 3.2×10^{16} cm^{-3},[30] very similar to Dennler's CV measurement of 5×10^{16} cm^{-3} for P3HT diodes.[7] Interestingly, considering the lower frequency employed in Dennler's work (1 kHz) as well as the CF theory outlined in Section 7.3.5, the deep trap is expected to be fully responding throughout the CV measurement. Subtracting the deep-band concentration (2.1×10^{16} cm^{-3})[30] from N_A^- (5×10^{16} cm^{-3}),[7] an ionized acceptor concentration of 2.9×10^{16} cm^{-3} is obtained. This is in almost perfect agreement with the aforementioned 3.2×10^{16} cm^{-3} and hence substantiates the applied theory. More importantly, this clearly points out that that $N_A^- \approx N_T$, signifying that N_T should have a significant impact on electrical characteristics.

Thus, one can expect that theoretical modeling employing similar trap distributions should be able to describe the performance characteristics of underlying OPV devices. Transport in OPVs is usually explained with SCL and trap-limited (TL) models. From such modeling, features of the dominant trap sites and their effect on transport can be studied (refer to Section 7.3.3 for details). Nikitenko et al. did just that with neat P3HT diodes.[48] From their TSC data (Figure 7.6), a multiple trap formalism was employed assuming a superposition of two Gaussian distributions for the DOS.[48]

TABLE 7.1
Summary of Fitting Parameters in Nikitenko et al.'s Dual Gaussian SCL Current Work

	E (meV)	N (cm^{-3})	σ (meV)
Shallow band	—	1.0×10^{18}	55
Deep band	500	1.5×10^{16}	80

$$g(E) = \sqrt{\frac{2}{\pi}} \frac{(N - N_T)}{\sigma} \exp\left(-\frac{E^2}{2\sigma^2}\right) + \frac{N_T}{\sqrt{2\pi}\sigma_T} \exp\left(-\frac{(E - E_T)}{2\sigma_T^2}\right) \quad (7.21)$$

Simulations showed high voltages or a zero trap concentration led to currents close to the trap-free law (Equation 7.11 with θ = 1), and variations of the N_T, E_T, and $σ_T$ showed changes in slope and shifts in voltage–current characteristics at intermittent biases.[48] The model was then applied to P3HT diodes.[48] The experimental data was well explained by the modeling parameters listed in Table 7.1.[48] This data supports the shallow-deep tDOS and quantifies the deep trap band concentration at 1.5×10^{16} cm^{-3}. N_T and $σ_T$ are in good agreement with the CF work discussed above; however, the activation energy is slightly higher. Nevertheless, this highlights that trap states have a notable effect on charge transport.

7.4.2.3 Electron Traps

Thus far, trap levels in the lower half of the gap of these neat materials have been discussed; however, traps above the midgap must also be visited. The discussion on electron traps in organic materials is relatively straightforward; a recent work by Nicolai et al. unifies the presence of such states across several OPV materials.[75] The authors used a numerical drift-diffusion model with a Gaussian distribution of states to study the current characteristics of electron-only organic diodes. The electron trap band was found to be fairly consistent for all materials that were investigated, with $E_T \approx -3.6$ eV below the vacuum level, a relatively high N_T at 3×10^{17} cm^{-3} and $σ_T \approx$ 100 meV.[75] Such a trap in P3HT resides at ca. 600 meV below the LUMO, which, when coupled with the works discussed above, depicts a trap profile much the same as the Figure 7.1 cartoon. Moving further, Nicolai et al.'s work implies that materials with an electron affinity larger than −3.6 eV should exhibit trap-free characteristics. Indeed, this was found to be true for PC$_{60}$BM (LUMO ≈ −3.9 eV) and poly[{N,N'-bis(2-octyldodecyl)-naphthalene-1,4,5,8-bis(dicarboximide)-2,6-diyl}-alt-5,5'-(2,2'-bithiophene)] (P(NDI2OD-T2); LUMO ≈ −4.0 eV), both of which gave trap-free SCL electron transport.[74] Of interest here is the former.

7.4.3 Defect States in Acceptor Materials

After discussing the donor material, we will now focus on acceptors. As mentioned, PC$_{60}$BM has been shown to exhibit trap-free electron transport[75]; however, some

Defect States in Organic Photovoltaic Materials, Thin Films, and Devices

have indicated electron trapping in other acceptors, such as higher adduct fullerenes may exist.[76] An example of this can be seen in the work of Schafferhans et al., in which Bis-PC$_{61}$BM-, PC$_{71}$BM-, and PC$_{61}$BM-based devices (ITO/PEDOT:PSS/PCBM/Lithium Fluoride [LiF]/Al) were studied via the conventional and $T_{start} - T_{stop}$ TSC measurements.[77] Figure 7.11 shows the resulting spectra, and quantification is summarized by Table 7.2.[77]

It is interesting to note that all three materials showed at least one distinct distribution of traps with bisPC$_{61}$BM and PC$_{71}$BM also showing a second, deeper level. Each fullerene derivative exhibited a notably different TSC spectrum. A common feature was observed between PC$_{61}$BM and bisPC$_{61}$BM, with which a low-temperature peak

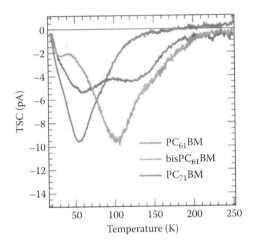

FIGURE 7.11 TSC spectra of PC$_{61}$BM, bisPC$_{61}$BM, and PC$_{71}$BM. (J. Schafferhans, C. Deibel and V. Dyakonov: *Advanced Energy Materials*, 2011, 1, 4, 655–660. Copyright Wiley-VCH Verlag GmbH & Co. KGaA. Reproduced with permission.)

TABLE 7.2
Lower Limit of the Trap Densities of PC$_{61}$BM, bisPC$_{61}$BM, and PC$_{71}$BM as well as the Temperatures of the TSC Maxima and the Corresponding Activation Energies Estimated by the T_{max} Method According to Equation 7.2

Sample	Trap Density (m^{-3})	T_{max} (K)	E_t (meV)
PC$_{61}$BM	≥1.7 × 10^{22}	54.9	86
bisPC$_{61}$BM	≥2.3 × 10^{22}	32.4	45
		103	184
PC$_{71}$BM	≥2.0 × 10^{22}	60.1	96
		121.4	223

Source: J. Schafferhans, C. Deibel and V. Dyakonov, *Advanced Energy Materials*, 2011, 1, 4, 655–660. Copyright Wiley-VCH Verlag GmbH & Co. KGaA. Reproduced with permission.

Note: For bisPC$_{61}$BM and PC$_{71}$BM, E_t and T_{max} values of both peaks are shown.

in bisPC$_{61}$BM (at 32 K or 45 meV) closely matched a slight PC$_{61}$BM TSC shoulder (Figure 7.11). This potentially signifies a common defect level.[77] For some substantiation, DLTS measurements on neat PC$_{61}$BM films also showed a single shallow trap band in this range, however, with a lower activation energy (21 meV).[69] At first, this shallow PC$_{61}$BM level seems to contrast the trap-free electron transport discussed earlier.[75] However, this can be reconciled if one considers the magnitude of the trap band. It is well known that for a Gaussian distribution of *shallow* traps, the SCL current is approximated by a modified Mott-Gurney law (see Equation 7.11). A quadratic current–voltage relationship with a reduced (effective) mobility is exhibited in such a case. The effective mobility is given by $\mu_p \theta$ and

$$\theta = \frac{\langle n_c \rangle}{\langle n_c \rangle + \langle n_t \rangle} \qquad (7.22)$$

where $\langle n_c \rangle$ is the average free carrier concentration and $\langle n_t \rangle$ is the average trapped charge with the trap-free case of $\theta = 1$. The free carrier concentration in PC$_{61}$BM is around 10^{17} to 10^{18} cm^{-3}. For a conservative estimate, if one considers the trapped charge concentration to be $1-2 \times 10^{16}$ cm^{-3}, one obtains a θ approaching 1, which indicates trap-free electron transport. Nevertheless, TSC of all three acceptors showed a continuous distribution of defect states throughout the measured energy regime. This is confirmed by fractional TSC measurements on PC$_{61}$BM.[77] Both bisPC$_{61}$BM and PC$_{71}$BM showed broadening in the TSC signal and slightly higher defect concentrations with significantly deeper energies.[77] This indicates toward higher energetic disorder as well as electron trapping and explains lowered dark currents in these two materials.[77]

7.4.4 Origins of Defect States: Oxygen, Moisture, Structural, and Synthesis Residuals

As discussed above, oxygen and moisture have been identified as main *p*-type dopants in donor materials. To further this discussion, we return to the P3HT TSC work of Schafferhans et al. (Figure 7.7).[68] In this work, the TSC measurement was conducted as a function of air and pure oxygen exposure. After 96 h of O$_2$ exposure, the magnitude of the 105 meV band tripled while the 50 meV did not change.[68] A similar observation was made in the case of air exposure. This shows that the 50 meV band is likely intrinsic or structural in nature; however, other chemical impurities originating from material synthesis cannot yet be ruled out. On the other hand, the 105 meV band is likely extrinsic in nature and closely related to oxygen. Photo-CELIV (charge extraction by linearly increasing voltage) measurements have also shown that mobility drops by nearly 50 times after 100 h of air exposure, further indicating that oxygen is altering states shallow within the gap.

An earlier work on the interaction of oxygen with polythiophene also explains this oxygen-related defect and its effect on carrier mobility.[3] Abdou et al. theorized that a charge transfer complex would form between oxygen and polymer with a binding energy on the order of 1.9 eV.[3] It was thought that this CTC state might facilitate charge transfer, thereby affecting electronic properties. To detect this band, they studied UV-Vis absorption of P3HT films in the presence of oxygen and nitrogen (Figure 7.12). As shown in the spectra, nitrogen did not affect the optical density, but oxygen did, especially near 630 nm or 1.97 eV, strongly implicating the predicted CTC.[3] Although the biggest change in optical density is near 630 nm, it can be seen that oxygen affects a range of subgap energies (Figure 7.12). This leads to the idea that oxygen itself affects and possibly generates a continuous distribution of states throughout the bandgap. This thought is also supported by a UV-Vis study of P3HT solution (Figure 7.13).[78] In this study, a broad subgap band around 810 nm was induced by oxidant treatments of molecular oxygen as well as Nitrosonium tetrafluoroborate. Interestingly, the absorption for wavelengths between 300 and 550 nm

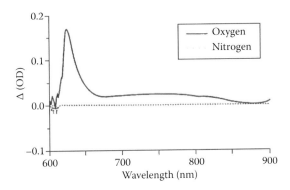

FIGURE 7.12 UV-Vis absorption spectra of a P3HT thin film (20 μm) in contact with oxygen (10 atm) and under oxygen-free N_2. Reference, O_2-free film. (Reprinted with permission from M. S. A. Abdou, F. P. Orfino, Y. Son, and S. Holdcroft, *Journal of the American Chemical Society*, 119, 4518–4524, 1997, Copyright 1997, American Chemical Society.)

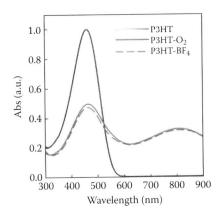

FIGURE 7.13 Comparison of UV-Vis spectra of P3HT in o-dichlorobenzene solution with the spectra of the P3HT solution after exposure to molecular oxygen (P3HT$^+$-O$_2^-$) and NOBF$_4$ (P3HT-BF$_4^-$). (Reprinted from *Solar Energy Materials and Solar Cells*, A. Guerrero, P. P. Boix, L. F. Marchesi, T. Ripolles-Sanchis, E. C. Pereira, and G. Garcia-Belmonte, Copyright 2012, with permission from Elsevier.)

was noticeably reduced. A similar decrease at moderate wavelengths along with an increase at higher wavelengths was also observed for treated polymer films.[78]

Abdou et al. also measured the effects of oxygen pressure on the conductivity and mobility of P3HT based thin-film transistors. With increasing oxygen pressure, conductivity increased, and mobility decreased.[3] Such opposite conductivity–mobility behavior is well known in inorganic semiconductors with increasing doping concentrations.[79] The above indicates that oxygen leads to extra carriers and explains the degraded mobility noticed in the work of Schafferhans et al. discussed earlier.[68] The presence of oxygen-induced charge carriers has been directly quantified by the work of Dennler et al. (Figure 7.8).[7] Through successive CV measurements in open air, it was found that the ionized acceptor density increased from 5.0×10^{16} to 1.0×10^{17} cm^{-3}.[7] This quantifies the Abdou et al.'s carrier concentration increase and again highlights that oxygen/moisture strongly affect the shallow impurity states.

Similarly, the electron traps noted by Nicolai et al. were also imputed, at least partly, to oxygen. As discussed earlier, the electron-trap parameters across several materials were found to be similar. This indicates a common origin and rules out material-specific structural disorder. Quantum-chemical calculations suggested hydrated oxygen complexes as the likely culprit.[75] This shows that oxygen and moisture together play an important role in the entire trap distribution of donor materials.

In addition to the donors, reports show that air/oxygen also affects the energetic tDOS of fullerene-based acceptors. In the case of C_{60}, exposure enhances the PDS spectrum of the deeper, localized states, and the shallow Urbach tail largely remains unaffected.[66] TSC experiments on C_{60}-based transistors further resolves this, pinpointing a trap band center on ca. 230 meV that was greatly affected by exposure and an even deeper band centered on ca. 370 to 420 meV, which remained unchanged.[9] It is worth noting that the deepest of these bands (370 and 420 meV) were dependent on the buffer layer used in the structure. Hence, this band may be interfacial (structural) in nature and not unique to the acceptor bulk. As such, it may not be surprising that they were unaffected by the exposure. Nonetheless, this again depicts a scenario in which structural disorder and/or chemical impurities from synthesis dominate a part of the defect distribution while the other is strongly influenced by oxygen and/or moisture. Similar to the case of donor materials, oxygen-induced modifications in C_{60} also lowers the carrier mobility.[9,80,81] Surprisingly, in C_{70}, the subgap absorption coefficient actually decreased with oxygen exposure, hinting that oxygen may be beneficial in mitigating some inherent defects.[67] Another interesting observation was that oxidant treatments of PCBM in solution led to no change in the optical density.[78]

Finally, in addition to structural and oxygen/moisture, defect origins from residuals left during material synthesis must also be considered. Although we could not find reports that link contaminant impurities to specific trap-band energies, it is well known that such impurities can negatively affect electronic properties. There are reports on contaminants from starting reagents, such as nickel, copper, rhodium, and palladium catalysts.[11,82] Most prominent in the literature seems to be residual palladium. Many conjugated polymers are synthesized via Stille or Suzuki coupling reactions, which are catalyzed by palladium complexes.[12,82] It has been reported that residuals of this palladium are linked to degradation of electronic properties and photovoltaic device performance.[11,12,82,83] Even trace amounts (for example, <0.1%

by weight) of contaminants can affect material properties and photovoltaic device performance.[11,12,82] The effects of residual catalysts on OPV device performance will be discussed further in Section 7.5.5.

From the discussion in this section, it is hopefully evident that common donor and acceptor OPV materials are rich in defects with said traps critically affecting the electronic properties. A depiction of energetic disorder with shallow and deep defect levels was given, and the possible origins of these traps were presented. Section 7.5 discusses how the defect profiles change in the popular bulk heterojunction OPV structures. The origins of these defects and their effects on device performance parameters will be discussed.

7.5 IDENTIFICATION AND CHARACTERIZATION OF DEFECT STATES IN DONOR–ACCEPTOR BLENDS

7.5.1 General Depiction

This section focuses on the tDOS of the donor–acceptor blends, which make up the contemporary OPV devices. First, we revisit the PDS data from the work of Goris et al.[65] As seen in Figure 7.5, when compared with the neat materials, the spectra of blended films showed a distribution that extended deeper into the gap.[65] This indicates a higher disorder in both PPV:PCBM and P3HT:PCBM devices. Stronger disorder was corroborated and further resolved by Neugebauer et al., who leveraged DLTS measurements to compare the defect bands of neat P3HT and PCBM films to that of blended systems.[69] The activation energies and trap concentrations are shown in Figure 7.14. This figure also includes a comparison to TSC measurements on neat

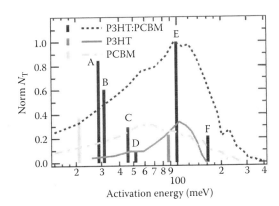

FIGURE 7.14 Bars: overview of the obtained activation energies in PCBM (light gray), P3HT (dark gray), and P3HT:PCBM blend (black) and relative maximum N_T in the spectrum found by I-DLTS. Lines: normalize activation energy spectra of earlier TSC measurements (colors are as above). (From J. Schafferhans et al.: *Organic Electron* 11 (10), 1693–1700, 2010; J. Schafferhans et al.: *Applied Physics Letters* 93, 093303, 2008; J. Schafferhans et al.: *Advanced Energy Materials* 1 (4), 655–660, 2011. Reprinted with permission from S. Neugebauer, J. Rauh, C. Deibel, and V. Dyakonov, *Applied Physics Letters*, 100, 26, 263304, Copyright 2012, American Institute of Physics.)

P3HT,[68] neat PCBM,[77] and their blend.[28] The neat-film DLTS data was mentioned earlier, and in it, both materials showed a single trap level with an activation energy of 21 meV for PCBM and 87 meV for P3HT.[69] The DLTS of blended films, however, indicated multiple trap levels. At the low end of the energy spectrum, a trap "A" with an activation energy around 28 meV was found. This corresponds quite well with the DLTS measured trap in neat $PC_{60}BM$ (21 meV).[69] The most concentrated band "E" has an activation energy of around 100 meV and correlates well with the neat P3HT trap at 105 meV as measured by TSC.[68,69] The other bands, "B–D" and "F," only appeared in the blend measurements and were missing in neat films. This supports the general depiction that greater energetic disorder results from mixing the materials in bulk heterojunction structures.

7.5.2 Defect States in the Donor and Acceptor Phases and New States in Blends

Trap bands in blended devices have also been examined through TSC measurements.[28] Investigation on an ITO/PEDOT:PSS/P3HT:PCBM/Ca/Al structure showed three TSC bands. Two (50 and 105 meV) matched the bands of the earlier neat P3HT TSC work,[68] and the third, centered deeper in the gap at 250 meV, was distinctive to the blended system.[28] Blended films also showed a substantial broadening and increase in concentration in the TSC spectrum.[28,68] This, coupled with the appearance of the unique deep band in the BHJ devices, again shows higher energetic disorder in blends. It is clear that the dominant band centered at 100 meV is common between the DLTS and TSC blend film work.[28,69] Interestingly, the TSC 50 meV band, attributed to a polymer contribution by the authors, correlates well with DLTS bands "C" and "D," which were considered unique to the blend.[28,69] Neglecting the origin aspect, DLTS bands "C" and "D" give more resolution to the TSC 50 meV band, indicating that two closely spaced trap sites make up this energy regime. The TSC 250 meV band is in qualitative agreement with the DLTS "F" band, albeit with a ca. 90 meV higher activation energy. The total trap density in the measured regime was estimated at 8×10^{16} cm^{-3}. This magnitude is eight times higher than that measured for neat P3HT, four times higher than that for neat $PC_{60}BM$, and nearly three times higher than the simple addition of the magnitudes of defects in neat films.[28,68,77]

In 2012, Yu et al. also studied the trap distribution of P3HT:PCBM BHJ films.[84] The authors utilized both conventional and $T_{start} - T_{stop}$ TSC measurements to probe ITO/P3HT:PCBM/Al structures.[84] Conventional TSC revealed two trap bands centered at 140 and 220 meV, similar to the DLTS (160 meV) and TSC (250 meV) bands, respectively.[28,69,84] The authors conjectured that more bands existed in this energy range but remained unresolved in the conventional TSC measurement. Thus, they performed fractional TSC along with numerical modeling for further elucidation. A total of five levels were cited (Table 7.3). Both exponential and Gaussian distributions were observed with concentrations between 8.0×10^{16} and 1.5×10^{19} cm^{-3}—much larger than those described by Schafferhans's work. Again, the N_3 and N_4 bands are in qualitative agreement with the DLTS "F" (160 meV) and TSC 250 meV bands, respectively. The newly resolved N_2 band seems similar to the dominant "E" band in Figure 7.14 whereas N_1 can be correlated to the "C–D" band. The authors attributed N_1–N_3 to the P3HT and

Defect States in Organic Photovoltaic Materials, Thin Films, and Devices

TABLE 7.3
Parameters of Electrically Active Trap Distributions Used in the Curve-Fitting Procedure

			Distribution			
			Gaussian		Exponential	
	E_a (eV)	Type	N_{g0} (cm^{-3})	σ (meV)	N_{t0} (cm^{-3})	E_t (eV)
N_1	0.06	Exp.	—	—	1.8×10^{17}	0.03
N_2	0.12	Exp.	—	—	2.5×10^{18}	0.03
N_3	0.14	Gauss.	8.2×10^{16}	52	—	—
N_4	0.20	Exp.	—	—	1.5×10^{19}	0.03
N_5	0.35	Gauss.	2.7×10^{18}	48	—	—

Source: Reprinted from *Journal of Non-Crystalline Solids*, P. Yu, A. Migan-Dubois, J. Alvarez, A. Darga, V. Vissac, D. Mencaraglia, Y. Zhou and M. Krueger, Copyright 2012, with permission from Elsevier.

N_4 to be resulting from the addition of PCBM—somewhat different than the interpretations of TSC and DLTS works. Perhaps the most interesting aspect is the deeper N_5 trap state. This band is vague in other TSC data but is strongly present here. The authors say that this type of data is often observed at the high-temperature edge and is possibly a result of weakening in the stimulated current.[84] However, a similar band has also been observed in measurements using other techniques, thus substantiating its existence. One can also recall that a similar deep band was seen in neat donor materials.

This deep band had been resolved by both CF measurements and current modeling methods.[30,59] Employing the CF measurement, Boix et al. revealed a Gaussian deep defect band centered at 380 meV above the HOMO with parameters $N_T = 1.2 \times 10^{16}$ cm^{-3} and $\sigma_T = 66$ meV (Figure 7.15), similar to that found in neat donor films.[30]

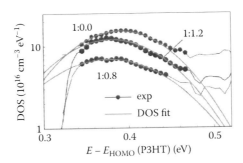

FIGURE 7.15 Density of defect states as a function of the energy with respect to the P3HT HOMO level (demarcation energy), $E - E_{HOMO}$, calculated using Equation 3 (of the source article) and the capacitance spectra in Figure 1a (of the source article). Gaussian DOS fits (Equation 5 of the source article) are also displayed. Composition of the blend is marked in each distribution. (Reprinted with permission from P. P. Boix, G. Garcia-Belmonte, U. Munecas, M. Neophytou, C. Waldauf, and R. Pacios, *Applied Physics Letters*, 95, 23, 233302–233303, Copyright 2009, American Institute of Physics.)

In these calculations, an attempt to escape frequency (v_0) of 1×10^{12} s^{-1} was used (see Equation 7.17). This band is in close proximity to the FTSC N_5 trap—although two orders of magnitude lower in density[84]—supporting the presence of this deep trap in blend films. Boix et al. also compared the energetic profile for films having different P3HT:PCBM loadings. It was found that the deep defect distribution is not significantly affected by the presence of PCBM, indicating that the band has its origin in the P3HT donor. Further, CV measurements showed that the ionized acceptor density in the blended films (ca. 4.3×10^{16} cm^{-3}) is only slightly higher than neat P3HT films (3.2×10^{16} cm^{-3}).[30] This indicates that the larger energetic disorder in the blends may not significantly affect the free carrier concentration.

In a subsequent work, Boix et al. revised the mean energy of this band to 35 meV, using a newly revealed attempt-to-escape frequency of 33.42 s^{-1} via a temperature-dependent CF technique.[85] Similar temperature-dependent CF measurements have also been performed on ITO/MDMO-PPV:PCBM/Al-[86] and ITO/PEDOT:PSS/MDMO-PPV:PCBM/Al-[57] based devices. The former report found a shallower trap between 24 to 34 meV with a v_0 in the range of 1×10^7 s^{-1} (estimated from presented data).[86] The latter report found two trap bands centered at 9 meV and 177 meV with attempt-to-escape frequencies of 1.3×10^7 s^{-1} and 7.6×10^9 s^{-1}, respectively.[57] Campbell et al. also found an attempt-to-escape frequency of 1×10^8 s^{-1} for PPV-based devices.[73] The sharp difference in v_0 between the P3HT and PPV seems implausible, and thus, it is important to obtain an accurate value to properly interpret these measurements. It is also worth noting that Dyakonov et al.'s temperature-dependent CF measurements shows a similar trap profile in PPV:PC$_{60}$BM compared to that of P3HT:PC$_{60}$BM devices. Bands with activation energies at 9, 22–34, and 177 meV were identified, and one may also expect the deeper, 750 meV level revealed by Campbell[72] in neat PPV, showing the qualitative resemblance to Figure 7.14.

The N_5 band in P3HT:PC$_{60}$BM was also revealed by Nam et al. using current modeling (SCL and Poole-Frenkel current methods)[59]; total concentration of midgap states in P3HT was calculated to be 6×10^{20} cm^{-3}.[59] This value is two to four orders of magnitude higher than what is typically reported for neat or blended cases. However, these devices were processed in open air, which can be expected to increase the impurity concentration. The effects of oxygen on trap states in blended systems will be further discussed in Section 7.5.5. Nevertheless, in Nam et al.'s work, the average trap activation energy was calculated to be 300–500 meV; this agrees with the TSC and original CF work.[30,59,84] Nam et al. also estimated relatively high electron trap density of ca. 1×10^{18} cm^{-3} in PCBM.[59] They suspected this high density to be oxygen-related due to fabrication processing in air. This thought is supported by the previously discussed optical measurement-related works centered on C$_{60}$, oxygen exposure and subgap states.[66,67] Interestingly, an annealing process significantly increased the degraded electron mobility and pushed the electron conduction towards trap-free transport.[59] This aligns their PCBM electron transport to that reported for inert processed diodes by Nicolai et al. and indicates that significant traps in PC$_{60}$BM may arise from appreciable oxygen exposure.[75] This also highlights the popular thought that thermal annealing can effectively remove structural defects and at least partially mitigated the trap states induced by oxygen.

In blends based on higher adduct fullerenes, such as bis-PC$_{60}$BM, higher disorder and electron trapping has been reported.[76] Lenes et al. investigated devices based on higher adduct blends and found reduced current–voltage characteristics in electron-only devices; enhanced energetic trapping in the higher adducts was found, akin to that expected from the TSC measurements on fullerene films discussed earlier (Section 7.4.3).[76,77] However, higher disorder did not have a negative effect on the photovoltaic performance, indicating that these traps are rapidly filled by photogenerated charge carriers.[76] This point naturally leads us into an important discussion on how OPV performance is affected by defects. This is the focus of Section 7.5.3.

7.5.3 Effects of Defects on the Performance of Organic Photovoltaic Devices

Continuing the discussion on the effects of disorder in the acceptor phase, it has been indicated that these states can indeed adversely alter OPV performance, even in PC$_{60}$BM-based BHJ devices. Garcia-Belmonte et al. have looked into these effects using the open-circuit impedance spectroscopy (IS) method.[26] In this work, BHJ cells based on a P3HT:PC$_{60}$BM system were compared with that of P3HT:DPM$_6$ (DPM$_6$: 4,4′-dihexyloxydiphenylmethano[60]fullerene).[26] DPM$_6$-based devices showed a significant positive shift in open-circuit voltage.[26] To elucidate its origins, the open-circuit impedance was measured at different illumination intensities and a profile of energetic DOS was generated (Figure 7.16).[26] Interestingly, the central energy of the DOS in PC$_{60}$BM-based devices was slightly shifted with a significantly larger concentration when compared to that of DPM$_6$. Consequently, it was concluded that the quasi Fermi level in PCBM-based devices is *pinned* at deeper energies within the

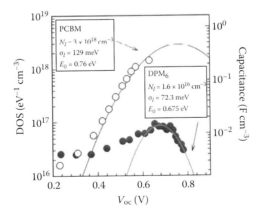

FIGURE 7.16 Capacitance values extracted from fits of the low-frequency arc of the impedance spectra as a function of V_{oc} reached under varying illumination levels. White dots correspond to PCBM-based solar cells and black dots to DPM$_6$-based solar cells. Gaussian DOS (solid lines) and distribution parameters resulting from fits. (Reprinted with permission from G. Garcia-Belmonte et al., *Solar Cells* 11, 14, Copyright 2010, American Chemical Society.)

DOS distribution. This brings the quasi electron Fermi level closer to the quasi hole Fermi level, thus reducing the achievable V_{oc} in PC$_{60}$BM-based devices.

Further, traps in both donor and acceptor phases may be detrimental to OPV device performance because of charge carrier recombination, especially if traps are at a donor–acceptor interface. Street et al. highlighted this by investigating the recombination kinetics in BHJ OPVs and carefully examining the possibility of geminate, nongeminate, interface state, and Auger recombination. Through light-intensity and temperature-dependent measurements coupled with modeling, the authors suggested recombination through interface states as a dominant recombination mechanism for the material system studied.[23] This study was performed on PCDTBT:PC$_{70}$BM BHJ devices; however, the authors also expressed the possibility of conclusions being generalized to other material systems. The model employed by Street et al. suggested that even an order of magnitude reduction of the interface states could potentially double the power conversion efficiency.[23] It is worth noting that some disagreement and further discussion on this work exists in the literature.[87,88]

There also exist other more empirical works on interface state or trap-assisted recombination.[22,24] Cowan et al. investigated such a case for PCDTBT:PC$_{60}$BM devices.[24] The introduction of PC$_{84}$BM impurities to the blend significantly reduced the efficiency; even impurity concentrations of 0.001% led to reduced efficiency.[24] By looking at the V_{oc} as a function of light intensity and trap density, the authors shed light on the thresholds after which trap-assisted losses become significant. Cowan et al. proposed that trap-assisted recombination becomes dominant at high trap densities and/or low light intensities. Mandoc et al. substantiate this further in their two reports.[22,89] In the first, the authors deliberately introduced electron traps into PPV:PC$_{60}$BM devices.[22] All photovoltaic parameters (V_{oc}, J_{sc}, FF) degraded, and the losses were attributed to higher trap-assisted recombination.[22] In their second work, Mandoc et al. investigated all polymer BHJ solar cells based on a blend of PPV and PCNEPV (poly-[oxa-1,4-phenylene-(1-cyano-1,2-vinylene)-(2-methoxy-5-(3′,7′-dimethyloctyloxy)-1,4-phenylene]-1,2-(2-cyanovinylene)-1,4-phenylene]).[89] No extra traps were added, but inherent electron traps in the acceptor polymer led to a dominant trap-assisted recombination loss, albeit only at low light intensities.[89] Clearly, these two works corroborate Cowan's proposal.

Aside from trap-assisted recombination, charge trapping can also lead to space charge effects.[90,91] McNeill et al. studied photocurrent transients in all polymer BHJ cells to reveal the trapping–detrapping of electrons.[91] The transients were induced by pulsed light exposure.[91] A sharp transient peak after *turn on* of the pulse with a long transient tail after *turn off* was revealed; the latter was attributed to the trapping–detrapping of electrons. The authors proposed that charges trapped near the electrode would lead to space charge effects, or in other words, perturb the internal electric field to (i) decrease the probability of charge separation and (ii) promote bimolecular recombination.[91]

7.5.4 Defect States at Organic–Electrode Interfaces

Electronic states at the organic–electrode interface layers have also been studied. Ecker et al. profiled trap bands near the anode using CF measurements on ITO\hole transport layer (HTL)\P3HT:PCBM\Ca\Al cells.[92] Devices using different HTLs, namely PEDOT:PSS and PANI:PSS based in water and alcohol solvents, were examined.[92] By conducting the CF measurement in illumination and at open-circuit conditions, two Gaussian trap distributions were revealed. The lower energy distribution was attributed to states in the bulk. An attempt-to-escape frequency v_0 was not mentioned; thus this work cannot be directly compared with other CF works, although the lower energy distribution is plausibly from the same deep states as those previously mentioned.[16,30] The higher energy Gaussian was attributed to the states at the HTL interface. The concentration of these states was significantly higher than that of the bulk states. Interestingly, unencapsulated devices with HTLs based in water solvents showed additional trap states at the blend-HTL interface.[92] This work highlights the importance of defect considerations when choosing these interface buffer layers.

Interface trap states have also been identified near the cathode.[93] Such states were revealed by CV measurements on an ITO/PEDOT:PSS/P3HT:PCBM/Al structure, measured in the dark as well as under illumination.[93] The ionized acceptor density was similar in both dark and illuminated conditions (ca. 4–5.0 × 10^{16} cm^{-3}). However, in the illuminated case, the built-in voltage was shifted negatively by 0.6 V. This shift is explained using a model that incorporated kinetically slow surface states at the organic–metal junction.[93] Under illumination, an accumulation of minority charges leads to charging of the interface states, which, in turn, induces band unpinning and the V_{bi} shift.[93] Although a quantification of these states and their affects on OPV performance has not yet been determined, this work highlights the importance of defect considerations at organic–electrode interfaces.

7.5.5 Origins of Defect States: Oxygen, Structural, and Synthesis Residuals

To look into the origins behind the traps in BHJ structures, we first revisit the 2010 TSC work on ITO/PEDOT:PSS/P3HT:PCBM/Ca/Al structures.[28] Upon exposure to air, in contrast to the case of neat P3HT, the magnitude of the 105 meV trap appeared to decrease in these P3HT:PCBM devices.[28,68] Yet, a new band at 142 meV emerged with a concentration increasing with exposure time.[28] Overall, the total trap density in the measured energy spectrum remained constant through 100 hours of exposure time.[28] Note that these measurements only provide a lower limit for the total trap density because charge-carrier recombination could be masking an increase in trap density and/or leading to the apparent reduction in the 105 meV band. Nonetheless, clearly oxygen is contributing to the deeper states. The above work also showed a decrease in carrier mobility and increase in charge density.[28] This is in line with the works and discussion on neat systems. Coupling their data with modeling, the authors assert that increased carrier densities degrade J_{sc}, and the increased deep(er) trap concentration lead to a lowering of V_{oc} and FF,[28] potentially through more trap-assisted recombination or space charge effects. The point of extra charge carriers degrading J_{sc} is supported by Seemann et al., who showed that oxygen-induced degradation of devices is ascribable to excess mobile holes and immobile anions.[38] Further, a direct correlation between generated photocurrent and free carrier concentration in P3HT:PCBM devices was shown by Guerrero et al., who found $J_{sc} \propto n^{-0.14}$.[78] It seems clear that defects originating from oxygen exposure significantly affect the long-term stability of organic devices. Of course, the topic of degradation is vast, and although aspects of it are related to defects, a comprehensive treatment of the topic is beyond the scope of this chapter. A full review on OPV degradation can be found in the literature.[94]

Turning to the yet deeper (ca. 380 meV) band discussed previously, its link to film morphology was elucidated to by Nalwa et al., who applied the CF method to P3HT:PCBM devices in which the active layers were spin coated at different rates.[16] Each spin rate produced a different film thickness, which greatly changes the film drying time, and therefore, ordering and chain packing within the layer.[72] It has been shown in several reports that better packing or self-organization leads to higher crystallinity and better OPV performance. Thus, Nalwa et al. set out to investigate how the deep-trap energetic profile changes with the film self-organization or growth rate.[16] They found that films spun at lower speeds (thicker, dried slowly) contained nearly a magnitude less deep traps than those spun at higher speeds (thinner, dried faster). Defect densities of 3.3×10^{15} and 2.1×10^{16} cm^{-3} and mean energies of 360 and 380 eV were found for devices with active layers spun at 400 and 1000 rpm, respectively.[16] If the assumed attempt-to-escape frequency ($v_0 = 1 \times 10^{12}$ s^{-1}) is accepted, this band closely matches the reports discussed above, and the data indicates that the deep defect band originates from intrinsic structural disorder or impurities. A similar conclusion was reached by Sharma et al. for CuPc devices.[62]

Seemann et al. also shed light on this deep band using impedance measurements.[38] As shown in Figure 7.17, the measured capacitance increased and parallel resistance decreased with exposure to synthetic air (20% O$_2$, 80% N$_2$), especially in the presence of light.[38] They attributed this to an increase in charge carriers induced by oxygen doping. It can also be seen that this response is frequency-dependent.

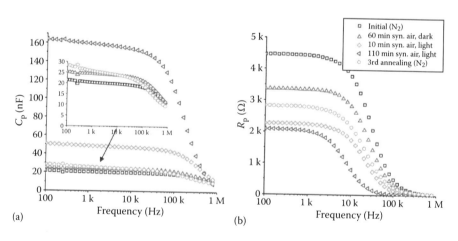

FIGURE 7.17 Spectrally resolved device capacitance (a) and device resistance (b) extracted from the admittance of a typical solar cell measured under nitrogen in the dark after exposure of the cell to different degradation conditions. The data are normalized to the respective values at 100 Hz before the first exposure of the device to synthetic air. (Reprinted from *Solar Energy*, 85, 6, A. Seemann, T. Sauermann, C. Lungenschmied, O. Armbruster, S. Bauer, H. J. Egelhaaf, and J. Hauch, 1238–1249, Copyright 2011, with permission from Elsevier.)

Defect States in Organic Photovoltaic Materials, Thin Films, and Devices 195

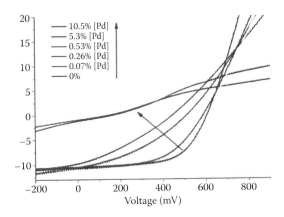

FIGURE 7.18 Effect of palladium impurities on the *I–V* characteristics of P3HT/PCBM organic solar cells. (P. A. Troshin, D. K. Susarova, Y. L. Moskvin, I. E. Kuznetsov, S. A. Ponomarenko, E. N. Myshkovskaya, K. A. Zakharcheva, A. A. Balakai, D. Babenko, and V. F. Razumov: *Advanced Functional Materials*, 20, 24, 4351–4357, 2010. Copyright Wiley-VCH Verlag GmbH & Co. KGaA. Reproduced with permission.)

According to the CF interpretation summarized in Section 7.3.5, this may indicate oxygen induced changes in the ca. 300–500 meV deep-defect distribution.

In addition to oxygen/humidity, trace impurities left from the material synthesis also play a role in the defect landscape. In the case of the commonly used palladium catalyst, this has been shown for P3HT:PCBM devices.[82,83] Troshin et al. investigated the effects of an intentional addition of tetrakis(triphenylphosphine)palladium(0) (Pd(PPh$_3$)$_4$) to P3HT:PCBM cells.[82] As can be seen in Figure 7.18, this intentional addition (from 0% to 10.5%) has a drastic effect on the OPV device performance parameters; performance reduction was noticeable even at a concentration of 0.07%.[82] In another study, similar results were found for a palladium addition in higher-efficiency organic solar cells based on PTB7: 5% (Pd(PPh$_3$)$_4$) was directly linked to greater trap-assisted recombination.[12] The negative impact of trace impurities is also seen in other works on nonpalladium residuals. Leong et al. showed that trace concentrations of (MePT)DTS(PTTh$_2$) in the molecular material *p*-DTS(PTTh$_2$)$_2$ originate from the synthesis process.[10] Through generation–recombination investigations, the authors found the presence of more energetic trap states even at very low contaminant levels. Consequently, the efficiency of BHJ structured cells (ITO/molybdenum oxide/*p*-DTS(PTTh$_2$)$_2$:PC$_{70}$BM.Al) was limited to 3%[10] whereas devices based on more purified *p*-DTS(PTTh$_2$)$_2$ exhibited efficiency of 6.5%. To date, there has been no direct link of synthesis residuals to specific defect energies/locations although it clearly appears that they play an important role in altering OPV performance.

7.6 CONCLUSIONS

In summary, this chapter discussed the current state of the identification and characterization defect levels in organic photovoltaic materials, thin films, and devices. Several techniques have been leveraged to study these midgap states. This list includes optical (PSD, UV-Vis), capacitance (CV, CF, DLTS), and current (TSC, SCL modeling, Poole-Frenkel modeling) measurements, and each has contributed important—complementary and supplementary—aspects to the overall theme. A compilation of the literature on defects depicts organic materials as disordered semiconductors with a distribution of both energetically shallow and deep trap bands within the bandgaps. Upon blending these materials to form the contemporary BHJ cells, energetic disorder increases, and new defect bands appear. These defect states have been shown to stem from both intrinsic (for example, structural disorder) and extrinsic (for example, oxygen, synthesis residuals) sources, and it is quite clear that such states can have profound effects on the electronic properties, efficiency, and stability of OPV devices or organic electronic devices in general. This is shown by several works highlighting the effects of traps on the OPV performance, citing enhanced trap-assisted recombination, Fermi-level pinning, space-charge effects, and others. Although these midgap traps have mostly a negative impact, it should also be remembered that they can give advantageous inherent doping, improving conductivity, and interfacial electric fields. It is clear that further progress in understanding the nature, sources, effects, and mitigation of the defects will be crucial to the optimization of OPVs and organic electronic device in general.

REFERENCES

1. T. Kirchartz, K. Taretto and U. Rau, *The Journal of Physical Chemistry C* 113 (41), 17958–17966 (2009).
2. L. Kaake, P. Barbara and X. Y. Zhu, *The Journal of Physical Chemistry Letters* 1 (3), 628–635 (2010).
3. M. S. A. Abdou, F. P. Orfino, Y. Son and S. Holdcroft, *Journal of the American Chemical Society* 119 (19), 4518–4524 (1997).
4. S. Hoshino, M. Yoshida, S. Uemura, T. Kodzasa, N. Takada, T. Kamata and K. Yase, *Journal of Applied Physics* 95, 5088 (2004).
5. M. M. Erwin, J. McBride, A. V. Kadavanich and S. J. Rosenthal, *Thin Solid Films* 409 (2), 198–205 (2002).
6. J. A. Carr, K. S. Nalwa, R. Mahadevapuram, Y. Chen, J. Anderegg and S. Chaudhary, *ACS Applied Materials and Interfaces* 4 (6), 2831–2835 (2012).
7. G. Dennler, C. Lungenschmied, N. S. Sariciftci, R. Schwödiauer, S. Bauer and H. Reiss, *Applied Physics Letters* 87, 163501 (2005).
8. R. F. Salzman, J. Xue, B. P. Rand, A. Alexander, M. E. Thompson and S. R. Forrest, *Organic Electronics* 6 (5), 242–246 (2005).
9. T. Matsushima, M. Yahiro and C. Adachi, *Applied Physics Letters* 91, 103505 (2007).
10. W. L. Leong, G. C. Welch, L. G. Kaake, C. J. Takacs, Y. Sun, G. C. Bazan and A. J. Heeger, *Chemical Science* 3, 2103–2109 (2012).
11. F. C. Krebs, R. B. Nyberg and M. Jørgensen, *Chemistry of Materials* 16 (7), 1313–1318 (2004).
12. M. P. Nikiforov, B. Lai, W. Chen, S. Chen, R. D. Schaller, J. Strzalka, J. Maser and S. B. Darling, *Energy and Environmental Science* 6 (5), 1513–1520 (2013).
13. D. Lang, X. Chi, T. Siegrist, A. Sergent and A. Ramirez, *Physical Review Letters* 93 (7), 76601 (2004).
14. D. Knipp and J. E. Northrup, *Advanced Materials* 21 (24), 2511–2515 (2009).
15. B. A. Gregg, S. G. Chen and R. A. Cormier, *Chemistry of Materials* 16 (23), 4586–4599 (2004).

16. K. S. Nalwa, R. C. Mahadevapuram and S. Chaudhary, *Applied Physics Letters* 98, 093306 (2011).
17. D. Wang, N. Kopidakis, M. O. Reese and B. A. Gregg, *Chemistry of Materials* 20 (20), 6307–6309 (2008).
18. Z. Liang, A. Nardes, D. Wang, J. J. Berry and B. A. Gregg, *Chemistry of Materials* 21 (20), 4914–4919 (2009).
19. A. Hepp, N. Von Malm, R. Schmechel and H. Von Seggern, *Synthetic Metals* 138 (1), 201–207 (2003).
20. D. Wang, M. Reese, N. Kopidakis and B. A. Gregg, presented at the Photovoltaic Specialists Conference, 2008, PVSC '08, 33rd IEEE, 2008 (unpublished).
21. B. A. Gregg, *Soft Matter* 5 (16), 2985–2989 (2009).
22. M. Mandoc, F. Kooistra, J. Hummelen, B. de Boer and P. Blom, *Applied Physics Letters* 91, 263505 (2007).
23. R. A. Street and M. Schoendorf, *Physical Review B* 81, 205307 (2010).
24. S. R. Cowan, W. L. Leong, N. Banerji, G. Dennler and A. J. Heeger, *Advanced Functional Materials* 21 (16), 3083–3092 (2011).
25. B. A. Gregg, *The Journal of Physical Chemistry C* 113 (15), 5899–5901 (2009).
26. G. Garcia-Belmonte, P. P. Boix, J. Bisquert, M. Lenes, H. J. Bolink, A. La Rosa, S. Filippone and N. Martín, *Solar Cells* 11, 14 (2010).
27. J. Bhattacharya, R. Mayer, M. Samiee and V. Dalal, *Applied Physics Letters* 100, 193501 (2012).
28. J. Schafferhans, A. Baumann, A. Wagenpfahl, C. Deibel and V. Dyakonov, *Organic Electron* 11 (10), 1693–1700 (2010).
29. G. Dicker, M. P. de Haas, J. M. Warman, D. M. de Leeuw and L. D. A. Siebbeles, *The Journal of Physical Chemistry B* 108 (46), 17818–17824 (2004).
30. P. P. Boix, G. Garcia-Belmonte, U. Munecas, M. Neophytou, C. Waldauf and R. Pacios, *Applied Physics Letters* 95 (23), 233302–233303 (2009).
31. B. A. Gregg, S. E. Gledhill and B. Scott, *Journal of Applied Physics* 99, 116104 (2006).
32. S. Jain, W. Geens, A. Mehra, V. Kumar, T. Aernouts, J. Poortmans, R. Mertens and M. Willander, *Journal of Applied Physics* 89 (7), 3804–3810 (2001).
33. H. Hintz, H. Peisert, H. J. Egelhaaf and T. Chassé, *The Journal of Physical Chemistry C* 115 (27), 13373–13376 (2011).
34. E. Meijer, A. Mangnus, B. H. Huisman, G. T. Hooft, D. De Leeuw and T. Klapwijk, *Synthetic Metals* 142 (1), 53–56 (2004).
35. H. H. Liao, C. M. Yang, C. C. Liu, S. F. Horng, H. F. Meng and J. T. Shy, *Journal of Applied Physics* 103 (10), 104506–104508 (2008).
36. C. K. Lu and H. F. Meng, *Physical Review B* 75 (23), 235206 (2007).
37. L. Lüer, H. J. Egelhaaf and D. Oelkrug, *Optical Materials* 9 (1), 454–460 (1998).
38. A. Seemann, T. Sauermann, C. Lungenschmied, O. Armbruster, S. Bauer, H. J. Egelhaaf and J. Hauch, *Solar Energy* 85 (6), 1238–1249 (2011).
39. A. Aguirre, S. Meskers, R. Janssen and H. J. Egelhaaf, *Organic Electronics* 12 (10), 1657–1662 (2011).
40. M. Chikamatsu, T. Taima, Y. Yoshida, K. Saito and K. Yase, *Applied Physics Letters* 84, 127 (2004).
41. V. Arkhipov, P. Heremans, E. Emelianova and H. Baessler, *Physical Review B* 71 (4), 045214 (2005).
42. A. Liu, S. Zhao, S. B. Rim, J. Wu, M. Könemann, P. Erk and P. Peumans, *Advanced Materials* 20 (5), 1065–1070 (2008).
43. A. Kadashchuk, R. Schmechel, H. Von Seggern, U. Scherf and A. Vakhnin, *Journal of Applied Physics* 98 (2), 024101–024109 (2005).
44. W. Graupner, G. Leditzky, G. Leising and U. Scherf, *Physical Review B* 54 (11), 7610 (1996).

45. Z. Fang, L. Shan, T. Schlesinger and A. Milnes, *Materials Science and Engineering: B* 5 (3), 397–408 (1990).
46. J. Steiger, R. Schmechel and H. Von Seggern, *Synthetic Metals* 129 (1), 1–7 (2002).
47. N. Von Malm, R. Schmechel and H. Von Seggern, *Synthetic Metals* 126 (1), 87–95 (2002).
48. V. Nikitenko, H. Heil and H. Von Seggern, *Journal of Applied Physics* 94, 2480 (2003).
49. V. Arkhipov, P. Heremans, E. Emelianova and G. Adriaenssens, *Applied Physics Letters* 79 (25), 4154–4156 (2001).
50. W. Hwang and K. Kao, *Solid-State Electronics* 15 (5), 523–529 (1972).
51. P. Mark and W. Helfrich, *Journal of Applied Physics* 33 (1), 205–215 (1962).
52. W. Hwang, *Solid-State Electronics* 19, 1045–1047 (1976).
53. J. Heath and P. Zabierowski, *Advanced Characterization Techniques for Thin Film Solar Cells* 81–105 (2011).
54. S. S. Hegedus and W. N. Shafarman, *Progress in Photovoltaics: Research and Applications* 12 (2–3), 155–176 (2004).
55. L. Kimerling, *Journal of Applied Physics* 45 (4), 1839–1845 (1974).
56. T. Walter, R. Herberholz, C. Muller and H. Schock, *Journal of Applied Physics* 80 (8), 4411–4420 (1996).
57. V. Dyakonov, I. Riedel, C. Deibel, J. Parisi, C. Brabec, N. Sariciftci and J. Hummelen, presented at the Materials Research Society Symposium Proceedings, 2002 (unpublished).
58. S. S. Hegedus and E. Fagen, *Journal of Applied Physics* 71 (12), 5941–5951 (1992).
59. C.-Y. Nam, D. Su and C. T. Black, *Advanced Functional Materials* 19 (22), 3552–3559 (2009).
60. D. Lang, *Journal of Applied Physics* 45 (7), 3023–3032 (1974).
61. D. V. Lang, J. D. Cohen and J. P. Harbison, *Physical Review B* 25 (8), 5285 (1982).
62. A. Sharma, P. Kumar, B. Singh, S. R. Chaudhuri and S. Ghosh, *Applied Physics Letters* 99, 023301 (2011).
63. G. Garcia-Belmonte, A. Munar, E. M. Barea, J. Bisquert, I. Ugarte and R. Pacios, *Organic Electronics* 9 (5), 847–851 (2008).
64. G. Garcia-Belmonte, P. P. Boix, J. Bisquert, M. Sessolo and H. J. Bolink, *Solar Energy Materials and Solar Cells* 94 (2), 366–375 (2010).
65. L. Goris, A. Poruba, L. Hodakova, M. Vanecek, K. Haenen, M. Nesladek, P. Wagner, D. Vanderzande, L. De Schepper and J. Manca, *Applied Physics Letters* 88, 052113 (2006).
66. T. Gotoh, S. Nonomura, S. Hirata and S. Nitta, *Applied Surface Science* 113, 278–281 (1997).
67. H. Habuchi, S. Nitta, D. Han and S. Nonomura, *Journal of Applied Physics* 87 (12), 8580–8588 (2000).
68. J. Schafferhans, A. Baumann, C. Deibel and V. Dyakonov, *Applied Physics Letters* 93, 093303 (2008).
69. S. Neugebauer, J. Rauh, C. Deibel and V. Dyakonov, *Applied Physics Letters* 100, 263304 (2012).
70. B. G. Streetman and S. Banerjee, *Solid State Electronic Devices*. (Prentice-Hall, USA 1995).
71. J. V. Li, A. M. Nardes, Z. Liang, S. E. Shaheen, B. A. Gregg and D. H. Levi, *Organic Electronics* 12 (11), 1879–1885 (2011).
72. J. A. Carr, Y. Chen, M. Elshobaki, R. C. Mahadevapuram and S. Chaudhary, *Nanomaterials and Energy* 1 (1), 8 (2012).
73. A. Campbell, D. Bradley, E. Werner and W. Brütting, *Synthetic Metals* 111, 273–276 (2000).
74. D. Q. Feng, A. Caruso, Y. B. Losovyj, D. Shulz and P. Dowben, *Polymer Engineering and Science* 47 (9), 1359–1364 (2007).

75. H. Nicolai, M. Kuik, G. Wetzelaer, B. de Boer, C. Campbell, C. Risko, J. Brédas and P. Blom, *Nature Materials* 11 (10), 882–887 (2012).
76. M. Lenes, S. W. Shelton, A. B. Sieval, D. F. Kronholm, J. C. K. Hummelen and P. W. M. Blom, *Advanced Functional Materials* 19 (18), 3002–3007 (2009).
77. J. Schafferhans, C. Deibel and V. Dyakonov, *Advanced Energy Materials* 1 (4), 655–660 (2011).
78. A. Guerrero, P. P. Boix, L. F. Marchesi, T. Ripolles-Sanchis, E. C. Pereira and G. Garcia-Belmonte, *Solar Energy Materials and Solar Cells* 100, 185–191 (2012).
79. R. F. Pierret and G. W. Neudeck, *Advanced Semiconductor Fundamentals*. (Addison-Wesley, Reading, MA, 1987).
80. R. Könenkamp, G. Priebe and B. Pietzak, *Physical Review B* 60 (16), 11804 (1999).
81. A. Tapponnier, I. Biaggio and P. Gunter, *Applied Physics Letters* 86, 112114 (2005).
82. P. A. Troshin, D. K. Susarova, Y. L. Moskvin, I. E. Kuznetsov, S. A. Ponomarenko, E. N. Myshkovskaya, K. A. Zakharcheva, A. A. Balakai, S. D. Babenko and V. F. Razumov, *Advanced Functional Materials* 20 (24), 4351–4357 (2010).
83. A. Saeki, M. Tsuji and S. Seki, *Advanced Energy Materials* 1 (4), 661–669 (2011).
84. P. Yu, A. Migan-Dubois, J. Alvarez, A. Darga, V. Vissac, D. Mencaraglia, Y. Zhou and M. Krueger, *Journal of Non-Crystalline Solids* 358, 2537–2540 (2012).
85. P. P. Boix, J. Ajuria, I. Etxebarria, R. Pacios and G. Garcia-Belmonte, *Thin Solid Films* 520, 2265–2268 (2011).
86. V. Dyakonov, D. Godovsky, J. Meyer, J. Parisi, C. Brabec, N. Sariciftci and J. Hummelen, *Synthetic Metals* 124 (1), 103–105 (2001).
87. C. Deibel and A. Wagenpfahl, *Physical Review B* 82 (20), 207301 (2010).
88. R. Street, *Physical Review B* 82 (20), 207302 (2010).
89. M. M. Mandoc, W. Veurman, L. J. A. Koster, B. de Boer and P. W. Blom, *Advanced Functional Materials* 17 (13), 2167–2173 (2007).
90. C. R. McNeill and N. C. Greenham, *Applied Physics Letters* 93, 203310 (2008).
91. C. R. McNeill, I. Hwang and N. C. Greenham, *Journal of Applied Physics* 106, 024507 (2009).
92. B. Ecker, J. C. Nolasco, J. Pallarés, L. F. Marsal, J. Posdorfer, J. Parisi and E. von Hauff, *Advanced Functional Materials* 21 (14), 2705–2711 (2011).
93. J. Bisquert, G. Garcia-Belmonte, A. Munar, M. Sessolo, A. Soriano and H. J. Bolink, *Chemical Physics Letters* 465 (1), 57–62 (2008).
94. S. K. Gupta, K. Dharmalingam, L. S. Pali, S. Rastogi, A. Singh and A. Garg, 2 (1), 42–58 (2012).

8 Interfacial Materials toward Efficiency Enhancement of Polymer Solar Cells

Zhiqiang Zhao, Wenfeng Zhang, Xuemei Zhao, and Shangfeng Yang

CONTENTS

8.1 Introduction 202
8.2 Anode Buffer Layers (ABLs) 202
    8.2.1 Poly(3,4-ethylenedioxithiophene):poly(styrene sulfonate) (PEDOT:PSS) 203
        8.2.1.1 Physical Treatments of PEDOT:PSS 205
        8.2.1.2 Additives in PEDOT:PSS 206
        8.2.1.3 ITO-Free PSCs Based on Modified PEDOT:PSS 210
    8.2.2 Metal Oxides 213
    8.2.3 Polymers and Small-Molecule Organic Materials 215
    8.2.4 Self-Assembled Buffer Layers 218
    8.2.5 Graphene Oxides 219
8.3 Cathode Buffer Layers (CBLs) 220
    8.3.1 Low Work Function Metals 220
    8.3.2 *n*-Type Metal Oxides 221
    8.3.3 Alkali Metal Compounds 223
    8.3.4 Organic Materials 225
        8.3.4.1 Small-Molecule Organic Materials 225
        8.3.4.2 Nonconjugated Polymers 227
        8.3.4.3 Water/Alcohol–Soluble Conjugated Polymers 228
        8.3.4.4 Fullerene Derivatives 228
    8.3.5 Self-Assembled Monolayers (SAMs) 230
    8.3.6 Graphenes 231
8.4 Summary and Outlook 232
Acknowledgments 233
References 233

8.1 INTRODUCTION

Polymer solar cells (PSCs) have been attracting considerable attention as a promising renewable energy source owing to their unique advantage of light weight, low cost, high flexibility, and easy roll-to-roll fabrication.[1-4] Since Yu et al. first reported a poly[2-methoxy-5-(2-ethylhexyloxy)-1,4-phenylenevinylene (MEH-PPV):[6,6]-phenyl-C61-butyric acid methyl ester (PC61BM) bulk heterojunction (BHJ) PSC structure in 1995,[5] significant progress in BHJ-PSC technology has been achieved.[5] During the past decade, the power conversion efficiency (PCE) of the champion small-area PSCs has been dramatically improved to 10.6%.[6-11] For PSCs, BHJ structure plays a determinative role in all of the major photoelectric conversion processes, including charge separation, which relies on effective donor/acceptor interfaces; charge transport, which is critically determined by the morphology of the thin film; and charge extraction, which can only occur at high-quality interfaces between donor(acceptor)/electrodes. In order to improve PCE of BHJ-PSCs so as to meet the requirement for commercial applications, a balanced consideration of photocurrent, photovoltage, and fill factor (FF), which are determined by the photoactive materials[12,13] and device architectures,[14-16] needs to be implemented.

Despite the recent advances in synthesizing novel conjugated low bandgap polymer donors with deep-lying highest occupied molecular orbital (HOMO) energies and fullerene acceptors leading to appreciably high PCE,[1-4,6,7,9,13,15] optimization of the device structure, particularly the interfaces between donor(acceptor)/electrodes, has been demonstrated to be so crucial that such interfaces play a determinative role on efficient charge transport and extraction.[1-4,17,18] Numerous studies of various donor(acceptor)/electrode interfaces demonstrate that a wide range of interfacial materials are effective in enhancing the PCE of BHJ-PSC devices, including organic molecules, inorganic dielectrics, and composites of organics with inorganic dopants etc.[17] Incorporation of such interfacial layers benefits the efficiency enhancement of PSCs because they can act as either a dedicated electron transporting layer or an exciton blocking layer, which allows the selective transport of only one type of charge carrier (hole or electron) but prevents the other one from reaching the electrode and quenching charge injection.[18-20] Besides, the interfacial layer generally plays the role of an isolating buffer layer capable of prohibiting the chemical reaction between the active layer and metal electrode as well as the diffusion of metal ions into the active layer.[19,21,22]

In this chapter, we review the recent state of the interfacial materials incorporated as buffer layers at the donor(acceptor)/electrode interfaces toward efficiency enhancement of PSCs, including both anode buffer layers (ABLs) sandwiched between the active layer/anode and cathode buffer layers (CBLs) applied between the active layer/cathode. The crucial role of the interfacial materials on the photovoltaic parameters of PSC devices with both conventional and inverted structures is discussed, and the correlation between the molecular structure of the interfacial materials and enhancement performance is emphasized.

8.2 ANODE BUFFER LAYERS (ABLs)

The typical BHJ-PSC device structure includes a transparent electrode based on indium tin oxide (ITO) and an ABL aiming to reduce the roughness of ITO and obtain an efficient hole extraction from the donors with deep-lying HOMO levels within the active layer. The role of ABL is thus to improve the anode efficiency in collecting and extracting holes.

The work function of ITO is usually reported at around 4.7 eV, which is aligned neither with the lowest unoccupied molecular orbital (LUMO) level of acceptor fullerenes and fullerene derivatives, nor with the HOMO level of most common donor polymers.[5,23,24] It has been reported that the Ohmic contact of the ITO/donor or ITO/acceptor interface is a critical factor for high-efficiency PSCs, affecting sensitively the maximum attainable open circuit voltage (V_{oc}) in conventional and inverted solar cells.[24] The maximum V_{oc} is correlated with the energy offset between the HOMO of the donor and the LUMO of the acceptor[25] in the case of Ohmic contacts.[20] In fact, both the efficiency and stability of a BHJ-PSC device have been shown to be strongly affected by the extraction of charge carriers at the donor(acceptor)/electrode interfaces.[20–22,26] On the other hand, ITO has no selectivity in collecting positive and negative carriers and cannot provide a required selectivity of carriers due to its high work function relative to the energy levels of the active materials. Thus, an ideal ABL incorporated between the active layer and ITO anode for the conventional-structure BHJ-PSCs should meet the following requirements: (1) high hole transporting mobility; (2) easy deposition onto anode surfaces via straightforward methods, such as spin coating, vapor deposition, or self-assembly; (3) good conformal matching to the substrate; (4) facile thickness control; and (5) being transparent with a well-defined microstructure free of pinhole defects.[17,27]

8.2.1 Poly(3,4-ethylenedioxithiophene):poly(styrene sulfonate) (PEDOT:PSS)

Poly(3,4-ethylenedioxithiophene) (PEDOT) doped with poly(styrene sulfonate) (PSS) (see Scheme 8.1 for the chemical structure) was first applied as an ABL to improve the performance of PSC in the late 1990s,[28] following the success in the similar application in the field of organic light-emitting diodes (OLEDs).[29] It was revealed that the incorporation of PEDOT:PSS as ABL was advantageous for the photodiode device because it improved the hole collection properties of the ITO anode.[28]

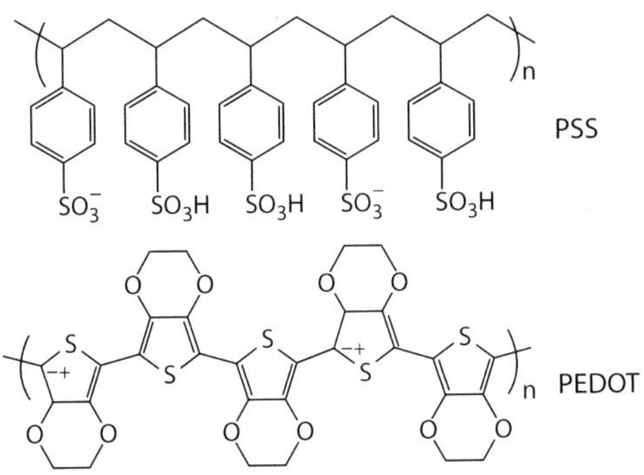

SCHEME 8.1 Chemical structure of PEDOT:PSS.

PEDOT is insoluble in most solvents but can be dispersed in water by using PSS as a counter ion, which also serves as an excellent oxidizing agent, as a charge compensator, and as a template for polymerization.[30] Different grades of PEDOT:PSS dispersions with different electric conductivities are now commercially available for low toxicity, antistatic coating, hole injection layer, and transparent electrode applications from H. C. Starck (Leverkusen, Germany).[18] A comparison of different poly(3-hexylthiophene-2,5-diyl) (P3HT):PC61BM BHJ-PSCs incorporating different types of PEDOT:PSS ABLs is summarized in Table 8.1, revealing the contribution of its conductivity to the short-circuit current (J_{sc}) and fill factor (FF).

PEDOT:PSS films have high transparency in the visible range, high mechanical flexibility, and excellent thermal stability and can be fabricated through conventional solution processing. PEDOT:PSS films can also be easily nanostructured to enhance the localized light intensity to the active layer and generate more power.[31] The other advantages of PEDOT:PSS ABL are its high work function (usually reported between 4.8 and 5.2 eV),[32] allowing the formation of an Ohmic contact with the most common donor polymers and its efficiency in transporting holes and blocking electrons to the anode.[33,34] Owing to its enormous advantages, PEDOT:PSS has been nowadays the most widely used ABL in conventional-structure BHJ-PSCs.

Considering the limited conductivity of PEDOT:PSS, the thickness-dependent conductivity is expected for the PEDOT:PSS ABL, and accordingly, the performance of PSC devices can be changed by varying the thickness of the PEDOT:PSS layer. A couple of studies have been reported concerning the thickness effect of the PEDOT:PSS layer on the performance of PSCs.[26,32,35–38] For instance, Kim et al. examined a broad thickness range (3–625 nm) of PEDOT:PSS ABLs in order to understand its thickness effect on P3HT:PC61BM PSCs,[32] revealing that the optical transmittance was gradually decreased with the PEDOT:PSS thickness, but the device performance was noticeably improved with only a 3-nm-thick PEDOT:PSS layer, and it was almost similar in the thickness range of 30–225 nm in the presence of a gradual decrease in the surface roughness (see Figure 8.1). This phenomena

TABLE 8.1
Photovoltaic Parameters of ITO/PEDOT:PSS/P3HT:PC61BM/Al Conventional Solar Cells for a Light Power Intensity of 100 mW cm², Made by Using Different Commercially Available PEDOT:PSS Formulations with Different Electric Conductivity

PEDOT:PSS	Conductivity (S cm⁻¹)	V_{oc} (V)	J_{sc} (mA cm⁻²)	Fill Factor (%)	PCE (%)	Reference
Baytron 4083 (H. C. Starck)	~10⁻³	0.61	8.37	60	3.09	39
Baytron P (H. C. Starck)	~0.4	0.61	10.0	54	3.29	40
Baytron PH 1000 (H. C. Starck)	~1	0.61	11.42	54	3.73	41

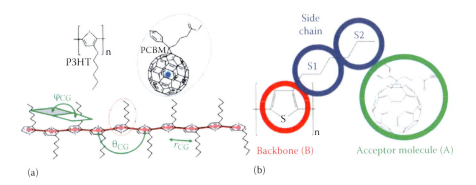

FIGURE 4.2 (a) Chemical structures of P3HT and PCBM are shown. Pink beads designate the CG representation of P3HT monomers, and a blue bead designates the CG representation of a PCBM molecule. The intramolecular degrees of freedom between the CG particles of P3HT monomers are highlighted, where r_{CG} represents bond length, θ_{CG} represents bond angle, and φ_{CG} represents planar angle. (Reproduced from C.-K. Lee et al., *Energy and Environmental Science* 2011, 4, 4124–4132, 2011, with permission from the Royal Society of Chemistry.) (b) Coarse-grained model of conjugated polymer with backbone beads (B) and side-chain beads (S1 and S2) and acceptor molecule (A) are shown. (Marsh, H. S.; Jayaraman, A., Morphological studies of blends of conjugated polymers and acceptor molecules using langevin dynamics simulations. *Journal of Polymer Science Part B-Polymer Physics* 51, 64–77, 2013. Copyright Wiley-VCH Verlag GmbH & Co. KGaA. Reproduced with permission.)

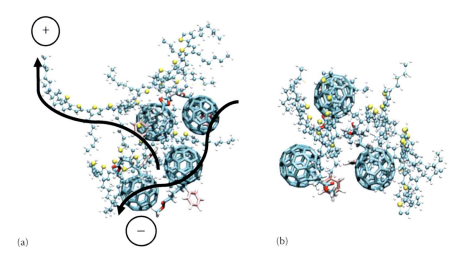

FIGURE 4.3 The molecular arrangement of P3HT and PCBM in systems of (a) 1:1 P3HT:PCBM and (b) 1:0.5 P3HT:PCBM at 310 K are shown. Although at a higher concentration of PCBM clear percolation networks are formed, it is relatively sparse at low concentrations.

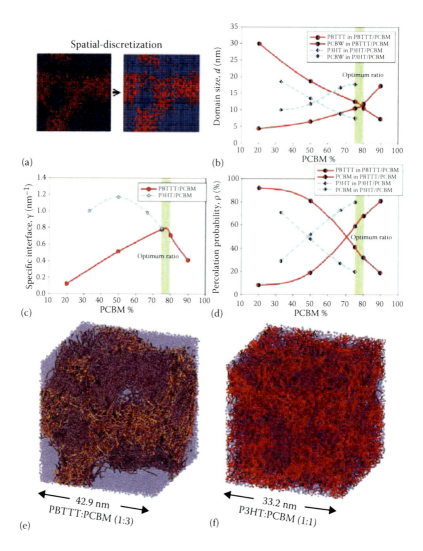

FIGURE 4.4 (a) A spatial-discretization scheme to evaluate the morphological properties of photoactive layers. Polymer and PCBM domains are colored in red and blue, respectively. (b) Domain size, (c) specific interfacial area, and (d) percolation probabilities are plotted with respect to the PCBM weight percentage of the PBTTT:PCBM blends (solid lines and symbols). The morphological properties of P3HT:PCBM blends are also plotted (dashed lines) for comparison purposes. (From Lee, C.-K. et al., *Energy and Environmental Science*, 4, 4124–4132, 2011.) The gray-shadowed areas highlight the optimal blending ratio for the PBTTT:PCBM blend from experiments. (e) Snapshots of PBTTT:PCBM (blending ratio 1:3) and (f) P3HT:PCBM (blending ratio 1:1) are shown, and PCBM molecules are transparent for clarity. (Reprinted with permission from Lee, C.-K. and Pao, C.-W., *Journal of Physical Chemistry C*, 116, 12455–12461. Copyright 2012, American Chemical Society; Hwang, I.-W. et al., *Journal of Physical Chemistry C*, 112, 7853–7857, 2008; Mayer, A. C. et al., *Advanced Functional Materials*, 19, 1173–1179, 2009; Cates, N. C. et al., *Nano Letters*, 9, 4153–4157, 2009; Rance, W. L. et al., *ACS Nano*, 5, 5635–5646, 2011.)

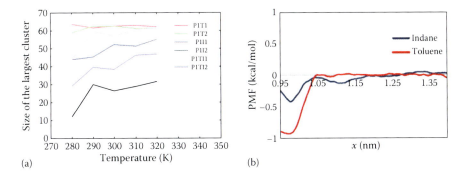

FIGURE 4.5 (a) Cluster size of PCBMs has been characterized by evaluating the size of the largest cluster with temperature at various concentrations. The notations P1T1, P1T2, P1I1, P1I2, P1TI1, and P1TI2 correspond to the simulated systems in the study of Mortuza et al.[26] in which systems with toluene, indane, and toluene-indane mixtures have the letters T, I, and TI in the names, respectively, and 1 and 2 refer to different concentrations. (b) Potential of mean force (PMF) for fullerene molecules in toluene and indane at 310 K are presented. (Reprinted with permission from S. M. Mortuza, S. Banerjee, *Journal of Chemical Physics*, 137. Copyright 2012, American Institute of Physics.)

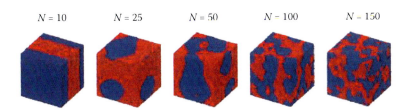

FIGURE 4.6 Effects of the P3HT degree of polymerization, N, on the morphology of P3HT:PCBM blends are presented at a weight ratio of 1:1. P3HT monomers and PCBM beads are colored in red and blue, respectively. (From J.-M. Y. Carrillo, R. Kumar, M. Goswami, B. G. Sumpter, W. M. Brown, *Physical Chemistry Chemical Physics*, 15, 17873–17882, 2013. Reproduced by permission of The Royal Society of Chemistry.)

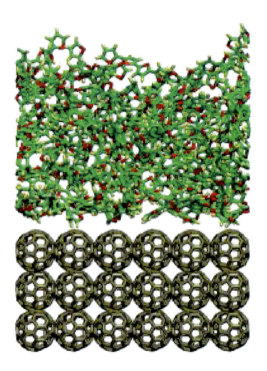

FIGURE 4.7 Molecular snapshot of the orientation of thiophene oligomers at the fullerene-oligothiophene interface is shown. Oligothiophene molecules are placed on top of fullerene molecules. (Reprinted with permission from Reddy, S. Y. and Kuppa, V. K., *Journal of Physical Chemistry C*, 116, 14873–14882. Copyright 2012, American Chemical Society.)

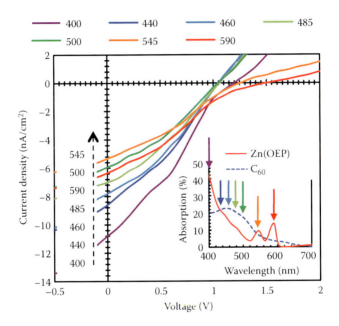

FIGURE 6.10 *J–V* characteristics of [ITO/20-nm-thick Zn(OEP)/30-nm-thick C$_{60}$/Al] cells under irradiation with monochromatic light of 400, 440, 460, 485, 500, 545, and 590 nm wavelengths. The inset shows the absorption spectra of 20-nm-thick Zn(OEP) (solid) and 30-nm-thick C$_{60}$ films.

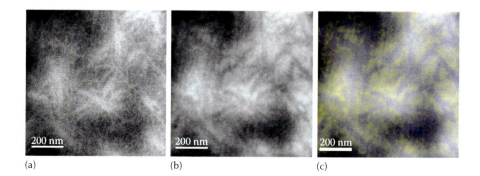

FIGURE 9.1 Low eV plasmon-based EFTEM images of P3HT/PCBM. (a) 19 eV-centered energy window highlights the P3HT-rich domains. (b) 30 eV-centered window highlights the PCBM-rich domains. (c) Artificially colored composite image shows the P3HT in blue and PCBM in yellow. (Light and dark areas correspond to sample thickness variations.)

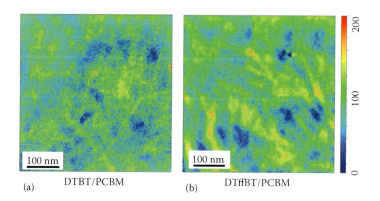

FIGURE 9.2 Sulfur maps of representative, high-performance, low-bandgap polymer (DTBT and DTffBT)-based OPV systems ([a] DTBT/PCBM, [b] DTffBT/PCBM). The details of the synthesis and molecular structure of DTBT and DTffBT polymers are reported elsewhere (Zhou et al. 2011). The higher sulfur content regions are shown with brighter colors. The corresponding thickness maps (not shown) are almost featureless.

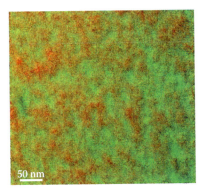

FIGURE 9.3 Combined sulfur-carbon maps of low-bandgap polymer PBDTTT-C-T/PC$_{70}$BM-based, high-performance, OPV systems. The sulfur map is shown in green and carbon in red. PCBM has a higher carbon density, corresponding to higher carbon content, and the polymer is rich in sulfur. (From P. Adhikary et al., *Nanoscale* 5(20): 10007–10013, 2013.)

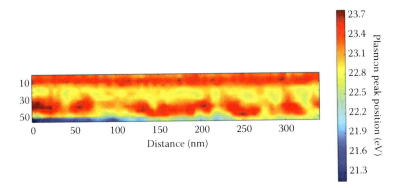

FIGURE 9.5 Bulk plasmon peak position reveals phase separation in a P3HT/PCBM BHJ cross-section. The higher energy plasmon peak (red) corresponds to the PCBM-rich phase, and the lower energy plasmon peak (yellow) corresponds to the P3HT-rich phase. A layer of platinum on top also shows a plasmon peak similar to that of PCBM. The silicon substrate shows the lowest energy plasmon peak (blue).

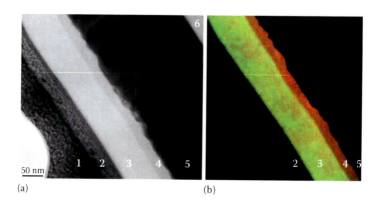

FIGURE 9.6 Elastic (0 eV) and composite energy-filtered TEM image of a high-performance OPV blend (PBDTTT-C-T/PC$_{70}$BM) in cross-sectional views. (a) The elastic image (0 eV) shows distinctive layer structures in a cross-sectional view of the OPV device (layers 1 and 2 represent ion and electron deposited Pt cap layer, 3 and 4 for active layer and PEDOT:PSS, and finally, 5 and 6 for ITO and glass). (b) Low-eV plasmon imaging of the same area with polymer-rich domains in red and fullerene-rich domains in green.

Interfacial Materials toward Efficiency Enhancement of Polymer Solar Cells 205

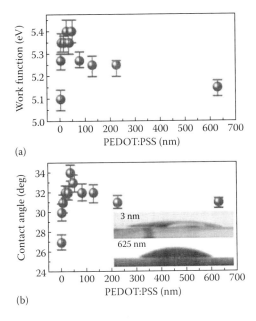

FIGURE 8.1 Work function (a) and contact angle (b) as a function of the thickness of the PEDOT:PSS layers; inset images in (b) show the water drops on the PEDOT:PSS layers. (The contact angles of the wet-cleaned ITO/glass substrates and the UV-ozone-treated ITO/glass substrates were ca. 10° and 70°, respectively.) The PEDOT:PSS layers that were coated on both glass substrates and ITO-coated glass substrates were examined, but almost similar results were obtained within given error bar ranges. (Reprinted with permission from Jeong, J. et al., *Org. Electron.*, *14* (11). Copyright 2013, Elsevier.)

was interpreted by the compensation effect between the reduced electrical resistance (increased conductivity) and the decreased optical transmittance as the thickness of the PEDOT:PSS layer increased.[32]

On the other hand, previous studies also revealed that PEDOT:PSS has some drawbacks, including (1) the extreme acidity of its dispersion (pH ~ 1–2) leads to indium loss from the ITO contact[42]; (2) it is highly hygroscopic because of the presence of water from both the dispersion and the humidity uptake in the ambient conditions[43]; and (3) the difficulty in controlling its morphology.[44] In order to improve the different commercially available types of PEDOT:PSS, many efforts have been made, focused on the treatment of PEDOT:PSS by either physical methods or chemical approaches involving chemical additives.[18]

8.2.1.1 Physical Treatments of PEDOT:PSS

Thermal treatment is the simplest physical treatment and has been widely adopted for PEDOT:PSS. In 2011, Kim et al.[45] found that PEDOT:PSS ABL incorporated between ITO and the P3HT:PC61BM active layer in the conventional BHJ-PSC devices resulted in dramatic enhancement of J_{sc} upon an annealing treatment in the 75°C–230°C temperature range for 15 min. A plausible explanation for the improved performance upon annealing treatment of PEDOT:PSS ABL was the change of its electrical conductivity induced by the change of PEDOT:PSS work function and the variation on its surface roughness due to the vertical distribution of PSS moiety.[18]

UV treatment of PEDOT:PSS has also been proved to be beneficial for the efficiency enhancement of PSCs by Lee et al. It was revealed that UV treatment did not damage PEDOT:PSS but increased its work function and, consequently, improving the Ohmic contact with the P3HT donor. The main advantage of the UV treatment is the high reproducibility and the facility of including it as a supplementary step in a continuous fabrication process. Besides, a similar treatment, UV-ozone-treated PEDOT:PSS film was used as the ABL in CuPc/C$_{60}$-based organic solar cell (OSCs).[46] The morphology of the PEDOT:PSS film was not affected by the UV-ozone treatment. However, the PCE was found to increase by about 20% compared to the reference device without UV-ozone treatment. The improved performance was attributed to the increased work function of the PEDOT:PSS layer, which increased the extraction efficiency of the photogenerated holes and decreased the recombination probability of holes and electrons in the active organic layers.

O$_2$ plasma-treated PEDOT:PSS ABLs were found to improve the hole collection as reflected by the improvement of FF in PSCs made of low-molecular-weight materials.[42,47]

8.2.1.2 Additives in PEDOT:PSS

Several different types of additives have been applied to improve the properties of PEDOT:PSS ABLs, specifically the conductivity, including (i) alcohols (mainly sorbitol, methanol, and glycerol); (ii) polar solvents, such as dimethyl sulfoxide (DMSO) and N,N-dimethylformamide (DMF); (iii) nanomaterials, such as gold nanoparticles, graphene, and metal nanoparticles; and (iv) salts, zwitterions, ionic liquids, acids, cosolvents, and surfactants.

Because the pristine PEDOT:PSS film spin-coated from the aqueous PEDOT:PSS solution usually has a conductivity lower than 1 S.cm^{-1}, an improvement of the conductivity of PEDOT:PSS is crucial for enhancing the performance of PSCs for which various methods have been investigated, for example, using salts,[48] zwitterions,[49] ionic liquids,[50] acids,[51] cosolvents,[52] and surfactants.[53] Alcohols, such as sorbitol, methanol, and glycerol, have been widely used as an additive of PEDOT:PSS to increase the conductivity of PEDOT:PSS ABL in PSCs.[43,54,55] For instance, upon doping PEDOT:PSS ABL by sorbitol, the BHJ-PSC devices based on a blend of a polyfluorene copolymer (DiO-PFDTBT) and PC61BM exhibited higher J_{sc} but lower V_{oc} compared to those with an unmodified PEDOT:PSS dispersion as ABL, and this phenomena was interpreted by the segregation of PSS on top of PEDOT.[54]

Recently, polar solvents, such as DMF, have been widely used to modify the PEDOT:PSS ABL.[56,57] DMF has the highest dipole moment (3.86 D) among common organic solvents,[58] and this can result in a large phase separation in PEDOT:PSS. DMF distributed the positively charged PEDOT segments over the entire thin film to the greatest extent and significantly increased the conductivity by increasing the probability that carriers would be transported through the conductive pathway without being recombined or trapped in the thin film. On the contrary, PEDOT:PSS thin film made from methanol solution did not show any noticeable reduction in resistivity because the dipole moment of methanol (1.68 D)[58] was not sufficiently high to change

Interfacial Materials toward Efficiency Enhancement of Polymer Solar Cells

the orientation in PEDOT:PSS.[56] The changes in the resistivity as a function of DMF content were investigated (see Figure 8.2). Even a small amount of DMF (that is, 10 wt%) lowered the resistivity of PEDOT:PSS from 3.51×10^{-1} Ω cm to 4.33×10^{-3} Ω cm. Minimum resistivities of DMF-treated PEDOT:PSS thin films were observed at 10–50 wt%. When 50 wt% DMF was used to dope PEDOT:PSS, the resistivity (4.89×10^{-3} Ω cm) of the film decreased by about 100 times compared to the pristine film (3.51×10^{-1} Ω cm). The conductive pathways were more extensive with solvents that have a large dipole moment and increase J_{sc}. In contrast, when larger amounts of DMF were used (for example, >60 wt%), the resistivity consequently increased to 2.57×10^{-2} Ω cm. In addition, the PEDOT:PSS ABL surface became more hydrophobic because the proportion of the insoluble PEDOT on the surface of the PEDOT:PSS thin film was higher when a solvent with a high dipole moment was used, following the changes in the resistivity.[56] The rough surface also increased the surface contact area between the active layer and the PEDOT:PSS ABL. As a result, the P3HT:PC61BM BHJ-PSCs incorporating PEDOT:PSS ABL doped by 50 wt% DMF had a PCE of 3.47%, much higher than that with the pristine PEDOT:PSS thin film as ABL (2.10%).[56]

In another similar report, the conductivity of PEDOT:PSS was systematically changed by polar solvent DMSO addition (up to 80% by weight) into the water dispersion, and its effect on the efficiency enhancement of P3HT:PC61BM BHJ-PSC devices was revealed.[59] It was revealed that the conductivity of PEDOT:PSS was enhanced dramatically by DMSO doping, and the maximum conductivity reached 1.25 S/cm when the doping concentration was 10 wt%, which increased about three orders of magnitude compared with that of the undoped PEDOT:PSS. Upon the incorporation of DMSO-doped PEDOT:PSS as an ABL, the BHJ-PSC device (ITO/PEDOT:PSS/P3HT:PC61BM/LiF/Al) showed an improvement of hole transport as well as an increase of J_{sc} and a decrease of series resistance, owing to

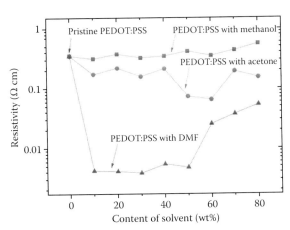

FIGURE 8.2 Resistivity of PEDOT:PSS thin films after treatment with different concentrations of various polar solvents ranging from 0 to 80 wt%. Polar solvents are methanol, acetone, and DMF. (Reprinted with permission from Yang, J. S. et al., *ACS Appl Mater Inter,* 4, 10, 5394–5398. Copyright 2012, American Chemical Society.)

the higher conductivity of the doped PEDOT:PSS. Consequently, the DMSO-doped PEDOT:PSS ABL improved the performance of the BHJ-PSC device with a PCE of 4.45%.[59]

More recently, tetramethylene sulfone (TMS), a solvent additive with a high boiling point and dielectric constant, was used as an additive to increase the conductivity of PEDOT:PSS ABL from 0.04 S/cm to 188.65 S/cm with 10 wt% doping ratio.[60] BHJ-PSC devices incorporating the TMS-doped PEDOT:PSS ABL were compared with those based on an unmodified PEDOT:PSS ABL, revealing higher J_{sc} and V_{oc} but lower FF.

The role of other additives, isopropanol and deionized water, in ink jet-printed PEDOT:PSS ABLs and their effect on P3HT:PC61BM BHJ-PSCs were also reported.[61] Depending on the layer deposition technique, the presence and role of additives can be determining for the processability of PEDOT:PSS dispersion.[18] To control film morphology, formation, and thickness, the optimized ink of 18:9:73 vol% (PEDOT:PSS:deionized water:isopropanol ratio) with the substrate at 30°C and a solution flow rate between 1.5 and 3.5 mL min^{-1} was determined in order to take advantage of surface tension gradients to create Marangoni flows that enhance the coverage of the substrate and reduce the roughness of the film. As a result, the highest PCE of spray-coated P3HT:PC61BM BHJ-PSC devices was 3.75% with J_{sc}, V_{oc}, and FF being 9.8 mA cm^{-2}, 550 mV, and 70%, respectively.[61]

Recently, the application of a localized surface plasmon resonance (LSPR) effect induced by addition of gold nanoparticles (Au NPs) has been attracting much attention as a means for increasing the photocurrent of PSCs via effective light trapping.[62–67] Enhanced J_{sc} and FF along with an 8% ~ 13% improvement in PCE were achieved with the incorporation of Au or Ag NPs in PEDOT:PSS ABL due to the LSPR effect.[37,40] Interestingly, PCE of the P3HT:PC61BM BHJ-PSC devices was dramatically enhanced to 3.80% with an enhancement ratio of ~16% upon the incorporation of Au@SiO$_2$ core/shell NPs (2 wt%) in PEDOT:PSS, which was higher than that obtained by incorporating bare Au NPs (~13%).[40] The Au@SiO$_2$ NP-incorporated device exhibited a much higher pure LSPR-induced enhancement on both light absorption and J_{sc}, and the good electrical-conducting property of Au NPs should be responsible for the enhancement of J_{sc} for the bare Au NP-incorporated device. The SiO$_2$ shell also provided an insulating barrier to limit the good electrical conducting properties of Au NPs and plays the role of "surfactant," fulfilling the good dispersion of Au NPs in the PEDOT:PSS layers. Moreover, upon a further DMF treatment of Au@SiO$_2$ NPs, PCE increased further to 4.22%.[40]

Hydrophilic grapheme oxide (GO) was used by Chen et al. as an additive of PEDOT:PSS in P3HT:PC61BM BHJ-PSCs, leading to an enhancement of PCE from 2.1% to 2.4% compared to a reference device made with pristine PEDOT:PSS ABL.[68] Furthermore, it was found that the preannealing treatment of GO resulted in further enhancement of PCE by 1.8 times of that of a reference cell made with pristine PEDOT:PSS ABL (see Table 8.2).[18]

Conductivity of PEDOT:PSS ABL was also found to be improved by adding a small amount of functionalized single-walled carbon nanotubes (SWCNTs) because of the facilitated carrier transport through SWCNTs.[69] Such a conductivity improvement of PEDOT:PSS ABLs contributed directly to the enhancement of PCE from 1.94% to 2.12%, which was primarily due to the improvement of both FF and J_{sc}.

TABLE 8.2
Nonexhaustive Survey of Conventional Polymer:Fullerene BHJ-PSCs, Including PEDOT:PSS Modified with Additives as ABLs

Additive	Cell Architecture	P_{in} (mW cm^{-2})	J_{sc} (mA cm^{-2})	V_{oc} (V)	FF (%)	PCE (%)	Ref.
TMS (10 wt%)	ITO/PEDOT:PSS:additive/ P3HT:PC61BM/TiO$_x$/Al	100	9.08 (5.77)	0.55 (0.53)	45 (51)	2.24 (1.48)	60
DMSO (15 wt%)	ITO/PEDOT:PSS:additive/ P3HT:PC61BM/TiO$_x$/Al	100	8.68 (5.77)	0.55 (0.53)	40 (51)	2.09 (1.48)	60
Au NPs (2 wt%)	ITO/PEDOT:PSS:additive/ P3HT:PC61BM/Al	100	10.6 (10.0)	0.62 (0.61)	56 (54)	3.69 (3.29)	40
Au@SiO$_2$ NPs (2 wt%)	ITO/PEDOT:PSS:additive/ P3HT:PC61BM/Al	100	10.6 (10.0)	0.62 (0.61)	57 (54)	3.80 (3.29)	40
Au@SiO$_2$ NPs (2 wt%) + DMF	ITO/PEDOT:PSS:additive/ P3HT:PC61BM/Al	100	11.8 (10.0)	0.63 (0.61)	57 (54)	4.22 (3.29)	40
SWCNTs	ITO/PEDOT:PSS:additive/ P3HT:PC61BM/Al	100	6.91 (6.68)	0.59 (0.61)	52 (48)	2.12 (1.94)	69
Fe$_2$O$_3$ (0.7 wt%)	ITO/PEDOT:PSS:additive/ P3HT:PC61BM/Al	100	11.1 (8.93)	0.63 (0.62)	52 (50)	3.68 (2.79)	70
Graphene oxide	ITO/PEDOT:PSS:additive/ P3HT:PC61BM/LiF/Al	100	10.2 (6.90)	0.56 (0.60)	42 (50)	2.40 (2.10)	68
Annealed GO	ITO/PEDOT:PSS:additive/ P3HT:PC61BM/LiF/Al	100	14.2 (6.90)	0.62 (0.60)	43 (50)	3.80 (2.10)	68
DMF (50 wt%)	ITO/PEDOT:PSS:additive/ P3HT:PC61BM/LiF/Al	100	12.0 (8.65)	0.60 (0.59)	48 (41)	3.47 (2.10)	56
Sorbitol (8 wt%)	ITO/PEDOT:PSS:additive/ P3HT:PC61BM/Al	100	11.27 (8.82)	0.53 (0.56)	49.1 (43)	2.93 (2.12)	55
DMSO (10 wt%)	ITO/PEDOT·PSS/P3HT: PC61BM/LiF/Al	100	11.09 (8.04)	0.63 (0.64)	63.7 (66.2)	4.45 (3.41)	59

Note: The photovoltaic parameters are compared to those measured for a reference cell (values in parentheses) in which the ABL was unmodified PEDOT:PSS.

Very recently, a solution-processed neutral ABL was developed by in situ formation of MoO$_3$ in aqueous PEDOT:PSS dispersion (MoO$_3$-PEDOT:PSS).[71] This MoO$_3$-PEDOT:PSS composite film took advantage of both the highly conductive PEDOT:PSS and the stability of MoO$_3$ in ambient conditions; consequently, it possessed a smooth surface and considerably reduced hygroscopicity. The resulting MoO$_3$-PEDOT:PSS ABL-incorporated BHJ-PSC devices based on poly[2,3-bis-(3-octyloxyphenyl)quinoxaline-5,8-diyl-alt-thiophene-2,5-diyl] (TQ1): [6,6]-phenyl-C71-butyric acid methyl ester (PC71BM) showed considerable improvement in PCE from 5.5% to 6.4% compared with the reference device made with pristine PEDOT:PSS ABL. More importantly, the device with MoO$_3$-PEDOT:PSS ABL showed considerably improved stability with the PCE remaining at 80% of its original value when stored in ambient air in the dark for 10 days, whereas the reference device made with pristine PEDOT:PSS ABL showed complete failure within 10 days.[71] Likewise, aqueous PEDOT:PSS dispersion doped by other metal oxides, such as V_2O_5[72] and Fe_2O_3 NPs,[70] was also investigated. Fe_2O_3 NPs contributed to increased light harvesting at an optimum doping concentration of 0.1 to 0.7 wt% in PEDOT:PSS aqueous dispersion, resulting in an enhanced external quantum efficiency (EQE). On the other hand, a higher concentration of Fe_2O_3 NPs led to decreased J_{sc} because Fe_2O_3 NPs hindered the hole transport. As a result, an optimized PCE of 3.68% with a 32% enhancement was obtained for the device incorporating Fe_2O_3 NP-doped PEDOT:PSS ABL.[70]

8.2.1.3 ITO-Free PSCs Based on Modified PEDOT:PSS

Given that a high conductivity of PEDOT:PSS can be achieved by additive doping as described above, a novel approach has been proposed to replace the ITO anode by modified PEDOT:PSS with high conductivity (more than 100 S cm^{-1}) as the transparent anode. Recent studies reveal that the addition of such solvents as dimethyl sulfoxide (DMSO),[73,74] sorbitol,[43,44] methanol,[75] and ethylene glycol (EG)[76] improves the conductivity of PEDOT:PSS by two to three orders of magnitude.[51,52] For instance, conductivity as high as 735 S cm^{-1} for PEDOT:PSS (PH 1000 from H. C. Starck) with the addition of 6 vol% EG solvent was obtained, and a solvent posttreatment remarkably increased the conductivity to 1418 S cm, which, however, decreased with extended posttreatment time of more than 2 hr.[45] In another study, the effect of the molecular weight/chain length of the EG and polyethylene glycol (PEG) additives on the conductivity and other properties of PEDOT:PSS film was systematically investigated.[77] The conductivity of PEDOT:PSS increased with increasing EG concentration and saturates at 6 vol% addition while the conductivity reached saturation at 2 vol% PEG concentration and even decreased afterward. The average conductivity was significantly improved from 0.3 S cm^{-1} to 805 S cm^{-1} with 2 vol% PEGs having molecular weights from 200 to 400, and PEGs with molecular weights higher than 600 still brought about conductivity enhancement. The authors suggested that the conductivity enhancement was attributed to the phase separation between PEDOT and PSS chains promoted by the additives, leading to better connected and bigger PEDOT chains. Consequently, BHJ-PSC devices based on a PEG-treated PEDOT:PSS anode showed a PCE of 3.62%, higher than that obtained by EG treatment (3.51%) and comparable to that based on an ITO counterpart (3.73%) (see Table 8.3).

Ouyang et al. used cosolvents, such as methanol/ethanol with water to treat PEDOT:PSS and succeeded in improving the conductivity of PEDOT:PSS by three orders of magnitude, and an alternative treatment of PEDOT:PSS with water or such a common organic solvent as ethanol, isopropyl alcohol, acetonitrile, acetone, or tetrahydrofuran alone did not change its conductivity remarkably. This phenomenon was interpreted by the preferential solvation of PEDOT:PSS with a cosolvent.[52] Chu et al. also found that the conductivity of PEDOT:PSS film was significantly improved by four orders of magnitude by either immersing the film in methanol or dropping a small amount of methanol onto the film or a combination of the two methods.[75]

TABLE 8.3
Photovoltaic Performances of PSCs with PEDOT:PSS Treated with Different Additives as Anode Electrodes

Anode Electrode	Cell Architecture	P_{in} (mW cm^{-2})	J_{sc} (mA cm^{-2})	V_{oc} (V)	FF (%)	PCE (%)	Ref.
PEDOT:PSS-EG 6 vol%	Glass/Anode/ P3HT:PC61BM/Ca/ Al	100	9.85 (9.38)	0.58 (0.58)	61 (68)	3.51 (3.73)	77
PEDOT:PSS-PEG200 2 vol%	Glass/Anode/ P3HT:PC61BM/Ca/ Al	100	9.80 (9.38)	0.58 (0.58)	63 (68)	3.62 (3.73)	77
PEDOT:PSS-PEG300 2 vol%	Glass/Anode/ P3HT:PC61BM/Ca/ Al	100	9.73 (9.38)	0.58 (0.58)	63 (68)	3.59 (3.73)	77
PEDOT:PSS-PEG400 2 vol%	Glass/Anode/ P3HT:PC61BM/Ca/ Al	100	9.22 (9.38)	0.58 (0.58)	64 (68)	3.43 (3.73)	77
PEDOT:PSS-PEG600 2 vol%	Glass/Anode/ P3HT:PC61BM/Ca/ Al	100	8.32 (9.38)	0.58 (0.58)	66 (68)	3.19 (3.73)	77
PEDOT:PSS/ methanol-drop	Glass/Anode/ P3HT:PC61BM/Ca/ Al	100	9.51 (9.18)	0.58 (0.58)	67 (70)	3.71 (3.77)	75
PEDOT:PSS/GMS 1 mg/mL	Glass/Anode/ P3HT:PC61BM/Al	100	12.16 (11.42)	0.61 (0.61)	53 (54)	3.9 (3.75)	41
PEDOT:PSS/GMS 1 mg/mL	Glass/Anode/ PTB7:PC71BM/Al	50	8.22 (8.24)	0.70 (0.72)	65 (67)	7.06 (7.77)	41
PEDOT:PSS/Ag nanowires	Plastic/Anode/ P3HT:PC61BM/Ca/ Al	100	9.74 (10.8)	0.61 (0.61)	52 (64)	3.8 (3.4)	36

Note: The photovoltaic parameters are compared to those measured for a reference cell (values in parentheses), in which the electrode was ITO/PEDOT:PSS without the additive.

They confirmed that PSS was washed away from the PEDOT:PSS film with methanol treatment (see Scheme 8.2). Methanol with a reasonably high dielectric constant induces a screening effect between the positively charged PEDOT chains and the negatively charged PSS chains, leading to a phase separation on the nanometer scale characterized by segregation of the excess PSS and facilitating PSS removal from the film and finally resulting in the improved conductivity of PEDOT:PSS. Furthermore, ITO-free P3HT:PC61BM BHJ-PSC devices using standalone PEDOT:PSS anodes treated by dropping 150 μL methanol solution onto the film showed a PCE of 3.71%, which is almost equal to that based on ITO anode (3.77%).

More recently, Yang and Wu et al. developed a new approach to improve the conductivity of PEDOT:PSS by formation of a PEDOT:PSS/surfactant bilayer film.

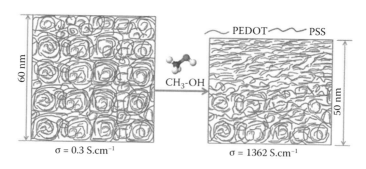

SCHEME 8.2 Schematic illustration of the mechanism of conductivity enhancement of PEDOT:PSS films by treatment with methanol. The core–shell structure is changed to a linear/coiled structure. The removal of PSS leads to the thickness reduction of the film and also brings about bigger and better-connected PEDOT chains. (Reprinted with permission from Alemu, D. et al., *Energy Environ. Sci.*, 5 (11). Copyright 2012, Royal Society of Chemistry.)

A surfactant layer, glycerol monostearate (GMS), was spin coated onto PEDOT:PSS film, leading to a remarkable improvement of the conductivity of PEDOT:PSS film from ~1 S.cm^{-1} to 1019 S.cm^{-1}. The conductivity improvement of PEDOT:PSS by GMS modification resulted from the GMS-induced segregation of PSS chains and the conformational change of the conductive PEDOT chains within PEDOT:PSS. Although the highly hydrophobic $-(CH_2)_{16}CH_3$ groups of GMS preferentially interacted with the hydrophobic PEDOT of PEDOT:PSS, the highly hydrophilic $-COOCH_2-CHOH-CH_2OH$ groups preferentially interacted with the hydrophilic PSS chains with the hydrozyl groups playing an important role in the consequent phase separation between the PEDOT and PSS chains. Using the PEDOT:PSS/GMS bilayer films as the transparent anodes replacing ITO, high-efficiency ITO-free BHJ-PSC devices based on a thieno[3,4-b]-thiophene/benzodithiophene (PTB7):PC71BM system exhibited a PCE of 7.06%, which was the highest one for ITO-free BHJ-PSC devices reported to date. Furthermore, the universality of the PEDOT:PSS/GMS bilayer anode for BHJ-PSC devices was confirmed by employing it in versatile active layer systems, for which comparable PCEs to the corresponding devices, based on the traditional ITO anode, were obtained in all cases, demonstrating its potential in the high-efficiency ITO-free BHJ-PSCs.[41]

An organic–inorganic composite embedding Ag nanowires into PEDOT:PSS was also applied as effective ITO replacement for ITO-free PSCs, resulting in high-quality devices with PCE being on par with or even better than those based on ITO electrodes due to a precise control of the nanoscale morphology of the composites.[36] The optimum PEDOT:PSS/Ag nanowires composites contributed to PCE of 4.2% and 3.8% for P3HT:PC61BM BHJ-PSC devices on glass and plastic substrate, respectively. Noteworthy is that the latter device showed an even higher PCE than that fabricated on ITO on plastic (3.4%), indicating its promise of high-efficiency, flexible, roll-to-roll PSCs. In particular, among the photovoltaic parameters, FF was revealed to be the most enhanced parameter on plastic, and the enhancement was attributed to the lower composite sheet resistance and the increased effective contact interfacial area with the active layer.[36]

8.2.2 Metal Oxides

Semiconducting metal oxides have been widely used to improve device performance as ABL substituting the PEDOT:PSS layer in both conventional and inverted PSCs. Numerous reports demonstrate that the devices with a metal oxide ABL showed performances similar to or better than devices with a PEDOT:PSS ABL. Such metal oxides as molybdenum oxide (MoO_3/MoO_x),[78] tungsten oxide (WO_3/WO_x),[79] and vanadium oxide (V_2O_5/VO_x),[80–82] etc., as n-type semiconductors have work functions exceeding 6 eV and could behave as strong acceptors with organic materials (see Figure 8.3). When they are used as anode interlayers, Fermi-level pinning occurs, and the built-in potential increases.[19] For instance, a P3HT:PC61BM BHJ-PSC device with a 5-nm-thick MoO_3 film as ABL showed the highest PCE of 3.3%, which was slightly higher than that with a PEDO:PSS ABL (3.18%).[83] On the other hand, p-type metal oxides, such as NiO, with large work functions also allow a good matching between the valence band and the HOMO level of most photoactive polymers and thus can be applied as ABLs, too.[33,84–86]

In terms of the fabrication technique, semiconducting metal oxides were typically deposited via vacuum evaporation in early studies, and the details can be found in previous dedicated reviews by So and Meyer et al.[87,88] For this technique, however, the metal oxide ABLs require relatively high-temperature treatment for getting good performance, rendering it incompatible with polymer substrates for future roll-to-roll manufacturing.[89] Therefore, a solution-processable technique with low-temperature annealing for incorporating metal oxide ABLs is in great demand for high performance PSCs, and recently fabrication of MoO_3, WO_3, V_2O_5, or NiO_x ABLs via

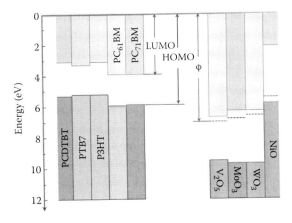

FIGURE 8.3 The energy level diagram of the state-of-the-art photovoltaic polymers, electron-accepting fullerene derivatives, and transition metal oxides. (Reprinted with permission from Chen, S. et al., *J. Mater. Chem.*, 22 (46). Copyright 2012, Royal Society of Chemistry.)

low-temperature solution-processable techniques have succeeded in several studies, which are reviewed in the following section.[78–82,84]

Recently, Choy et al. developed a simple one-step method to synthesize low-temperature solution-processed MoO_3^- and $V_2O_5^-$ with oxygen vacancies as n-dopants for good ABL performance.[90] Sub-bandgap states could be generated by the oxygen vacancies as indicated by the fluctuated features, which may push the Fermi level close to the conduction band in the MoO_3^- and $V_2O_5^-$ films. Using MoO_3^- and $V_2O_5^-$ as ABLs, PCE of PBDTTT-C-T:PC71BM BHJ-PSC devices improved to 7.75% and 7.62%, respectively, which were better than that of the control devices using PEDOT:PSS as ABL (7.24%). Noteworthy, is that the oxygen level of MoO_3^- and $V_2O_5^-$ can be controlled by this method for achieving high-performance PSCs.[91]

Li et al. prepared CuO_x ABL through a low-temperature and inexpensive method of spin coating a copper acetylacetonate ($Cu[acac]_2$) solution in 1,2-dichlorobenzene onto ITO followed by an annealing treatment at 80°C. The as-prepared CuO_x ABL showed high light transmittance and an excellent hole extraction property due to its high work function. Upon incorporating CuO_x as ABL, an Ohmic contact formed, and the resistance of the entire device decreased, and this benefited the enhancement of PCE by reducing the separated charge recombination losses. The BHJ-PSC devices incorporating CuO_x ABL exhibited enhanced performance over the PEDOT:PSS-based devices, approaching as high as 7.14% for the device based on a PBDTTT-C:PC71BM active layer.[92] In another work by the same group, a nickel acetate (NiAc) layer was prepared as ABL by spin coating its aqueous solution and then it was thermally annealed at 150°C for 30 min.[89] The resultant NiAc layer was highly transparent in the visible range and showed an effective hole collection property. The PCE of the P3HT:ICBA BHJ-PSC device with NiAc ABL reached 6.08%, which was among the best efficiencies in the P3HT-based PSCs.[89]

Similarly to MoO_3, molybdenum sulfides (MoS_x) were fabricated using a solution-processed method followed by thermal decomposition of $(NH_4)_2MoS_4$.[93] The calculated S/Mo ratios were 3.0, 2.85, 2.33, and 1.80 for the MoS_x film annealed at 150°C, 200°C, 300°C, and 400°C, respectively. This trend indicated that, with increasing temperature, MoS_3 in MoS_x films slowly decreased and finally converted to MoS_2 at 400°C completely. Both MoS_3 and MoS_2 were found beneficial to the device performance as ABLs due to the low-lying valence band (from 5.64 eV at 150°C to 5.81 eV at 400°C). Although the carrier collection of MoS_x processed at high temperature was improved, the chance of recombination at the interface was also increased because of the lower conduction band (CB), which was a reason for the lower V_{oc} of the device with MoS_x processed at high temperature. Thus, MoS_3 could result in higher V_{oc}, and MoS_2 could lead to higher J_{sc}. When the components of MoS_3 and MoS_2 reached a relatively appropriate balance (300°C), both V_{oc} and J_{sc} were improved, and the highest PCE of 3.90% for P3HT:PC61BM BHJ-PSC devices was obtained.[93]

In addition to metal oxides and sulfides, a metal halide BiI_3 was also reported very recently to function as ABL in PSCs. Chu et al. reported that the P3HT:PC61BM BHJ-PSC devices incorporating a novel solution-processable BiI_3 nanosheets as ABL, which were prepared by hydrothermal processing, followed by a simple and cost-effective wet milling method, exhibited a PCE of 3.5%, comparable to that based on PEDOT:PSS ABL.[94]

8.2.3 POLYMERS AND SMALL-MOLECULE ORGANIC MATERIALS

Many organic materials have been successfully applied as ABLs in PSCs, including both polymers[95–100] and small-molecule organic materials, which are in some cases in combination with PEDOT:PSS. These include PEDOT:PSS/pentacene,[95] PEDOT:PSS/acid-doped poly(aniline) nanotubes (a-PANIN),[101,102] PEDOT:PSS/poly[(9,9-dioctylfluorene)-co-N-(4-(1-methylpropyl) phenyl) diphenylamine] (PFT):4,4′-bis[(p-trichlorosilylpropylphenyl) phenylamino]biphenyl (TSPP),[103] copper phthalocyanine (CuPc),[104] Poly(styrenesulfonic acid)-graft-poly(aniline) (PSSA-g-PANI),[105] PFT:TSPP,[106] PFT: 5,5′-bis[(p-trichlorosilylpropylphenyl) phenylamino]-2,2′-bithiophene (TSPT),[107] etc.

One of the most popular polymeric ABLs is poly(aniline) (PANI) in different forms, which was used as a sole ABL or in a combination with PEDOT:PSS.[96–100] Recently, Ecker et al. reported that PANI blended with PSS incorporated as ABL into PSCs enhanced hole injection from low work function metals, such as Al (work function of 4.3 eV), and consequently can be used to establish high-efficiency and air-stable PSCs.[96] In addition, Chen et al. showed that efficient hole injection can be realized with a high work function metal, for example, Au (5.2 eV), due to the presence of a strong interface dipole.[97] In a more recent study reported by the same group, they employed PANI as an ABL in an inverted poly[(4,40-bis(3-(2-ethylhexyl)dithieno[3,2-b:30-d]silole)-2,6-diyl-alt-(2,5-(3-(2-ethylhexyl)-thiophene-2-yl) thiazolo[5,4-d]thiazole)]:indene-C60 bisadduct (PSEHTT:ICBA) solar cell.[98] PANI ABL can block electrons and transport holes to the appropriate electrode due to the higher LUMO of PANI (2.6 eV) relative to that of ICBA (3.4 eV) and the same HOMO (5.1 eV) of PANI and ICBA. The device performance as a function of PANI layer thickness was investigated, revealing that a 18 nm thickness was optimum for the PANI layer in a transparent flexible PSC. The optimal device showed a J_{sc} of 11.60 mA cm^{-2}, a V_{oc} of 0.89 V, a FF of 66.87%, and thus, a PCE of 6.87%, which was improved by 14.5% over their previously reported record efficiency of 6%.[99] Kim et al. added a self-assembled structure composed of terephthalic acid and long alkyl chains to PANI for inducing a high crystallinity of PANI film through the improved orientation of extended PANI chains.[90] The electrical and optical properties of blended PANI film were improved compared with the conventional PANI film. The optimized flexible PSCs fabricated with the modified PANI film as the anode and P3HT:PC61BM as the active layer showed a PCE of 1.02%.[100]

A new class of conductive polyelectrolyte films with tunable work function and hydrophobicity based on a copolymer of ethylenedioxythiophene (EDOT) was developed by Nakamura et al. as a novel ABL in PSC devices.[108] Several advantages were claimed for this copolymerization strategy: (i) The polymers were more hydrophobic, making film formation from aqueous solution possible; (ii) the variation of the copolymer ratio allowed easy control of the distribution of the counteranions along the polymer backbone; and (iii) the use of monomers with a variety of functional groups can influence the surface energy of the films. As a result, the work function of these films was found to be easily tunable over a range of almost 1 eV and reached values as high as those of PEDOT:PSS.[109,110] The new ABL materials did not need the addition of any insulating or acidic material that might limit the film conductivity or device lifetime. BHJ-PSC devices built with these ABLs showed improved V_{oc} over those of the known PSS-free conductive EDOT-based polymers with values as high as that obtained for PEDOT:PSS.[108]

Tris-(8-hydroxyquinoline)aluminum (Alq3) and its derivatives, which were commonly used as light-emitting materials in OLEDs,[111–116] were recently applied also as ABLs in inverted P3HT:PC61BM BHJ-PSCs.[117,118] Because of the positions of the HOMO levels of P3HT and Alq3, an extraction barrier for the photogenerated holes to escape the device was created. To reduce the height of such a barrier, the position of the HOMO level of the Alq3 can be elevated by attaching different substituents on the 8-hydroxyquinoline ligand. Three new Alq3 complexes with electron-donating amino substituents were synthesized successfully, and the performed spectroscopic and electrochemical characterizations confirmed that 5-amino substitution of the hydroxyquinoline ligand was directly correlated with the position of HOMO levels in the complexes while the LUMO levels remained unaffected. Although the complexes exhibited extremely low emission properties compared to the parent Alq3, they performed nicely as ABLs in the inverted P3HT:PC61BM BHJ-PSC devices with improved FF and PCE.[119]

Very recently, Ma et al. designed and synthesized a series of small-molecule organic materials, TPDA, TPDB, and TPDH, which comprise the same backbone TPD (N,N′-diphenyl-N,N′-bis(3-methylphenyl)-(1,1′-biphenyl)-4,4′-diamine) and different carboxyl side chains (see Scheme 8.3) as new ABLs in PSCs.[120] Because of the introduction of the weakly acidic carboxyl groups, TPDA, TPDB, and TPDH exhibited improved solubility in polar solvents. Besides, TPDA, TPDB, and TPDH showed

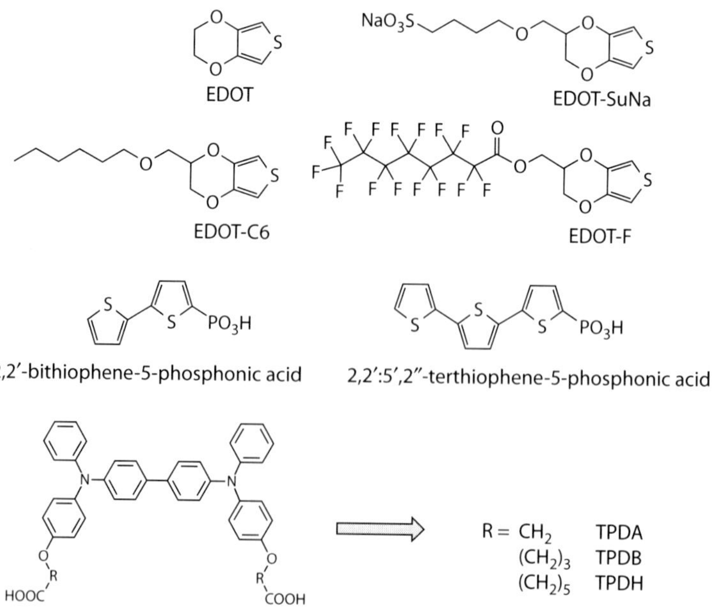

SCHEME 8.3 Molecular structures of selected organic materials used as ABLs.

higher transmittance in the visible range compared to the conventional PEDOT:PSS ABL. More importantly, absorption and UPS spectroscopic studies confirmed the desirable energy-level alignment for efficient hole-transporting and electron-blocking ability of TPDA, TPDB, and TPDH. As a result, a highest PCE of 6.51% was obtained by a TPDB ABL-incorporated device based on a poly(benzo[1,2-b:4,5-b']-dithiophenethieno[3,4-c]pyrrole-4,6-dione) (PBDTTPD):PC61BM active layer, which showed ~15% enhancement over the control devices with the conventional PEDOT:PSS ABL. Moreover, a better device stability was also achieved by using TPDB ABL (Table 8.4).[120]

TABLE 8.4
Nonexhaustive Survey of Other Organic Materials as ABLs in Conventional Solar Cells

ABL	Thickness (nm)	Cell Architecture	P_{in} (mW cm^{-2})	J_{sc} (mA cm^{-2})	V_{oc} (V)	FF (%)	PCE (%)	Ref.
Polyaniline (PANI)	18	ITO/CBL/PSEHTT:ICBA/ABL/Al	100	11.6	0.89	67	6.87	98
Alq3	4.3	ITO/CBL/P3HT:PC61BM/ABL/Al	50	2.91	0.51	60	2.36	119
Al(5-pipq)$_3$	3.66	ITO/CBL/P3HT:PC61BM/ABL/Al	50	3.20	0.48	64	2.63	119
PEDOT:PSS/pentacene	3–5	ITO/ABL/P3HTV/C$_{60}$/BCP/Al	100	3.42 (5.25)	0.63 (0.26)	55 (53)	1.19 (0.73)	95
CuPc	10	ITO/ABL/P3OT:C$_{60}$/BCP/LiF/Al	1.3	0.17 (0.17)	0.46 (0.40)	35 (37)	2.1 (1.8)	104
PEDOT:PSS/a-PANIN	30/50	ITO/ABL/P3HT:PCBM/Al	100	10.96 (8.83)	0.64 (0.64)	60 (55)	4.26 (3.39)	101
PEDOT:PSS/a-PANIN	40/50	ITO/ABL/P3HT:PCBM/Al	100	8.43 (7.33)	0.57 (0.56)	58 (53)	2.78 (2.17)	102
PFT:TSPT	25	ITO/ABL/P3HT:PCBM/LiF/Al	100	9.31 (8.04)	0.54 (0.52)	63 (64)	3.14 (2.69)	107
PEDOT:PSS/PFT:TSPP	—	ITO/ABL/MDMO-PPV:PCBM/LiF/Al	100	4.97 (5.80)	0.91 (0.47)	37 (41)	1.67 (0.74)	103
PFT:TSPP	10	ITO/ABL/MDMO-PPV:PCBM/Al	100	4.62 (3.75)	0.89 (0.47)	54 (41)	2.23 (0.73)	106
PSSA-g-PANI	40	ITO/ABL/P3HT:PCBM/Al	100	10.9 (9.7)	0.59 (0.56)	62 (61)	3.99 (3.31)	105

Note: The photovoltaic parameters are compared to those measured for a reference cell (values in parentheses), when available, in which the ABL was PEDOT:PSS.

8.2.4 Self-Assembled Buffer Layers

Many approaches, such as oxygen-plasma etching, UV-ozone treatments[121] and self-assembed monolayers (SAMs) of dipolar molecules,[122] have been developed to improve the properties of ITO substrates. As one of the most promising methods, SAM-modified ITO electrodes have also been employed for PSCs, for which a less air-sensitive buffer layer was inserted between the active layer and ITO electrode in order to reduce the amount of degradation and the amount of oxygen and moisture diffusion inside the active layer.[123]

Cho et al. modified ITO substrate using SAMs with different terminal groups, −CH3, −CF3, −NH2, which were based on *N*-propyltriethoxysilane, trichloro(3,3,3-trifluoropropylsilane), and aminopropyl triethoxysilane, respectively, and managed to control the hole injection barrier of the ITO closer to the HOMO energy level of the P3HT:PC61BM active layer and the surface energy of the ITO substrate. The authors found that SAMs with electron-withdrawing groups (−CF$_3$) increased the work function of the ITO/active layer interface (from 4.7 to 5.16 eV) more than SAMs with electron-donating groups (−CH3), and the phase separation of the P3HT:PC61BM active layer was a more important factor than the matching between the injection barrier and the active layer's HOMO level. Under the optimized condition, the device

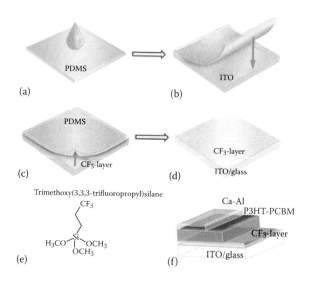

SCHEME 8.4 Schematic representation of the formation of self-assembled multilayers by the soft-imprinting method. (a) The CF$_3$–silane solution was deposited on the surface of the PDMS film. (b) The CF$_3$–silane-coated PDMS film was inversely attached onto the ITO surface. (c) The top transparent PDMS film is peeled off from the ITO substrate. (d) After the removal of the stamp, the self-assembled multilayers of CF$_3$ were formed onto the ITO surface. (e) Chemical structure of the CF$_3$–silane material. (f) The polymer solar cell device structure consisting of ITO/CF$_3$-layer/P3HT:PCBM/Ca/Al. (Reprinted with permission from Huang, L. C. et al., *Soft Matter*, 8 (5). Copyright 2012, Royal Society of Chemistry.)

based on the CF3 SAM-treated ITO afforded the highest PCE of 3.15%, which was comparable to that with PEDOT:PSS ABL.[80–82,124]

Chao et al. employed self-assembled multilayers with terminal $-CF_3$ groups to modify the surface of ITO substrates by the soft-imprinting method (see Scheme 8.4) and investigated the effect of such treatments on the morphology of the P3HT:PC61BM active layer. The CF_3-layer-modified ITO exhibited a work function of 5.12 eV, which was 0.42 eV higher than that of the plasma-treated ITO and close to the HOMO level of P3HT (−5.2 eV). Thus, the hole injection barrier from the active layer to the ITO anode was reduced. Accordingly, the PCE of the P3HT:PC61BM BHJ-PSC device in the absence of a PEDOT:PSS layer was 3.42% for a CF_3-silane-treated ITO anode, which is higher than that of the corresponding device with PEDOT:PSS ABL (3.18%). The enhanced PCE resulting from CF_3-silane modification was attributed to the increase of both J_{sc} and FF, and a significant decrease in series resistance was also observed in the device.[125]

8.2.5 GRAPHENE OXIDES

Graphene is considered to be a potential candidate for application as an electrode in various organic electronic devices because of its high electrical and thermal conductivities, transparency, and flexibility.[126–130] However, fine patterning of graphene is not easy because of its small thickness and high sensitivity to contamination from residue of photoresists.[131–134] Nevertheless, upon oxidation forming GO, cyclohexane-like units decorated with hydroxyl, ether, carbonyl, and carboxyl groups would be attached onto the graphene planar.[135] These functional groups enable water dispersibility of GO sheets, allowing them to be facilely spin coated on the patterned ITO surface. In the past few years, comprehensive works have been done on the application of GO in organic electronic devices.[136] In particular, its application as an efficient ABL in BHJ-PSCs leading to enhancement of both efficiency and stability has been extensively reported.[137–141] In 2011, Na et al. reported that, by replacing the PEDOT:PSS ABL with a GO layer modified by p-toluenesulfonyl hydrazide, the P3HT:PC61BM BHJ-PSC device exhibited not only a comparable PCE to that based on PEDOT:PSS ABL, but also a considerably enhanced shelf lifetime.[135] In a further study by the same group, the application of a moderately reduced GO prepared by a simple and fast thermal treatment of solution-processed GO as an efficient ABL in P3HT:PC61BM BHJ-PSC device was reported, and the device based on a reduced GO thermally treated at ~250°C exhibited a highest PCE of 3.98%, quite comparable to that based on conventional PEDOT:PSS ABL (3.85%). The enhancement was attributed to the better conducting property of moderately reduced GO than that of the insulating GO.[142] Furthermore, the BHJ-PSC device with thermally treated GO showed superior stability compared to that based on conventional PEDOT:PSS ABL under the atmospheric condition without any encapsulation process.[142]

The dramatic improvement in the stability of the BHJ-PSC device with a GO ABL compared to that with a PEDOT:PSS ABL was also reported by several other groups.[137,138,140,143,144] For instance, Lee et al. found that PCE of the BHJ-PSC device fabricated on a GO-buffered ITO substrate can maintain 84% of its initial value after 132 days of operation-storage cycles whereas PCE based on bare ITO substrate dropped below 1/1000 of its original value after 60 days.[139]

Application of GO as the ABL in inverted-structure BHJ-PSCs was investigated as well.[141] Yusoff et al. incorporated two different solution-processed aluminum-doped zinc oxide (AZO) and trilayer GO as CBL and ABL, respectively, in inverted PCDTBT:PC71BM PSC devices.[140] The fabricated inverted PSC devices showed superior performance compared to the conventional devices, with a PCE being higher than 5%, while that based on the conventional PSCs has a PCE of around 2.81% only.

Recently, Chao et al. found that a composite of solution-processed GO and vanadium oxide (GO/VO$_x$) as ABL exhibited a significant enhancement in their electron-blocking properties and sol–gel-precursor blocking abilities, compared to the buffer layer of VO$_x$ only. As a result, the GO/VO$_x$ composite afforded a high V_{oc} and FF as well as a PCE of 6.7% for the inverted poly{(5,6-difluorobenzo-2,1,3-thiadiazole-4,7-diyl)-alt-(3′,4″-di-(2-octyldodecyl)-2, 2′;5′,2″;5″,2‴-quaterthiophene-5,5‴-diyl)} (PTh4FBT)):PC71BM BHJ-PSC device.[144]

8.3 CATHODE BUFFER LAYERS (CBLS)

For a given polymer:fullerene-derivative conventional-structure BHJ-PSC device, generally there is an energy level offset between the work function of the cathode (typically 4.3 eV for Al cathode) and the LUMO level of the fullerene-derivative acceptor (for example, −4.0 eV for PC61BM), resulting in unfavorable electron extraction due to charge accumulation and, consequently, a recombination loss of charge carriers. Therefore, to decrease such an energy level offset is of high importance for the optimization of the electrical contact between the active layer and the cathode. CBLs incorporated between the active layer and the cathode have been extensively investigated in recent years and found to improve the cathode efficiency in electron collection and extraction effectively. Following the pioneering work applying the low work function metal Ca as an efficient CBL,[145] so far, the reported CBL materials include other low work function metals, such as Ba;[146] n-type metal oxides, such as TiO$_x$ and ZnO; alkali metal compounds, such as LiF; organic materials, including small-molecule organic materials; nonconjugated polymers and water/alcohol–soluble conjugated polymers; SAMs; and graphenes.[6–11] For these versatile CBL materials, the common requirements are threefold besides the stability: (1) providing an Ohmic contact with the acceptor material, such as PC61BM; (2) transporting electrons efficiently; and (3) blocking holes.[18–20,147–151]

8.3.1 Low Work Function Metals

Stimulated by the wide application of the low work function metals, such as calcium (Ca) as the interfacial materials for polymeric light-emitting diodes (PLEDs), Yang et al. first applied Ca as a CBL inserted between the Al cathode and organic active layer and improved successfully the device performance via the formation of ideal Ohmic contact at the interface as confirmed by the increase of V_{oc}.[152] Brabec et al. studied systematically the influence of the work function of a cathode on V_{oc} of the poly(2-methoxy-5-(3′,7′-dimethyloctyloxy)-1,4-phenylenevinylene) (MDMO-PPV):PC61BM BHJ-PSC devices and claimed that a Fermi level-pinning between the cathode and the fullerene reduction potential via charged interfacial states would occur, accounting for the observation that a variation of the cathode work function influenced the open circuit voltage in only a minor way.[152] In a more recent study by Chen and Wu et al., ultraviolet and x-ray photoemission spectroscopy (XPS and UPS) results revealed that chemical reactions occurred at the P3HT/Ca interface, and a 0.8 eV downward shift of the HOMO level of P3HT was observed while the energy levels of PC61BM remained unchanged, leading to the increase of the energy level difference between the HOMO of P3HT and LUMO of PC61BM and consequently the increase of V_{oc}. The authors further claimed that the better contact properties with Ca cathodes reduced the series resistance and leakage current, contributing to the increase of FF.[153] However, the very low work function of Ca (2.9 eV) enables Ca CBL quite reactive, and thus the Ca-incorporated PSC devices are sensitive to the moisture in the ambient atmosphere and suffer from poor stability.[154]

Very recently, Heeger et al. introduced a 10–20 nm thick barium (Ba) CBL in a small-molecule BHJ solar cells composed by p-DTS(FBTTh$_2$)$_2$:PC71BM active layer, and increased the FF up to 75.1%, which is one of the highest values reported for an organic solar cell, and the PCE reached 8.6% with an increase of 46% compared to the device without Ba CBL. The role of Ba was proposed to prevent trap-assisted Shockley-Read-Hall (SRH) recombination at the interface and to increases in shunt resistance with the decreases in the series resistance. These resulted in an increase in the charge collection probability leading to high FF.[146]

8.3.2 N-Type Metal Oxides

The LUMO energy levels of *n*-type semiconducting metal oxides are generally close to that of PC61BM and the work function of Al cathode and thus are favorable for electron extraction. In particular, titanium oxide (TiO$_x$) thin films prepared from a sol–gel process have attracted extensive attentions as a CBL (see Scheme 8.5).[155,156] This solution-processed buffer layer is multifunctional as both an electron-transporting/hole-blocking layer and an oxygen barrier and optical spacer, leading to significant enhancements in the device's efficiency and stability. For instance, in a recent report by Park et al., a high PCE of 6% was achieved for the PCDTBT:PC$_{71}$BM BHJ-PSC device by incorporating a TiO$_x$ CBL, which functioned as an electron extraction layer and optical spacer.[157] To further improve the performance of a TiO$_2$ CBL, Park et al. doped Cs into the TiO$_2$ layer by mixing Cs$_2$CO$_3$ solution with a nanocrystalline TiO$_2$ solution and successfully lowered the work function of TiO$_2$ for more efficient electron extraction. As a result, the P3HT:PC61BM solar cells showed further enhancement in PCE compared to that using only pure TiO$_2$ buffer layers.[158]

Inspired by the superior performance of TiO$_x$ as a CBL, Jen et al. incorporated a solution-processed thin film of ZnO nanoparticles as a CBL to improve the efficiency of P3HT:PC61BM BHJ-PSCs.[159,160] The Fermi level of ZnO (4.3 eV) matches well with the LUMO of PC61BM (4.3 eV), and this facilitated efficient electron transfer and extraction from the P3HT:PC61BM active layer to the Al cathode. In addition to the optical spacer effect, the low-lying valence band of ZnO also prevented hole carriers in the active layer from reaching the cathode. A buffer layer of a thin film of ZnO nanoparticles was advantageous because of its high electron mobility (2.5 cm^2 V^{-1} s^{-1})

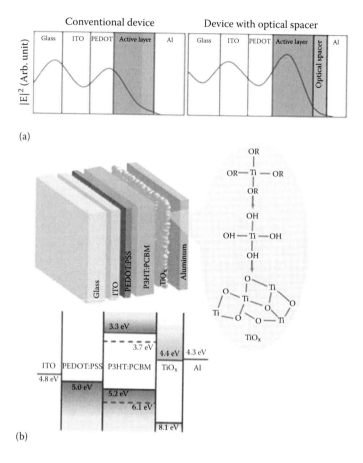

SCHEME 8.5 (a) Schematic representation of the spatial distribution of the squared optical electric field strength E^2 inside the devices with a structure of ITO/PEDOT/active layer/Al (left) and ITO/PEDOT/active layer/optical spacer/Al (right). (b) Schematic illustration of the device structure with a brief flow chart of the steps involved in the preparation of the TiO$_x$ layer. The energy levels of the single components of photovoltaic cell are also shown. (Reprinted with permission from Kim, J. Y. et al., *Adv. Mater.*, 18, 572. Copyright 2006, Wiley-VCH Verlag GmbH & Co. KGaA.)

and solution processability.[161,162] Furthermore, the electrical and electronic properties of ZnO can be easily tuned by modifying the ZnO surface with a self-assembled (SAM) functional molecular layer. This provided a versatile way to engineer the contact between the ZnO CBL and different metal cathodes (Table 8.5).[163,164]

Al$_2$O$_3$ was also applied as a CBL in inverted-structure P3HT:PC61BM solar cells by Kippelen et al. using the atomic layer deposition technique. The ultrathin Al$_2$O$_3$ CBL was found to reduce the work function of ITO and turned ITO into an electron-collecting electrode. Interestingly, the measured J–V curves of the unexposed devices showed an s-shape kink, which was eliminated upon exposure to UV illumination.

Interfacial Materials toward Efficiency Enhancement of Polymer Solar Cells

TABLE 8.5
Nonexhaustive Survey of Conventional and Inverted Solar Cells, Including Metal Oxide Cathode Buffer Layers

CBL	Cell Architecture	P_{in} (mW cm^{-2})	J_{sc} (mA cm^{-2})	V_{oc} (V)	FF (%)	PCE (%)	Ref.
TiO$_x$	ITO/PEDOT:PSS/ P3HT:PC61BM/CBL/Al	90	11.10 (7.50)	0.61 (0.51)	66 (54)	5.00 (2.30)	155
TiO$_x$	ITO/PEDOT:PSS/ MEHPPV:PC61BM/CBL/Al	100	6.10 (5.10)	0.81 (0.81)	(42)	2.10 (1.70)	156
TiO$_x$	ITO/PEDOT:PSS/ PCDTBT:PC71BM/CBL/Al	100	10.60	0.88	66	6.10	157
TiO$_2$:Cs	ITO/PEDOT:PSS/ P3HT:PC61BM/CBL/Al	100	10.76	0.58	67	4.2	158
ZnO	ITO/CBL/P3HT:PC61BM/ MoO$_3$/Ag	100	11.90 (8.50)	0.59 (0.33)	60 (21)	4.18 (0.57)	161
ZnO NPs	ITO/CBL/P3HT:PC61BM/ PEDOT:PSS/Ag	100	10.69	0.62	54	3.61	159
ZnONPs/ C$_{60}$-SAM	ITO/CBL/P3HT:PC61BM/ PEDOT:PSS/Ag	100	12.00	0.63	61	4.54	160
Al$_2$O$_3$	ITO/CBL/P3HT:PC61BM/ PEDOT:PSS/Ag	67	10.10 (10.40)	0.57 (0.63)	64 (58)	5.34 (5.49)	165
CaO	ITO/PEDOT:PSS/ P3HT:PC61BM/CBL/Al	100	10.08 (10.23)	0.78 (0.63)	66 (59)	5.19 (3.83)	166

Note: The photovoltaic parameters are compared to those measured for a reference cell (values in parentheses), when available, without the buffer layer.

This phenomenon was interpreted by the change in the conductivity of ITO arising from oxygen desorption/absorption. Consequently, the Al$_2$O$_3$ CBL-incorporated device exhibited a FF of 0.64 and a PCE of about 2.8%.[165]

Xie et al. used calcium oxide (CaO) as a CBL in conventional P3HT:PC61BM BHJ-PSC devices and compared its performance with that of LiF. The authors found that the V_{oc} and PCE were both increased upon the introduction of both CaO and LiF CBLs (1 nm thick) in preannealing devices, and a postthermal annealing after cathode deposition resulted in further enhancement of PCE incorporating a LiF CBL, but that of the device with CaO CBL decreased. They also found that, when the postannealed cell with Ca/Al cathode was exposed in air, V_{oc} and PCE (4.60%) were largely enhanced probably due to the conversion of Ca to CaO.[166]

8.3.3 ALKALI METAL COMPOUNDS

Inorganic fluorides have been widely used as CBL materials in PSCs as in OLEDs. Among them, lithium fluoride (LiF) is the most commonly used, and the optimal thickness of LiF is limited to less than 1 nm because thicker LiF layers were found to be unfavorable for efficient electron collection. The formation of dipoles at the LiF interfacial layer was reported so that the control of the work function of the cathode contact is evident in PSCs.[167,168] Brabec et al. revealed the insertion of thin interlayers of LiF under the cathode (Al or Au) for the MDMO-PPV:PC61BMBHJ-PSCs significantly enhanced the FF and afforded high V_{oc}. Compared to devices without the LiF CBL, the PCE increased by more than 20% up to 3.3%. Substitution of the LiF by another insulating interlayer of SiO_x resulted in lower overall PCE. In the case of a LiF/Au cathode, substantial efficiency enhancement was observed compared to a pristine Au cathode and PCE reached up to 2.3%. The formation of a dipole moment across the junction, due to either orientation of the LiF or chemical reactions leading to charge transfer across the interface, was suggested as the mechanism for the enhancement by LiF CBL (Table 8.6).[169]

Other metal fluorides, including NaF, KF, and CsF, were also applied as CBLs, leading to improved performance of BHJ-PSC devices. Compared to LiF, NaF exhibited an improved PCE, FF, and V_{oc} with an ultrathin layer thickness (<0.2 nm).[170] Alternatively, using CsF as CBL, the PCE of MDMO-PPV:PC61BM BHJ-PSC devices was higher than with LiF CBL as well. These results could be explained by the lower series resistance of CsF, which was almost constant in the range of the CsF thicknesses studied (0.4–3 nm).[171]

Another alkali metal compound, caesium carbonate (Cs_2CO_3), which was first proposed for OLED devices, was successfully applied as a CBL in both conventional and inverted PSCs.[34,172,173] For instance, Yang et al. demonstrated a highly efficient inverted P3HT:PC61BM BHJ-PSC with a low temperature annealed Cs_2CO_3 as CBL.

TABLE 8.6
Nonexhaustive Survey of Conventional and Inverted Solar Cells Including an Alkaline-Based CBL

CBL	Cell Architecture	P_{in} (mW cm^{-2})	J_{sc} (mA cm^{-2})	V_{oc} (V)	FF (%)	PCE (%)	Ref.
LiF	ITO/PEDOT:PSS/ MDMOPPV:PC61BM/CBL/Al	80	5.25	0.82 (0.76)	61 (53)	3.30	169
CsF	ITO/PEDOT:PSS/ MEHPPV:PC61BM/CBL/Al	100	5.26	0.72	37	2.20 (1.40)	171
Cs_2CO_3	ITO/PEDOT:PSS/ P3HT:PC61BM/CBL/Al	130	5.95 (7.44)	0.52 (0.42)	66 (52)	1.55 (1.25)	34
Cs_2CO_3	ITO/CBL/P3HT:PC61BM/ V_2O_5/Al	130	8.42 (6.97)	0.56 (0.30)	62 (41)	2.25 (0.66)	34
Cs_2CO_3	ITO/PEDOT:PSS/ P3HT:PC61BM/CBL/Al	100	9.50 (11.20)	0.56 (0.41)	60 (50)	3.10 (2.30)	173
Cs_2CO_3 anneal	ITO/CBL/P3HT:PC61BM/ V_2O_5/Al	100	11.13	0.59	64	4.19	172

Note: The photovoltaic parameters are compared to those measured for a reference cell (values in parentheses), when available, without the buffer layer.

Interfacial Materials toward Efficiency Enhancement of Polymer Solar Cells

By this approach, the PCE of the inverted device increased dramatically from 2.3% to 4.2%, with J_{sc} of 11.17 mA/cm^2, V_{oc} of 0.59 V, and FF of 63% under AM1.5G 100 mW/cm^2 irradiation. Ultraviolet photoelectron spectroscopy (UPS) showed that the work function of the annealed Cs$_2$CO$_3$ layer decreased from 3.45 to 3.06 eV. Further x-ray photoelectron spectroscopy (XPS) results revealed that Cs$_2$CO$_3$ could decompose into a low work function, doped cesium oxide Cs$_2$O upon annealing, which accounted for the work function decrease of Cs$_2$CO$_3$ and device efficiency improvement upon Cs$_2$CO$_3$ CBL incorporation.[172]

8.3.4 Organic Materials

8.3.4.1 Small-Molecule Organic Materials

A variety of organic materials, including small-molecule organic materials, non-conjugated polymers and water/alcohol–soluble conjugated polymers, and fullerene derivatives have been successfully applied in BHJ-PSCs via the facile solution processing approach in recent years (see Scheme 8.6). Among them, small-molecule organic material is more readily available than polymers and advantageous in terms of its certain molecular weight and assembly structure. Several groups used a thin layer (1–10 nm) of bathocuproine (BCP) as CBL.[95,174,175] For instance, Feng et al. investigated the effects of BCP CBL on the P3HT:PC61BM BHJ-PSCs, and found that, obvious enhancements of V_{oc} (from 0.38 to 0.65 V) and FF (from 44% to 63%)

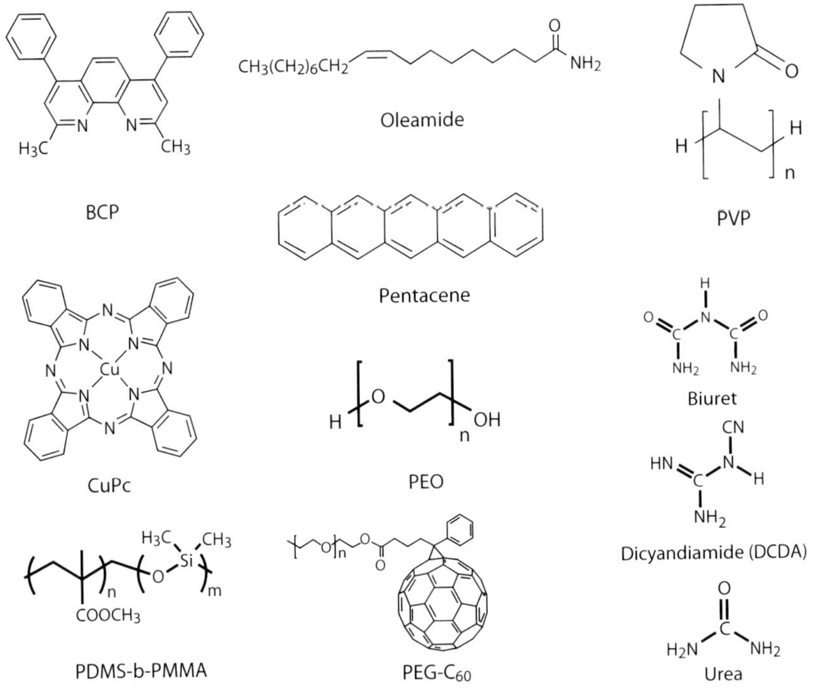

SCHEME 8.6 Chemical structure of some organic materials used as CBLs.

were achieved by using a 2-nm BCP layer compared with the device without a BCP CBL. As a result, the PCE of the device improved significantly from 1.63% to 4.11%. They studied the underlying mechanisms of BCP CBL for BHJ-PSCs utilizing XPS, which revealed that the major role of BCP is to prohibit aluminum carbide formation and energy level shift at the interfaces.[174]

Copper phthalocyanine (CuPc) and pentacene as p-type organic semiconductors have also been applied as effective CBLs in P3HT:PC61BM BHJ-PSCs in spite of their poor hole-blocking properties.[175] Hou et al. compared the CBL performance of CuPc and pentacene with respect to BCP, and found that the PCE of the CuPc CBL-based devices was only slightly worse than that based on a BCP CBL whereas the PCE of the device incorporating a 1-nm-thick pentacene CBL even surpassed that with a LiF CBL.[176] More recently, Kim et al. demonstrated that introducing a thin pentacene layer as a CBL resulted in a better than 50% improvement in the PCE of P3HT:PC61BM BHJ-PSCs. An additional path for electron transfer provided by the thin pentacene layer appeared to be mainly responsible for the improvement along with the increase in the V_{oc} that occurred with increasing pentacene thickness.[177]

Several small-molecule electron transporting materials, including 2-(4-biphenyl)-5-(4-tert-butylphenyl)1,3,4-oxidiazole (PBD), tris-8-hydroxy-quinolinato aluminum (Alq$_3$), and Bis[2-(2-benzothiazoly)phenolato]zinc(II) (Zn(BTZ)$_2$ were incorporated as a CBL in P3HT:PC61BM BHJ-PSCs via thermal evaporation, and it was found that the insertion of these CBLs improved V_{oc} and PCE owing to the increased built-in potential at the interface between the photovoltaic active layer and the Al electrode.[178] Ma et al. synthesized a new alcohol-soluble electron-transporting material, 4,7-diphenyl-1,10-phenanthroline-2,9-dicarboxylicacid (DPPA), which was used to modify ZnO and then incorporated as a CBL into inverted P3HT:PC61BM BHJ-PSCs, resulting in an enhancement of about 10% compared with the device without DPPA. Moreover, DPPA was also used as the sole CBL, affording PCE enhancement as well due to the improved interfacial contact between the substrate and active layer.[179]

Li et al. applied a solution processable titanium chelate, titanium (diisopropoxide) bis(2,4-pentanedionate) (TIPD), as a CBL in poly (2-methoxy-5-(2′-ethylhexyloxy)-1,4-phenylenevinylene) (MEH-PPV):PC61BM BHJ-PSCs, and found that introducing a TIPD buffer layer reduced the interface resistance between the active layer and Al electrode, and consequently, the PCE with TIPD CBL reached 2.52%, increased by 51.8% compared with that without a TIPD CBL (1.66%).[180] More recently, the same group further applied TIPD as a CBL in inverted-structure poly(4,8-bis-alkyloxybenzo(1,2-b:4,5-b′)dithiophene-2,6-diyl-alt-(alkyl thieno(3,4-b) thiophene-2-carboxylate)-2,6-diyl) (PBDTTT-C):PC$_{71}$BM BHJ-PSCs, achieving a high PCE of 7.4%, which was enhanced by 16%. The authors proposed that the high PCE benefited from the hydrophobic surface and the suitable electronic energy levels of the TIPD CBL.[181]

Very recently, Yang et al. applied three amino-containing small-molecule organic materials—biuret, dicyandiamide (DCDA), and urea—as CBLs in P3HT:PC61BM BHJ-PSCs, resulting in PCE enhancements of 3.84%, 4.25%, and 4.39% for biuret, DCDA, and urea, which were enhanced by ~15%, ~27%, and ~31%, respectively, compared to that without any CBL. The authors proposed that the efficiency enhancement was due to the conjunct effects of the formation of a dipole layer between the P3HT:PC61BM active layer and the Al electrode, which decreased effectively the energy level offset between the work function of Al and the LUMO level of the PC61BM acceptor, and the coordination interaction between the lone-pair electrons on the nitrogen atoms of the amino group with Al atoms, which prohibited the interaction between Al and the thiophene rings of P3HT. The latter effect accounted partially for the dependence of the CBL performance of these three amino-containing CBLs on their chemical structures specifically the number of the amino groups.[182]

Amphiphilic surfactant was also incorporated as CBL in BHJ-PSCs by Yang et al. Oleamide was initially doped in the P3HT:PC61BM solution and then migrated onto the P3HT:PC61BM active layer surface via self-assembly, resulting in a significant efficiency enhancement by ~28% at the optimum oleamide doping ratio of 2.5%. The PCE enhancement was found to be primarily due to the increase of FF by ~22%, and this resulted from the formation of an oleamide interfacial dipole layer, which could lower the work function of Al, and consequently, facilitated the electron extraction by the Al cathode because of the decreased energy level offset between the work function of Al and the LUMO level of the PC61BM acceptor.[183]

8.3.4.2 Nonconjugated Polymers

Several nonconjugated insulating polymers used as CBLs in PSCs have been reported. Zhang et al. reported the first study of application of nonconjugated polymers as CBLs by inserting a thin poly(ethylene oxide) (PEO) interlayer between the P3HT:PC61BM active layer and the Al cathode via spin coating. The V_{oc} was dramatically enhanced up to 200 mV and noticeable enhancements of FF and J_{sc} were observed as well, resulting in the enhancement of PCE by 50%. They proposed that PEO had a similar function as LiF, that is, the built-in potential was increased upon inserting the PEO interfacial layer, thus improving charge transportation.[184] Chen et al. applied a thin poly(ethylene glycol) (PEG) CBL by adding up to 5 wt% of PEG to the P3HT:PC61BM solution followed by the spontaneous migration of PEG to the surface of the active layer. The PEG CBL reduced the contact resistance after undergoing chemical reactions with the Al atoms of the cathode, leading to the improved efficiency of electron collection and, consequently, the enhancement of PCE.[185] A similar self-organization behavior was also reported in the BHJ-PSC device containing poly(dimethylsiloxane)-block-poly(methyl methacrylate) (PDMS-*b*-PMMA) as CBL, which was at first mixed into the P3HT:PC61BM active layer solution and then surface-segregated from the active layer due to the low surface energy of the PDMS block. A plausible interpretation of the enhancement of the PCE was that the PDMS-*b*-PMMA interface layer suppressed charge carrier recombination at the organic–metal interface, but the detailed mechanism was not yet clear.[186] Besides, fullerene end-capped PEG was also reported by two groups independently to form a CBL via self-assembly, resulting in not only the significant enhancement of the P3HT:PC61BM device performance, but also the improved thermal stability. The enhancement of the device performance upon addition of fullerene end-capped PEG was mainly attributed to the increase of V_{oc} and FF, which is due to the generation of the interfacial dipole moment and the improved vertical morphology of the P3HT:PC61BM active layer with the uniform distribution of PC61BM crystallites in the P3HT matrix, respectively.[187,188] Recently, Yang et al. applied a poly(vinylpyrrolidone) (PVP) CBL in P3HT:PC61BM BHJ-PSCs by means of either spin coating or self-assembly, resulting in significant efficiency enhancement. The PCE of the device with PVP CBL (3.90%) was enhanced by 29% under the optimum PVP spin-coating speed of 3000 rpm, which led to the optimum thickness of the PVP layer of about 3 nm. The incorporation of a PVP CBL led to not only the formation of a dipole layer between the P3HT:PC61BM active layer and Al electrodes, but also the chemical reactions of PVP with Al atoms, and the overall effect of PVP CBL was the increase of the charge carriers collected by the Al cathode, and thus an overwhelming increase of J_{sc} was observed.[189]

8.3.4.3 Water/Alcohol–Soluble Conjugated Polymers

Water/alcohol–soluble conjugated polymers can be processed from water or other polar solvents, which offer good opportunities to avoid interfacial mixing upon fabrication of multilayer polymer solar cells by solution processing and can dramatically improve charge injection from a high work function metal cathode resulting in great enhancement of the device performance.[190,191] A series of amino N-oxide functionalized polyfluorene homopolymers and copolymers (PNOs) were synthesized by oxidizing their amino functionalized precursor polymers (PNs) with hydrogen peroxide. Excellent solubility in polar solvents and good electron injection from high work function metals made PNOs good candidates for interfacial modification of PSCs. Both PNOs and PNs were used as CBLs in BHJ-PSCs and the resulting devices showed much better performance than devices based on a bare Al cathode. Based on the comparison to PNs, the amino N-oxide groups in the PNO structure not only enhanced alcohol solubility, but also significantly enhanced the ability of interlayers to reduce electron injection barriers. As a result, BHJ-PSCs based on PNOs/Al cathode also exhibited higher V_{oc}, J_{sc}, and FF than those of PNs/Al cathode devices.[192] Wu et al. applied alcohol/water–soluble poly [(9,9-bis(3′-(N,N-dimethylamino) propyl)-2,7-fluorene)-alt-2,7-(9,9-dioctylfluorene)] (PFN) as a conjugated polymer CBL in the conventional BHJ-PSC device based on thieno[3,4-b]-thiophene/benzodithiophene (PTB7) and PC71BM and achieved a certified PCE as high as 8.37%.[193] Later on, the same group used PFN CBL in an inverted-structure BHJ-PSC device and achieved a certified efficiency of 9.2% (see Scheme 8.7). The role of the PFN CBL on the efficiency enhancement was interpreted by the conjunct effects, including an enhanced built-in potential across the device due to the existence of an interface dipole, improved charge-transport properties, elimination of the buildup of space charge, and reduced recombination loss due to the increase in built-in field and charge carrier mobility.[194]

8.3.4.4 Fullerene Derivatives

Some fullerene C_{60} derivatives have been investigated as CBLs by different groups. For instance, Wei et al. synthesized a novel fullerene derivative with a fluorocarbon chain (F-PC61BM) and applied it in P3HT:PC61BM BHJ-PSCs. In F-PC61BM, the C_{60} moiety worked as an electron-accepting and transporting material, and the perfluoroalkyl group gave a low surface energy to the material. When a small amount of F-PC61BM was mixed in the P3HT:PC61BM solution for spin coating, it was

Interfacial Materials toward Efficiency Enhancement of Polymer Solar Cells 229

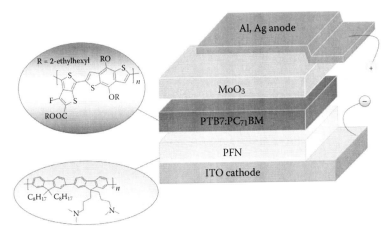

SCHEME 8.7 Schematic of the inverted-type PSCs, in which the photoactive layer is sandwiched between a PFN-modified ITO cathode and an Al, Ag-based top anode. Insets: chemical structures of the water/alcohol–soluble conjugated polymer and electron donor materials used in the study. (Reprinted with permission from He, Z. et al., *Nat. Photon.*, 6, 591. Copyright 2012, Nature Publishing Group.)

expected that F-PC61BM spontaneously migrated to the surface of the organic layer during spin casting, owing to the low surface energy of the fluorocarbon, and formed a very thin layer of F-PC61BM in a single step. Indeed, F-PC61BM CBL-incorporated P3HT:PC61BM BHJ-PSCs exhibited a PCE of 3.79%, much higher than that without CBL (3.09%). It was found that the surface dipole moment induced by the F-PC61BM CBL could be the origin of the improvement in the device performance.[195] Hsieh et al. applied a CBL of a cross-linked fullerene derivative (C-PCBSD) on the top of an ITO/ZnO cathode in inverted P3HT:PC61BM BHJ-PSCs. In situ cross-linking of PCBSD was carried out by heating at a low temperature of 160°C for 30 min to generate a robust, adhesive, and solvent-resistant thin film. This cross-linked network enabled a sequential active layer to be successfully deposited on top of this interlayer to overcome the problem of interfacial erosion and realized a multilayer inverted device by all-solution processing. This C-PCBSD interlayer exerted multiple positive effects on both P3HT/C-PCBSD and PC61BM/C-PCBSD localized heterojunctions at the interface of the active layer, including improved exciton dissociation efficiency, reduced charge recombination, decreased interface contact resistance, and induction of vertical phase separation to reduce the bulk resistance of the active layer as well as passivation of the local shunts at the ZnO interface.[196]

Very recently, Li et al. developed several new functionalized fullerene derivatives, namely [6,6]-phenyl-C61-butyricacid 2-((2-(dimethylamino)ethyl)(methyl)amino) ethyl ester (PCBDAN), [6,6]-phenyl-C61-butyric acid 2-((2-(trimethylammonium)ethyl)(dimethyl)ammonium) ethyl ester diiodonium (PCBDANI), and an amine group functionalized fullerene derivative DMAPA-C$_{60}$, as CBLs in versatile BHJ-PSC devices based on different photoactive layers.[197,198] For PCBDAN and PCBDANI, the polar side chain with the amine end group made the fullerene derivatives alcohol-soluble and able to form a dipole interfacial layer between the active layer and Al electrode. Both PCBDAN and PCBDANI were applied as CBLs in different PSCs based on P3HT:PC61BM, P3HT:ICBA, and PBDTTT-C-T:PC71BM, affording enhanced PCE in all cases. Thus, PCBDAN and PCBDANI can be used as universal CBLs in PSCs.[197] In another report by the same group, DMAPA-C$_{60}$ was applied as a CBL in different BHJ-PSCs based on P3HT:PC61BM, PBDTTT-C:PC71BM, and PBDTTT-CT:PC71BM. It was revealed that PCE of the three systems with DMAPA-C$_{60}$ CBL reached 3.88%, 6.29%, and 7.42%, respectively, which were much higher than those of the corresponding PSCs with the Al-only cathode and even slightly higher than those of the corresponding Ca/Al devices of these systems.[198]

8.3.5 Self-Assembled Monolayers (SAMs)

SAMs, which were mainly used to modify CBLs made of zinc oxide (ZnO), have also been used as CBLs in PSCs (see Scheme 8.8). Jen et al. applied different ZnO-based CBLs, which were modified by SAMs of different saturated carboxylic acids with various dipoles and terminal groups R–COOH (R = –C$_{11}$H$_{22}$SH, –C$_{11}$H$_{23}$, or –C$_{14}$F$_{29}$), and found that carboxylic acid could tune the contact property between ZnO and the cathode due to different interactions of the hydrophobic terminal groups with metal. With the appropriate choice of SAM modifiers, the BHJ-PSC devices showed dramatically improved PCE, and even high work function metals, such as Ag and Au, could be used as efficient cathodes.[199]

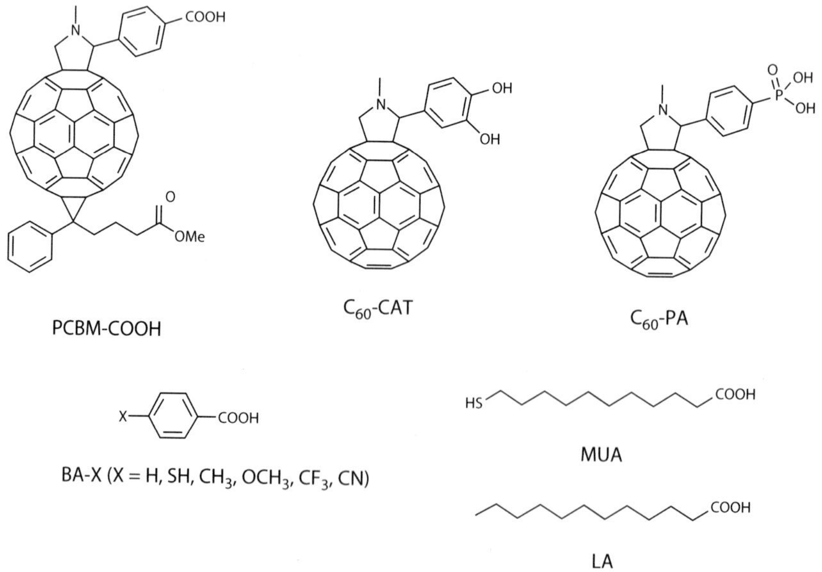

SCHEME 8.8 Compounds able to form self-assembled monolayers used to modify cathode buffer layers. (Reprinted with permission from Po, R. et al., *Energy Environ. Sci.* 4 (2). Copyright 2011, Royal Society of Chemistry.)

Interfacial Materials toward Efficiency Enhancement of Polymer Solar Cells 231

The study on SAMs was extended to C_{60} derivatives with different anchoring groups (carboxylic, phosphonic, catechol). The formation of the C_{60} SAM onto the surface of ZnO was investigated by processing the SAM through either a solution immersion technique or a solution spin-coating method. It was found that C_{60} SAMs with the carboxylic acid and catechol termination could be formed onto the surface of ZnO by a simple solution spin-coating process, whereas all three anchoring groups could be formed by a solution immersion technique. P3HT:PC61BM BHJ-PSCs were fabricated under different processing conditions to form SAM leading to twofold, 75%, and 30% efficiency improvement with the carboxylic acid, catechol, and phosphonic acid C_{60} SAMs, respectively. The main contribution to the variation of efficiency from different SAMs was due to the V_{oc} affected by different anchoring groups and functionalization of the C_{60} SAM. The SAM formation condition influenced the device performance. Because of the strong acidic nature of the phosphonic acid anchoring group, immersing the ZnO substrate into a solution containing the C_{60} phosphonic acid SAM for an extended period of time led to degradation of the ZnO surface. This, in turn, led to devices without any photovoltaic activity whereas weaker acids, such as carboxylic acid, and catechol-based C_{60} SAMs can be assembled onto ZnO, leading to devices with an average PCE of 4.4% and 4.2%, respectively.[200]

8.3.6 Graphenes

Application of graphene derivatives as CBLs has been much less reported compared to GO ABLs as discussed above, and up to now there are only two successes. As the first success, Dai et al. applied cesium-neutralized graphene oxide (GO-Cs), which was prepared through simple charge neutralization of the peripheral carboxylic acid groups of GO with Cs_2CO_3, as an excellent CBL for P3HT:PC61BM BHJ-PSCs. PSCs with a GO-Cs CBL exhibited a PCE of 3.67%, much higher than that of corresponding devices with state-of-the-art ABL and CBLs (3.15%) (see Scheme 8.9).[201] Chen et al. also reported that a GO/GO-Cs (cesium neutralized GO) bilayer modified

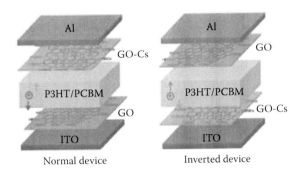

SCHEME 8.9 Device structures of (left) normal and (right) inverted BHJ-PSCs with GO as ABL and cesium-neutralized graphene oxide (GO-Cs) as CBL. (Liu, J.; Xue, Y. H.; Gao, Y. X.; Yu, D. S.; Durstock, M.; Dai, L. M.: Hole and Electron Extraction Layers Based on Graphene Oxide Derivatives for High-Performance Bulk Heterojunction Solar Cells. *Adv. Mater.* 2012. 24. 2228. Copyright Wiley-VCH Verlag GmbH & Co. KGaA. Reproduced with permission.)

with ultrathin Al and MoO$_3$ could function as an efficient interconnecting layer in tandem PSCs to achieve a significantly increased V_{oc}, reaching almost 100% of the sum of the subcell V_{oc} under standard AM 1.5 conditions.[202]

Very recently, Yang et al. reported a new graphene-fullerene composite (rGO-pyrene-PC61BM), in which [6,6]-phenyl-C61-butyric acid methyl ester (PC61BM) was attached onto reduced graphene oxide (rGO) via the noncovalent functionalization approach, as CBL. When rGO-pyrene-PC61BM was applied as CBL in P3HT:PC61BM BHJ-PSC devices, an improved PCE of 3.89% with an enhanced ratio of ca. 15% compared to that of the reference device without CBL (3.39%) was observed. Contrarily, the comparative devices incorporating the pure rGO or pyrene-PC61BM component as CBL showed dramatically decreased PCE, indicating the importance of composite formation between rGO and pyrene-PC61BM components for its electron extraction property.[203]

8.4 SUMMARY AND OUTLOOK

In the past two decades, the efficiency of PSCs has been dramatically increasing to 10.6%, and the interfacial materials, including both ABLs and CBLs, play determinative roles in efficient charge transport and extraction. A wide range of interfacial materials have been developed and demonstrated to be effective in enhancing the PCE of BHJ-PSC devices, including organic molecules, inorganic dielectrics, and composites of organics with inorganic dopants, etc. For ABL materials, PEDOT:PSS are the most commonly used and are proved to be effective in smoothing the surface of ITO and modulating the work function of ITO, which however, suffers from some drawbacks, such as acidity, hygroscopy, and sensitivity to oxygen exposure. Semiconducting p-type metal oxides, such as MoO$_3$, NiO, WO$_3$, V$_2$O$_5$, etc., represent alternative ABL materials, demonstrating similar or better performances than devices with a PEDOT:PSS ABL. Although semiconducting metal oxides were typically deposited via vacuum evaporation in some early studies, a solution-processable technique with low-temperature annealing for incorporating metal oxide ABLs has been successfully developed recently but only for limited types of metal oxides; thus further development of other metal oxide ABL materials and/or new solution-processable techniques is needed.

On the other hand, incorporation of CBL appears more facile and flexible because it can be fabricated via either thermal deposition or a solution-processable technique. Therefore, many different types of CBL materials have been developed in recent years as reviewed here, including low work function metals, such as Ca and Ba; n-type metal oxides, such as TiO and ZnO; alkali metal compounds, such as LiF; organic materials, including small-molecule organic materials; nonconjugated polymers and water/alcohol–soluble conjugated polymers; self-assembled monolayers (SAMs); and graphenes, etc. In spite of the great progress in CBL research, the solution-processable technique of CBLs needs to be optimized further to improve the reproducibility of the CBL performance, especially for metal oxides. Besides, for the newly developed organic CBL materials, which usually have large dipole moments beneficial for electron transport and lowering the interfacial barrier for charge carrier collection by cathode, their stability toward post-treatment and the correlation between their molecular structures and CBL performance are yet to be investigated further.

Up to now, the record PCE of 10.6% was achieved by tandem PSCs, for which the interconnection layers play a determinative role as well. The interfacial materials developed for single-junction BHJ-PSC devices may function as ideal interconnection layers for tandem PSCs. Besides, recently, the transparent PSCs have drawn great attention from researchers, and some of the interfacial materials for single-junction BHJ-PSC devices may be suitable for a transparent anode or cathode as well. Finally, all-solution-processed PSCs are very important for commercial application of PSCs, which places a high demand on the stability of PSCs, and this can be improved by interfacial modification via solution-processable interfacial material incorporation. Thus, the promising progress achieved for interfacial materials provides a versatile and practical solution for commercialization of the low-cost PSCs.

ACKNOWLEDGMENTS

We thank the present and former members of our group for the contributions as well as our collaborators on the published works cited in this review. S. Y. thanks the National Basic Research Program of China (2010CB923300, 2011CB921401) National Natural Science Foundation of China (Nos. 21132007, 21371164), and Key Project of Hefei Center for Physical Science and Technology (No. 2012FXZY006) for financial support.

REFERENCES

1. Heeger, A. J., 25th Anniversary Article: Bulk Heterojunction Solar Cells: Understanding the Mechanism of Operation. *Adv. Mater.* **2014**, *26* (1), 10–28.
2. Dou, L. T.; You, J. B.; Hong, Z. R.; Xu, Z.; Li, G.; Street, R. A.; Yang, Y., 25th Anniversary Article: A Decade of Organic/Polymeric Photovoltaic Research. *Adv. Mater.* **2013**, *25* (46), 6642–6671.
3. He, Z.; Wu, H.; Cao, Y., Recent Advances in Polymer Solar Cells: Realization of High Device Performance by Incorporating Water/Alcohol-Soluble Conjugated Polymers as Electrode Buffer Layer. *Adv. Mater.* **2014**, *26* (7), 1006–1024.
4. Li, Y. F., Fullerene-Bisadduct Acceptors for Polymer Solar Cells. *Chem. Asian J.* **2013**, *8* (10), 2316–2328.
5. Yu, G.; Gao, J.; Hummelen, J. C.; Wudl, F.; Heeger, A. J., Polymer Photovoltaic Cells—Enhanced Efficiencies Via a Network of Internal Donor-Acceptor Heterojunctions. *Science* **1995**, *270* (5243), 1789–1791.
6. Chu, T.-Y.; Lu, J.; Beaupre, S.; Zhang, Y.; Pouliot, J.-R.; Wakim, S.; Zhou, J. et al., Bulk Heterojunction Solar Cells Using Thieno 3,4-c pyrrole-4,6-dione and Dithieno 3,2-b:2′,3′-d silole Copolymer with a Power Conversion Efficiency of 7.3%. *J. Am. Chem. Soc.* **2011**, *133* (12), 4250–4253.
7. Liang, Y.; Xu, Z.; Xia, J.; Tsai, S.-T.; Wu, Y.; Li, G.; Ray, C.; Yu, L., For the Bright Future-Bulk Heterojunction Polymer Solar Cells with Power Conversion Efficiency of 7.4%. *Adv. Mater.* **2010**, *22* (20), E135.
8. Chen, S.; Small, C. E.; Amb, C. M.; Subbiah, J.; Lai, T.-H.; Tsang, S.-W.; Manders, J. R.; Reynolds, J. R.; So, F., Inverted Polymer Solar Cells with Reduced Interface Recombination. *Adv. Energy Mater.* **2012**, *2* (11), 1333–1337.
9. Dou, L.; You, J.; Yang, J.; Chen, C.-C.; He, Y.; Murase, S.; Moriarty, T.; Emery, K.; Li, G.; Yang, Y., Tandem Polymer Solar Cells Featuring a Spectrally Matched Low-Bandgap Polymer. *Nat. Photon.* **2012**, *6* (3), 180–185.

10. Small, C. E.; Chen, S.; Subbiah, J.; Amb, C. M.; Tsang, S.-W.; Lai, T.-H.; Reynolds, J. R.; So, F., High-Efficiency Inverted Dithienogermole-Thienopyrrolodione-Based Polymer Solar Cells. *Nat. Photon.* **2012**, *6* (2), 115–120.
11. You, J. B.; Dou, L. T.; Yoshimura, K.; Kato, T.; Ohya, K.; Moriarty, T.; Emery, K.; Chen, C. C.; Gao, J.; Li, G.; Yang, Y., A Polymer Tandem Solar Cell with 10.6% Power Conversion Efficiency. *Nat. Commun.* **2013**, *4*, 1446.
12. Shaheen, S. E.; Brabec, C. J.; Sariciftci, N. S.; Padinger, F.; Fromherz, T.; Hummelen, J. C., 2.5% Efficient Organic Plastic Solar Cells. *Appl. Phys. Lett.* **2001**, *78* (6), 841–843.
13. Blouin, N.; Michaud, A.; Leclerc, M., A Low-Bandgap Poly(2,7-Carbazole) Derivative for Use in High-Performance Solar Cells. *Adv. Mater.* **2007**, *19* (17), 2295.
14. Amb, C. M.; Chen, S.; Graham, K. R.; Subbiah, J.; Small, C. E.; So, F.; Reynolds, J. R., Dithienogermole As a Fused Electron Donor in Bulk Heterojunction Solar Cells. *J. Am. Chem. Soc.* **2011**, *133* (26), 10062–10065.
15. Hou, J.; Chen, H.-Y.; Zhang, S.; Chen, R. I.; Yang, Y.; Wu, Y.; Li, G., Synthesis of a Low Band Gap Polymer and Its Application in Highly Efficient Polymer Solar Cells. *J. Am. Chem. Soc.* **2009**, *131* (43), 15586.
16. Zhou, H.; Yang, L.; You, W., Rational Design of High Performance Conjugated Polymers for Organic Solar Cells. *Macromolecules* **2012**, *45* (2), 607–632.
17. Turak, A., Interfacial Degradation in Organic Optoelectronics. *RSC Adv.* **2013**, *3* (18), 6188–6225.
18. Po, R.; Carbonera, C.; Bernardi, A.; Camaioni, N., The Role of Buffer Layers in Polymer Solar Cells. *Energy Environ. Sci.* **2011**, *4* (2), 285–310.
19. Park, J. H.; Lee, T.-W.; Chin, B.-D.; Wang, D. H.; Park, O. O., Roles of Interlayers in Efficient Organic Photovoltaic Devices. *Macromol. Rapid Commun.* **2010**, *31* (24), 2095–2108.
20. Ratcliff, E. L.; Garcia, A.; Paniagua, S. A.; Cowan, S. R.; Giordano, A. J.; Ginley, D. S.; Marder, S. R.; Berry, J. J.; Olson, D. C., Investigating the Influence of Interfacial Contact Properties on Open Circuit Voltages in Organic Photovoltaic Performance: Work Function Versus Selectivity. *Adv. Energy Mater.* **2013**, *3* (5), 647–656.
21. Jorgensen, M.; Norrman, K.; Gevorgyan, S. A.; Tromholt, T.; Andreasen, B.; Krebs, F. C., Stability of Polymer Solar Cells. *Adv. Mater.* **2012**, *24* (5), 580–612.
22. Lloyd, M. T.; Olson, D. C.; Lu, P.; Fang, E.; Moore, D. L.; White, M. S.; Reese, M. O.; Ginley, D. S.; Hsu, J. W. P., Impact of Contact Evolution on the Shelf Life of Organic Solar Cells. *J. Mater. Chem.* **2009**, *19* (41), 7638–7642.
23. Park, Y.; Choong, V.; Gao, Y.; Hsieh, B. R.; Tang, C. W., Work Function of Indium Tin Oxide Transparent Conductor Measured by Photoelectron Spectroscopy. *Appl. Phys. Lett.* **1996**, *68* (19), 2699–2701.
24. Tang, Z.; Andersson, L. M.; George, Z.; Vandewal, K.; Tvingstedt, K.; Heriksson, P.; Kroon, R.; Andersson, M. R.; Inganas, O., Interlayer for Modified Cathode in Highly Efficient Inverted ITO-Free Organic Solar Cells. *Adv. Mater.* **2012**, *24* (4), 554.
25. Mihailetchi, V. D.; Blom, P. W. M.; Hummelen, J. C.; Rispens, M. T., Cathode Dependence of the Open-Circuit Voltage of Polymer:Fullerene Bulk Heterojunction Solar Cells. *J. Appl. Phys.* **2003**, *94* (10), 6849–6854.
26. Voroshazi, E.; Verreet, B.; Buri, A.; Mueller, R.; Di Nuzzo, D.; Heremans, P., Influence of Cathode Oxidation via the Hole Extraction Layer in Polymer:Fullerene Solar Cells. *Org. Electron.* **2011**, *12* (5), 736–744.
27. Cui, J.; Huang, Q. L.; Veinot, J. C. G.; Yan, H.; Wang, Q. W.; Hutchison, G. R.; Richter, A. G.; Evmenenko, G.; Dutta, P.; Marks, T. J., Anode Interfacial Engineering Approaches to Enhancing Anode/Hole Transport Layer Interfacial Stability and Charge Injection Efficiency in Organic Light-Emitting Diodes. *Langmuir* **2002**, *18* (25), 9958–9970.
28. Roman, L. S.; Mammo, W.; Pettersson, L. A. A.; Andersson, M. R.; Inganas, O., High Quantum Efficiency Polythiophene/C-60 Photodiodes. *Adv. Mater.* **1998**, *10* (10), 774–777.

29. Scott, J. C.; Malliaras, G. G.; Chen, W. D.; Breach, J. C.; Salem, J. R.; Brock, P. J.; Sachs, S. B.; Chidsey, C. E. D., Hole Limited Recombination in Polymer Light-Emitting Diodes. *Appl. Phys. Lett.* **1999**, *74* (11), 1510–1512.
30. Groenendaal, B. L.; Jonas, F.; Freitag, D.; Pielartzik, H.; Reynolds, J. R., Poly(3,4-ethylenedioxythiophene) and Its Derivatives: Past, Present, and Future. *Adv. Mater.* **2000**, *12* (7), 481–494.
31. Wei, H.-Y.; Huang, J.-H.; Hsu, C.-Y.; Chang, F.-C.; Ho, K.-C.; Chu, C.-W., Organic Solar Cells Featuring Nanobowl Structures. *Energy Environ. Sci.* **2013**, *6* (4), 1192–1198.
32. Jeong, J.; Woo, S.; Park, S.; Kim, H.; Lee, S. W.; Kim, Y., Wide Range Thickness Effect of Hole-Collecting Buffer Layers for Polymer:Fullerene Solar Cells. *Org. Electron.* **2013**, *14* (11), 2889–2895.
33. Irwin, M. D.; Buchholz, B.; Hains, A. W.; Chang, R. P. H.; Marks, T. J., p-Type Semiconducting Nickel Oxide as an Efficiency-Enhancing Anode Interfacial Layer in Polymer Bulk-Heterojunction Solar Cells. *Proc. Natl. Acad. Sci. U.S.A.* **2008**, *105* (8), 2783–2787.
34. Li, G.; Chu, C. W.; Shrotriya, V.; Huang, J.; Yang, Y., Efficient Inverted Polymer Solar Cells. *Appl. Phys. Lett.* **2006**, *88* (25), 253503.
35. Kim, Y.; Ballantyne, A. M.; Nelson, J.; Bradley, D. D. C., Effects of Thickness and Thermal Annealing of the PEDOT:PSS Layer on the Performance of Polymer Solar Cells. *Org. Electron.* **2009**, *10* (1), 205–209.
36. Gaynor, W.; Burkhard, G. F.; McGehee, M. D.; Peumans, P., Smooth Nanowire/Polymer Composite Transparent Electrodes. *Adv. Mater.* **2011**, *23* (26), 2905.
37. Kang, J.-W.; Kang, Y.-J.; Jung, S.; Song, M.; Kim, D.-G.; Kim, C. S.; Kim, S. H., Fully Spray-Coated Inverted Organic Solar Cells. *Sol. Energy Mater. Sol. Cells* **2012**, *103*, 76–79.
38. Kettle, J.; Waters, H.; Horie, M.; Chang, S. W., Effect of Hole Transporting Layers on the Performance of PCPDTBT:PCBM Organic Solar Cells. *J. Phys. D: Appl. Phys.* **2012**, *45* (12), 125102.
39. Kim, W.; Kim, N.; Kim, J. K.; Park, I.; Choi, Y. S.; Wang, D. H.; Chae, H.; Park, J. H., Polymer Bulk Heterojunction Solar Cells with PEDOT:PSS Bilayer Structure as Hole Extraction Layer. *ChemSusChem* **2013**, *6* (6), 1070–1075.
40. Chen, B. X.; Zhang, W. F.; Zhou, X. H.; Huang, X.; Zhao, X. M.; Wang, H. T.; Liu, M.; Lu, Y. L.; Yang, S. F., Surface Plasmon Enhancement of Polymer Solar Cells by Penetrating Au/SiO$_2$ Core/Shell Nanoparticles into All Organic Layers. *Nano Energy* **2013**, *2* (5), 906–915.
41. Zhang, W. F.; Zhao, B. F.; He, Z. C.; Zhao, X. M.; Wang, H. T.; Yang, S. F.; Wu, H. B.; Cao, Y., High-Efficiency ITO-Free Polymer Solar Cells Using Highly Conductive PEDOT:PSS/Surfactant Bilayer Transparent Anodes. *Energy Environ. Sci.* **2013**, *6* (6), 1956–1964.
42. Sha, W.; Shaohu, H.; Yina, Z.; Hua, Z.; Nanliu, L.; Lei, W.; Yong, C.; Jian, W., pH-neutral PEDOT:PSS as Hole Injection Layer in Polymer Light Emitting Diodes. *Org. Electron.* **2011**, *12* (3), 504–508.
43. Nardes, A. M.; Kemerink, M.; de Kok, M. M.; Vinken, E.; Maturova, K.; Janssen, R. A. J., Conductivity, Work Function, and Environmental Stability of PEDOT:PSS Thin Films Treated with Sorbitol. *Org. Electron.* **2008**, *9* (5), 727–734.
44. Jonsson, S. K. M.; Birgerson, J.; Crispin, X.; Greczynski, G.; Osikowicz, W.; van der Gon, A. W. D.; Salaneck, W. R.; Fahlman, M., The Effects of Solvents on the Morphology and Sheet Resistance in Poly(3,4-ethylenedioxythiophene)-Polystyrenesulfonic Acid (PEDOT-PSS) Films. *Synth. Met.* **2003**, *139* (1), 1–10.
45. Kim, Y. H.; Sachse, C.; Machala, M. L.; May, C.; Muller-Meskamp, L.; Leo, K., Highly Conductive PEDOT:PSS Electrode with Optimized Solvent and Thermal Post-Treatment for ITO-Free Organic Solar Cells. *Adv. Funct. Mater.* **2011**, *21* (6), 1076–1081.

46. Su, Z. S.; Wang, L. D.; Li, Y. T.; Zhao, H. F.; Chu, B.; Li, W. L., Ultraviolet-Ozone-Treated PEDOT:PSS as Anode Buffer Layer for Organic Solar Cells. *Nanoscale Res. Lett.* **2012**, *7*, 1–6.
47. Petti, L.; Rippa, M.; Capasso, R.; Nenna, G.; De Girolamo Del Mauro, A.; Pandolfi, G.; Maglione, M. G.; Minarini, C., Fabrication of Novel Two-Dimensional Nanopatterned Conductive PEDOT:PSS Films for Organic Optoelectronic Applications. *ACS Appl. Mater. Interfaces* **2013**, *5* (11), 4777–4782.
48. Montibon, E.; Jarnstrom, L.; Lestelius, M., Characterization of Poly(3,4-ethylene-dioxythiophene)/Poly(Styrene Sulfonate) (PEDOT:PSS) Adsorption on Cellulosic Materials. *Cellulose* **2009**, *16* (5), 807–815.
49. Xia, Y. J.; Zhang, H. M.; Ouyang, J. Y., Highly Conductive PEDOT:PSS Films Prepared through a Treatment with Zwitterions and Their Application in Polymer Photovoltaic Cells. *J. Mater. Chem.* **2010**, *20* (43), 9740–9747.
50. Dobbelin, M.; Marcilla, R.; Salsamendi, M.; Pozo-Gonzalo, C.; Carrasco, P. M.; Pomposo, J. A.; Mecerreyes, D., Influence of Ionic Liquids on the Electrical Conductivity and Morphology of PEDOT:PSS Films. *Chem. Mater.* **2007**, *19* (9), 2147–2149.
51. Xia, Y. J.; Ouyang, J. Y., Significant Conductivity Enhancement of Conductive Poly(3,4-ethylenedioxythiophene):Poly(styrenesulfonate) Films through a Treatment with Organic Carboxylic Acids and Inorganic Acids. *ACS Appl. Mater. Interfaces* **2010**, *2* (2), 474–483.
52. Xia, Y. J.; Ouyang, J. Y., PEDOT:PSS Films with Significantly Enhanced Conductivities Induced by Preferential Solvation with Cosolvents and Their Application in Polymer Photovoltaic Cells. *J. Mater. Chem.* **2011**, *21* (13), 4927–4936.
53. Fang, G.; Wu, S. P.; Xie, Z. Y.; Geng, Y. H.; Wang, L. X., Enhanced Performance for Polymer Solar Cells by Using Surfactant-Modified PEDOT:PSS as the Anode Buffer Layer. *Macromol. Chem. Phys.* **2011**, *212* (17), 1846–1851.
54. Zhang, F. L.; Gadisa, A.; Inganas, O.; Svensson, M.; Andersson, M. R., Influence of Buffer Layers on the Performance of Polymer Solar Cells. *Appl. Phys. Lett.* **2004**, *84* (19), 3906–3908.
55. Li, J.; Liu, J. C.; Gao, C. J., Influence of PEDOT:PSS Film Doped with Sorbitol on Performances of Organic Solar Cells. *Acta Phys. Sin.* **2011**, *60* (7), 125102.
56. Yang, J. S.; Oh, S. H.; Kim, D. L.; Kim, S. J.; Kim, H. J., Hole Transport Enhancing Effects of Polar Solvents on Poly(3,4-ethylenedioxythiophene):Poly(styrene sulfonic acid) for Organic Solar Cells. *ACS Appl. Mater. Interfaces* **2012**, *4* (10), 5394–5398.
57. Wang, H. T.; Zhang, W. F.; Chen, B. X.; Yang, S. F., Enhancing Power Conversion Efficiency of Polymer Solar Cells via Treatment of PEDOT: PSS Anode Buffer Layer Using DMF Solvent. *J. Univ. Sci. Technol. China* **2012**, *42* (10), 775–784.
58. Lide, D. R., *CRC Handbook of Chemistry and Physics*, 85th Edition. CRC Press, Boca Raton, FL, **2004**, pp. 15–21.
59. Hu, Z. Y.; Zhang, J. J.; Hao, Z. H.; Zhao, Y., Influence of Doped PEDOT:PSS on the Performance of Polymer Solar Cells. *Sol. Energy Mater. Sol. Cells* **2011**, *95* (10), 2763–2767.
60. Keawprajak, A.; Koetniyom, W.; Piyakulawat, P.; Jiramitmongkon, K.; Pratontep, S.; Asawapirom, U., Effects of Tetramethylene Sulfone Solvent Additives on Conductivity of PEDOT:PSS Film and Performance of Polymer Photovoltaic Cells. *Org. Electron.* **2013**, *14* (1), 402–410.
61. Girotto, C.; Moia, D.; Rand, B. P.; Heremans, P., High-Performance Organic Solar Cells with Spray-Coated Hole-Transport and Active Layers. *Adv. Funct. Mater.* **2011**, *21* (1), 64–72.
62. Choi, H.; Lee, J.-P.; Ko, S.-J.; Jung, J.-W.; Park, H.; Yoo, S.; Park, O.; Jeong, J.-R.; Park, S.; Kim, J. Y., Multipositional Silica-Coated Silver Nanoparticles for High-Performance Polymer Solar Cells. *Nano Lett.* **2013**, *13* (5), 2204–2208.

63. Liu, S.; Meng, F.; Xie, W.; Zhang, Z.; Shen, L.; Liu, C.; He, Y.; Guo, W.; Ruan, S., Performance Improvement of Inverted Polymer Solar Cells by Doping Au Nanoparticles into TiO2 Cathode Buffer Layer. *Appl. Phys. Lett.* **2013**, *103* (23), 233303.
64. Lu, Q.; Lu, Z.; Lu, Y.; Lv, L.; Ning, Y.; Yu, H.; Hou, Y.; Yin, Y., Photocatalytic Synthesis and Photovoltaic Application of Ag-TiO$_2$ Nanorod Composites. *Nano Lett.* **2013**, *13* (11), 5698–5702.
65. Xu, X.; Kyaw, A. K. K.; Peng, B.; Zhao, D.; Wong, T. K. S.; Xiong, Q.; Sun, X. W.; Heeger, A. J., A Plasmonically Enhanced Polymer Solar Cell with Gold-Silica Core-Shell Nanorods. *Org. Electron.* **2013**, *14* (9), 2360–2368.
66. Kozanoglu, D.; Apaydin, D. H.; Cirpan, A.; Esenturk, E. N., Power Conversion Efficiency Enhancement of Organic Solar Cells by Addition of Gold Nanostars, Nanorods, and Nanospheres. *Org. Electron.* **2013**, *14* (7), 1720–1727.
67. Tan, Z. F.; Imae, I.; Ooyama, Y.; Komaguchi, K.; Ohshita, J.; Harima, Y., Low Bandgap Polymers with Benzodithiophene and Bisthienylacrylonitrile Units for Photovoltaic Applications. *Eur. Polym. J.* **2013**, *49* (6), 1634–1641.
68. Yin, B.; Liu, Q.; Yang, L.; Wu, X.; Liu, Z.; Hua, Y.; Yin, S.; Chen, Y., Buffer Layer of PEDOT:PSS/Graphene Composite for Polymer Solar Cells. *J. Nanosci. Nanotechnol.* **2010**, *10* (3), 1934–1938.
69. Kishi, N.; Kato, S.; Saito, T.; Hayashi, J.; Ito, D.; Hayashi, Y.; Soga, T.; Jimbo, T., Poly(3,4-ethylenedioxythiophene): Poly(Styrenesulfonate)/Single-Wall Carbon Nanotube Composite Film for the Hole Transport Layer in Polymer Solar Cells. *Nano* **2011**, *6* (6), 583–588.
70. Park, E. K.; Choi, M.; Jeun, J. H.; Lim, K. T.; Kim, J. M.; Kim, Y. S., The Effect of Metal Oxide Nanoparticle Concentrations in PEDOT:PSS Layer on the Performance of P3HT:PCBM Organic Solar Cells. *Microelectron. Eng.* **2013**, *111*, 166–169.
71. Shao, S.; Liu, J.; Bergqvist, J.; Shengwei, S.; Veit, C.; Wurfel, U. Z. X.; Zhang, F., In Situ Formation of MoO$_3$ in PEDOT:PSS Matrix: A Facile Way to Produce a Smooth and Less Hygroscopic Hole Transport Layer for Highly Stable Polymer Bulk Heterojunction Solar Cells. *Adv. Energy Mater.* **2013**, *3* (3), 349–355.
72. Lee, S. J.; Kim, H. P.; Yusoff, A. R. b. M.; Jang, J., Organic Photovoltaic with PEDOT:PSS and V$_2$O$_5$ Mixture as Hole Transport Layer. *Sol. Energy Mater. Sol. Cells* **2014**, *120*, 238–243.
73. Cruz-Cruz, I.; Reyes-Reyes, M.; Aguilar-Frutis, M. A.; Rodriguez, A. G.; Lopez-Sandoval, R., Study of the Effect of DMSO Concentration on the Thickness of the PSS Insulating Barrier in PEDOT:PSS Thin Films. *Synth. Met.* **2010**, *160* (13–14), 1501–1506.
74. Dimitriev, O. P.; Grinko, D. A.; Noskov, Y. V.; Ogurtsov, N. A.; Pud, A. A., PEDOT:PSS Films-Effect of Organic Solvent Additives and Annealing on the Film Conductivity. *Synth. Met.* **2009**, *159* (21–22), 2237–2239.
75. Alemu, D.; Wei, H. Y.; Ho, K. C.; Chu, C. W., Highly Conductive PEDOT:PSS Electrode by Simple Film Treatment with Methanol for ITO-Free Polymer Solar Cells. *Energy Environ. Sci.* **2012**, *5* (11), 9662–9671.
76. Crispin, X.; Jakobsson, F. L. E.; Crispin, A.; Grim, P. C. M.; Andersson, P.; Volodin, A.; van Haesendonck, C.; Van der Auweraer, M.; Salaneck, W. R.; Berggren, M., The Origin of the High Conductivity of Poly(3,4-ethylenedioxythiophene)-Poly(Styrenesulfonate) (PEDOT- PSS) Plastic Electrodes. *Chem. Mater.* **2006**, *18* (18), 4354–4360.
77. Mengistie, D. A.; Wang, P. C.; Chu, C. W., Effect of Molecular Weight of Additives on the Conductivity of PEDOT: PSS and Efficiency for ITO-Free Organic Solar Cells. *J. Mater. Chem. A* **2013**, *1* (34), 9907–9915.
78. Zilberberg, K.; Gharbi, H.; Behrendt, A.; Trost, S.; Riedl, T., Low-Temperature, Solution-Processed MoO$_x$ for Efficient and Stable Organic Solar Cells. *ACS Appl. Mater. Interfaces* **2012**, *4* (3), 1164–1168.

79. Choi, H.; Kim, B.; Ko, M. J.; Lee, D.-K.; Kim, H.; Kim, S. H.; Kim, K., Solution Processed WO$_3$ Layer for the Replacement of PEDOT:PSS Layer in Organic Photovoltaic Cells. *Org. Electron.* **2012**, *13* (6), 959–968.
80. Zilberberg, K.; Trost, S.; Meyer, J.; Kahn, A.; Behrendt, A.; Luetzenkirchen-Hecht, D.; Frahm, R.; Riedl, T., Inverted Organic Solar Cells with Sol-Gel Processed High Work-Function Vanadium Oxide Hole-Extraction Layers. *Adv. Funct. Mater.* **2011**, *21* (24), 4776–4783.
81. Meyer, J.; Zilberberg, K.; Riedl, T.; Kahn, A., Electronic Structure of Vanadium Pentoxide: An Efficient Hole Injector for Organic Electronic Materials. *J. Appl. Phys.* **2011**, *110* (3), 033710.
82. Laubach, S.; Schmidt, P. C.; Thissen, A.; Fernandez-Madrigal, F. J.; Wu, Q.-H.; Jaegermann, W.; Klemm, M.; Horn, S., Theoretical and Experimental Determination of the Electronic Structure of V$_2$O$_5$, Reduced V$_2$O$_{5-x}$ and Sodium Intercalated NaV$_2$O$_5$. *Phys. Chem. Chem. Phys.* **2007**, *9* (20), 2564–2576.
83. de Jong, M. P.; van Ijzendoorn, L. J.; de Voigt, M. J. A., Stability of the Interface between Indium-Tin-Oxide and Poly(3,4-ethylenedioxythiophene)/Poly(Styrenesulfonate) in Polymer Light-Emitting Diodes. *Appl. Phys. Lett.* **2000**, *77* (14), 2255–2257.
84. Steirer, K. X.; Ndione, P. F.; Widjonarko, N. E.; Lloyd, M. T.; Meyer, J.; Ratcliff, E. L.; Kahn, A. et al., Enhanced Efficiency in Plastic Solar Cells via Energy Matched Solution Processed NiO$_x$ Interlayers. *Adv. Energy Mater.* **2011**, *1* (5), 813–820.
85. Steirer, K. X.; Chesin, J. P.; Widjonarko, N. E.; Berry, J. J.; Miedaner, A.; Ginley, D. S.; Olson, D. C., Solution Deposited NiO Thin-Films as Hole Transport Layers in Organic Photovoltaics. *Org. Electron.* **2010**, *11* (8), 1414–1418.
86. Berry, J. J.; Widjonarko, N. E.; Bailey, B. A.; Sigdel, A. K.; Ginley, D. S.; Olson, D. C., Surface Treatment of NiO Hole Transport Layers for Organic Solar Cells. *IEEE J. Sel. Topics Quantum Electron.* **2010**, *16* (6), 1649–1655.
87. Chen, S.; Manders, J. R.; Tsang, S.-W.; So, F., Metal Oxides for Interface Engineering in Polymer Solar Cells. *J. Mater. Chem.* **2012**, *22* (46), 24202–24212.
88. Meyer, J.; Hamwi, S.; Kroger, M.; Kowalsky, W.; Riedl, T.; Kahn, A., Transition Metal Oxides for Organic Electronics: Energetics, Device Physics and Applications. *Adv. Mater.* **2012**, *24* (40), 5408–5427.
89. Tan, Z. A.; Zhang, W. Q.; Qian, D. P.; Cui, C. H.; Xu, Q.; Li, L. J.; Li, S. S.; Li, Y. F., Solution-Processed Nickel Acetate as Hole Collection Layer for Polymer Solar Cells. *Phys. Chem. Chem. Phys.* **2012**, *14* (41), 14217–14223.
90. Xie, F. X.; Choy, W. C. H.; Wang, C. D.; Li, X. C.; Zhang, S. Q.; Hou, J. H., Low-Temperature Solution-Processed Hydrogen Molybdenum and Vanadium Bronzes for an Efficient Hole-Transport Layer in Organic Electronics. *Adv. Mater.* **2013**, *25* (14), 2051–2055.
91. Greiner, M. T.; Helander, M. G.; Tang, W.-M.; Wang, Z.-B.; Qiu, J.; Lu, Z.-H., Universal Energy-Level Alignment of Molecules on Metal Oxides. *Nat. Mater.* **2012**, *11* (1), 76–81.
92. Xu, Q.; Wang, F. Z.; Tan, Z. A.; Li, L. J.; Li, S. S.; Hou, X. L.; Sun, G.; Tu, X. H.; Hou, J. H.; Li, Y. F., High-Performance Polymer Solar Cells with Solution-Processed and Environmentally Friendly CuO$_x$ Anode Buffer Layer. *ACS Appl. Mater. Interfaces* **2013**, *5* (21), 10658–10664.
93. Li, X.; Zhang, W.; Wu, Y.; Min, C.; Fang, J., Solution-Processed MoS(x) as an Efficient Anode Buffer Layer in Organic Solar Cells. *ACS Appl. Mater. Interfaces* **2013**, *5* (18), 8823–8827.
94. Boopathi, K. M.; Raman, S.; Mohanraman, R.; Fang-Cheng, C.; Yang-Yuang, C.; Chih-Hao, L.; Feng-Chih, C.; Chih-Wei, C., Solution-Processable Bismuth Iodide Nanosheets as Hole Transport Layers for Organic Solar Cells. *Sol. Energy Mater. Sol. Cells* **2014**, *121*, 35–41.

95. Stevens, D. M.; Qin, Y.; Hillmyer, M. A.; Frisbie, C. D., Enhancement of the Morphology and Open Circuit Voltage in Bilayer Polymer/Fullerene Solar Cells. *J. Phys. Chem. C* **2009**, *113* (26), 11408–11415.
96. Ecker, B.; Posdorfer, J.; von Hauff, E., Influence of Hole Extraction Efficiency on the Performance and Stability of Organic Solar Cells. *Sol. Energy Mater. Sol. Cells* **2013**, *116*, 176–181.
97. Chen, H. Y.; Hou, J. H.; Zhang, S. Q.; Liang, Y. Y.; Yang, G. W.; Yang, Y.; Yu, L. P.; Wu, Y.; Li, G., Polymer Solar Cells with Enhanced Open-Circuit Voltage and Efficiency. *Nat. Photon.* **2009**, *3* (11), 649–653.
98. da Silva, W. J.; Kim, H. P.; Yusoff, A. R. B.; Jang, J., Transparent Flexible Organic Solar Cells with 6.87% Efficiency Manufactured by an All-Solution Process. *Nanoscale* **2013**, *5* (19), 9324–9329.
99. Chen, K. S.; Salinas, J. F.; Yip, H. L.; Huo, L. J.; Hou, J. H.; Jen, A. K. Y., Semi-Transparent Polymer Solar Cells with 6% PCE, 25% Average Visible Transmittance and a Color Rendering Index Close to 100 for Power Generating Window Applications. *Energy Environ. Sci.* **2012**, *5* (11), 9551–9557.
100. Lim, T. H.; Oh, K. W.; Kim, S. H., Self-Assembly Supramolecules to Enhance Electrical Conductivity of Polyaniline for a Flexible Organic Solar Cells Anode. *Sol. Energy Mater. Sol. Cells* **2012**, *101*, 232–240.
101. Chang, M. Y.; Wu, C. S.; Chen, Y. F.; Hsieh, B. Z.; Huang, W. Y.; Ho, K. S.; Hsieh, T. H.; Han, Y. K., Polymer Solar Cells Incorporating One-Dimensional Polyaniline Nanotubes. *Org. Electron.* **2008**, *9* (6), 1136–1139.
102. Han, Y. K.; Lee, Y. J.; Huang, P. C., Regioregularity Effects in Poly(3-hexylthiophene): PCBM-Based Solar Cells Incorporating Acid-Doped Polyaniline Nanotubes as an Interfacial Layer. *J. Electrochem. Soc.* **2009**, *156* (4), K37–K43.
103. Hains, A. W.; Liu, J.; Martinson, A. B. F.; Irwin, M. D.; Marks, T. J., Anode Interfacial Tuning via Electron-Blocking/Hole-Transport Layers and Indium Tin Oxide Surface Treatment in Bulk-Heterojunction Organic Photovoltaic Cells. *Adv. Funct. Mater.* **2010**, *20* (4), 595–606.
104. Yoo, I.; Lee, M.; Lee, C.; Kim, D. W.; Moon, I. S.; Hwang, D. H., The Effect of a Buffer Layer on the Photovoltaic Properties of Solar Cells with P3OT: Fullerene Composites. *Synth. Met.* **2005**, *153* (1–3), 97–100.
105. Jung, J. W.; Lee, J. U.; Jo, W. H., High-Efficiency Polymer Solar Cells with Water-Soluble and Self-Doped Conducting Polyaniline Graft Copolymer as Hole Transport Layer. *J. Phys. Chem. C* **2010**, *114* (1), 633–637.
106. Hains, A. W.; Marks, T. J., High-Efficiency Hole Extraction/Electron-Blocking Layer to Replace Poly(3,4-ethylenedioxythiophene): Poly(Styrene Sulfonate) in Bulk-Heterojunction Polymer Solar Cells. *Appl. Phys. Lett.* **2008**, *92* (2), 023504.
107. Hains, A. W.; Ramanan, C.; Irwin, M. D.; Liu, J.; Wasielewski, M. R.; Marks, T. J., Designed Bithiophene-Based Interfacial Layer for High-Efficiency Bulk-Heterojunction Organic Photovoltaic Cells. Importance of Interfacial Energy Level Matching. *ACS Appl. Mater. Interfaces* **2010**, *2* (1), 175–185.
108. Lacher, S.; Obata, N.; Luo, S. C.; Matsuo, Y.; Zhu, B.; Yu, H. H.; Nakamura, E., Electropolymerized Conjugated Polyelectrolytes with Tunable Work Function and Hydrophobicity as an Anode Buffer in Organic Optoelectronics. *ACS Appl. Mater. Interfaces* **2012**, *4* (7), 3396–3404.
109. Nasybulin, E.; Wei, S.; Cox, M.; Kymissis, I.; Levon, K., Morphological and Spectroscopic Studies of Electrochemically Deposited Poly(3,4-ethylenedioxythiophene) (PEDOT) Hole Extraction Layer for Organic Photovoltaic Device (OPVd) Fabrication. *J. Phys. Chem. C* **2011**, *115* (10), 4307–4314.
110. Rider, D. A.; Harris, K. D.; Wang, D.; Bruce, J.; Fleischauer, M. D.; Tucker, R. T.; Brett, M. J.; Buriak, J. M., Thienylsilane-Modified Indium Tin Oxide as an Anodic Interface in Polymer/Fullerene Solar Cells. *ACS Appl. Mater. Interfaces* **2009**, *1* (2), 279–288.

111. Bhagat, S. A.; Raut, S. B.; Dhoble, S. J., Study of Photophysical Properties of Different Metal Complexes of Alq(3). *Luminescence* **2013**, *28* (5), 755–759.
112. Kim, K. S.; Hwang, Y. W.; Won, T. Y., Numerical Study on Exciton Transport and Light Emission for Organic Light Emitting Diodes with an Emission Layer. *J. Nanosci. Nanotechnol.* **2013**, *13* (12), 8050–8054.
113. Jin, J.; Lee, J.; Jeong, S.; Yang, S.; Ko, J. H.; Im, H. G.; Baek, S. W.; Lee, J. Y.; Bae, B. S., High-Performance Hybrid Plastic Films: A Robust Electrode Platform for Thin-Film Optoelectronics. *Energy Environ. Sci.* **2013**, *6* (6), 1811–1817.
114. Yun, S. O.; Hwang, Y.; Park, J.; Jeong, Y.; Kim, S. H.; Noh, B. I.; Jung, H. S. et al., Sticker-Type Alq(3)-Based OLEDs Based on Printable Ultrathin Substrates in Periodically Anchored and Suspended Configurations. *Adv. Mater.* **2013**, *25* (39), 5626–5631.
115. Vivo, P.; Jukola, J.; Ojala, M.; Chukharev, V.; Lemmetyinen, H., Influence of Alq(3)/Au Cathode on Stability and Efficiency of a Layered Organic Solar Cell in Air. *Sol. Energy Mater. Sol. Cells* **2008**, *92* (11), 1416–1420.
116. Tolkki, A.; Kaunisto, K.; Heiskanen, J. P.; Omar, W. A. E.; Huttunen, K.; Lehtimaki, S.; Hormi, O. E. O.; Lemmetyinen, H., Organometallic Tris(8-Hydroxyquinoline) Aluminum Complexes as Buffer Layers and Dopants in Inverted Organic Solar Cells. *Thin Solid Films* **2012**, *520* (13), 4475–4481.
117. Guan, Z.-Q.; Yu, J.-S.; Zang, Y.; Zeng, X.-X., Inverted Organic Solar Cells with Improved Performance using Varied Cathode Buffer Layers. *Chinese J. Chem. Phys.* **2012**, *25* (5), 625–630.
118. Guo-Fu, M.; Hao-Jun, X.; Pan-Pan, C.; Yan-Qing, L.; Jian-Xin, T., Performance Enhancement of Polymer Solar Cells with Luminescent Down-Shifting Sensitizer. *Appl. Phys. Lett.* **2013**, *103* (4), 043302.
119. Manninen, V. M.; Omar, W. A. E.; Heiskanen, J. P.; Lemmetyinen, H. J.; Hormi, O. E. O., Synthesis and Characterization of Tris-(5-Amino-8-Hydroxyquinoline) Aluminum Complexes and Their Use as Anode Buffer Layers in Inverted Organic Solar Cells. *J. Mater. Chem.* **2012**, *22* (43), 22971–22982.
120. Lu, K. Y.; Yuan, J. Y.; Peng, J.; Huang, X. D.; Cui, L. S.; Jiang, Z. Q.; Wang, H. Q.; Ma, W. L., New Solution-Processable Small Molecules as Hole-Transporting Layer in Efficient Polymer Solar Cells. *J. Mater. Chem. A* **2013**, *1* (45), 14253–14261.
121. Lee, H. K.; Kim, J. K.; Park, O. O., Effects of UV Light-Irradiated Buffer Layer on the Performance of Polymer Solar Cells. *Org. Electron.* **2009**, *10* (8), 1641–1644.
122. Vaynzof, Y.; Kabra, D.; Zhao, L. H.; Ho, P. K. H.; Wee, A. T. S.; Friend, R. H., Improved Photoinduced Charge Carriers Separation in Organic–Inorganic Hybrid Photovoltaic Devices. *Appl. Phys. Lett.* **2010**, *97* (3), 033309.
123. Wong, K. W.; Yip, H. L.; Luo, Y.; Wong, K. Y.; Lau, W. M.; Low, K. H.; Chow, H. F.; Gao, Z. Q.; Yeung, W. L.; Chang, C. C., Blocking Reactions between Indium-Tin Oxide and Poly (3,4-Ethylene Dioxythiophene): Poly(Styrene Sulphonate) with a Self-Assembly Monolayer. *Appl. Phys. Lett.* **2002**, *80* (15), 2788–2790.
124. Kim, J. S.; Park, J. H.; Lee, J. H.; Jo, J.; Kim, D. Y.; Cho, K., Control of the Electrode Work Function and Active Layer Morphology via Surface Modification of Indium Tin Oxide for High Efficiency Organic Photovoltaics. *Appl. Phys. Lett.* **2007**, *91* (11), 112111.
125. Huang, L. C.; Liu, H. W.; Liang, C. W.; Chou, T. R.; Wang, L.; Chao, C. Y., Self-Assembled Multilayers of Modified ITO in Polymer Solar Cells by Soft-Imprinting. *Soft Matter* **2012**, *8* (5), 1467–1472.
126. Li, T.; Hauptmann, J. R.; Wei, Z.; Petersen, S.; Bovet, N.; Vosch, T.; Nygard, J. et al., Solution-Processed Ultrathin Chemically Derived Graphene Films as Soft Top Contacts for Solid-State Molecular Electronic Junctions. *Adv. Mater.* **2012**, *24* (10), 1333–1339.
127. Li, S.-S.; Tu, K.-H.; Lin, C.-C.; Chen, C.-W.; Chhowalla, M., Solution-Processable Graphene Oxide as an Efficient Hole Transport Layer in Polymer Solar Cells. *ACS Nano* **2010**, *4* (6), 3169–3174.

128. Matyba, P.; Yamaguchi, H.; Eda, G.; Chhowalla, M.; Edman, L.; Robinson, N. D., Graphene and Mobile Ions: The Key to All-Plastic, Solution-Processed Light-Emitting Devices. *ACS Nano* **2010,** *4* (2), 637–642.
129. Lee, Y.-Y.; Tu, K.-H.; Yu, C.-C.; Li, S.-S.; Hwang, J.-Y.; Lin, C.-C.; Chen, K.-H.; Chen, L.-C.; Chen, H.-L.; Chen, C.-W., Top Laminated Graphene Electrode in a Semitransparent Polymer Solar Cell by Simultaneous Thermal Annealing/Releasing Method. *ACS Nano* **2011,** *5* (8), 6564–6570.
130. Pang, S.; Hernandez, Y.; Feng, X.; Muellen, K., Graphene as Transparent Electrode Material for Organic Electronics. *Adv. Mater.* **2011,** *23* (25), 2779–2795.
131. Teweldebrhan, D.; Balandin, A. A., Modification of graphene properties due to electron-beam irradiation. *Appl. Phys. Lett.* **2009,** *94* (1), 013101.
132. Fan, J.; Michalik, J. M.; Casado, L.; Roddaro, S.; Ibarra, M. R.; De Teresa, J. M., Investigation of the Influence on Graphene by Using Electron-Beam and Photo-Lithography. *Solid State Commun.* **2011,** *151* (21), 1574–1578.
133. Becerril, H. A.; Stoltenberg, R. M.; Tang, M. L.; Roberts, M. E.; Liu, Z.; Chen, Y.; Kim, D. H.; Lee, B.-L.; Lee, S.; Bao, Z., Fabrication and Evaluation of Solution-Processed Reduced Graphene Oxide Electrodes for p- and n-Channel Bottom-Contact Organic Thin-Film Transistors. *ACS Nano* **2010,** *4* (11), 6343–6352.
134. Lee, B. R.; Kim, J.-W.; Kang, D.; Lee, D. W.; Ko, S.-J.; Lee, H. J.; Lee, C.-L.; Kim, J. Y.; Shin, H. S.; Song, M. H., Highly Efficient Polymer Light-Emitting Diodes Using Graphene Oxide as a Hole Transport Layer. *ACS Nano* **2012,** *6* (4), 2984–2991.
135. Yun, J.-M.; Yeo, J.-S.; Kim, J.; Jeong, H.-G.; Kim, D.-Y.; Noh, Y.-J.; Kim, S.-S.; Ku, B.-C.; Na, S.-I., Solution-Processable Reduced Graphene Oxide as a Novel Alternative to PEDOT:PSS Hole Transport Layers for Highly Efficient and Stable Polymer Solar Cells. *Adv. Mater.* **2011,** *23* (42), 4923–4928.
136. Pang, S.; Tsao, H. N.; Feng, X.; Muellen, K., Patterned Graphene Electrodes from Solution-Processed Graphite Oxide Films for Organic Field-Effect Transistors. *Adv. Mater.* **2009,** *21* (34), 3488.
137. Kim, H. P.; Yusoff, A. R. b. M.; Ryu, M. S.; Jang, J., Stable Photovoltaic Cells Based on Graphene Oxide/Indium Zinc Oxide Bilayer Anode Buffer. *Org. Electron.* **2012,** *13* (12), 3195–3202.
138. Stubhan, T.; Krantz, J.; Li, N.; Guo, F.; Litzov, I.; Steidl, M.; Richter, M.; Matt, G. J.; Brabec, C. J., High Fill Factor Polymer Solar Cells Comprising a Transparent, Low Temperature Solution Processed Doped Metal Oxide/Metal Nanowire Composite Electrode. *Sol. Energy Mater. Sol. Cells* **2012,** *107*, 248–251.
139. Yang, Q. D.; Ng, T. W.; Lo, M. F.; Wong, N. B.; Lee, C. S., Enhanced Storage/Operation Stability of Small Molecule Organic Photovoltaics Using Graphene Oxide Interfacial Layer. *Org. Electron.* **2012,** *13* (12), 3220–3225.
140. Yusoff, A. R. b. M.; Kim, H. P.; Jang, J., Comparison of Organic Photovoltaic with Graphene Oxide Cathode and Anode Buffer Layers. *Org. Electron.* **2012,** *13* (11), 2379–2385.
141. Yusoff, A. R. b. M.; Kim, H. P.; Jang, J., Inverted Organic Solar Cells with TiO$_x$ Cathode and Graphene Oxide Anode Buffer Layers. *Sol. Energy Mater. Sol. Cells* **2013,** *109*, 63–69.
142. Jeon, Y. J.; Yun, J. M.; Kim, D. Y.; Na, S. I.; Kim, S. S., High-Performance Polymer Solar Cells with Moderately Reduced Graphene Oxide as an Efficient Hole Transporting Layer. *Sol. Energy Mater. Sol. Cells* **2012,** *105*, 96–102.
143. Pan, Z.; Gu, H. L.; Wu, M. T.; Li, Y. X.; Chen, Y., Graphene-Based Functional Materials for Organic Solar Cells [Invited]. *Opt. Mater. Express* **2012,** *2* (6), 814–824.
144. Chao, Y. H.; Wu, J. S.; Wu, C. E.; Jheng, J. F.; Wang, C. L.; Hsu, C. S., Solution-Processed (Graphene Oxide)-(d(0) Transition Metal Oxide) Composite Anodic Buffer Layers toward High-Performance and Durable Inverted Polymer Solar Cells. *Adv. Energy Mater.* **2013,** *3* (10), 1279–1285.

145. Li, G.; Shrotriya, V.; Huang, J. S.; Yao, Y.; Moriarty, T.; Emery, K.; Yang, Y., High-Efficiency Solution Processable Polymer Photovoltaic Cells by Self-Organization of Polymer Blends. *Nat. Mater.* **2005**, *4*, 864.
146. Gupta, V.; Kyaw, A. K. K.; Wang, D. H.; Chand, S.; Bazan, G. C.; Heeger, A. J., Barium: An Efficient Cathode Layer for Bulk-Heterojunction Solar Cells. *Sci. Rep.* **2013**, *3*, 1965.
147. Steim, R.; Kogler, F. R.; Brabec, C. J., Interface Materials for Organic Solar Cells. *J. Mater. Chem.* **2010**, *20*, 2499.
148. Chen, L. M.; Xu, Z.; Hong, Z. R.; Yang, Y., Interface Investigation and Engineering-Achieving High Performance Polymer Photovoltaic Devices. *J. Mater. Chem.* **2010**, *20*, 2575.
149. Ma, H.; Yip, H. L.; Huang, F.; Jen, A. K. Y., Interface Engineering for Organic Electronics. *Adv. Funct. Mater.* **2010**, *20*, 1371.
150. Lee, T. W.; Lim, K. G.; Kim, D. H., Approaches toward Efficient and Stable Electron Extraction Contact in Organic Photovoltaic Cells: Inspiration from Organic Light-Emitting Diodes. *Electron. Mater. Lett.* **2010**, *6*, 41.
151. Yip, H. L.; Jen, A. K., Recent Advances in Solution-Processed Interfacial Materials for Efficient and Stable Polymer Solar Cells. *Energy Environ. Sci.* **2012**, *5*, 5994.
152. Brabec, C. J.; Cravino, A.; Meissner, D.; Sariciftci, N. S.; Fromherz, T.; Rispens, M. T.; Sanchez, L.; Hummelen, J. C., Origin of the Open Circuit Voltage of Plastic Solar Cells. *Adv. Funct. Mater.* **2001**, *11*, 374.
153. Tseng, W. H.; Chen, M. H.; Wang, J. Y.; Tseng, C. T.; Lo, H.; Wang, P. S.; Wu, C., Investigations of Efficiency Improvements in Poly(3-Hexylthiophene) Based Organic Solar Cells Using Calcium Cathodes. *Sol. Energy Mater. Sol. Cells* **2011**, *95*, 3424.
154. Jorgensen, M.; Norrman, K.; Krebs, F. C., Stability/Degradation of Polymer Solar Cells. *Sol. Energy Mater. Sol. Cells* **2008**, *92*, 686.
155. Kim, J. Y.; Kim, S. H.; Lee, H. H.; Lee, K.; Ma, W.; Gong, X.; Heeger, A. J., New Architecture for High-Efficiency Polymer Photovoltaic Cells Using Solution-Based Titanium Oxide as an Optical Spacer. *Adv. Mater.* **2006**, *18*, 572.
156. Kim, S. H.; Park, S. H.; Lee, K., Efficiency Enhancement in Polymer Optoelectronic Devices by Introducing Titanium Sub-Oxide Layer. *Curr. Appl. Phys.* **2010**, *10*, S528.
157. Park, S. H.; Roy, A.; Beaupre, S.; Cho, S.; Coates, N.; Moon, J. S.; Moses, D.; Leclerc, M.; Lee, K.; Heeger, A. J., Bulk Heterojunction Solar Cells with Internal Quantum Efficiency Approaching 100%. *Nat. Photon.* **2009**, *3*, 297.
158. Park, M. H.; Li, J. H.; Kumar, A.; Li, G.; Yang, Y., Doping of the Metal Oxide Nanostructure and its Influence in Organic Electronics. *Adv. Funct. Mater.* **2009**, *19*, 1241.
159. Hau, S. K.; Yip, H. L.; Baek, N. S.; Zou, J. Y.; O'Malley, K.; Jen, A. K. Y., Air-Stable Inverted Flexible Polymer Solar Cells Using Zinc Oxide Nanoparticles as an Electron Selective Layer. *Appl. Phys. Lett.* **2008**, *92*, 253301.
160. Hau, S. K.; Yip, H. L.; Ma, H.; Jen, A. K. Y., High Performance Ambient Processed Inverted Polymer Solar Cells through Interfacial Modification with a Fullerene Self-Assembled Monolayer. *Appl. Phys. Lett.* **2008**, *93*, 233304.
161. Wang, J. C.; Weng, W. T.; Tsai, M. Y.; Lee, M. K.; Horng, S. F.; Perng, T. P.; Kei, C. C.; Yu, C. C.; Meng, H. F., Highly Efficient Flexible Inverted Organic Solar Cells Using Atomic Layer Deposited ZnO as Electron Selective Layer. *J. Mater. Chem.* **2010**, *20*, 862.
162. Gilot, J.; Wienk, M. M.; Janssen, R. A. J., Double and Triple Junction Polymer Solar Cells Processed from Solution. *Appl. Phys. Lett.* **2007**, *90*, 143512.
163. Salomon, A.; Berkovich, D.; Cahen, D., Molecular Modification of an Ionic Semiconductor–Metal Interface: ZnO/Molecule/Au Diodes. *Appl. Phys. Lett.* **2003**, *82*, 1051.
164. Monson, T. C.; Lloyd, M. T.; Olson, D. C.; Lee, Y. J.; Hsu, J. W. P., Photocurrent Enhancement in Polythiophene- and Alkanethiol-Modified ZnO Solar Cells. *Adv. Mater.* **2008**, *20*, 4755.

165. Zhou, Y.; Cheun, H.; Postcavage, Jr., W. J.; Fuentes-Hernandez, C.; Kim, S. J.; Kippelen, B., Inverted Organic Solar Cells with ITO Electrodes Modified with an Ultrathin Al2O3 Buffer Layer Deposited by Atomic Layer Deposition. *J. Mater. Chem.* **2010,** *20,* 6189.
166. Zhao, Y.; Xie, Z.; Qin, C.; Qu, Y.; Geng, Y.; Wang, L., Effects of Thermal Annealing on Polymer Photovoltaic Cells with Buffer Layers and In Situ Formation of Interfacial Layer for Enhancing Power Conversion Efficiency. *Synth. Met.* **2008,** *158,* 908.
167. Lee, J.; Park, Y.; Kim, D. Y.; Chu, H. Y.; Lee, H.; Do, L. M., High Efficiency Organic Light-Emitting Devices with Al/NaF Cathode. *Appl. Phys. Lett.* **2003,** *82,* 173.
168. Shaheen, S. E.; Jabbour, G. E.; Morrell, M. M.; Kawabe, Y.; Kippelen, B.; Peyghambarian, N.; Nabor, M. F.; Schlaf, R.; Mash, E. A.; Armstrong, N. R., Bright Blue Organic Light-Emitting Diode with Improved Color Purity Using a LiF/Al Cathode. *J. Appl. Phys.* **1998,** *84,* 2324.
169. Brabec, C. J.; Shaheen, S. E.; Winder, C.; Sariciftci, N. S.; Denk, P., Effect of LiF/Metal Electrodes on the Performance of Plastic Solar Cells. *Appl. Phys. Lett.* **2002,** *80,* 1288.
170. Ahlswede, E.; Hanish, J.; Powalla, M., Comparative Study of the Influence of LiF, NaF, and KF on the Performance of Polymer Bulk Heterojunction Solar Cells. *Appl. Phys. Lett.* **2007,** *90,* 163504.
171. Jiang, X.; Xu, H.; Yang, L.; Shi, M.; Wang, M.; Chen, H., Effect of CsF Interlayer on the Performance of Polymer Bulk Heterojunction Solar Cells. *Sol. Energy Mater. Sol. Cells* **2009,** *93,* 650.
172. Liao, H. H.; Chen, L. M.; Xu, Z.; Li, G.; Yang, Y., Highly Efficient Inverted Polymer Solar Cell by Low Temperature Annealing of Cs_2CO_3 Interlayer. *Appl. Phys. Lett.* **2008,** *92,* 173303.
173. Chen, F. C.; Wu, J. L.; Yang, S. S.; Hsieh, K. H.; Chen, W. C., Cesium Carbonate as a Functional Interlayer for Polymer Photovoltaic Devices. *J. Appl. Phys.* **2008,** *103,* 103721.
174. Chang, C. C.; Lin, C. F.; Chiou, J. M.; Ho, T. H.; Tai, Y.; Lee, J. H.; Chen, Y. F.; Wang, J. K.; Chen, L. C.; Chen, K. H., Effects of Cathode Buffer Layers on the Efficiency of Bulk-Heterojunction Solar Cells. *Appl. Phys. Lett.* **2010,** *96,* 263506.
175. Sahin, Y.; Alem, S.; de Bettignies, R.; Nunzi, J. M., Development of Air Stable Polymer Solar Cells Using an Inverted Gold on Top Anode Structure. *Thin Solid Films* **2005,** *476,* 340.
176. Feng, Z.; Hou, Y.; Lei, D., The Influence of Electrode Buffer Layers on the Performance of Polymer Photovoltaic Devices. *Renew. Energy* **2010,** *35,* 1175.
177. Kim, J. H.; Huh, S. Y.; Kim, T. I.; Lee, H. H., Thin Pentacene Interlayer for Polymer Bulk-Heterojunction Solar Cell. *Appl. Phys. Lett.* **2008,** *93,* 143305.
178. Du, H.; Deng, Z.; Lü, Z.; Chen, Z.; Zou, Y.; Yin, Y.; Xu, D.; Wang, Y., The Effect of Small-Molecule Electron Transporting Materials on the Performance of Polymer Solar Cells. *Thin Solid Films* **2011,** *519,* 4357.
179. Li, J.; Huang, X.; Yuan, J.; Lu, K.; Yue, W.; Ma, W., A New Alcohol-Soluble Electron-Transporting Molecule for Efficient Inverted Polymer Solar Cells. *Org. Electron.* **2013,** *14,* 2164.
180. Tan, Z.; Yang, C.; Zhou, E.; Wang, X.; Li, Y., Performance Improvement of Polymer Solar Cells by Using a Solution Processable Titanium Chelate as Cathode Buffer Layer. *Appl. Phys. Lett.* **2007,** *91,* 023509.
181. Tan, Z.; Zhang, W.; Zhang, Z.; Qian, D.; Huang, Y.; Hou, J.; Li, Y., High-Performance Inverted Polymer Solar Cells with Solution Processed Titanium Chelate as Electron-Collecting Layer on ITO Electrode. *Adv. Mater.* **2012,** *24,* 1476.
182. Zhao, X. M.; Xu, C. H.; Wang, H. T.; Chen, F.; Zhang, W. F.; Zhao, Z. Q.; Chen, L. W.; Yang, S. F., Application of Biuret, Dicyandiamide, or Urea as a Cathode Buffer Layer toward the Efficiency Enhancement of Polymer Solar Cells. *ACS Appl. Mater. Interfaces* **2014,** *6,* 4329.

183. Zhang, W.; Wang, H.; Chen, B.; Bi, X.; Venkatesan, S.; Qiao, Q.; Yang, S., Oleamide as a Self-Assembled Cathode Buffer Layer for Polymer Solar Cells: The Role of the Terminal Group on the Function of the Surfactant. *J. Mater. Chem.* **2012**, *22*, 24067.
184. Zhang, F. L.; Ceder, M.; Inganäs, O., Enhancing the Photovoltage of Polymer Solar Cells by Using a Modified Cathode. *Adv. Mater.* **2007**, *19*, 1835.
185. Chen, F. C.; Chien, S. C., Nanoscale Functional Interlayers Formed through Spontaneous Vertical Phase Separation in Polymer Photovoltaic Devices. *J. Mater. Chem.* **2009**, *19*, 6865.
186. Yamakawa, S.; Tajima, K.; Hashimoto, K., Buffer Layer Formation in Organic Photovoltaic Cells by Self-Organization of Poly(dimethylsiloxane)s. *Org. Electron.* **2009**, *10*, 511.
187. Jung, J. W.; Jo, J. W.; Jo, W. H., Enhanced Performance and Air Stability of Polymer Solar Cells by Formation of a Self-Assembled Buffer Layer from Fullerene-End-Capped Poly(ethylene glycol). *Adv. Mater.* **2011**, *23*, 1782.
188. Tai, Q. D.; Li, J. H.; Liu, Z. K.; Sun, Z. H.; Zhao, X. Z.; Yan, F., Enhanced Photovoltaic Performance of Polymer Solar Cells by Adding Fullerene End-Capped Polyethylene Glycol. *J. Mater. Chem.* **2011**, *21*, 6848.
189. Wang, H.; Zhang, W.; Xu, C.; Bi, X.; Chen, B.; Yang, S., Efficiency Enhancement of Polymer Solar Cells by Applying Poly(Vinylpyrrolidone) as a Cathode Buffer Layer via Spin Coating or Self-Assembly. *ACS Appl. Mater. Interfaces* **2013**, *5*, 26.
190. Huang, F.; Wu, H. B.; Cao, Y., Water/Alcohol Soluble Conjugated Polymers as Highly Efficient Electrontransporting/Injection Layer in Optoelectronic Devices. *Chem. Soc. Rev.* **2010**, *39*, 2500.
191. Huang, F.; Wu, H. B.; Peng, J. B.; Yang, W.; Cao, Y., Polyfluorene Polyelectrolytes and Their Precursors Processable from Environment-Friendly Solvents (Alcohol or Water) for PLED Applications. *Curr. Org. Chem.* **2007**, *11*, 1207.
192. Guan, X.; Zhang, K.; Huang, F.; Bazan, G. C.; Cao, Y., Amino N-Oxide Functionalized Conjugated Polymers and Their Amino-Functionalized Precursors: New Cathode Interlayers for High-Performance Optoelectronic Devices. *Adv. Funct. Mater.* **2012**, *22*, 2846.
193. He, Z. C.; Zhong, C. M.; Huang, X.; Wong, W.-Y.; Wu, H. B.; Chen, L. W.; Su, S. J.; Cao, Y., Simultaneous Enhancement of Open-Circuit Voltage, Short-Circuit Current Density, and Fill Factor in Polymer Solar Cells. *Adv. Mater.* **2011**, *23*, 4636.
194. He, Z.; Zhong, C.; Su, S.; Xu, M.; Wu, H.; Cao, Y., Enhanced Power-Conversion Efficiency in Polymer Solar Cells Using an Inverted Device Structure. *Nat. Photon.* **2012**, *6*, 591.
195. Wei, Q. H.; Nishizawa, T.; Tajima, K.; Hashimoto, K., Self-Organized Buffer Layers in Organic Solar Cells. *Adv. Mater.* **2008**, *20*, 2211.
196. Hsieh, C. H.; Cheng, Y. J.; Li, P. J.; Chen, C. H.; Dubosc, M.; Liang, R. M.; Hsu, C. S., Highly Efficient and Stable Inverted Polymer Solar Cells Integrated with a Cross-Linked Fullerene Material as an Interlayer. *J. Am. Chem. Soc.* **2010**, *132*, 4887.
197. Li, S.; Lei, M.; Lv, M.; Watkins, S. E.; Tan, Z.; Zhu, J.; Hou, J.; Chen, X.; Li, Y., [6,6]-Phenyl-C61-Butyric Acid Dimethylamino Ester as a Cathode Buffer Layer for High-Performance Polymer Solar Cells. *Adv. Energy Mater.* **2013**, *3*, 1569.
198. Zhang, Z. G.; Li, H.; Qi, B.; Chi, D.; Jin, Z.; Qi, Z.; Hou, J.; Li, Y.; Wang, J., Amine Group Functionalized Fullerene Derivatives as Cathode Buffer Layers for High Performance Polymer Solar Cells. *J. Mater. Chem. A* **2013**, *1*, 9624.
199. Yip, H. L.; Hau, S. K.; Baek, N. S.; Jen, A. K. Y., Self-Assembled Monolayer Modified ZnO/Metal Bilayer Cathodes for Polymer/Fullerene Bulk-Heterojunction Solar Cells. *Appl. Phys. Lett.* **2008**, *92*, 193313.
200. Hau, S. K.; Cheng, Y. J.; Yip, H. L.; Zhang, Y.; Ma, H.; Jen, A. K. Y., Effect of Chemical Modification of Fullerene-Based Self-Assembled Monolayers on the Performance of Inverted Polymer Solar Cells. *ACS Appl. Mater. Interfaces* **2010**, *2*, 1892.

201. Liu, J.; Xue, Y. H.; Gao, Y. X.; Yu, D. S.; Durstock, M.; Dai, L. M., Hole and Electron Extraction Layers Based on Graphene Oxide Derivatives for High-Performance Bulk Heterojunction Solar Cells. *Adv. Mater.* **2012**, *24*, 2228.
202. Chen, Y. H.; Lin, W. C.; Liu, J.; Dai, L. M., Graphene Oxide-Based Carbon Interconnecting Layer for Polymer Tandem Solar Cells. *Nano Lett.* **2014**, *14*, 1467.
203. Qu, S.; Li, M.; Xie, L.; Huang, X.; Yang, J.; Wang, N.; Yang, S., Noncovalent Functionalization of Graphene Attaching [6,6]-PhenylC61-Butyric Acid Methyl Ester (PCBM) and Application as Electron Extraction Layer of Polymer Solar Cells. *ACS Nano* **2013**, *7*, 4070.

9 Nanophase Separation in Organic Solar Cells

*Wei Chen, Feng Liu, Ondrej E. Dyck,
Gerd Duscher, Huipeng Chen,
Mark D. Dadmun, Wei You, Qiquan Qiao,
Zhengguo Xiao, Jinsong Huang, Wei Ma,
Jong K. Keum, Adam J. Rondinone,
Karren L. More, and Jihua Chen*

CONTENTS

9.1	Introduction	248
	9.1.1 Organic Photovoltaics and Their Nanophase Separation	248
	9.1.2 Manipulation of Nanophase Separations in OPVs	249
	9.1.2.1 Solvent Selection	250
	9.1.2.2 Thermal Annealing	250
	9.1.2.3 Solvent Annealing	251
	9.1.2.4 Blending Solvent Additive	251
9.2	Characterizing Vertical and Planar Nanophase Separation	252
	9.2.1 Scanning Probes	252
	9.2.2 Techniques Using Electrons	253
	9.2.3 Techniques Using Ions	258
	9.2.4 Techniques Using X-Rays	259
	9.2.5 Techniques Using Neutrons	263
9.3	Case Studies	264
	9.3.1 Nanophase Separation in OPV with the Second-Generation Semiconducting Polymers	265
	9.3.2 Nanophase Separation in OPV with the Third-Generation Semiconducting Polymers	268
	9.3.3 Nanophase Separation in OPV with Small-Molecule Semiconductors	271
	9.3.4 Nanophase Separation in Advanced Ternary OPVs	271
9.4	Conclusion	272
Acknowledgment		272
References		272

9.1 INTRODUCTION

9.1.1 Organic Photovoltaics and Their Nanophase Separation

Organic photovoltaics (OPVs) are cost-effective thin-film devices that can be printed onto flexible substrates and integrated into building materials (Brabec et al. 2001; Thompson and Fréchet 2008). They are regarded as one of the most promising renewable energy sources for the near future (Mayer et al. 2007; Nelson 2011). In the past two decades, tremendous efforts from both academia and industry have been devoted to this quickly evolving field, leading to breathtaking improvement of the device power conversion efficiencies (PCEs) (Bian et al. 2012; Brabec et al. 2010; Dennler et al. 2008; Inganäs et al. 2009; Liang and Yu 2010). Up to now, a record efficiency of 10.6% has been obtained from polymer-based bulk heterojunction (BHJ) tandem devices (You et al. 2013). These achievements arose from improvement of materials and device optimization. However, further developments will rely more and more on a sophisticated interdisciplinarity of material science, device physics, interface science, and optical science. Such a highly interdisciplinary research frontier calls for more insightful contributions, for instance, further development and application of cutting-edge nanostructure characterization techniques, to address fundamental challenges in OPV research (Chen et al. 2012c; DeLongchamp et al. 2012; Liu et al. 2013a). Those characterization techniques will largely complement input from manufacturing engineering (such as innovative production instruments, processes, and ink formulas) to realize a path toward possible commercialization of OPVs (Krebs et al. 2010a,b; Søndergaard et al. 2012).

Fundamental mechanisms of OPVs can be found in other chapters of this book or from reviews in this field (Brabec 2004; Brabec et al. 2001; Coakley and McGehee 2004; Mayer et al. 2007; Nuzi 2002; Spanggaard and Krebs 2004). In order to get an OPV device to function, an electron-donating component and an electron-accepting component are used in couple (Brabec et al. 2001). Under the concept of a BHJ, donors and acceptors are blended intimately to increase the interfacial area (Halls et al. 1995; Yu et al. 1995). Both donor and acceptor materials should absorb sunlight efficiently to generate excitons, which will subsequently diffuse to domain interfaces and split into charge carriers. Subsequent transport and collection of the charge carriers at their respective electrodes leads to an internal current. The energy offset of the frontier orbitals of the donor/acceptor materials and the work functions of asymmetric electrode materials provides voltage to drive an outside circuit (Thompson and Fréchet 2008). To obtain high PCE, simultaneous improvement of open circuit voltage (V_{oc}), short circuit current (J_{sc}), and fill factor (FF) need to be achieved (He et al. 2011). Although V_{oc} is more related to the intrinsic properties of the active layer materials and electrodes (Brabec et al. 2002; Graham et al. 2013; Vandewal et al. 2009; Yamamoto et al. 2011), J_{sc} and FF are factors that can be manipulated by varied processing methods and post-treatments through the kinetic control of the morphology or nanophase separation of the active layer mixture (Liu et al. 2013a). For example, BHJ films with vertical nanophase separation (or a gradient BHJ), with donor enriched at the anode and acceptor enriched at the cathode side, have many advantages over the uniform BHJ films, which leads to efficiency enhancement. A gradient BHJ film with a donor-rich anode side and an acceptor-rich cathode side provides not only better pathways for hole and electron transport to their corresponding electrode, but also a better match of the distribution of currents in the devices. Therefore, gradient BHJ films may reduce the piling up of charges in the active layer and the leakage of photogenerated charges to the wrong electrodes, thus increasing the charge collection and diminishing charge recombination. Several groups have used thermal evaporation of small molecules SubPc and C_{60} to accurately control the vertical profile of donors and acceptors (Pandey and Holmes 2010; Xue et al. 2005; Yang et al. 2004). These studies demonstrate that the gradient BHJ devices have larger J_{sc} than the donor/acceptor planar structure due to the increased interface area. Meanwhile, the vertical phase separation improves charge carrier transport compared to a uniformly mixed heterojunction leading to an enhancement in both the J_{sc} and FF (Pandey and Holmes 2010).

Solution-processed BHJ solar cell devices are currently the most efficient type of OPV devices. The fundamental benefit is that the nanosized phase separation of donor and acceptor results in a large interfacial area in the blended active layer (Halls et al. 1995; Yu et al. 1995). Currently, the state-of-the-art BHJ thin film consists of a bicontinuous, interpenetrating network of donor/acceptor domains with length scales commensurate with the exciton diffusion length. Devices based on this concept intrinsically bear a large J_{sc}. Because J_{sc} and FF are strongly affected by the details of the internal nanophase separated structure, a systematic study of the structure–property relationship, or nanostructure-performance correlation, is of paramount importance (Dang et al. 2013; Hoppe and Sariciftci 2006; Liu et al. 2012a; Ruderer and Müller-Buschbaum 2011). Up to now, although tremendous efforts have been made to reveal a clear nanomorphology of BHJ OPVs, no cohesive picture has yet been obtained (Liu et al. 2013a). This is attributed to the limitation of characterization methods and the complication of the BHJ structure itself. In real-world scenarios, multilength scale morphology is commonly observed, which adds difficulties in their resolution and interpretation (Chen et al. 2011b; Gu et al. 2012; Yin and Dadmun 2011). In addition, although chemically modified fullerenes (such as PCBM) are often used as acceptors, the vast variety of donor materials also brings about tremendous challenges in morphology characterization and the realization of a generalized correlation between morphology and performance (Bian et al. 2012; Chochos and Choulis 2011). In BHJ blends, structural order of the materials, interactions between and miscibility of the components, as well as surface and vertical segregation, are key factors that determine the morphology of spin-coated blends with and without postdeposition treatment (Brabec 2004; Coakley and McGehee 2004; Hegde et al. 2012; Liu et al. 2012a). Some OPV blend systems show a uniform mixing in as-spun films due to the fast solvent evaporation, which require additional posttreatment to enhance the nanophase separation of active layers (Liu et al. 2012a). In other cases, however, as-spun OPV films naturally demonstrate large-sized phase segregation, far beyond the suitable length scale for efficient charge generation, which also makes novel film-processing techniques necessary in order to reduce the size of these phase domains (Liu et al. 2013a).

9.1.2 Manipulation of Nanophase Separations in OPVs

Before we elaborate on characterization methods and case studies of OPV nanophase separation, herein we briefly discuss various treatments for manipulating and controlling the nanophase separation in polymer:fullerene BHJs, such as solvent selection, solvent drying, rate/solvent vapor annealing, thermal annealing, and processing additives (Campoy-Quiles et al. 2008).

9.1.2.1 Solvent Selection

Both polymer:fullerene film morphology and device performance are highly dependent on the solvents from which the films are deposited (Park et al. 2009). The degree of phase separation between donor and acceptor generally varies with different solvents. On one hand, the time for preparing a dry film through spin coating depends on the evaporation rate (vapor pressure) of the depositing solvent. A longer drying time generally results in a larger domain size. On the other hand, the phase separation between donor and acceptor is influenced by their solubility in the processing solvent (Hoppe et al. 2004; Zhang et al. 2006). For a poly(*p*-phenylene vinylene) (PPV):PCBM system, the PCBM domains increased from tens of nanometers using chlorobenzene to hundreds of nanometers using toluene as the depositing solvent even with the same bending ratio of PPV to PCBM and the same solution concentration (Hoppe et al. 2004; Shaheen et al. 2001). Such a difference in domain size was attributed to the PCBM solubility in the processing solvent (Hoppe et al. 2004).

Recently, several groups have reported using solvent mixtures to modify the nanophase separation of solution-processed BHJs. Typically, there are two types of solvent mixtures. One consists of solvent pairs in which both polymers and fullerenes are considerably soluble, such as dichlorobenzene (DCB) and chlorobenzene (CB). Park et al. reported that, for the poly[2,6-(4,4-bis(2-ethylhexyl)-4H-cyclopenta[2,1-b;3,4-b′]-dithiophene)-alt-4,7-(2,1,3-benzothiadiazole)] (PCDTBT):PCBM system, the phase separation between polymer donor and fullerene in the deposited film becomes finer by increasing the amount of DCB in the solvent, leading to larger photocurrent (Park et al. 2009). In addition, Zhang et al. observed that, for the APFO-3:PCBM system, mixing CB with chloroform as the depositing solvent resulted in smoother film surfaces and smaller domains, which translate to larger J_{sc} (Zhang et al. 2006). The other option involves solvent pairs in which either the polymer or fullerenes is much less soluble in one solvent than in the other. Moule et al. revealed that introducing a polar solvent nitrobenzene (NtB) into CB as the processing solvent controls the aggregation of P3HT in the deposited film (Moulé and Meerholz 2008). Moreover, Chu et al. demonstrate that, after adding 13% (v%) of the solvent dimethyl sulfoxide (DMSO) into the DCB processing solvent, the PCE of PCDTBT:PCBM-based devices increases from 6.0% to 7.1%. This increase in PCE was also ascribed to the aggregation and improved packing of PCDTBT molecules, resulting in enhanced charge mobility (Chu et al. 2011).

9.1.2.2 Thermal Annealing

Most as-prepared polymer:fullerene (such as P3HT:PCBM) films are less crystalline, for spin coated P3HT:PCBM blends also show peaks in GIXD, which limits their charge carrier mobility. Thermal annealing, either postthermal annealing (on real device) or prethermal annealing (on the active layer only), is one of the most effective methods to improve the crystallinity of polymers and/or fullerenes. The enhanced crystallinity, as evidenced by XRD, increases the near-infrared region absorption and the hole/electron mobility (Chu et al. 2011; Moulé and Meerholz 2008) and reduces charge recombination as a result of the improved percolation pathway (Chen et al. 2009b). Both electron and hole mobility of thermal annealed films could be increased by orders of magnitude (Mihailetchi et al. 2006). As reported by Campoy-Quiles et al., thermal annealing could also induce a vertical phase separation in some polymer/fullerene systems. For example, upon annealing of P3HT/PCBM blends, P3HT chains were initially crystallized, followed by diffusion of PCBM molecules to nucleation sites and then PCBM growth to form aggregates at the surface of the blend films (Campoy-Quiles et al. 2008). For the PCDTBT:PCBM system, a thin PCDTBT layer can be formed at the cathode side after annealing at 140°C for a few minutes, but this thin PCDTBT layer hampers the charge transportation and collection, which is detrimental for the device efficiency (Synooka et al. 2013). In addition, thermal annealing could enhance the charge collection efficiency, arising from an increase in the surface roughness of the OPV active layer and, thereby, the contact areas with metal cathode (Li et al. 2005). However, the thermal annealing process is not compatible to many other low bandgap polymer systems because it can cause overly large PCBM and/or polymer domains. In those systems, the interfacial area between donors and acceptors is decreased upon thermal annealing, which results in much lower J_{sc}.

9.1.2.3 Solvent Annealing

Solvent annealing is a versatile method in which a thin film is exposed to an atmosphere of solvent vapors that diffuse into the deposited layer, the extent of which is dependent on, and thus controllable by, exposure time. Solvent annealing has effects similar to thermal annealing to some extent and results in an increase in crystallinity and change in nanophase separation (Campoy-Quiles et al. 2008; Hegde et al. 2012).

The solvent vapor increases molecular mobility of the components in the active layer, allowing the system to evolve toward a more thermodynamically stable morphology. Moreover, compared to thermal annealing, with which structure develops in seconds, solvent annealing provides more precise morphological control (Chen et al. 2013b). Most importantly, solvent annealing is performed at room temperature that is very suitable for low-bandgap polymers, which usually have high glass transition temperatures (T_g) or no observable T_g (Blouin et al. 2008; Lu et al. 2011) and are therefore difficult to modify by temperature change.

9.1.2.4 Blending Solvent Additive

Adding a solvent additive to the depositing solution is another effective method to control the nanophase separation of the deposited polymer/fullerene film (Lee et al. 2008; Liang et al. 2010; Peet et al. 2007). Generally, two criteria for incorporating additives to control the blend-film morphology were proposed: (i) selective solubility of the fullerene acceptor and (ii) a higher boiling point (lower vapor pressure) than the process solvent (Lee et al. 2008; Lou et al. 2011). Difunctionalized alkanes, such as alkanedithiols or diiodooctane (DIO), are widely used additives to enhance the PCE of as-cast BHJ active layers. The additive can decrease the size of the polymer and fullerene domains and form better interpenetrating networks, contributing to larger J_{sc} and FF (Liang et al. 2010). Lou et al. found that the active layer of a Poly({4,8-bis[(2-ethylhexyl)oxy]benzo[1,2-b:4,5-b']dithiophene-2,6-diyl}

{3-fluoro-2-[(2-ethylhexyl)carbonyl]thieno[3,4-b]thiophenediyl}) (or PTB7):PCBM system is highly aggregated without an additive (Lou et al. 2011). The level of aggregation was significantly decreased after adding 3% v/v DIO in the processing solvent, facilitating the intercalation of PCBM into the polymer networks. Su et al. found the solvent additive (1,6-diiodohexane or DIH) can decrease the average size of fullerene aggregation from 150 nm to around 35 nm (Su et al. 2011). The PCE of a polymer/$PC_{71}BM$-based device increased from 5% to 7.3% due to the increased photocurrent. On the contrary, Lee et al. found that, for another polymer:PCBM system, solvent additives result in larger polymer/fullerene domains relative to those formed without solvent additives. In addition, connectivity of domains appears to improve with the additive, which results in improved charge transport and collection (Lee et al. 2008).

The presence of a solvent additive has also been shown to affect the vertical distribution of the polymer donor and acceptor (Yao et al. 2008). For example, Yao et al. found that there was a vertical phase separation in a P3HT/PCBM film when 1,8-octanedithiol OT was added into DCB as the processing solvent (Yao et al. 2008). The vapor pressure of DCB is 200 times higher than that of OT so that OT concentrates in the film during deposition. Due to a lower solubility of PCBM in OT, PCBM therefore forms clusters and readily precipitates in the presence of OT. Considering the higher surface energy of PCBM, PCBM therefore enriches the bottom of the deposited film.

The additive method was also applicable to small molecular molecule solar cells to increase the small molecule donor crystallinity (Perez et al. 2013; Sun et al. 2012). The efficiency was very sensitive to the additive ratio. A small additive concentration of 0.25% was used to enhance the 5,5′-bisf(4-(7-hexylthiophen-2-yl)thiophen-2-yl)-[1,2,5]thiadiazolo[3,4-c]pyridineg-3,3′-di-2-ethylhexylsilylene-2,2′-bithiophene, $(DTS(PTTh_2)_2)$:PCBM device performance (Sun et al. 2012). Increasing the additive volume ratio to 1% drastically decrease the PCE from 6.7% to 0.43% (Sun et al. 2012).

9.2 CHARACTERIZING VERTICAL AND PLANAR NANOPHASE SEPARATION

9.2.1 SCANNING PROBES

Scanning probe microscopy captures surface images of a sample by using a physical probe that scans the samples, in which atomic force microscopy (AFM) is the most widely used scanning probe technique in characterizing phase separation of the polymer:fullerene BHJ organic solar cells.

Binnig, Quate, and Gerber invented the AFM in 1986 (Binnig et al. 1986). It is a very high-resolution scanning probe microscopy with demonstrated resolution of fractions of a nanometer, which provide sufficient resolution to resolve length scales comparable to the exciton diffusion length in organic solar cells. The basic principle of operation is as follows: the tip, mounted on the cantilever, is approached to the sample surface and is subject to an attractive or repulsive interaction with the sample. This leads to a bending of the cantilever, which is detected and converted into an electronic signal.

For OPV studies, scanning probe methods have the powerful potential to correlate local device performance to local phase structure, which is rather unique as compared to many other nanophase detection methods (Giridharagopal and Ginger 2010). Variants of scanning probe microscopies that are applied in OPV research include photoconducting AFM (pcAFM), time-resolved electrostatic force microscopy (trEFM), scanning Kelvin probe microscopy (SKPM), and near-field scanning optical microscopy (NSOM) (Giridharagopal and Ginger 2010).

9.2.2 Techniques Using Electrons

Scanning electron microscopy (SEM) and transmission electron microscopy (TEM) are two well-known electron microscopy techniques. SEM has its advantages because of simple sample preparation requirements, and TEM provides significantly higher resolution. Detecting nanophase separation of OPV materials in TEM can be experimentally challenging for a couple of reasons: (1) Conventional TEM studies of organic or hybrid films suffer significantly from poor contrast caused by their low atomic numbers, and (2) in addition, organic compounds are notorious for their electron beam sensitivity and difficulties in high-resolution TEM studies (Martin et al. 2005).

To overcome these obstacles, several TEM techniques have emerged as promising and powerful approaches. These include low-dose high-resolution TEM imaging, low eV plasmon imaging of energy-filtered TEM (EFTEM), core-loss elemental mapping in EFTEM, electron energy-loss spectroscopy (EELS) imaging, 3-D tomography, and cross-sectional EFTEM. Among these approaches, EFTEM is among the most widely used because of its versatility and superior compatibility, and tomography produces the most revealing data in three dimensions.

Low-electron-dose high-resolution TEM imaging enables direct imaging of lattice fringes of organic and polymer crystallites in OPV active layers. For example, Sun et al. reported crystal lattice imaging and crystallite orientation variation in small molecule donor/PCBM high-performance OPV blends (Sun et al. 2012). Drummy et al., achieved low-dose lattice imaging of P3HT crystallites in P3HT/PCBM films (Drummy et al. 2011). Local crystalline structure, crystal defects, and crystal orientation distribution of donors or acceptors can be then correlated with device performance although a direct correlation can be extremely complex. Therefore, there is a need to carefully examine structures over a broad range of length scales.

EFTEM imaging uses selected energy ranges in the EELS spectrum to produce contrast in TEM images and therefore requires the TEM to be equipped with an EELS spectrometer and energy filter. Often the bulk plasmon peak in the low loss portion of the spectrum is used to distinguish between the donor- and acceptor-rich domains. This low eV plasmon peak occurs at different energies for the two components as it depends on the material's electronic structure. For example this plasmon peak is at ~22 eV for pure P3HT and at ~25 eV for pure PCBM (Herzing et al. 2010; Pfannmoeller et al. 2011). Selecting an energy window of ~8 eV and centering it at 19 eV will preferentially allow electrons that have interacted with the P3HT to result in P3HT-rich regions appearing brighter in the EFTEM image, and repeating the same procedure with a window centered at 30 eV will preferentially allow electrons that have interacted with the PCBM to form the image (Drummy et al. 2011). If there is sharp phase separation, these two types of images will be complementary and can be overlaid to form a composite image showing distinct phase positions (Figure 9.1). This technique has relatively low beam damage with image exposure time only on the order of seconds or less under proper imaging conditions and can be easily applicable to other donor-acceptor systems (Ajuria et al. 2013).

Core-loss EFTEM may also be used when distinctive elements are found in the phases, such as sulfur in P3HT (Domanski et al. 2013). Representative sulfur maps of two high-performance, low-bandgap polymer (DTBT and DTffBT)/PCBM blends (Zhou et al. 2011) are shown in Figure 9.2, obtained using a conventional three-window method. Figure 9.3 shows a combined sulfur-carbon map of a benchmark, high-performance, low-bandgap polymer PBDTTT-C-T/PC$_{70}$BM blend, in which the sulfur map is shown in green, indicating the polymer-rich domains, and the carbon

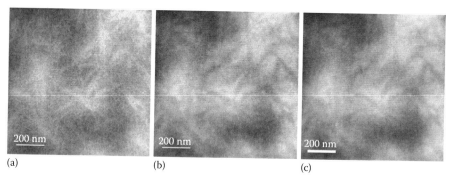

FIGURE 9.1 (See color insert.) Low eV plasmon-based EFTEM images of P3HT/PCBM. (a) 19 eV-centered energy window highlights the P3HT-rich domains. (b) 30 eV-centered window highlights the PCBM-rich domains. (c) Artificially colored composite image shows the P3HT in blue and PCBM in yellow. (Light and dark areas correspond to sample thickness variations.)

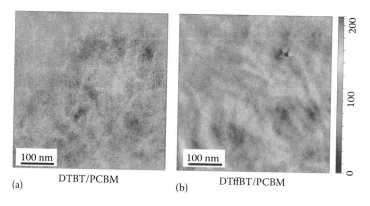

FIGURE 9.2 (See color insert.) Sulfur maps of representative, high-performance, low-bandgap polymer (DTBT and DTffBT)-based OPV systems ([a] DTBT/PCBM, [b] DTffBT/PCBM). The details of the synthesis and molecular structure of DTBT and DTffBT polymers are reported elsewhere (Zhou et al. 2011). The higher sulfur content regions are shown with brighter colors. The corresponding thickness maps (not shown) are almost featureless.

Nanophase Separation in Organic Solar Cells

FIGURE 9.3 (See color insert.) Combined sulfur-carbon maps of low-bandgap polymer PBDTTT-C-T/PC$_{70}$BM-based, high-performance, OPV systems. The sulfur map is shown in green and carbon in red. PCBM has a higher carbon density, corresponding to higher carbon content, and the polymer is rich in sulfur. (From P. Adhikary et al., *Nanoscale* 5(20): 10007–10013, 2013.)

map is shown in red, revealing the PCBM-rich regions (Adhikary et al. 2013). To confirm the consistency between core-loss imaging and the low-eV plasmon imaging method described above, both methods were used to image an identical area of a PBDTTT-C-T/PC$_{70}$BM film. In Figure 9.4, the dotted circles highlight an acceptor-rich dot-shaped area, which is brighter in the carbon map and 30 ± 4 eV image. The solid-line circles highlight a donor-rich Y-shaped area, which is darker in the carbon map and 30 ± 4 eV image. Both the dot- and Y-shaped domains show distinctive contrast reverse in the sulfur map and 19 ± 4 eV image. The results suggest that our low-eV plasmon imaging and core-loss elemental mapping have a high degree of consistency in terms of nanophase identification. Although the elemental mapping takes significantly longer exposure time (up to 10 s for each window and 30 s per elemental maps or longer), it contains quantitative information about elemental distribution. In contrast, low-eV plasmon imaging typically has complicated background signals that are not easily correctable.

Another analytical TEM technique available for OPV characterization is EELS imaging. In this technique, the beam is scanned across the sample while an EELS spectrum is recorded at each pixel. Unlike EFTEM imaging, the advantage of this technique is that the spectra are saved for later analysis. The resultant spectra can be fit with an empirical model and feature changes with position on the sample recorded quantitatively (Hunt and Williams 1991). Figure 9.5 shows color-coded changes in position of the bulk plasmon peak in a P3HT/PCBM BHJ cross-section, which enables phase identification. It is important to point out that care must be taken when employing this technique because the beam is concentrated on a relatively small area. To avoid beam damage, close monitoring of the imaged areas is necessary; otherwise, beam-induced morphological changes can easily occur (Herzing et al. 2010).

Tomography is a 3-D imaging technique during which images are taken at various sample tilt angles. A computer program is then used to track positions of distinctive features in the sample and to reconstruct a 3-D model (Midgley and Weyland

256   Organic Solar Cells

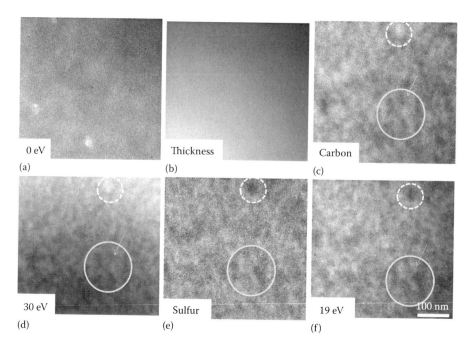

FIGURE 9.4 Energy-filtered TEM of a benchmark low-bandgap, polymer-based (PBDTTT-C-T/PC$_{70}$BM) OPV layer in planar views. For elemental mapping, a typical energy slit window is 10 eV with a conventional "three-window" method; for 0 eV and plasmon imaging (19 and 30 eV), a typical energy window of 8 eV (or ±4 eV) is used. Panels (a) to (f) represent 0 eV image, carbon map, sulfure map, thickness map, 30 eV as well as 19 eV image of the PBDTTT-C-T/PC$_{70}$BM system at an identical imaging area.

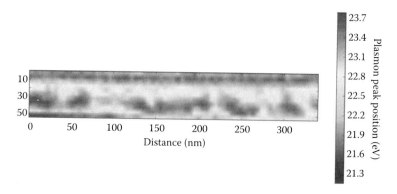

FIGURE 9.5 (See color insert.) Bulk plasmon peak position reveals phase separation in a P3HT/PCBM BHJ cross-section. The higher energy plasmon peak (red) corresponds to the PCBM-rich phase, and the lower energy plasmon peak (yellow) corresponds to the P3HT-rich phase. A layer of platinum on top also shows a plasmon peak similar to that of PCBM. The silicon substrate shows the lowest energy plasmon peak (blue).

Nanophase Separation in Organic Solar Cells

2003). Although this is a much more time-consuming technique than EFTEM, it offers much more information than simple 2-D projection images. A drawback for using tomography in OPVs is that there is limited contrast between phases within the blends, and these phases must be visible for proper reconstruction (Andersson et al. 2009). In addition, the sample must also be robust enough to withstand very many exposures to the electron beam. Nevertheless Loos and coworkers successfully applied this technique in a few organic or hybrid photovoltaic systems (van Bavel et al. 2009, 2010). In order to address the issue of contrast in tomography, one can combine tomography with EFTEM imaging (EFTEM tomography) to identify phases and produce contrast (Herzing et al. 2010; Midgley and Weyland 2003). In this way, a 3-D reconstruction with phase identification can be produced.

As revealed by tomography results, the vertical phase morphology is a crucial aspect in determining the charge transport and power conversion efficiency of BHJ OPVs because it determines both exciton generation and the charge transport pathway (van Bavel et al. 2010). Alternatively, cross-sectional TEM can be used to complement or substitute for the often-tedious tomography experiments. Figure 9.6 shows the vertical phase separation of a high-performance PBDTTT-C-T/$PC_{70}BM$ organic photovoltaic film with energy-filtered TEM in its cross-sectional views. The sample preparation involves focused ion beam techniques, which will be elaborated in Section 9.2.3. In Figure 9.6, the elastic image (0 eV) shows a distinctive layer structure in a cross-sectional view of the OPV device. Layers 1 and 2 represent ion and electron deposited Pt cap layers, 3 and 4 are the active layer and PEDOT:PSS, and finally, 5 and 6 are the ITO and glass. The low-eV plasmon imaging of the same area is also shown in Figure 9.6 with polymer-rich domains in red and fullerene-rich domains in green. The combination of cross-sectional TEM and energy-filtered TEM present a powerful way to monitor process conditions and correlate with device performance.

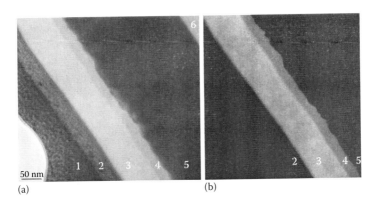

FIGURE 9.6 (See color insert.) Elastic (0 eV) and composite energy-filtered TEM image of a high-performance OPV blend (PBDTTT-C-T/$PC_{70}BM$) in cross-sectional views. (a) The elastic image (0 eV) shows distinctive layer structures in a cross-sectional view of the OPV device (layers 1 and 2 represent ion and electron deposited Pt cap layer, 3 and 4 for active layer and PEDOT:PSS, and finally, 5 and 6 for ITO and glass). (b) Low-eV plasmon imaging of the same area with polymer-rich domains in red and fullerene-rich domains in green.

9.2.3 Techniques Using Ions

A well-established technique for preparing cross-section TEM samples employs the use of a focused ion beam (FIB), typically Ga^+. The instrument is often equipped with a scanned electron probe as well as the ion beam, which allows dual function as a SEM. Preparation using a FIB is accomplished by focusing the ion beam onto various areas of the sample so that the material is sputtered away. This is done in such a way as to leave a thin lamella of untouched sample that can be removed and thinned further to produce an electron-transparent lamella suitable for examination in a TEM. Using a FIB for OPV sample preparation has several advantages over other techniques. The FIB technique allows the user to precisely select the location for sample extraction. It also produces a very uniform thickness sample (usually less than 100 nm), which is very helpful for proper interpretation of TEM generated data, such as image contrast and electron energy loss spectra. Soft organic materials are often prepared for TEM using an ultramicrotome that slices thin sections from the sample. When an ultramicrotome is used in OPV applications, the OPV material must be transferred from the original substrate (that is, silicon or glass) to a softer material that can be sliced. This process may introduce contamination, and the slicing procedure may introduce morphological changes (for example shearing and tearing) not present in the virgin specimen (Bassim et al. 2012; White et al. 2001). For these reasons, FIB is an attractive alternative to prepare cross-section samples.

The main drawback with the FIB technique is that the very nature of the method causes the destruction of the sample. It is, therefore, necessary to take precautions to limit beam damage brought on by sample heating, knock-on damage, and radiolysis (Bassim et al. 2012). The first precautionary measure is the deposition of a protective platinum layer over the area of interest before milling. This limits direct exposure to the beam during milling; however, the deposition of this Pt layer is accomplished by either electron beam or ion beam exposure. Bassim and De Gregorio (Bassim et al. 2012) showed in several coal samples prepared via FIB (for x-ray studies) that exposure to the electron beam was more damaging to the sample than exposure to the ion beam. It was suggested that less than 1 keV electrons be used for imaging and exposure time minimized. The implications of this are unclear for TEM studies because the final sample will be exposed to the electron beam in the TEM at much higher energies. It is assumed that if the sample is resilient under the TEM beam, the SEM beam should negligibly affect it. Current practice is to deposit an initial layer of platinum with the electron beam and a subsequent ion beam deposited layer. The ion beam is preferable for deposition when possible because deposition occurs more rapidly. Sample heating under the ion beam is an issue in organic materials due to the lower thermal conductivity (compared to inorganic materials). This heating may be mitigated during the final thinning by leaving a frame of native material around the thinned area. This reinforces the mechanical structure while providing a heat sink. Additionally, the beam current and energy should be reduced during the final thinning process. The local heating due to an incident ion beam is given by (Bassim et al. 2012)

$$\Delta T = \frac{JV}{2k} r_0 \qquad (9.1)$$

Nanophase Separation in Organic Solar Cells

where ΔT is the change in temperature, J is the beam current density, V is the accelerating voltage, k is the thermal conductivity, and r_0 is the beam radius. The current density can be found by $J = I/(\pi r_0^2)$ where I is the beam current. A final polish with beam currents ≤50 pA and accelerating voltages ≤10 kV should be adequate to ensure limited beam-induced heating. It has been shown in crystalline specimens (Kato 2004) that the amorphized surface layer (due to beam damage) decreases approximately linearly with beam energy, and for beam energies of 10 keV, the amorphized layer is ~10 nm (in silicon). While this does not translate directly to organic materials, it is worth noting that there will always be some amount of surface damage, and the lower the beam energy, the thinner the damaged layer will be. It should be noted here that this carbon layer likely has a higher thermal conductivity than the polymers of interest and may actually decrease beam heating, thereby (somewhat) protecting the underlying polymers, although we know of no direct study to verify this. Unfortunately, there is uncertainty of the amount of damage introduced by the FIB from one sample to the next. In order to address this inconsistency, it is possible to prepare a single sample with multiple layers of interest. For instance, Moon et al. demonstrated this technique (Moon et al. 2011) in which layers of P3HT, PCBM, and two BHJs were each separated vertically by layers of ITO and PEDOT:PSS. A single FIB sample was then able to capture the cross-section of each material, and the milling conditions for each were, necessarily, identical, eliminating any question about differences introduced by the FIB fabrication.

In addition, secondary ion mass spectroscopy (SIMS) and helium ion microscopy (HIM) are among other useful ion-related tools to determine the nanoscale morphology of OPV active layers. SIMS can be used, coupled with isotopic labeling, to extract depth profiles of conventional and inverted OPV devices based on a given ion species or to generate ion maps with moderate resolution (Andreasen et al. 2012; Norrman et al. 2009, 2010). HIM (Ramachandra et al. 2009) utilizes the helium ion as a probe source, which provides high surface contrast, high spatial resolution (up to 4 angstrom), and large depth of view. HIM has the potential for nanoscale chemical analysis and is compatible with soft materials without the tedious sample preparation required for TEM. Pearson et al. reported the first HIM study of the P3HT/PCBM system, in which the internal structures of the BHJ devices were accessed via HIM after controlled plasma etching (Pearson et al. 2011). Figure 9.7 is a HIM image representative of the DTffBT/PCBM (Zhou et al. 2011) OPV system, which shows large surface contrast that is not available in conventional TEM/SEM techniques. Moreover sample preparation of HIM is much less tedious than that of TEM, which includes floating the film on water and picking up with a grid.

9.2.4 Techniques Using X-Rays

X-ray-based characterization techniques are important in studying OPV nanophase separation, including grazing incidence small angle x-ray scattering (GISAXS), resonant soft x-ray scattering (RSoXS), x-ray reflectivity (XRR), and x-ray photon spectroscopy (XPS). In combination, they contribute toward a better understanding of the 3-D nanophase morphology of the active layer in a real device or device-like scenarios.

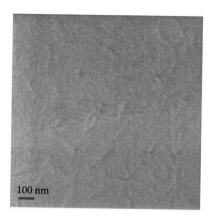

FIGURE 9.7 Representative HIM image of DTffBT/PCBM based OPV systems, which shows large surface contrast that is not available in conventional TEM techniques. (From H. Zhou et al., *Angewandte Chemie-International Edition* 50(13): 2995–2998, 2011.)

GISAXS is a nondestructive scattering probe suitable to characterizing the lateral phase inhomogeneities of OPV active layers. In a GISAXS experiment, a thoroughly collimated x-ray beam strikes the flat surface of a film at a fixed incidence angle (α_i), from which the scattered x-ray beams, with in-plane and out-of-plane exit angles $2\theta_f$ and α_f, are recorded on a 2-D detector for nanophase analysis (inset of Figure 9.8). Shao et al. monitored the lateral domain morphology of PCBM dispersed in various deuterated P3HT matrices using GISAXS (Shao et al. 2014). In these studies, the in-plane GISAXS profiles (Intensity vs. q_y) were fit using a scattering model of that combined the Guinier-Porod model (Hammouda 2010) and the polydisperse hard-sphere model with a Schulz distribution (Griffith et al. 1987) as shown in Figure 9.8a through f. In Figure 9.8, q_y and q_z refer to the in-plane and out-of-plane scattering wave vectors, $q_y = k_0[\sin(2\theta_f)\cos(\alpha_f)]$ and $q_z = k_0[\sin(\alpha_i) + \sin(\alpha_f)]$, respectively, and k_0 is the elastically conserved wave vector modulus. In the model fits, the Guinier-Porod and polydisperse hard-sphere models account for the phase-separated PCBM domains with two different ranges of sizes; that is, the Guinier-Porod model characterizes the larger domains (low-q_y region) and the hard-sphere model characterizes the smaller domains (high-q_y region). The radii of gyration (R_gs) of the large and small PCBM domains in the annealed P3HT/PCBM films (150°C) were ~5.5 and ~10.0 nm, respectively. Liu et al. and Wu et al. also reported similar GISAXS results, in which PCBM domains with $R_g \approx 2.7$ nm and 7.0 nm were observed in the as-spun and annealed (150°C) P3HT:PCBM film, respectively (Liu et al. 2011; Wu et al. 2011).

The recent development of RSoXS has made it possible to enhance electron density contrast by tuning x-ray photon energy close to the absorption edges of constituent materials (Ade 2012). This has been particularly powerful in enabling RSoXS to probing OPV blends with relatively low electron density contrasts (Ade 2012; Chen et al. 2011b; Lu et al. 2013; Swaraj et al. 2010). Indeed, RSoXS is an excellent alternative to GISAXS and transmission SAXS, which use hard x-rays, and

Nanophase Separation in Organic Solar Cells

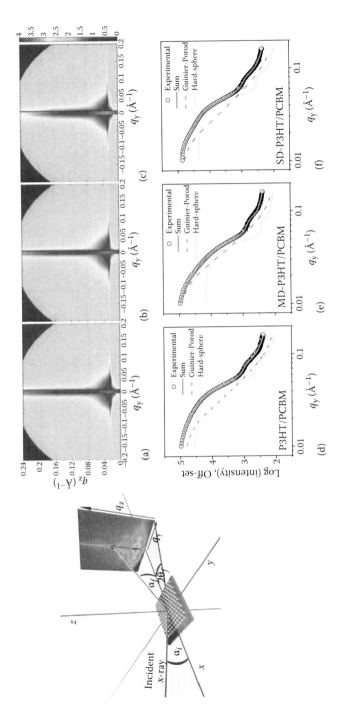

FIGURE 9.8 Two-dimensional GISAXS images and in-plane scattering curves extracted from the 2-D images for P3HT/PCBM (a) and (d), main-chain deuterated P3HT/PCBM (b) and (e), and side-chain deuterated P3HT/PCBM (c) and (f). GISAXS geometry is shown in the left inset. (Reprinted by permission from Macmillan Publishers Ltd., *Nature Communications*, M. Shao et al., The Isotopic Effects of Deuteration on Optoelectronic Properties of Conducting Polymers. Copyright 2014. The inset figure was kindly provided by Dr. Andreas Meyer of the University of Hamburg.)

small-angle neutron scattering (SANS). Contrary to conventional SAXS in which the samples must be either freestanding films or prepared on x-ray transparent substrate with considerable thickness, RSoXS samples can be quite thin, making it suitable to study the in-plane structure of a device in transmission geometry. Also, RSoXS provides tunable electron density contrast by simply varying the x-ray photon energy. Recently, Chen et al. (2011b) and Swaraj et al. (2010) have successfully applied RSoXS to investigate the lateral phase morphologies of OPV active layers, in which the authors quantitatively analyzed the RSoXS curves by calculating pair distribution function $P(r)$ from the measured scattering curves and obtained the size and purity of phase domains (Figure 9.9a and b).

XRR has been shown to be viable in the investigation of the composition profiles of the donor and acceptor in the direction perpendicular to the surface plane of OPV blend films (Kirschner et al. 2012; Liu et al. 2011; Ruderer et al. 2009). However, electron density contrast between the organic donor and acceptor is often very low, and hence, XRR has often been used in conjunction with neutron reflectivity in order to complement each other. In standard x-ray/neutron reflectometry measurement, a slit-collimated x-ray or neutron beam impinges onto the flat surface and interface of the film at shallow incidence angles, α_is. At each α_i, only the reflected specular beam that reflects at angle (α_f), which is identical to α_i, is recorded to explore the vertical depth profile of the sample film. As is typical in any scattering experiment, the phase information is lost; therefore, the analysis of the measured reflectivity curve is generally carried out by constructing an electron density model or scattering length density (SLD) profile from which the calculated reflectivity profile using Parratt formalism (Parratt 1954) is compared with experimental reflectivity data until a goodness of fit with at least χ^2 is attained. Using the obtained electron density or SLD depth profiles, the composition depth profiles of constituents are readily determined by expressions described in the literature (Keum et al. 2013; Liu et al. 2011).

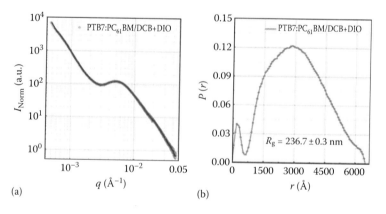

FIGURE 9.9 (a) RRSoXS curve for film spun-cast from PTB7/PCBM/1,2-dichlorobenzene/diiodooctane solution and (b) the pair distribution function calculated from the RSoXS curve in (a). (Reprinted with permission from W. Chen et al. 2011, 3707–3713, Copyright 2011, American Chemical Society.)

In addition, XPS offers a unique opportunity for a quantitative analysis of the elemental composition near the surface and interfaces of an OPV blend film (Orimo et al. 2010; Xu et al. 2009). In XPS measurements, films are irradiated by a series of focused x-ray beams, and the total numbers of electrons that escape from the film at different excitation energies are counted on the detector to identify the elemental composition of the sample surface. Xu et al. (2009) have recently investigated the interfacial compositions of P3HT and PCBM in P3HT/PCBM OPV films by peeling off the films from the substrates using a water floatation method and measuring the XPS spectra of the top and bottom surfaces. It was shown that P3HT tends to enrich at the free surface of the film due to its lower surface-free energy than PCBM whereas PCBM was found to prefer the film/substrate interface due to its interaction with the substrate. Orimo et al. have studied the effects of thermal annealing on the composition of P3HT and PCBM at the Al electrode/active layer interface using XPS (Orimo et al. 2010). In this study, two devices, preannealed and postannealed devices in which the active layers were annealed prior to and after Al deposition, respectively, were examined using XPS. The XPS results revealed that in the preannealed device P3HT is enriched in the Al interface, and in the postannealed device PCBM is accumulated at the Al interface. The accumulated PCBM at the Al interface of the postannealed device was found to serve as a hole-blocking layer, resulting in an improved open circuit voltage and hence PCE.

Other x-ray-related techniques in the nanophase characterization of OPV include near edge x-ray absorption fine structure (NEXAFS) spectroscopy and scanning transmission x-ray microscopy (STXM) (Giridharagopal and Ginger 2010).

9.2.5 Techniques Using Neutrons

Neutron techniques have been instrumental in developing a recent paradigm shift in the understanding of the structure and phase separation of conjugated polymer/fullerene mixtures (Chen et al. 2011a, 2012a, 2013a,b; Kiel et al. 2010a,b; Lee et al. 2011; Lu et al. 2011; Parnell et al. 2010; Yin and Dadmun 2011). Among the neutron techniques, SANS is a powerful technique to study the planar morphology of conjugated polymer/fullerene mixtures on length scales from 1 nm to more than 100 nm, providing detailed information about the morphology, miscibility, and interfaces (Chen et al. 2011a, 2013a,b; Kiel et al. 2010a; Yin and Dadmun 2011). The experimental determination of the details of this morphology is difficult due to minimal contrast between the C_{60} fullerene and the carbon-based polymer in electron microscopy and the thin film nature of the materials studied in photovoltaics. Fortunately, SANS provides sufficient contrast to unequivocally determine the structure of fullerenes in an organic medium due to the fact that the neutron SLD of a protonated polymer matrix (that is, SLD of P3HT is 0.7×10^{-6} cm^{-2}) differs significantly from that of the fullerene (that is, SLD of PCBM is 4.4×10^{-6} cm^{-2}) (Chen et al. 2013a), which is very similar to that of deuterated organic compounds. In neutron scattering, the coherent scattering intensity can be modeled as

$$\frac{d\Sigma(Q)}{d\Omega} = \left(\frac{N}{V}\right) V_p^2 \Delta\rho^2 P(Q) S(Q) \quad (9.2)$$

where N/V is the number density of scattering particles, V_p is the scattering particle volume, and $\Delta\rho^2$ is the square of the difference of the scattering length density of phases. Usually a significant increase of scattering intensity is associated with fullerene phase separation, which dramatically increases the SLD contrast between phases. $P(Q)$ is the form factor, which is associated with the shape and size of domains. $S(Q)$ is the interparticle structure factor. A careful analysis of the scattering curves can provide information about domain shape and size, concentration of fullerene aggregates, interfacial area between the fullerene-rich phase and polymer-rich phase, and miscibility of the fullerene in polymer (Chen et al. 2011a, 2013a,b; Kiel et al. 2010a; Yin and Dadmun 2011).

Neutron reflectometry is another powerful and essential technique to study the vertical morphology of polymer/fullerene mixtures with high spatial resolution of about a few angstroms (Chen et al. 2012a; Kiel et al. 2010b; Lee et al. 2011; Lu et al. 2011; Parnell et al. 2010). In neutron reflectometry, a well-collimated beam is incident on the smooth surface of a sample with a small incident grazing angle, and the intensity of the reflected beam is obtained as a function of neutron transfer momentum, $Q = 4\pi\sin\theta/\lambda$, where θ is the incident angle, and λ is the wavelength of the neutron. The measured reflectivity can then be analyzed to extract the scattering length density profile of the sample (Nelson 2006). The experimentally determined SLD profile is then analyzed to determine the concentration depth profile of each component in the system. For instance, the fullerene concentration depth profile is determined using the equation

$$\phi(z)_{\text{fullerene}} = \frac{\rho(z) - \rho_{\text{Polymer}}}{\rho_{\text{fullerene}} - \rho_{\text{Polymer}}} \tag{9.3}$$

where $\phi(z)_{\text{fullerene}}$ is the volume fraction of fullerene at depth z, $\rho(z)$ is the experimentally determined scattering length density at depth z, and ρ_{Polymer} and $\rho_{\text{fullerene}}$ are the SLD of the conjugated polymer and fullerene, respectively. A careful analysis of neutron reflectometry curves provides information about the interface, miscibility, and composition depth profile perpendicular to the sample surface (Chen et al. 2012a; Kiel et al. 2010b; Lee et al. 2011; Lu et al. 2011; Parnell et al. 2010).

9.3 CASE STUDIES

One of the core structural aspects of the active layer of an OPV is the 3-D nanophase morphology of the BHJ blends. In an attempt to understand the nanophase morphology buried in the active layer and its correlation to the PCE of OPV, various combined techniques have been actively applied (Chen 2011b; Liu et al. 2011; Lu et al. 2011, 2013; Ruderer et al. 2009; Shao et al. 2014). For example, Li et al. have successfully employed bright field TEM and cross-sectional TEM to map out the BHJ morphology embedded in P3HT/PCBM OPV (Li et al. 2012a). Liu et al. have utilized GISAXS and XRR and neutron reflectometry to obtain the information about the nanophase domains and distributions of blending components in a direction parallel and perpendicular to the surface plane of P3HT/PCBM blend film

(Liu et al. 2011). Shao et al. have combined real and reciprocal space measurement techniques, that is, energy-filtered TEM, GISAXS, and NR to systematically study the lateral and vertical nanophase separations of P3HT and PCBM occurring in the active layer (Shao et al. 2014).

9.3.1 Nanophase Separation in OPV with the Second-Generation Semiconducting Polymers

Cis- and *trans*-polyacetylene have been known as the first generation of semiconducting polymers since their discovery in the 1970s (Heeger 2010). Among the second-generation semiconducting polymers, poly(3-hexyl thiophene) (P3HT) has been shown to be one of the most successful and widely studied materials in organic electronics (Heeger 2010). In OPVs based on a binary blend of electron-donating P3HT and electron-accepting [6,6]-phenyl C_{61} butyric acid methyl ester (PCBM), PCE as high as 4%–5% has been achieved through the manipulation of kinetic nanophase separation between donor and acceptor (Brabec et al. 2010; Li et al. 2005; Reyes-Reyes et al. 2005).

Due to the chemical dissimilarity, P3HT and PCBM that are homogeneously dissolved in good solvent are phase separated during the film formation process. Indeed, the phase separation of donor and acceptor is greatly affected by the rate of solvent evaporation, and if the rate of solvent evaporation is faster than or comparable to that of phase separation, the resultant morphology is trapped in a state far from the equilibrium. Hence, thermal or solvent annealing has typically been applied for more evolved phase separation. Also, if one or both components preferentially segregate to the air/film, bulk, or film/substrate interfaces, the phase separation is distorted dramatically from that in the bulk. In a photoactive layer of OPV, the ideal phase morphology is known to be BHJ, in which the dissimilar electron donor and acceptor are phase separated and form a nanoscale bicontinuous network as shown in Figure 9.10 (Li et al. 2007, 2012a; Yang et al. 2005). In such a BHJ layer, the

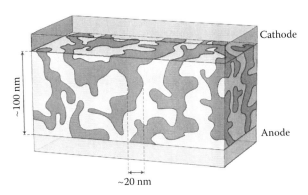

FIGURE 9.10 Conceptual BHJ morphology model with bicontinuous interpenetration network of electron-donating polymer and electron acceptor. (Reprinted by permission from Macmillan Publishers Ltd., *Nature Photonics*, G. Li et al., Polymer Solar Cells, 6, 153–161, Copyright 2012.)

unique nanophase morphology delivers many advantages in the device performance. The nanoscale phase network ensures a large interfacial area between the donor and acceptor and, hence, effective charge separation, and the bicontinuity of donor and acceptor phase facilitates efficient charge transport and collection. In practice, however, current OPV systems, for example, consisting of a blend of P3HT and PCBM, still suffer from many drawbacks in terms of the nanophase morphology (Keum et al. 2013; Kiel et al. 2010a–c; Lee et al. 2011; Orimo et al. 2010; Parnell et al. 2010; Tsai et al. 2010). Spin casting the P3HT/PCBM blend solution onto a substrate to prepare a thin film usually generates stratified morphology perpendicular to the surface plane of the substrate due to the fundamental interfacial effect existing in the thin film. It has been shown that PCBM tends to accumulate on the top and/or bottom interfaces depending on the fabrication condition. Also, further stratification and the growth of phase domains are inevitable as heat is generated during device operation, in effect annealing the active layer. As a result, the domains become much larger than the maximum length of exciton diffusion (<10 nm), the interfacial area between donor and acceptor is reduced, and hence the device performance degrades (Markov et al. 2005). For the origin of the stratification, it has been hypothesized that the relatively hydrophilic PCBM tends to migrate toward the hydrophilic substrates, such as PEDOT:PSS (Oh et al. 2011), $CsCO_3$ (Xu et al. 2009), and SiO_x (Björström et al. 2005), resulting in the PCBM accumulation near the substrates. de Villers et al. have also discussed the origin for the interfacial accumulation of PCBM in terms of the solubility difference between P3HT and PCBM as follows (Tremolet de Villers et al. 2009). During the film-forming process, solvent evaporates first from the top surface, resulting in a solvent-rich region near the substrate. In this case, more soluble PCBM in solvent as compared to P3HT tends to migrate toward the solvent-rich region near the substrate leading to the film stratification. More recently, using neutron reflectometry, Keum et al. have demonstrated that the film stratification occurring during film formation and thermal annealing can be due to the unique crystallization habit of P3HT in thin film and the partial miscibility between amorphous P3HT and PCBM (Keum et al. 2013). Figure 9.11a and b shows measured and modeled neutron reflectivity curves for as-spun and annealed P3HT/PCBM films deposited on Si-substrate and the volume fraction profiles of PCBM, v_{PCBM}s obtained from the neutron reflectivity modeling in Figure 9.11a. The reflectivity modeling was based on the Parratt formalism (Parratt 1954). From the changes in v_{PCBM}, it is evident that spin casting P3HT/PCBM onto Si-substrate results in stratified morphology with PCBM accumulations near the free surface and Si interface. It is also seen that increasing annealing temperature accompanies increased stratification. For the film stratification, it is proposed that, because P3HT remains less crystalline in the two interfacial regions (free surface and substrate interface) than in the bulk region and PCBM is miscible with amorphous P3HT, PCBM preferentially diffuses to the two interfacial regions and accumulates, resulting in the stratification. Recently, Wu et al. have examined the effect of thermal annealing on the size of PCBM domains using GISAX (Wu et al. 2011). Figure 9.11c and d shows in-plane GISAXS profiles for the P3HT/PCBM film measured at 30°C, 60°C, 100°C, 130°C, and 150°C and during the subsequent isothermal annealing at 150°C, when the GISAXS profiles were fit using a polydisperse sphere model with Schulz distribution (Griffith et al. 1987). As

Nanophase Separation in Organic Solar Cells

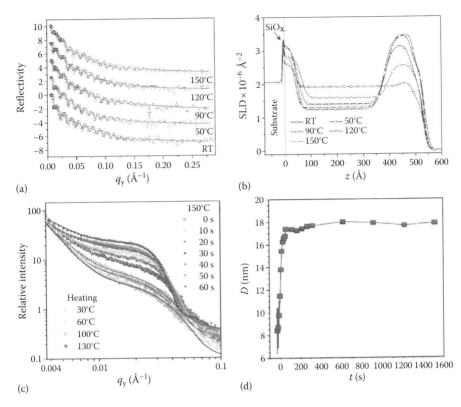

FIGURE 9.11 (a) Measured and model fit NR profiles of as-cast and annealed P3HT/PCBM film and (b) volume fraction profiles of PCBM (v_{PCBM}) extracted from the NR data model fit shown in (a). The annealing temperatures were as indicated and the annealing time at each temperature was 90 min. (c) GISAXS profiles for the P3HT/PCBM film measured at 30°C, 60°C, 100°C, 130°C, and 150°C and during the subsequent isothermal annealing at 150°C. The data were fit (solid curves) using a polydisperse sphere model with the Schultz size distribution. (From W.L. Griffith et al., *Physical Review A* 35: 2200–2206, 1987.) (d) Time evolution of the diameter (*D*) of PCBM domains at 150°C. (a and b reprinted with permission from J.K. Keum et al., Morphological Origin for the Stratification of P3HT:PCBM Blend Film Studied by Neutron Reflectometry, *Applied Physics Letters*, 103, 223301. Copyright 2013, American Institute of Physics.) (c and d reprinted with permission from W.-R. Wu et al. 2011, 6233–6243. Copyright 2011, American Chemical Society.)

shown in Figure 9.11d, the PCBM domains grew quickly from ~7.0 to ~18.0 nm in diameter at 150°C within 60 s due to phase separation.

In an effort to optimize the phase morphology, various diblock copolymer surfactants have been used as a compatibilizer or interfacial modifier between donor- and acceptor-rich domains. For example, Sun et al. have shown that a P3HT-*b*-polystyrene (P3HT-*b*-PS) diblock copolymer can act as an interfacial compatibilizer at the interface between P3HT and PCBM, where the added small amount of P3HT-*b*-PS preferentially stays at the interface between P3HT and PCBM and stabilizes the domains while reducing the film stratification and domain growth during thermal annealing (Sun et al. 2012). Chen et al. have also reported reduced domain size by the added small amount of P3HT-*b*-polyethylene oxide (P3HT-*b*-PEO) to the P3HT/PCBM blend (Chen et al. 2012b). The reduced stratification and size of phase domains might result in the improved charge transport and collection. However, the existence of P3HT-*b*-PS and P3HT-*b*-PEO at the donor–acceptor interface could negatively affect the open circuit voltage of the devices due to the existence of a nonconductive PS- and PEO-block in the donor–acceptor interface. It has been reported that the presence of nonconductive DIO in the interface between donor and acceptor can result in the reduced open circuit voltage of the OPV (Su et al. 2011). In this case, the average distance between donor and acceptor increases by the DIO, and hence, the probability of exciton dissociation at the donor–acceptor interface is reduced, resulting in the reduction of open circuit voltage.

9.3.2 NANOPHASE SEPARATION IN OPV WITH THE THIRD-GENERATION SEMICONDUCTING POLYMERS

The majority of OPV research efforts gradually shifted to the third generation of semiconducting polymers or the low-bandgap (LBG) polymers due to their improved light absorption and performances (Bian et al. 2012; Zhou et al. 2012). A 9.2% PCE has been achieved for single-layer low-bandgap polymer devices (He et al. 2012). LBG polymers are usually constructed by donor–acceptor hybridization or quinoid resonance (Zhou et al. 2012). The prosperity of LBG polymers has also enriched the content of nanophase separation study. It was shown that both chemical structure and processing conditions had profound effects in affecting the resultant BHJ thin-film morphology. It should also be noted that a chemical additive is widely used in optimizing the morphology of LBG polymers (Liu et al. 2013a). In the following discussions, we will focus on a few well-studied LBG polymers and their nanophase morphologies.

PCPDTBT is one of the most well-studied low-bandgap polymers. It has a bandgap of 1.46 eV and absorbs deeply in spectrum. By blending PCPDTBT with PCBM and fabricating solar cells through a conventional approach, a moderate efficiency ~3% is recorded (Peet et al. 2008). In the earlier stage of research, PCPDTBT was thought to be low in crystallinity because it mixes well with PCBM. Such intimate mixing between the donor and acceptor makes it difficult to phase separate into bicontinuous morphology, which causes the serious charge recombination. For this reason, high PCBM loading is required in its device fabrication. Posttreatments, such as thermal annealing were applied to the system, yet no improvement was seen. Importantly, Heeger and coworkers developed a very effective method of adding a high boiling point bad solvent into the BHJ solution to modulate the morphology formation of PCPDTBT blends in processing (Peet et al. 2007). Although only a tiny amount of additive was added (usually ~3 v%), the PCE of the devices was increased to above 5%. This observation opened a new route to optimize the morphology, and it quickly became a standard procedure to fabricate LBG polymer OPVs. After testing a variety of additives and combinations of the additive and the host solvent, herein the requirements for processing additives have been summarized as follows: (i) There must be a selective solubility for PCBM, and (ii) it must have a higher boiling point than the host solvent (Lee et al. 2008). Detailed morphology characterization revealed that phase separation between PCPDTBT and PCBM by using an additive was the main factor for the increased efficiency, demonstrated by AFM and TEM images (Lee et al. 2008). GIXD characterization showed the ordered packing of PCPDTBT chains, and the chain packing took the edge-on orientation similar to that of P3HT (Agostinelli et al. 2011; Rogers et al. 2011). In situ UV-vis and GIXD were used to monitor the chain packing or crystal evolution process during solvent evaporation (Peet et al. 2008). Russell and coworkers characterized the morphology of PCPDTBT:PCBM blends in detail (Gu et al. 2012). By using x-ray/neutron scattering and diffraction methods in combination with TEM, they demonstrated a multilength scale morphology, which comprised domains of pure, crystalline PCPDTBT fibrils and another domain of a PCBM-rich mixture with noncrystalline PCPDTBT aggregations. The fibril-to-fibril spacing forms the major length scale of phase-separated morphology and determines the performance of the blends.

Besides the processing optimization and morphology tuning, synthetic effort was also employed to improve the performance of the material. Yang and coworkers synthesized poly{[4,40-bis(2-ethylhexyl)dithieno(3,2-b;20,30-d)silole]-2,6-diyl-alt-(2,1,3-benzothidiazole)-4,7-diyl} (PSBTBT), a silole-containing polymer (Hou et al. 2008). By replacing the five-position carbon of PCPDTBT with a silicon atom, the stacking of the polymer chain is improved, and thus the crystallinity of the polymer was largely improved also. GIXD showed a strong edge-on orientation of the crystallites (Chen et al. 2010; Scharber et al. 2010). The higher crystallinity of the polymer is thought to be responsible for the better charge transport and reduced bimolecular recombination. Thermal annealing was demonstrated to be effective to increase PCE from 3.8% to 5.6%. GIXD and SANS showed that the crystallization of PSBTBT and segregation of PCBM occurred during spin coating, and thermal annealing increased the ordering of PSBTBT and enhanced the segregation of the PCBM, forming domains ~10 nm in size, leading to an improvement in photovoltaic performance (Lu et al. 2011).

Conjugated polymers based on diketopyrrolopyrrole (DPP) have also been widely used in OPVs (Li et al. 2013). DPP has a planar conjugated bicyclic structure, which leads to strong π–π interaction and high crystallinity. Its strong electron withdrawing property also leads to a low bandgap when polymerized with proper comonomers. Janssen and coworkers reported a DPP-quarter thiophene polymer, which showed 4% PCE when processed from a chloroform-dichlorobenzene solvent mixture (Wienk et al. 2008). The low-bandgap property of this polymer effectively extended the EQE to around 900 nm. Other polymers were also developed, and enhanced performance was recorded (Bijleveld et al. 2009, 2010; Bronstein et al. 2011; Li et al. 2012b; Meager et al. 2013; Ye et al. 2012; Zhang et al. 2011; Zhou et al. 2009). Although DPP polymers demonstrate excellent absorption, its miscibility with PCBM is poor. When the BHJ blends were directly casted from chloroform, large (>200 nm wide) fullerene clusters dominate the morphology. This macroscopic phase separation leads to poor efficiency around 2% (Bijleveld et al. 2010; Wienk et al. 2008). In morphology optimization, it was seen that by using either solvent mixtures or adding additives, the phase became much more uniform, and fibril-like polymer domains were well developed (Bijleveld et al. 2010; Wienk et al. 2008). Frechet and coworkers studied the effect of a bridge molecule of DPP-based polymer (Woo et al. 2010). As the furan was incorporated into the conjugated polymer as a bridge molecule, the solubility of the polymer was dramatically increased. They also used a processing additive to develop the BHJ morphology with nanoscale phase separation. When the mixture of polymer and PCBM was spin coated with chlorobenzene, macroscopic phase separation was observed. However, after addition of 9% of 1-chloronaphthalene to the chlorobenzene, the nanoscale phase separation was achieved, and 5% efficiency was achieved. Russell and coworkers characterized the morphology details of solvent mixture-processed BHJ blends and proposed a poor solvent-induced crystallization to explain the morphology formation mechanism (Liu et al. 2012b). In solvent mixture processing, 1,2-dichlorobenzene was a poor solvent for DPP polymer; its presence promotes the crystallization of DPP polymer into fibrils with smooth distribution of PCBM. A fibril network morphology was observed for this system, and the interfibril spacing determined the length scale of phase separation as supported by TEM and resonant soft x-ray scattering. In a following study, they compared the morphology of a series of DPP polymers with varied chemical structures and showed that slight variation of chemical structures had a profound effect on the properties of the materials, such as absorption, crystallization, and phase separation when blended with PCBM. It was shown that solvent mixture processing and solvent–additive processing achieved the same results in terms of nanomorphological control, and the interfibril spacing was shown to be a key parameter in determining the device performance (Liu et al. 2013b; Ferdous et al. 2014).

Low-bandgap polymers combining thieno[3,4-b]thiophene and benzodithiophene are another category of high-performance LBG materials. Although their absorption is not as deep as PCPDTBT or DPP polymers, the suitable energy levels led to enhanced open circuit voltage and thus increased PCE (Zhou et al. 2012). Yu and coworkers developed a series of thienothiophene-benzodithiophene polymers (PTB1 to PTB7) (Liang et al. 2009, 2010; Liang and Yu 2010). The PCE of 7.4% was achieved based on BHJ solar cells of PTB7/PC$_{71}$BM blends (Chen et al. 2009a; Liang et al. 2010). The PTBx series showed interesting crystal packing: the chains were found to order on the substrate at the face-on orientation. GIXD images revealed the intensive out-of-plane (OOP) (010) peaks for the neat polymer, which is rather different from many other conjugated polymers used in organic solar cells (Chen et al. 2011b; Collins et al. 2013; Hammond et al. 2011). In terms of their device preparations, an additive is also necessary to reduce the PCBM aggregation and optimize the nanophase morphology (Liang et al. 2010). The combined results of resonant soft x-ray scattering and energy-filtered TEM microscopy yielded a hierarchical morphology picture of PTB7:PCBM blends (Chen et al. 2011b). Characterizations using scanning transmission x-ray microscopy showed that PCBM aggregation is critical in realizing optimal nanomorphology (Collins et al. 2013). Better resolved TEM combined with RSoXS, GISAXS, and GIXD showed that PTB7:PCBM blends had a multilength scale morphology, originating from the unique structure order of PTB7 materials (Liu et al. 2014). More specifically, PTB7 can form two different types of ordered structures; the larger sized aggregation forms the 140-nm phase separation, and another size-scaled phase separation around 20–30 nm (as revealed by scattering methods) is responsible for the large current of the devices.

9.3.3 NANOPHASE SEPARATION IN OPV WITH SMALL-MOLECULE SEMICONDUCTORS

Small molecule-based BHJ (SM-BHJ) solar cells are based on blends of small molecules and fullerenes, and there is a dramatic growth in these promising systems in recent years (Garcia et al. 2012; Liu et al. 2013c; Love et al. 2013; Perez et al. 2013; Sun et al. 2012; van der Poll et al. 2012; Walker et al. 2011; Wang et al. 2013). Recently, a PCE of 10.1% for a SM-BHJ based solar cell has been demonstrated (Liu et al. 2013c), which is comparable to polymer-based BHJ solar cells. Additionally, SM-BHJ based solar cells offer some advantages relative to polymer-based solar cells, such as the fact that their well-defined structure readily allows their functionalization and purification. Similar to a polymer-based solar cell, morphology and nanophase separation of a SM-BHJ solar cell is crucial for the optimization of an OPV system. Love and coauthors demonstrated the morphological control of a SM-BHJ system (p-DTS[FBTTh$_2$]$_2$ and PC$_{70}$BM) using thermal annealing and solvent additive (Love et al. 2013). The as-cast sample shows a relatively modest PCE of 1.8%, however, thermal annealing and solvent additive processing lead to PCE up to 5.8% and 7.0%, which is associated with morphology change. The bright field TEM images shows that there is no significant crystallization or phase separation observed for the as-cast film. Thermal annealing results in the crystallization of p-DTS(FBTTh$_2$)$_2$, which shows wire-like structure with 40–50 nm width in TEM. A different image is observed for the film processed with 0.4% 1,8-diiodooctane (DIO), which shows a continuous network with smaller width (30–40 nm). These results show that the formation of interconnected network of crystalline small molecule domains is crucial for the optimization for the SM-BHJ system.

Besides the crystallization of SM, the aggregates of fullerene have also been found to be crucial in the optimization of a SM-BHJ device. Wang and coauthors studied the impact of DIO on the morphology and device performance of a DPP-based SM-BHJ device (Wang et al. 2013). The grazing incidence x-ray diffraction results shows the ratio of the PCBM aggregates to SM crystallites is strongly impacted by the concentration of DIO. The results shows that a smaller concentration of DIO (below 0.5%) mainly enhanced the crystallization of SM, and a higher concentration of DIO induced the aggregates of PCBM. The best PCE was found with 3% DIO. The morphology of the SM/fullerene mixture consists of the SM crystalline phase, miscible phase of SM and fullerene, and a pure fullerene phase. A delicate balance among these phases is the key for the optimization of morphology of the active layer of a SM-BHJ device.

9.3.4 NANOPHASE SEPARATION IN ADVANCED TERNARY OPVS

Nanophase separation is equally or even more important for other types of advanced organic solar cells, for example, ternary blend BHJ organic solar cells. Ternary blend BHJ solar cells are made from the blends of two donors and one acceptor or two acceptors and one donor in order to enlarge the absorption window. As the fundamental device configuration of a ternary blend BHJ is identical to the binary solar cells, the fabrication process of a ternary BHJ is simple (Ameri et al. 2012; Khlyabich et al. 2011, 2012). Two donors and one acceptor ternary blends are particularly interesting as donor materials can have red-shifted absorption more readily (Ameri et al. 2012; Yang et al. 2012), and some improvement of device performance can be achieved (Ameri et al. 2012; Khlyabich et al. 2011, 2012; Yang et al. 2012). For example, You and coworkers showed in thin devices that the efficiency of poly (benzodithiophene-dithienylbezotriazole) TAZ and poly (benzodithiophene-dithienylbezothiadiazole) DTBT blended with PCBM was improved compared to binary blends (Yang et al. 2012). Brabec and coworkers showed that average performing P3HT/PCBM binary solar cells can be enhanced by introducing the low-bandgap material Si-PCPDTBT (Ameri et al. 2012).

The BHJ active layer consists of an interpenetrating network (~10 nm) of a hole conductor and an electron acceptor. The prerequisite of improving photocurrent using the ternary components strategy is to keep efficient charge separation and transport properties of single-layer devices. The phase separation of a ternary BHJ organic solar cell may negatively impact transport; for example, if the higher bandgap donor material surrounds the PCBM, the holes or/and electrons produced in the low-bandgap material will never contribute to the current due to energy transfer restriction. Similarly, the polymer with the higher LUMO might be dispersed in a matrix of polymer with the lower LUMO, thus trapping holes. This indicates that the morphology in the ternary blend device is even more critical than in binary blends. Furthermore, many factors, including the material–material interactions, play an important role in the final morphology. For example, it is likely that PCBM has different miscibility in the two donor materials, which could induce a different PCBM ratio in each donor-rich phase (Ma et al. 2013).

9.4 CONCLUSION

Interfacial arrangements are well known as an important factor in determining the local properties and performance of polymer-based hybrid systems (Kobayashi et al. 2011; Song et al. 2012, 2013). In this chapter, we explored, in detail, the nanophase separation of cutting-edge OPV systems and demonstrated that the fundamental understanding and controllable manipulation of nanophase separation are both necessary and urgent in order to further improve these novel energy conversion devices.

ACKNOWLEDGMENT

This research was conducted at the Center for Nanophase Materials Sciences, which is a DOE Office of Science User Facility.

REFERENCES

Ade, H. (2012). "Characterization of Organic Thin Films with Resonant Soft X-Ray Scattering and Reflectivity Near the Carbon and Fluorine Absorption Edges." *European Physical Journal: Special Topics* **208**: 305–318.

Adhikary, P. et al. (2013). "Enhanced Charge Transport and Photovoltaic Performance of PBDTTT-C-T/PC70BM Solar Cells via UV-Ozone Treatment." *Nanoscale* **5**(20): 10007–10013.

Agostinelli, T. et al. (2011). "The Role of Alkane Dithiols in Controlling Polymer Crystallization in Small Band Gap Polymer: Fullerene Solar Cells." *Journal of Polymer Science Part B: Polymer Physics* **49**(10): 717–724.

Ajuria, J. et al. (2013). "Nanomorphology Influence on the Light Conversion Mechanisms in Highly Efficient Diketopyrrolopyrrole Based Organic Solar Cells." *Organic Electronics* **14**(1): 326–334.

Ameri, T. et al. (2012). "Performance Enhancement of the P3HT/PCBM Solar Cells through NIR Sensitization Using a Small-Bandgap Polymer." *Advanced Energy Materials* **2**: 1198–1202.

Andersson, B. V. et al. (2009). "Imaging of the 3D Nanostructure of a Polymer Solar Cell by Electron Tomography." *Nano Letters* **9**(2): 853–855.

Andreasen, B. et al. (2012). "TOF-SIMS Investigation of Degradation Pathways Occurring in a Variety of Organic Photovoltaic Devices—The ISOS-3 Inter-Laboratory Collaboration." *Physical Chemistry Chemical Physics* **14**(33): 11780–11799.

Bassim, N. D. et al. (2012). "Minimizing Damage during FIB Sample Preparation of Soft Materials." *Journal of Microscopy* **245**(3): 288–301.

Bian, L. et al. (2012). "Recent Progress in the Design of Narrow Bandgap Conjugated Polymers for High-Efficiency Organic Solar Cells." *Progress in Polymer Science* **37**(9): 1292–1331.

Bijleveld, J. C. et al. (2009). "Poly(Diketopyrrolopyrrole–terthiophene) for Ambipolar Logic and Photovoltaics." *Journal of the American Chemical Society* **131**(46): 16616–16617.

Bijleveld, J. C. et al. (2010). "Efficient Solar Cells Based on an Easily Accessible Diketopyrrolopyrrole Polymer." *Advanced Materials* **22**(35): E242–E246.

Binnig, G. et al. (1986). "Atomic Force Microscope." *Physical Review Letters* **56**: 930–933.

Björström, C. M. et al. (2005). "Multilayer Formation in Spin-Coated Thin Films of Low-Bandgap Polyfluorene:PCBM Blends." *Journal of Physics: Condensed Matter* **17**: L529–L534.

Blouin, N. et al. (2008). "Towards a Rational Design of Poly(2,7-carbazole) Derivatives for Solar Cells." *Journal of the American Chemical Society* **130**: 732–742.

Brabec, C. J. (2004). "Organic Photovoltaics: Technology and Market." *Solar Energy Materials and Solar Cells* **83**(2–3): 273–292.

Brabec, C. J. et al. (2001). "Plastic Solar Cells." *Advanced Functional Materials* **11**: 15–26.

Brabec, C. J. et al. (2002). "Effect of LiF/Metal Electrodes on the Performance of Plastic Solar Cells." *Applied Physics Letters* **80**(7): 1288.

Brabec, C. J. et al. (2010). "Polymer–Fullerene Bulk-Heterojunction Solar Cells." *Advanced Materials* **22**(34): 3839–3856.

Bronstein, H. et al. (2011). "Thieno[3,2-b]thiophene–Diketopyrrolopyrrole-Containing Polymers for High-Performance Organic Field-Effect Transistors and Organic Photovoltaic Devices." *Journal of the American Chemical Society* **133**(10): 3272–3275.

Campoy-Quiles, M. et al. (2008). "Morphology Evolution via Self-Organization and Lateral and Vertical Diffusion in Polymer: Fullerene Solar Cell Blends." *Nature Materials* **7**(2): 158–164.

Chen, D. A. et al. (2011a). "P3HT/PCBM Bulk Heterojunction Organic Photovoltaics: Correlating Efficiency and Morphology." *Nano Letters* **11**(2): 561–567.

Chen, H. P. et al. (2012a). "The Miscibility and Depth Profile of PCBM in P3HT: Thermodynamic Information to Improve Organic Photovoltaics." *Physical Chemistry Chemical Physics* **14**(16): 5635–5641.

Chen, H. P. et al. (2013a). "Correlation of Polymeric Compatibilizer Structure to Its Impact on the Morphology and Function of P3HT:PCBM Bulk Heterojunctions." *Journal of Materials Chemistry A* **1**(17): 5309–5319.

Chen, H. P. et al. (2013b). "Precise Structural Development and Its Correlation to Function in Conjugated Polymer: Fullerene Thin Films by Controlled Solvent Annealing." *Advanced Functional Materials* **23**(13): 1701–1710.

Chen, H.-Y. et al. (2009a). "Polymer Solar Cells with Enhanced Open-Circuit Voltage and Efficiency." *Nature Photonics* **3**(11): 649–653.

Chen, H.-Y. et al. (2010). "Silicon Atom Substitution Enhances Interchain Packing in a Thiophene-Based Polymer System." *Advanced Materials* **22**(3): 371–375.

Chen, J. et al. (2012b). "Ternary Behavior and Systematic Nanoscale Manipulation of Domain Structures in P3HT/PCBM/P3HT-b-PEO Films." *Journal of Materials Chemistry* **22**: 13013–13022.

Chen, L. M. et al. (2009b). "Recent Progress in Polymer Solar Cells: Manipulation of Polymer: Fullerene Morphology and the Formation of Efficient Inverted Polymer Solar Cells." *Advanced Materials* **21**(14–15): 1434–1449.

Chen, W. et al. (2011b). "Hierarchical Nanomorphologies Promote Exciton Dissociation in Polymer/Fullerene Bulk Heterojunction Solar Cells." *Nano Letters* **11**(9): 3707–3713.

Chen, W. et al. (2012c). "Morphology Characterization in Organic and Hybrid Solar Cells." *Energy & Environmental Science* **5**(8): 8045–8074.

Chochos, C. L. and S. A. Choulis (2011). "How the Structural Deviations on the Backbone of Conjugated Polymers Influence Their Optoelectronic Properties and Photovoltaic Performance." *Progress in Polymer Science* **36**(10): 1326–1414.

Chu, T.-Y. et al. (2011). "Morphology Control in Polycarbazole Based Bulk Heterojunction Solar Cells and Its Impact on Device Performance." *Applied Physics Letters* **98**: 253301.

Coakley, K. M. and M. D. McGehee (2004). "Conjugated Polymer Photovoltaic Cells." *Chemistry of Materials* **16**(23): 4533–4542.

Collins, B. A. et al. (2013). "Absolute Measurement of Domain Composition and Nanoscale Size Distribution Explains Performance in PTB7:PC 71BM Solar Cells." *Advanced Energy Materials* **3**(3): 65–74.

Dang, M. T. et al. (2013). "Controlling the Morphology and Performance of Bulk Heterojunctions in Solar Cells. Lessons Learned from the Benchmark Poly(3-hexylthiophene):[6,6]-Phenyl-C 61-butyric Acid Methyl Ester System." *Chemical Reviews* **113**(5): 3734–3765.

DeLongchamp, D. M. et al. (2012). "Nanoscale Structure Measurements for Polymer-Fullerene Photovoltaics." *Energy & Environmental Science* **5**(3): 5980–5993.

Dennler, G. et al. (2008). "Design Rules for Donors in Bulk-Heterojunction Tandem Solar Cells? Towards 15% Energy-Conversion Efficiency." *Advanced Materials* **20**(3): 579–583.

Domanski, A. L. et al. (2013). "Effect of Morphological Changes on Presence of Trap States in P3HT:PCBM Solar Cells Studied by Cross-Sectional Energy Filtered TEM and Thermally Stimulated Current Measurements." *Journal of Physical Chemistry C* **117**(45): 23495–23499.

Drummy, L. F. et al. (2011). "Molecular-Scale and Nanoscale Morphology of P3HT:PCBM Bulk Heterojunctions: Energy-Filtered TEM and Low-Dose HREM." *Chemistry of Materials* **23**(3): 907–912.

Ferdous, S. et al. (2014). "Solvent-Polarity-Induced Active Layer Morphology Control in Crystalline Diketopyrrolopyrrole-Based Low Band Gap Polymer Photovoltaics." *Advanced Energy Materials* **4**: 1300834.

Garcia, A. et al. (2012). "Improvement of Interfacial Contacts for New Small-Molecule Bulk-Heterojunction Organic Photovoltaics." *Advanced Materials* **24**(39): 5368–5373.

Giridharagopal, R. and D. S. Ginger (2010). "Characterizing Morphology in Bulk Heterojunction Organic Photovoltaic Systems." *Journal of Physical Chemistry Letters* **1**(7): 1160–1169.

Graham, K. R. et al. (2013). "Re-Evaluating the Role of Sterics and Electronic Coupling in Determining the Open-Circuit Voltage of Organic Solar Cells." *Advanced Materials* **25**(42): 6076–6082.

Griffith, W. L. et al. (1987). "Analytical Scattering Function of a Polydisperse Percus-Yevick Fluid with Schulz- (Γ-) Distributed Diameters." *Physical Review A* **35**: 2200–2206.

Gu, Y. et al. (2012). "Multi-Length-Scale Morphologies in PCPDTBT/PCBM Bulk-Heterojunction Solar Cells." *Advanced Energy Materials* **2**(6): 683–690.

Halls, J. J. M. et al. (1995). "Efficient Photodiodes from Interpenetrating Polymer Networks." *Nature* **376**(6540): 498–500.

Hammond, M. R. et al. (2011). "Molecular Order in High-Efficiency Polymer/Fullerene Bulk Heterojunction Solar Cells." *ACS Nano* **5**(10): 8248–8257.

Hammouda, B. (2010). "A New Guinier-Porod Model." *Journal of Applied Crystallography* **43**: 716–719.

He, Z. et al. (2011). "Simultaneous Enhancement of Open-Circuit Voltage, Short-Circuit Current Density, and Fill Factor in Polymer Solar Cells." *Advanced Materials* **23**(40): 4636–4643.

He, Z. et al. (2012). "Enhanced Power-Conversion Efficiency in Polymer Solar Cells Using an Inverted Device Structure." *Nature Photonics* **6**(9): 593–597.

Heeger, A. J. (2010). "Semiconducting Polymers: The Third Generation." *Chemical Society Reviews* **39**(7): 2354–2371.

Hegde, R. et al. (2012). "The Impact of Controlled Solvent Exposure on the Morphology, Structure and Function of Bulk Heterojunction Solar Cells." *Solar Energy Materials and Solar Cells* **107**: 112–124.

Herzing, A. A. et al. (2010). "3D Nanoscale Characterization of Thin-Film Organic Photovoltaic Device Structures via Spectroscopic Contrast in the TEM." *Journal of Physical Chemistry C* **114**(41): 17501–17508.

Hoppe, H. et al. (2004). "Nanoscale Morphology of Conjugated Polymer/Fullerene Based Bulk Heterojunction Solar Cells." *Advanced Functional Materials* **14**(10): 1005–1011.

Hoppe, H. and N. S. Sariciftci (2006). "Morphology of Polymer/Fullerene Bulk Heterojunction Solar Cells." *Journal of Materials Chemistry* **16**(1): 45–61.

Hou, J. et al. (2008). "Synthesis, Characterization, and Photovoltaic Properties of a Low Band Gap Polymer Based on Silole-Containing Polythiophenes and 2,1,3-Benzothiadiazole." *Journal of the American Chemical Society* **130**(48): 16144–16145.

Hunt, J. A. and D. B. Williams (1991). "Electron Energy-Loss Spectrum-Imaging." *Ultramicroscopy* **38**(1): 47–73.

Inganäs, O. et al. (2009). "Alternating Polyfluorenes Collect Solar Light in Polymer Photovoltaics." *Accounts of Chemical Research* **42**(11): 1731–1739.

Kato, N. I. (2004). "Reducing Focused Ion Beam Damage to Transmission Electron Microscopy Samples." *Journal of Electron Microscopy* **53**(5): 451–458.

Keum, J. K. et al. (2013). "Morphological Origin for the Stratification of P3HT:PCBM Blend Film Studied by Neutron Reflectometry." *Applied Physics Letters* **103**: 223301.

Khlyabich, P. P. et al. (2011). "Efficient Ternary Blend Bulk Heterojunction Solar Cells with Tunable." *Journal of the American Chemical Society* **133**: 14534–14537.

Khlyabich, P. P. et al. (2012). "Compositional Dependence of the Open-Circuit Voltage in Ternary Blend Bulk Heterojunction Solar Cells Based on Two Donor Polymers." *Journal of the American Chemical Society* **134**: 9074–9077.

Kiel, J. W. et al. (2010a). "Nanoparticle Agglomeration in Polymer-Based Solar Cells." *Physical Review Letters* **105**(16): 4.

Kiel, J. W. et al. (2010b). "Nanoparticle Concentration Profile in Polymer-Based Solar Cells." *Soft Matter* **6**(3): 641–646.

Kiel, J. W. et al. (2010c). "Phase-Sensitive Neutron Reflectometry Measurements Applied in the Study of Photovoltaic Films." *The Journal of Chemical Physics* **133**: 074902.

Kirschner, S. B. et al. (2012). "X-Ray and Neutron Reflectivity and Electronic Properties of PCBM-Poly(bromo)styrene Blends and Bilayers with Poly(3-hexylthiophene)." *Journal of Materials Chemistry* **22**: 4364–4370.

Kobayashi, S. et al. (2011). "Amino-Functionalized Polyethylene for Enhancing the Adhesion between Polyolefins and Polyurethanes." *Industrial & Engineering Chemistry Research* **50**(6): 3274–3279.

Krebs, F. C. et al. (2010a). "Product Integration of Compact Roll-to-Roll Processed Polymer Solar Cell Modules: Methods and Manufacture Using Flexographic Printing, Slot-Die Coating and Rotary Screen Printing." *Journal of Materials Chemistry* **20**(41): 8994.

Krebs, F. C. et al. (2010b). "Upscaling of Polymer Solar Cell Fabrication Using Full Roll-to-Roll Processing." *Nanoscale* **2**(6): 873.

Lee, J. K. et al. (2008). "Processing Additives for Improved Efficiency from Bulk Heterojunction Solar Cells." *Journal of the American Chemical Society* **130**(11): 3619–3623.

Lee, K. H. et al. (2011). "Morphology of All-Solution-Processed 'Bilayer' Organic Solar Cells." *Advanced Materials* **23**(6): 766–770.

Li, G. et al. (2005). "High-Efficiency Solution Processable Polymer Photovoltaic Cells by Self-Organization of Polymer Blends." *Nature Materials* **4**: 864–868.

Li, G. et al. (2007). "'Solvent Annealing' Effect in Polymer Solar Cells Based on Poly (3-hexylthiophene) and Methanofullerenes." *Advanced Functional Materials* **17**(10): 1636–1644.

Li, G. et al. (2012a). "Polymer Solar Cells." *Nature Photonics* **6**: 153–161.

Li, W. et al. (2012b). "Enhancing the Photocurrent in Diketopyrrolopyrrole-Based Polymer Solar Cells via Energy Level Control." *Journal of the American Chemical Society* **134**(33): 13787–13795.

Li, Y. et al. (2013). "High Mobility Diketopyrrolopyrrole (DPP)-Based Organic Semiconductor Materials for Organic Thin Film Transistors and Photovoltaics." *Energy & Environmental Science* **6**(6): 1684.

Liang, Y. et al. (2009). "Development of New Semiconducting Polymers for High Performance Solar Cells." *Journal of the American Chemical Society* **131**(1): 2008–2009.

Liang, Y. et al. (2010). "For the Bright Future—Bulk Heterojunction Polymer Solar Cells with Power Conversion Efficiency of 7.4%." *Advanced Materials* **22**(20): E135–E138.

Liang, Y. and L. Yu (2010). "A New Class of Semiconducting Polymers for Bulk Heterojunction Solar Cells with Exceptionally High Performance." *Accounts of Chemical Research* **43**(9): 1227–1236.

Liu, F. et al. (2012a). "On the Morphology of Polymer-Based Photovoltaics." *Journal of Polymer Science Part B: Polymer Physics* **50**(15): 1018–1044.

Liu, F. et al. (2012b). "Efficient Polymer Solar Cells Based on a Low Bandgap Semi-Crystalline DPP Polymer-PCBM Blends." *Advanced Materials* **24**(29): 3947–3951.

Liu, F. et al. (2013a). "Characterization of the Morphology of Solution-Processed Bulk Heterojunction Organic Photovoltaics." *Progress in Polymer Science* **38**(12): 1990–2052.

Liu, F. et al. (2013b). "Relating Chemical Structure to Device Performance via Morphology Control in Diketopyrrolopyrrole-Based Low Band Gap Polymers." *Journal of the American Chemical Society* **135**(51): 19248–19259.

Liu, F. et al. (2014). "Understanding the Morphology of PTB7:PCBM Blends in Organic Photovoltaics." *Advanced Energy Materials* **4**: 1301377.

Liu, H.-J. et al. (2011). "Surface and Interface Porosity of Polymer/Fullerene-Derivative Thin Films Revealed by Contrast Variation of Neutron and X-Ray Reflectivity." *Soft Matter* **7**: 9276–9282.

Liu, Y. S. et al. (2013c). "Solution-Processed Small-Molecule Solar Cells: Breaking the 10% Power Conversion Efficiency." *Scientific Reports* **3**: 8.

Lou, S. J. et al. (2011). "Effects of Additives on the Morphology of Solution Phase Aggregates Formed by Active Layer Components of High-Efficiency Organic Solar Cells." *Journal of the American Chemical Society* **133**(51): 20661–20663.

Love, J. A. et al. (2013). "Film Morphology of High Efficiency Solution-Processed Small-Molecule Solar Cells." *Advanced Functional Materials* **23**(40): 5019–5026.

Lu, H. et al. (2011). "Morphological Characterization of a Low-Bandgap Crystalline Polymer:PCBM Bulk Heterojunction Solar Cells." *Advanced Energy Materials* **1**(5): 870–878.

Lu, L. et al. (2013). "The Role of N-Doped Multiwall Carbon Nanotubes in Achieving Highly Efficient Polymer Bulk Heterojunction Solar Cells." *Nano Letters* **13**(6): 2365–2369.

Ma, W. et al. (2013). "Competition between Morphological Attributes in the Thermal Annealing and Additive Processing of Polymer Solar Cells." *Journal of Materials Chemistry* **1**: 5023–5030.

Markov, D. E. et al. (2005). "Simultaneous Enhancement of Charge Transport and Exciton Diffusion in Poly(p-Phenylene Vinylene) Derivatives." *Physical Review B* **72**: 045217.

Martin, D. et al. (2005). "High Resolution Electron Microscopy of Ordered Polymers and Organic Molecular Crystals: Recent Developments and Future Possibilities." *Journal of Polymer Science Part B-Polymer Physics* **43**(14): 1749–1778.

Mayer, A. C. et al. (2007). "Polymer-Based Solar Cells." *Materials Today* **10**(11): 28–33.

Meager, I. et al. (2013). "Photocurrent Enhancement from Diketopyrrolopyrrole Polymer Solar Cells through Alkyl-Chain Branching Point Manipulation." *Journal of the American Chemical Society* **135**(31): 11537–11540.

Midgley, P. A. and M. Weyland (2003). "3D Electron Microscopy in the Physical Sciences: The Development of Z-contrast and EFTEM Tomography." *Ultramicroscopy* **96**(3–4): 413–431.

Mihailetchi, V. D. et al. (2006). "Charge Transport and Photocurrent Generation in Poly (3-hexylthiophene): Methanofullerene Bulk-Heterojunction Solar Cells." *Advanced Functional Materials* **16**(5): 699–708.

Moon, J. S. et al. (2011). "Spontaneous Formation of Bulk Heterojunction Nanostructures: Multiple Routes to Equivalent Morphologies." *Nano Letters* **11**(3): 1036–1039.

Moulé, A. J. and K. Meerholz (2008). "Controlling Morphology in Polymer–Fullerene Mixtures." *Advanced Materials* **20**(2): 240–245.

Nelson, A. (2006). "Co-refinement of Multiple-Contrast Neutron/X-Ray Reflectivity Data Using MOTOFIT." *Journal of Applied Crystallography* **39**: 273–276.

Nelson, J. (2011). "Polymer: Fullerene Bulk Heterojunction Solar Cells." *Materials Today* **14**(10): 462–470.

Norrman, K. et al. (2009). "Water-Induced Degradation of Polymer Solar Cells Studied by (H2O)-O-18 Labeling." *ACS Applied Materials & Interfaces* **1**(1): 102–112.

Norrman, K. et al. (2010). "Degradation Patterns in Water and Oxygen of an Inverted Polymer Solar Cell." *Journal of the American Chemical Society* **132**(47): 16883–16892.

Nuzi, J. M. (2002). "Organic Photovoltaic Materials and Devices." *Comptes Rendus Physique* **3**: 523–542.

Oh, J. Y. et al. (2011). "Driving Vertical Phase Separation in a Bulk-Heterojunction by Inserting a Poly (3-hexylthiophene) Layer for Highly Efficient Organic Solar Cells." *Applied Physics Letters* **98**: 023303.

Orimo, A. et al. (2010). "Surface Segregation at the Aluminum Interface of Poly(3-hexylthiophene)/ Fullerene Solar Cells." *Applied Physics Letters* **96**: 043305.

Pandey, R. and R. J. Holmes (2010). "Graded Donor-Acceptor Heterojunctions for Efficient Organic Photovoltaic Cells." *Advanced Materials* **22**(46): 5301–5305.

Park, S. H. et al. (2009). "Bulk Heterojunction Solar Cells with Internal Quantum Efficiency Approaching 100%." *Nature Photonics* **3**(5): 297–302.

Parnell, A. J. et al. (2010). "Depletion of PCBM at the Cathode Interface in P3HT/PCBM Thin Films as Quantified via Neutron Reflectivity Measurements." *Advanced Materials* **22**(22): 2444–2447.

Parratt, L. G. (1954). "Surface Studies of Solids by Total Reflection of X-Rays." *Physical Review* **95**(2): 359–369.

Pearson, A. J. et al. (2011). "Imaging the Bulk Nanoscale Morphology of Organic Solar Cell Blends Using Helium Ion Microscopy." *Nano Letters* **11**(10): 4275–4281.

Peet, J. et al. (2007). "Efficiency Enhancement in Low-Bandgap Polymer Solar Cells by Processing with Alkane Dithiols." *Nature Materials* **6**(7): 497–500.

Peet, J. et al. (2008). "Transition from Solution to the Solid State in Polymer Solar Cells Cast from Mixed Solvents." *Macromolecules* **41**(22): 8655–8659.

Perez, L. A. et al. (2013). "Solvent Additive Effects on Small Molecule Crystallization in Bulk Heterojunction Solar Cells Probed During Spin Casting." *Advanced Materials* **25**(44): 6380–6384.

Pfannmoeller, M. et al. (2011). "Visualizing a Homogeneous Blend in Bulk Heterojunction Polymer Solar Cells by Analytical Electron Microscopy." *Nano Letters* **11**(8): 3099–3107.

Ramachandra, R. et al. (2009). "A Model of Secondary Electron Imaging in the Helium Ion Scanning Microscope." *Ultramicroscopy* **109**(6): 748–757.

Reyes-Reyes, M. et al. (2005). "High-Efficiency Photovoltaic Devices Based on Annealed Poly(3-hexylthiophene) and 1-(3-methoxycarbonyl)-Propyl-1-(6,6)C61 Blends." *Applied Physics Letters* **87**(8): 083506.

Rogers, J. T. et al. (2011). "Structural Order in Bulk Heterojunction Films Prepared with Solvent Additives." *Advanced Materials* **23**(20): 2284–2288.

Ruderer, M. A. et al. (2009). "Thin Films of Photoactive Polymer Blends." *ChemPhysChem* **10**: 664–671.

Ruderer, M. A. and P. Müller-Buschbaum (2011). "Morphology of Polymer-Based Bulk Heterojunction Films for Organic Photovoltaics." *Soft Matter* **7**(12): 5482.

Scharber, M. C. et al. (2010). "Influence of the Bridging Atom on the Performance of a Low-Bandgap Bulk Heterojunction Solar Cell." *Advanced Materials* **22**(3): 367–370.

Shaheen, S. E. et al. (2001). "2.5% Efficient Organic Plastic Solar Cells." *Applied Physics Letters* **78**: 841.

Shao, M. et al. (2014). "The Isotopic Effects of Deuteration on Optoelectronic Properties of Conducting Polymers." *Nature Communications* **5**: 3180.

Søndergaard, R. et al. (2012). "Roll-to-Roll Fabrication of Polymer Solar Cells." *Materials Today* **15**(1–2): 36–49.

Song, J. et al. (2012). "Blends of Polyolefin/PMMA for Improved Scratch Resistance, Adhesion and Compatibility." *Polymer* **53**(16): 3636–3641.

Song, J. et al. (2013). "Reactive Coupling between Immiscible Polymer Chains: Acceleration by Compressive Flow." *AIChE Journal* **59**(9): 3391–3402.

Spanggaard, H. and F. C. Krebs (2004). "A Brief History of the Development of Organic and Polymeric Photovoltaics." *Solar Energy Materials and Solar Cells* **83**(2–3): 125–146.

Su, M.-S. et al. (2011). "Improving Device Efficiency of Polymer/Fullerene Bulk Heterojunction Solar Cells through Enhanced Crystallinity and Reduced Grain Boundaries Induced by Solvent Additives." *Advanced Materials* **23**(29): 3315–3319.

Sun, Y. et al. (2012). "Solution-Processed Small-Molecule Solar Cells with 6.7% Efficiency." *Nature Materials* **11**(1): 44–48.

Swaraj, S. et al. (2010). "Nanomorphology of Bulk Heterojunction Photovoltaic Thin Films Probed with Resonant Soft X-Ray Scattering." *Nano Letters* **10**(8): 2863–2869.

Synooka, O. et al. (2013). "Influence of Thermal Annealing on PCDTBT:PCBM Composition Profiles." *Advanced Energy Materials* **4**: 1300981.

Thompson, B. C. and J. M. J. Fréchet (2008). "Polymer–Fullerene Composite Solar Cells." *Angewandte Chemie International Edition* **47**(1): 58–77.

Tremolet de Villers, B. et al. (2009). "Improving the Reproducibility of P3HT:PCBM Solar Cells by Controlling the PCBM/Cathode Interface." *The Journal of Physical Chemistry C* **113**(44): 18978–18982.

Tsai, J.-H. et al. (2010). "Enhancement of P3HT/PCBM Photovoltaic Efficiency Using the Surfactant of Triblock Copolymer Containing Poly(3-hexylthiophene) and Poly(4-vinyltriphenylamine) Segments." *Macromolecules* **43**(14): 6085–6091.

van Bavel, S. S. et al. (2009). "Three-Dimensional Nanoscale Organization of Bulk Heterojunction Polymer Solar Cells." *Nano Letters* **9**(2): 507–513.

van Bavel, S. S. et al. (2010). "P3HT/PCBM Bulk Heterojunction Solar Cells: Impact of Blend Composition and 3D Morphology on Device Performance." *Advanced Functional Materials* **20**(9): 1458–1463.

van der Poll, T. S. et al. (2012). "Non-Basic High-Performance Molecules for Solution-Processed Organic Solar Cells." *Advanced Materials* **24**(27): 3646–3649.

Vandewal, K. et al. (2009). "On the Origin of the Open-Circuit Voltage of Polymer—Fullerene Solar Cells." *Nature Materials* **8**(11): 904–909.

Walker, B. et al. (2011). "Small Molecule Solution-Processed Bulk Heterojunction Solar Cells." *Chemistry of Materials* **23**(3): 470–482.

Wang, H. et al. (2013). "The Role of Additive in Diketopyrrolopyrrole-Based Small Molecular Bulk Heterojunction Solar Cells." *Advanced Materials* **25**(45): 6519–6525.

White, H. et al. (2001). "Focused Ion Beam/Lift-Out Transmission Electron Microscopy Cross Sections of Block Copolymer Films Ordered on Silicon Substrates." *Polymer* **42**(4): 1613–1619.

Wienk, M. M. et al. (2008). "Narrow-Bandgap Diketo-Pyrrolo-Pyrrole Polymer Solar Cells: The Effect of Processing on the Performance." *Advanced Materials* **20**(13): 2556–2560.

Woo, C. H. et al. (2010). "Incorporation of Furan into Low Band-Gap Polymers for Efficient Solar Cells." *Journal of the American Chemical Society* **132**(44): 15547–15549.

Wu, W.-R. et al. (2011). "Competition between Fullerene Aggregation and Poly(3-hexylthiophene) Crystallization upon Annealing of Bulk Heterojunction Solar Cells." *ACS Nano* **5**(8): 6233–6243.

Xu, Z. et al. (2009). "Vertical Phase Separation in Poly(3-hexylthiophene): Fullerene Derivative Blends and its Advantage for Inverted Structure Solar Cells." *Advanced Functional Materials* **19**(8): 1227–1234.

Xue, J. et al. (2005). "Mixed Donor-Acceptor Molecular Heterojunctions for Photovoltaic Applications. II. Device Performance." *Journal of Applied Physics* **98**(12): 124903.

Yamamoto, S. et al. (2011). "Molecular Understanding of the Open-Circuit Voltage of Polymer:Fullerene Solar Cells." *Advanced Energy Materials* **2**(2): 229–237.

Yang, F. et al. (2004). "Controlled Growth of a Molecular Bulk Heterojunction Photovoltaic Cell." *Nature Materials* **4**(1): 37–41.

Yang, L. et al. (2012). "Parallel-like Bulk Heterojunction Polymer Solar Cells." *Journal of the American Chemical Society* **134**: 5432–5435.

Yang, X. et al. (2005). "Nanoscale Morphology of High-Performance Polymer Solar Cells." *Nano Letters* **5**(4): 579–583.

Yao, Y. et al. (2008). "Effects of Solvent Mixtures on the Nanoscale Phase Separation in Polymer Solar Cells." *Advanced Functional Materials* **18**(12): 1783–1789.

Ye, L. et al. (2012). "From Binary to Ternary Solvent: Morphology Fine-Tuning of D/A Blends in PDPP3T-based Polymer Solar Cells." *Advanced Materials* **24**: 6335–6341.

Yin, W. and M. Dadmun (2011). "A New Model for the Morphology of P3HT/PCBM Organic Photovoltaics from Small-Angle Neutron Scattering: Rivers and Streams." *ACS Nano* **5**(6): 4756–4768.

You, J. et al. (2013). "A Polymer Tandem Solar Cell with 10.6% Power Conversion Efficiency." *Nature Communications* **4**(1446): 1441–1410.

Yu, G. et al. (1995). "Polymer Photovoltaic Cells: Enhanced Efficiencies via a Network of Internal Donor-Acceptor Heterojunctions." *Science* **270**(5243): 1789–1791.

Zhang, F. et al. (2006). "Influence of Solvent Mixing on the Morphology and Performance of Solar Cells Based on Polyfluorene Copolymer/Fullerene Blends." *Advanced Functional Materials* **16**(5): 667–674.

Zhang, X. et al. (2011). "Molecular Packing of High-Mobility Diketo Pyrrolo-Pyrrole Polymer Semiconductors with Branched Alkyl Side Chains." *Journal of the American Chemical Society* **133**(38): 15073–15084.

Zhou, E. et al. (2009). "Synthesis and Photovoltaic Properties of Diketopyrrolopyrrole-Based Donor–Acceptor Copolymers." *Chemistry of Materials* **21**(17): 4055–4061.

Zhou, H. et al. (2011). "Development of Fluorinated Benzothiadiazole as a Structural Unit for a Polymer Solar Cell of 7% Efficiency." *Angewandte Chemie-International Edition* **50**(13): 2995–2998.

Zhou, H. et al. (2012). "Rational Design of High Performance Conjugated Polymers for Organic Solar Cells." *Macromolecules* **45**(2): 607–632.

10 Engineering of Active Layer Nanomorphology via Fullerene Ratios and Solvent Additives for Improved Charge Transport in Polymer Solar Cells

*Swaminathan Venkatesan,
Evan Ngo, and Qiquan Qiao*

CONTENTS

10.1 Introduction 282
10.2 Effect of Fullerene Ratio and Solvent Additive on the Nanomorphology
and Charge Transport in PBT-T1-Based Bulk Heterojunction Solar Cells....283
    10.2.1 UV-Visible Absorption of PBT-T1 283
    10.2.2 UV-Visible Absorption and XRD Spectra of PBT-T1:$PC_{60}BM$
        Films with Different Fullerene Ratios and Solvent Additive 284
    10.2.3 AFM Topography and Phase Images of PBT-T1:$PC_{60}BM$ Films
        with Different Fullerene Ratios and Solvent Additive......................285
    10.2.4 Current–Voltage Characteristics of PBT-T1:$PC_{60}BM$ Solar Cells
        with Different Fullerene Ratio...286
    10.2.5 Current–Voltage Characteristics of PBT-T1:$PC_{70}BM$ Solar Cells
        Processed with Additive 288
    10.2.6 Photo-CELIV and Extracted Charge Carrier Density versus
        Delay Time of PBT-T1 Solar Cells 290
10.3 Effect of Solvent Additive on the Nanomorphology and Charge
Transport in PDPP3T:PCBM Solar Cells 291
    10.3.1 EFTEM Images of PDPP3T:PCBM Films Modified with
        Solvent Additives 292

 10.3.2 Selected Area Electron Diffraction and Raman Spectra of
PDPP3T:PCBM Thin Films..293
 10.3.3 Effect of Solvent Additives on the Surface Morphology of
PDPP3T:PCBM Films..294
 10.3.4 Current–Voltage Characteristics of PDPP3T:PCBM Solar Cells.....296
 10.3.5 Effect of Solvent Additive on Charge Transport and
Recombination Dynamics of PDPP3T:PCBM Solar Cells...............297
10.4 Summary..300
Acknowledgments...300
References...300

10.1 INTRODUCTION

Polymer solar cells (PSCs) are promising candidates for future photovoltaic devices [1–8] owing to their low cost of material production and device processing, low temperature processing for flexible devices, and ease of fabrication [7,9–11]. However, due to their lower lifetime and efficiency, PSCs require deeper understanding to eradicate the aforementioned drawbacks. One of the most important parameters that influence charge transport and recombination dynamics in bulk heterojunction solar cells is the nanomorphology [12–15]. In the past, several methods, such as polymer:fullerene weight ratio, thickness of active layer, charge-selective interfacial layers, choice of solvents [16], solvent additives [17], solvent annealing, and thermal annealing [7] have been utilized to obtain optimized morphology to attain higher power conversion efficiency. Also, over the last decade, several groups have been working on defining and resolving different aspects of morphology, such as domain size, purity, crystallinity, miscibility, interface sharpness/roughness, lateral and vertical phase separation, or aggregation [12,18–22]. In most cases, such morphological attributes are correlated with device performance in steady-state illumination, and there is still a lack of understanding of the effect of domain size, purity, and their spatial distribution on the charge transport and recombination behavior. Bimolecular recombination has widely been shown to be the dominant recombination mechanism in bulk heterojunction solar cells [23]. Hence, there is a need to understand the role of processing conditions in order to achieve optimal nanomorphology to suppress the bimolecular recombination in bulk heterojunction solar cells.

 Lyons et al. [20] performed a rather interesting morphological alteration using Cahn–Hilliard modeling and correlated it with solar cell performance as predicted by Monte Carlo simulations. In the paper, domain purity and interface sharpness were emphasized, assuming domain sizes were below 10 nm. However, the authors neglected the effect of domain aggregation, which can lead to lower donor–acceptor interface and cause charge accumulation at the donor–donor interface. Another important study is by Spoltore et al. [24] in which the authors report the role of polymer crystallinity on the bimolecular recombination order and defect density; however, the authors failed to analyze the role of domain size and crystallinity on the device efficacies. Most studies on recombination kinetics of polymer bulk heterojunction solar cells provide lesser insight on the nanomorphological parameters,

Engineering of Active Layer Nanomorphology

such as domain purity, domain size, and interfaces; hence, it is hard to visualize the role of morphological parameters on charge transport.

In this chapter, we have sequentially studied and analyzed the role of different fullerene ratios and solvent additives on morphological and structural features, such as domain purity (crystallinity), size, and interfaces, after which we have correlated nanomorphology with charge carrier lifetime and carrier concentration, which was derived from small perturbation transient optoelectronic measurements. Finally, the effects of nanomorphology on steady-state current voltage characteristics were discussed.

10.2 EFFECT OF FULLERENE RATIO AND SOLVENT ADDITIVE ON THE NANOMORPHOLOGY AND CHARGE TRANSPORT IN PBT-T1-BASED BULK HETEROJUNCTION SOLAR CELLS

10.2.1 UV-Visible Absorption of PBT-T1

Figure 10.1a shows the chemical structure of poly{Thiophene-2,5-diyl-alt-[5,6-bis(dodecyloxy)-benzo[c][1,2,5]thiadiazole]-4,7-diyl (PBT-T1). The difference between the previously reported polymer [25] and the PBT-T1 polymer is the change in the solubilizing side chain alkoxy group from $C_{14}H_{29}O$ to $C_{12}H_{25}O$ [26]. Synthesized PBT-T1 was found to be soluble in a wide range of organic solvents, such as xylene, anisole, chloroform, chlorobenzene, and dichlorobenzene. The UV-visible optical absorption of PBT-T1 in chloroform solvent and as a film is shown in Figure 10.1b. When cast as film, a red shift of absorption onset from 640 nm in chloroform solution to 714 nm in solid film was observed. The optical bandgap of PBT-T1 calculated from the onset of optical absorption in solution and solid film was 1.92 eV and 1.73 eV, respectively. Two vibronic peaks at 612 nm and 661 nm were observed in PBT-T1 film's absorption, which indicated structural order within the polymer when cast as a film.

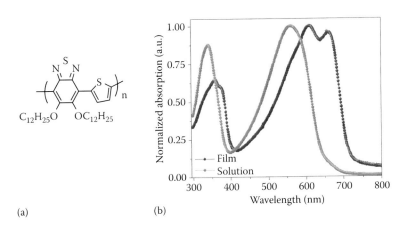

FIGURE 10.1 (a) Chemical structure of PBT-T1 polymer. (b) Normalized UV-visible absorption of PBT-T1 in chloroform solution and solid film. (Reproduced from Venkatesan, S. et al., *Nanoscale*, 2014. With permission.)

Cyclic voltammetry measurements were performed on PBT-T1 to calculate the HOMO and LUMO levels of the PBT-T1. The HOMO and LUMO levels of PBT-T1 are −5.44 eV and −3.12 eV, respectively. Hence, PBT-T1 possessed optimal bandgap and energy levels for photovoltaic applications.

10.2.2 UV-Visible Absorption and XRD Spectra of PBT-T1:PC$_{60}$BM Films with Different Fullerene Ratios and Solvent Additive

PBT-T1 was blended with PC$_{60}$BM in different weight ratios, and solvent additive 1,8-diiodooctane (DIO) was added to a 1:2 (PBT-T1:PC$_{60}$BM) blend solution. The reason that we added DIO to a 1:2 (PBT-T1:PC$_{60}$BM) blend solution is that this ratio provided us higher photovoltaic performance as shown in Figure 10.5. To study the role of fullerene ratio and solvent additive on the optical properties and polymer crystallinity, UV-visible spectroscopy and x-ray diffraction measurements were done as shown in Figure 10.2a and b, respectively.

The absorption spectrum of 1:1 blend film had the same characteristic vibronic peaks as that of the pure polymer film shown in Figure 10.1, which indicated that the structural order was maintained within the polymer phase when blended with PC$_{60}$BM in a 1:1 weight ratio. When the PBT-T1:PC$_{60}$BM blending ratios were increased to 1:2 and 1:3, a blue shift in the onset of optical absorption was observed. The absorption peak blue shifted from 616 nm in 1:1 to 598 nm for 1:2 and to 585 nm for 1:3 blend films. Broadening of the vibronic shoulder was also observed, which suggested a higher degree of disorder in films with 1:2 and 1:3 weight ratios. A similar trend was also observed for poly (3-hexylthiophene) (P3HT); the structural order of P3HT was disrupted when higher PCBM ratios were used [27–29]. Hence, increasing PCBM blending ratios led to breaking of the polymer chain packing, which resulted in higher amorphous phase of polymer. This was further supported by in-plane XRD measurements shown in Figure 10.2b. A sharp peak at 2θ = 4.4° was found in both the PBT-T1:PC$_{60}$BM 1:1 blend ratio film without an additive and 1:2 blend ratio films

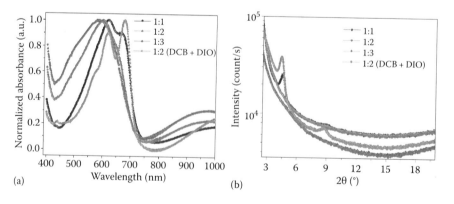

FIGURE 10.2 (a) UV-visible optical absorption of PBT-T1:PC$_{60}$BM thin films and (b) parallel beam XRD measurements of PBT-T1:PC$_{60}$BM thin films. (Reproduced from Venkatesan, S. et al., *Nanoscale*, 2014. With permission.)

Engineering of Active Layer Nanomorphology

with a DIO additive. As seen in 1:2 blend films cast from DCB with DIO additive, the shoulders in absorption spectra (Figure 10.2a) were distinctively highlighted, which suggested a higher degree of structural order within the polymer phase. XRD spectra also showed that crystalline chain packing within the polymer phase was not disrupted by the addition of PC$_{60}$BM when DIO was used as a solvent additive. However, the peak was completely suppressed in 1:2 blend films cast without DIO additive. XRD results proved that the loss of interchain packing order is induced with a higher ratio of PCBM. A small peak at 2θ = 8.8° was also observed for films processed with DIO additive, which corresponded to second-order (200) diffraction. The (200) diffraction is suppressed for 1:1 blend films, which indicated that films cast from DIO possessed higher structural order.

10.2.3 AFM Topography and Phase Images of PBT-T1:PC$_{60}$BM Films with Different Fullerene Ratios and Solvent Additive

Figure 10.3 shows the AFM topography (Figure 10.3a through d and phase 10.3e through h) images of PBT-T1:PCBM thin films in which (Figure 10.3a and e) 1:1, (Figure 10.3b and f) 1:2, and (Figure 10.3c and g) 1:3 were cast from DCB without

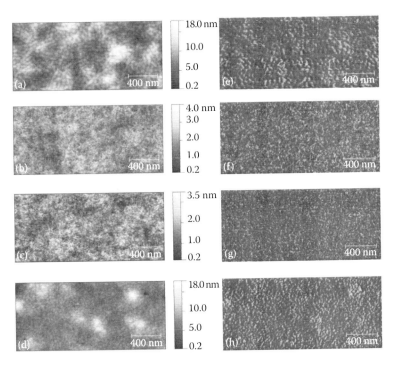

FIGURE 10.3 AFM (a–d) topography and (e–h) phase images of PBT-T1:PC$_{60}$BM thin films. (a, e) 1:1, (b, f) 1:2, and (c, g) 1:3 films cast from DCB without additive, and (d, h) 1:2 cast from DCB with DIO additive. Dimension of AFM images are 1 μm by 2 μm. (Reproduced from Venkatesan, S. et al., *Nanoscale*, 2014. With permission.)

additive, and (Figure 10.3d and h) 1:2 were cast from DCB with 2.5 v% DIO as a solvent additive.

Large fiber-type domains with width ranging from 40 to 50 nm were observed in 1:1 blend film, which showed a root mean square (RMS) roughness of 2.34 nm. When the PCBM weight ratios were increased to 1:2 and 1:3, shorter and thinner fiber-like domains were observed. The RMS roughness measured for 1:2 and 1:3 blend films were 0.375 nm and 0.407 nm, respectively. The 1:2 blend film exhibited a thinner network of fibers whose widths ranged from 15 nm to 20 nm. The widths of the fiber were measured from average values obtained in several line profiles on the topography images and were cross-verified with their corresponding phase images. As the PCBM weight ratio increased to 1:3, shorter and thinner fiber-shaped domains were observed. AFM results correlated with absorption and XRD spectra showed that higher fullerene loading not only led to disruption of interchain packing, but also led to formation of smaller domains. When the $PC_{60}BM$ ratio was low, the polymer phase aggregated into larger domains as seen in 1:1 cast films. When the $PC_{60}BM$ ratio was increased to 1:2 or 1:3, ratio polymer aggregation was prevented, which led to formation of smaller amorphous phases. When DIO was added to a PBT-T1:$PC_{60}BM$ (1:2) blend solution, it led to formation of nanocrystalline polymer domains confirmed by XRD and as seen in Figure 10.3d and h. These domains were of the similar shape as in the 1:1 cast films but had smaller dimensions. The average widths of fiber-type domains were about 30 nm, and lengths ranged from 30 nm to 35 nm. Films cast with DIO additive had RMS roughness of 1.75 nm.

10.2.4 Current–Voltage Characteristics of PBT-T1:PC$_{60}$BM Solar Cells with Different Fullerene Ratio

PBT-T1 was blended with $PC_{60}BM$ in different weight ratios in dichlorobenzene (DCB) to determine the optimum mixing ratio for photovoltaic devices. The current voltage characteristics of PBT-T1:$PC_{60}BM$ with different weight ratios of PBT-T1 to PCBM is shown in Figure 10.4.

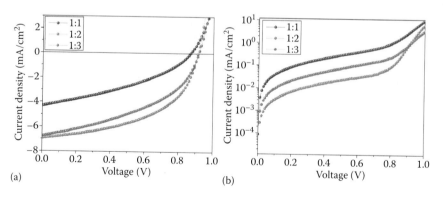

FIGURE 10.4 (a) Illuminated current–voltage characteristics and (b) semilog plot of dark current–voltage characteristics of PBT-T1:$PC_{60}BM$ solar cells. (Reproduced from Venkatesan, S. et al., *Nanoscale*, 2014. With permission.)

Solar cells having a 1:1 blended active layer (PBT-T1:PC$_{60}$BM) showed the lowest current density, fill factor, and open circuit voltage with respect to active layers with higher PCBM weight ratios. Comparing the best devices in 1:2 and 1:3 blended solar cells, it is observed that J_{sc} is comparable between both; however, on average, 1:3 devices showed lower FF and J_{sc} consistently. The low J_{sc} and FF in 1:1 blend devices is attributed to large aggregated domains observed in the AFM images. When PCBM loading was increased to 1:2 and 1:3, finer phase separation was observed, which led to improvement in J_{sc} and FF. The solar cell parameters are summarized in Table 10.1.

From the dark J–V curves shown in Figure 10.2b, it is seen that, among all blending ratios, the 1:2 (PBT-T1:PC$_{60}$BM) showed the lowest dark current below threshold voltage (0.82 V), which indicated the lowest recombination current among all devices cast with only DCB as solvent. The diode parameters calculated from dark current–voltage characteristics are given in Table 10.2.

Shunt resistance was the highest for cells processed in the 1:2 weight ratio. 1:2 cells also possessed lower series resistance compared to cells with 1:3 blend active layers. The reverse saturation current (J_0) and ideality factor (n) were measured from the exponential region of the dark current–voltage regions. A significant decrease in both J_0 and n were observed for 1:2 blend solar cells. J_0 decreased by almost four

TABLE 10.1
Photovoltaic Characteristics of PBT-T1:PC$_{60}$BM Solar Cells

	V_{oc} (V)	J_{sc} (mA/cm²)	FF	η (%)
1:1 (PC$_{60}$BM)	0.87 (0.82)	4.29 (4.26)	0.36 (0.36)	1.36 (1.25)
1:2 (PC$_{60}$BM)	0.93 (0.92)	7.29 (7.11)	0.47 (0.46)	3.18 (3.02)
1:3 (PC$_{60}$BM)	0.91 (0.93)	7.26 (6.92)	0.43 (0.41)	2.82 (2.65)

Note: Values in parentheses are average values of six to eight devices.

TABLE 10.2
Dark Current–Voltage Parameters of PBT-T1:PCBM Solar Cells

	R_{sh} (Ωcm²)[a]	R_s (Ωcm²)[a]	J_0 (mA/cm²)[b]	n[b]
1:1 (PC$_{60}$BM)	2.55 × 10³	9.9	1.78 × 10⁻⁵	2.91
1:2 (PC$_{60}$BM)	3.19 × 10⁴	12.1	2.07 × 10⁻⁹	1.76
1:3 (PC$_{60}$BM)	1.05 × 10⁴	27.4	8.46 × 10⁻⁷	2.52

[a] R_{sh} and R_s values are calculated by inverse slopes of linear fits to dark J–V curves near 0 V and 1 V, respectively.

[b] J_0 and n values are calculated from exponential fit to the linear portion of semilog plot of dark J–V curves as per equation $J = J_0 \left(e^{\frac{qV}{nkT}} - 1 \right)$ where J_0 is the intercept and $\frac{qV}{nkT}$ is the slope.

orders of magnitude from 1:1 to 1:2 blend devices and is further increased by 24 times when PC$_{60}$BM loading was increased from 1:2 to 1:3. The ideality factor of the diode was observed to be lowest in 1:2 blend solar cells. Low J_0 and n were indicative of reduced trap-assisted recombination [30] in the solar cells blended in a 1:2 weight ratio. Hence, 1:2 was found to be the optimum mixing ratio for PBT-T1:PCBM solar cells.

10.2.5 Current–Voltage Characteristics of PBT-T1:PC$_{70}$BM Solar Cells Processed with Additive

To further enhance the power conversion efficiency in 1:2 blend devices, additive DIO was used in small volume concentration (2 or 2.5 v%). The current–voltage characteristics of PBT-T2 (1:2) solar cells with and without DIO is shown in Figure 10.5.

A significant improvement in fill factor and power conversion efficiency were observed for solar cells processed with additive DIO. For 1:2 blended devices with 2.5 v% DIO, the average FF increased from 0.46 to 0.65. Average J_{sc} increased slightly from 7.12 mA/cm^2 to 7.53 mA/cm^2. A 60 mV decrease in V_{oc} is observed for

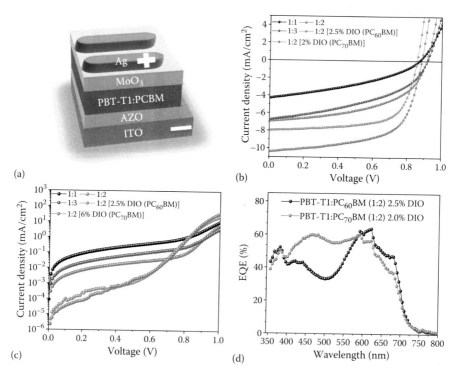

FIGURE 10.5 (a) Device schematic of inverted PBT-T1:PCBM solar cells, (b) illuminated current–voltage, (c) semilog plot of dark current–voltage characteristics, and (d) EQE spectra of PBT-T1:PC$_{60}$BM and PBT-T1:PC$_{70}$BM devices blended in a 1:2 weight ratio. (Reproduced from Venkatesan, S. et al., *Nanoscale*, 2014. With permission.)

Engineering of Active Layer Nanomorphology

devices cast with DIO additive, which was due to a narrowing of bandgap that led to a slightly higher HOMO level. The higher HOMO level was caused by improved crystallinity induced by lower solubility of polymer in DIO leading to an upward shift in the HOMO level of PBT-T1 as observed in other polymers [31,32]. The solar cell parameters are tabulated in Table 10.3.

In devices processed with additive DIO, J_0 and n were suppressed as seen in the dark current–voltage curves (Figure 10.5b). The diode characteristics derived from dark are tabulated in Table 10.4. J_0 reduced in PBT-T1:$PC_{60}BM$ solar cells from 2.07×10^{-9} mA/cm² to 4.66×10^{-10} mA/cm², which indicated the lowest charge carrier recombination in active layers processed with DIO. Inclusion of DIO as a solvent additive led to decrease in R_s and more significantly enhanced R_{sh} by almost two orders of magnitude. Due to reduced R_s and increased R_{sh}, the fill factor improved from 0.47 to 0.66. Improvement in FF led to an increase in device efficiency from 3.18% to 4.53% (42% enhancement).

The power conversion efficiency is further enhanced by employing $PC_{70}BM$ as an acceptor, which led to the highest power conversion efficiency of 5.65%.

Improvement in J_{sc} using $PC_{70}BM$ was due to increased photocurrent generation in 400 nm to 600 nm as seen in external quantum efficiency (EQE) spectra (Figure 10.5c). Enhancement of photocurrent in the 400–600 nm spectral region was attributed to the enhanced absorption coefficient of $PC_{70}BM$ in visible spectra. For $PC_{70}BM$, a slightly lower concentration of DIO (2 v%) was required compared to $PC_{60}BM$ (2.5 v%) due to higher solubility of $PC_{70}BM$ in DCB [33].

TABLE 10.3
Solar Cell Parameters for PBT-T1:PCBM (1:2) Solar Cells

	V_{oc} (V)	J_{sc} (mA/cm²)	FF	η (%)	Enhancement (%)
1:2 ($PC_{60}BM$) (No DIO)	0.93 (0.92)	7.29 (7.11)	0.47 (0.46)	3.18 (3.02)	—
1:2 ($PC_{60}BM$ with 2.5% DIO)	0.86 (0.86)	7.97 (7.53)	0.66 (0.65)	4.53 (4.22)	42.45
1:2 ($PC_{70}BM$ with 2% DIO)	0.88 (0.87)	10.39 (9.84)	0.62 (0.62)	5.65 (5.35)	77.67

TABLE 10.4
Diode Parameters of PBT-T1:PCBM (1:2) Solar Cells Derived from Dark Current–Voltage Characteristics

	R_{sh} (Ωcm²)[a]	R_s (Ωcm²)[a]	J_0 (mA/cm²)	n
1:2 ($PC_{60}BM$) (No DIO)	3.19×10^4	12.1	2.07×10^{-9}	1.76
1:2 ($PC_{60}BM$ with 2.5% DIO)	9.43×10^5	5.2	4.66×10^{-10}	1.47
1:2 ($PC_{70}BM$ with 2% DIO)	2.80×10^6	2.9	1.27×10^{-10}	1.39

[a] R_{sh} and R_s values are calculated by inverse slopes of linear fits to dark J–V curves near 0 V and 1 V, respectively.

10.2.6 Photo-CELIV and Extracted Charge Carrier Density versus Delay Time of PBT-T1 Solar Cells

Photo-CELIV measurements were done with varying delay times to find the mobility and to measure extracted charge carrier density as a function of delay time. Figure 10.6a serves as a visual guide to the reader showing a typical photo-CELIV signal obtained along with indications of t_{max}, j_0, and Δj. The shaded portion in the graph shows the extracted charges (Q_{ext}) from the photo-CELIV signal. Figure 10.6b shows the photo-CELIV curves at 4 µs delay time for PBT-T1:PC$_{60}$BM solar cells with different PC$_{60}$BM weight ratios and 1:2 solar cells processed with 2.5 v% DIO additive. The results from Figure 10.6a are summarized in Table 10.5.

From Figure 10.6b, higher Δj was observed for 1:2 solar cells compared to 1:1 and 1:3 ratios, indicating higher charge carrier density. The t_{max} observed for 1:2 was 3.5 µs. For the 1:1 blend devices, a slightly shorter t_{max} of 3.05 µs was observed, and the 1:3 device showed the longest t_{max} of 5.25 µs. t_{max} indicated the voltage required to extract the photogenerated trapped charge carriers, and hence, for 1:1 and 1:2

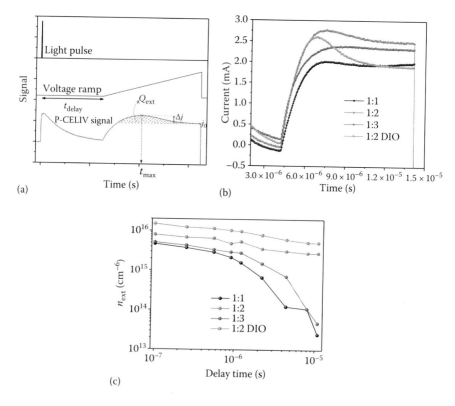

FIGURE 10.6 (a) Typical photo-CELIV curve, (b) photo-CELIV measurements of PBT-T1:PC$_{60}$BM solar cells at 4 µs delay time between light pulse and voltage ramp, and (c) extracted charge carrier density from photo-CELIV curves with varying delay times. (Reproduced from Venkatesan, S. et al., *Nanoscale*, 2014. With permission.)

TABLE 10.5
Charge Carrier Mobility, t_{max}, and Extracted Charge Carrier Density (n_{ext}) Values Estimated from the Photo-CELIV Measurements at 4 µs Delay Time along with Thickness Values

	t_{max} (µs)	Mobility (cm²/Vs)	Thickness (nm)	n_{ext} (cm⁻³)
1:1	3.05	3.06 × 10⁻⁵	80	1.26 × 10¹⁴
1:2	3.5	4.04 × 10⁻⁵	110	3.14 × 10¹⁵
1:3	5.25	1.86 × 10⁻⁵	120	7.15 × 10¹⁴
1:2 (DCB + 2.5% DIO)	2.85	6.98 × 10⁻⁵	110	6.16 × 10¹⁵

blend devices, lower bias voltage was required to extract the maximum charges. Higher bias voltage was required for 1:3 devices owing to the amorphous nature of the polymer that led to charges being occupied in deeper defect states. However, the extracted charge carrier density of 1:3 was higher than that of the 1:1 blend devices due to finer phase separation as observed in AFM images. The 1:2 blend devices showed the highest mobility of 4.04 × 10⁻⁵ cm² V⁻¹ s⁻¹ among all blend devices processed from DCB. For 1:2 devices processed from a DCB and DIO mixture, the mobility was further enhanced to 6.98 × 10⁻⁵ cm² V⁻¹ s⁻¹ due to the shorter t_{max} and higher extracted charge carrier density (n_{ext}). Figure 10.6c shows the variation of extracted charge carrier density (n_{ext}) with delay time used in photo-CELIV measurements. The n_{ext} reduced with higher delay times due to higher recombination of charge carriers before the voltage ramp is applied. The log–log plot of n_{ext} versus delay time showed power law decay typical of bimolecular recombination [34]. The decay of n_{ext} in the 1:1 and 1:3 blend devices was found to be much faster than that in the 1:2 blend devices attributed to higher recombination in 1:1 and 1:3 solar cells. The 1:2 ratio devices with and without DIO additive showed reduced recombination decay with delay time, which led to higher extracted carriers.

10.3 EFFECT OF SOLVENT ADDITIVE ON THE NANOMORPHOLOGY AND CHARGE TRANSPORT IN PDPP3T:PCBM SOLAR CELLS

Figure 10.7a shows the device schematic used to study the effect of solvent additive on the nanomorphology and charge transport in PDPP3T:PCBM solar cells [35]. Here, the inverted devices were specifically chosen due to higher stability. The devices studied showed minimal (<5%) degradation after complete testing. Zinc oxide (ZnO) acts as an electron transport layer and molybdenum oxide (MoO₃) as a hole transport layer. ITO and silver act as a cathode and anode, respectively. Active-layer morphology was altered with different commonly used additives for polymer bulk heterojunction solar cells, namely CN, DIO, and ODT. For comparison, active layers without any solvent additives (pristine) were also spin casted. Chemical structures of the additives are shown in Figure 10.7b.

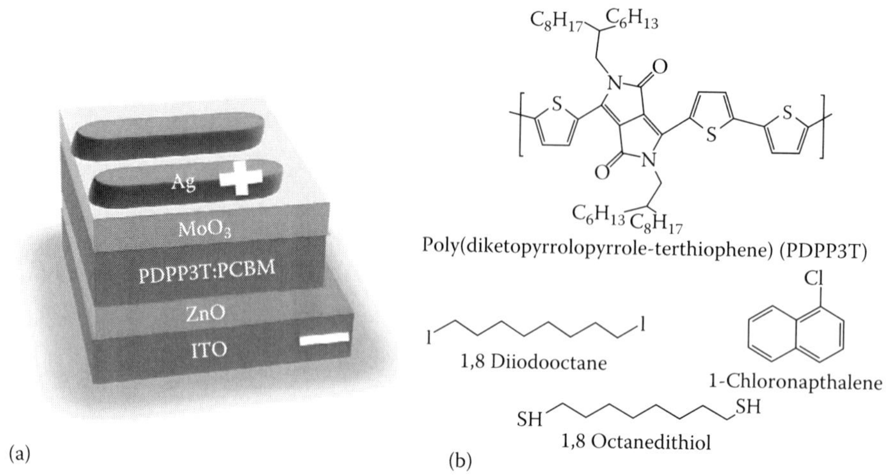

FIGURE 10.7 (a) Device structure of inverted PDPP3T:PC$_{60}$BM solar cells, and (b) chemical structures of PDPP3T, 1,8 diiodooctane, 1-chloronapthalene, and 1,8 octanedithiol. (Reproduced from Venkatesan, S. et al., *Nanoscale*, 6, 2, 1011–1019, 2014. With permission.)

10.3.1 EFTEM Images of PDPP3T:PCBM Films Modified with Solvent Additives

The active layer morphology of PDPP3T:PCBM bulk heterojunction was modified using different solvent additives, namely 1-chloronapthalene (CN), 1,8 diiodooctane (DIO), and 1,8 octanedithiol (ODT).

To investigate the effect of solvent additive on the bulk morphology and donor purity of these films, energy-filtered TEM (EFTEM) was used to map the donor purity as shown in Figure 10.8. For comparison, films without additive were also studied and termed as "pristine" in future sections.

In the pristine films, no contrast inversion is observed between the donor and acceptor maps. The bright regions in the donor map are also bright regions in the acceptor map, which suggested that donor (PDPP3T) and acceptor (PCBM) are highly intermixed without clearly separated donor-rich or acceptor-rich domains in pristine films. It has also been reported by others that spatial randomness in maps of films cast without additives is due to formation of a mixed phase in which the fullerene molecules intercalate between an amorphous or semicrystalline matrix of polymer chains [36,37]. Therefore, the brighter and darker spots in pristine films were only caused by the thickness variation. However, in the DIO, CN, and ODT additive-processed films, a contrast inversion between donor and acceptor maps is observed; that is, brighter regions in donor maps corresponded to darker regions in acceptor maps. The bright regions indicated high-purity donor domains in the donor maps, which correspond to the dark regions that represented acceptor-poor domains in the acceptor maps in DIO, CN, and ODT films. Such spatial correlation indicated that high-purity donor-rich and acceptor-rich domains were formed in the blend films when cast with solvent additives. In the case of DIO and CN films,

Engineering of Active Layer Nanomorphology

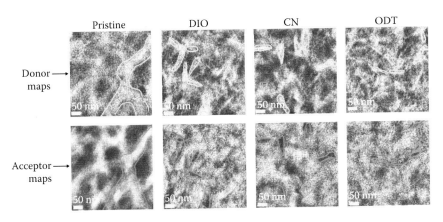

FIGURE 10.8 EFTEM donor and acceptor maps of PDPP3T:PCBM films cast with different solvent additives. (Reproduced from Venkatesan, S. et al., *Nanoscale*, 6, 2, 1011–1019, 2014. With permission.)

tripod-shaped and rod-like domains of polymer-rich regions were observed. Most of these domains were 10–20 nm in width. These domains were much larger than the domain sizes observed in films cast with ODT additive; however, ODT films showed higher aggregation of donor domains possibly due to the high boiling point of ODT.

10.3.2 Selected Area Electron Diffraction and Raman Spectra of PDPP3T:PCBM Thin Films

To analyze the crystallinity of the polymer, selected area electron diffraction (SAED) and Raman spectroscopy were performed on PDPP3T:PCBM thin films shown in Figure 10.9. The grayness intensity of SAED images shown in Figure 10.9a was plotted versus the diffraction radius as shown in Figure 10.9b.

The peak at 50–65 pixels is due to the d-spacing of PCBM [38]. The peak was observed at a radius of 120 pixels, which is attributed to the d-spacing of PDPP3T. A broad peak was observed for pristine films. However, there was negligible or no change in domain crystallinity when PDPP3T:PCBM films were cast with different solvent additives, implying no change in domain crystallinity. Similar behavior was also obtained in the Raman spectra shown in Figure 10.9c in which a negligible change in full width half maximum (FWHM) at 1421 cm^{-1} was observed for all films processed with additives. This was attributed to the C=C stretching in the polymer's molecular structure related to π electron delocalization arising from chain packing as emphasized in previous reports [17,39,40]. As seen in the Raman spectra, the FWHM at 1421 cm^{-1} was the highest for pristine samples, indicating poor chain packing compared to films processed with solvent additives. Hence, use of different solvent additives did not change or had negligible impact on the domain crystallinity or chain packing in the donor-rich domains; however, they affected the donor domain size as observed from EFTEM maps.

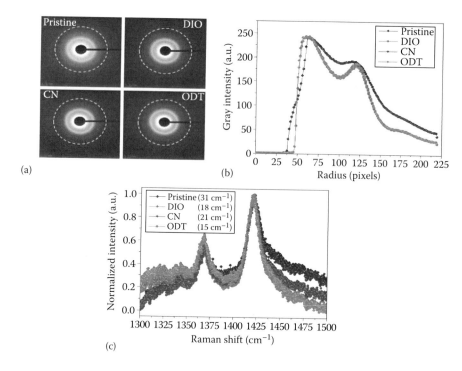

FIGURE 10.9 (a) Selected area electron diffraction (SAED) pattern of PDPP3T:PC$_{60}$BM thin films. (b) Radial grayness profile of SAED images. (c) Raman spectra of PDPP3T:PC$_{60}$BM thin films. The numbers in the legend show FWHM at 1421 cm^{-1} denoting the C=C bond stretching extracted from Lorentz Fit of Raman peaks. (Reproduced from Venkatesan, S. et al., *Nanoscale*, 6, 2, 1011–1019, 2014. With permission.)

10.3.3 Effect of Solvent Additives on the Surface Morphology of PDPP3T:PCBM Films

Tapping mode atomic force microscopic (AFM) imaging was performed to analyze the surface morphology and phase separation of PDPP3T:PCBM films. Figure 10.10 shows the topographic (a–d) and phase (e–h) images of PDPP3T:PCBM films cast with different solvent additives.

The topographic images of all films exhibited the fiber type of morphology. These fibers showed widths of 25 nm to 40 nm. With the addition of solvent additives, formation of longer fibers was observed for PDPP3T:PCBM films attributed to enhancement in the structural order of the polymer. Additives serve as good solvents for fullerenes but poor solvents for the polymer phase, leading to polymer aggregation into more crystalline domains. Such structural ordering led to larger polymer domains, which thereby resulted in higher RMS roughness of the film's surface. The RMS roughness increased from 2.67 nm for pristine films to 5.13 nm, 4.18 nm, and 5.53 nm for films cast with DIO, CN, and ODT additives, respectively. Shorter fiber-like domains were observed for pristine films, which indicated higher density of domain boundaries. Domain boundaries can act as traps for charge recombination. On the other hand, the

Engineering of Active Layer Nanomorphology

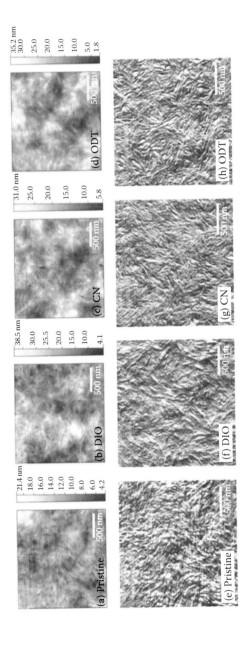

FIGURE 10.10 AFM topography images (a–d) and AFM phase images (e–h) of PDPP3T:PCBM films cast with different solvent additives. (Reproduced from Venkatesan, S. et al., *Nanoscale*, 6, 2, 1011–1019, 2014. With permission.)

films cast with additives show longer fiber-like domains whose lengths varied from 80 nm to 120 nm. Films cast with ODT additive showed aggregation of polymer fibers and also possessed the longest fibers with lengths of 100 nm to 150 nm.

10.3.4 CURRENT–VOLTAGE CHARACTERISTICS OF PDPP3T:PCBM SOLAR CELLS

The effect of solvent additive on the photovoltaic performance of PDPP3T:PCBM solar cells was evaluated by performing current versus voltage spectroscopy as shown in Figure 10.11.

The J_{sc} increased significantly for all solar cells processed with additives compared to pristine devices. The solar cell parameters extracted from the current–voltage measurements are summarized in Table 10.6.

Among all additives, the cells processed with DIO showed the highest J_{sc} of 10.75 mA/cm² and the highest power conversion efficiency (η) of 4.77% having a V_{oc} of 0.68 V and FF of 0.65. Pristine solar cells showed the highest fill factor of 0.68 due to high shunt resistance (R_{sh}) of 2345 Ωcm². The high fill factor of PDPP3T:PCBM solar cells is attributed to high ambipolar mobility of PDPP3T [41,42]. Cells processed

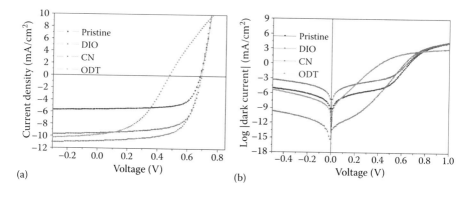

FIGURE 10.11 (a) Current density–voltage characteristics of PDPP3T:PC$_{60}$BM solar cells under AM1.5 illumination. (b) Semilog plot of dark current density–voltage of PDPP3T:PCBM solar cells. (Reproduced from Venkatesan, S. et al., *Nanoscale*, 6, 2, 1011–1019, 2014. With permission.)

TABLE 10.6
Photovoltaic Properties of PDPP3T:PC$_{60}$BM Solar Cells Using Different Solvent Additives

	J_{sc} (mA/cm²)	V_{oc} (V)	FF	η (%)	R_{sh} (Ωcm²)	R_s (Ωcm²)
Pristine	5.55	0.68	0.68	2.58	2345	6.5
DIO	10.75	0.68	0.65	4.77	1541	5.8
CN	9.50	0.67	0.68	4.33	1533	5.6
ODT	9.54	0.46	0.46	2.01	780	23.8

with additive ODT showed s-shaped current voltage characteristics, which also possessed the lowest power conversion efficiency. ODT-processed solar cells showed similar J_{sc} (9.54 mA/cm^2) to CN- and DIO-processed cells but possessed lowest FF and V_{oc} compared to all the other cells. The higher series resistance (23.8 Ωcm^2) and lower shunt resistance (780 Ωcm^2) in the ODT devices than those in the DIO or CN cells led to a lower FF. The ODT cell also showed the lowest threshold voltage; this can be seen as an earlier onset of dark current in the medium-bias voltage range as compared to the pristine, DIO, and CN cells, which led to a corresponding lower V_{oc} in the ODT cells.

Domain impurity in pristine cells gave rise to a high density of traps that acted as recombination centers. Solvent additives enhanced domain purity, which led to vast improvement in short-circuit current density. However, ODT additive led to s-shaped J–V behavior due to charge accumulation, leading to low FF. Accumulation of charges is attributed to inefficient bicontinuous transport pathways due to aggregation of pure polymer domains [43]. Aggregation also led to more polymer–polymer interfaces, which reduced the built-in potential resulting in lower V_{oc}. DIO and CN additives gave rise to desirable distribution consisting of crystalline polymer domains evenly distributed throughout the bulk of the film. This led to reduced recombination within the polymer domains and efficient charge transport due formation of bicontinuous pathways.

10.3.5 Effect of Solvent Additive on Charge Transport and Recombination Dynamics of PDPP3T:PCBM Solar Cells

To study the charge collection and recombination dynamics in PDPP3T:PCBM solar cells, transient photocurrent (TPC) and transient photovoltage (TPV) spectroscopy were performed. The TPC plot is shown in Figure 10.12a, and the recombination lifetime and coefficient extracted from TPV spectroscopy is shown in Figure 10.12b and c, respectively.

As seen in Figure 10.12a, cells processed with CN and DIO additive showed faster decay of photocurrent than those processed without additive and cells processed with ODT additive. By fitting the TPC curves with mono-exponential decay functions, the charge collection time was calculated to be 220.6 ns and 246.4 ns for CN- and DIO-processed cells, respectively. Slow photocurrent decay in ODT-processed cells was observed with decay lifetime of 342.6 ns, which indicated charge accumulation as supported by the s-shaped behavior in the J–V results in Figure 10.12a.

Figure 10.12b shows the variation of charge carrier lifetime on the charge carrier concentration for the pristine and different additives PDPP3T:PCBM solar cells. Charge carrier lifetime decreased and bimolecular recombination coefficient increased with an increase in carrier concentration, which is attributed to bimolecular recombination from distribution of tail trap states present within the bandgap of the polymer [24]. For ODT-processed devices, a higher charge carrier lifetime at higher charge carrier densities is observed, which indicated lower bimolecular recombination order. The bimolecular recombination order (λ) is derived from the slope of the charge carrier lifetime versus charge carrier density plot. The bimolecular recombination order was 4.13 for pristine solar cells. The bimolecular recombination order

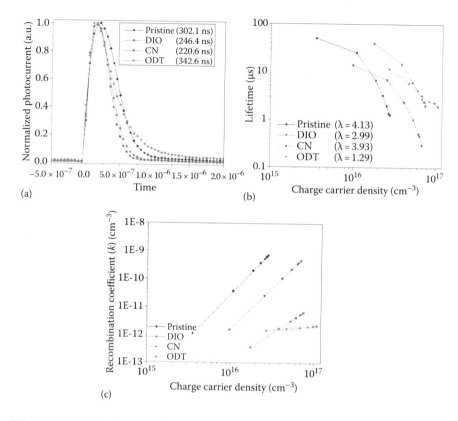

FIGURE 10.12 (a) Transient photocurrent decay of PDPP3T:PCBM solar cells. Numbers indicated in parentheses in the legend show charge carrier lifetime in the short-circuit condition obtained by fitting the decay function with the mono-exponential equation. (b) Charge carrier lifetime versus charge carrier density plot of PDPP3T:PC$_{60}$BM solar cells with apparent bimolecular recombination order (λ) and (c) bimolecular recombination coefficient versus charge carrier density of PDPP3T:PC$_{60}$BM solar cells. (Reproduced from Venkatesan, S. et al., *Nanoscale*, 6, 2, 1011–1019, 2014. With permission.)

decreased from 4.13 to 2.99 and 3.93 for DIO- and CN-processed devices, respectively, which indicated a lower trap-assisted recombination in DIO and CN cells. The ODT-processed solar cells had the lowest bimolecular recombination order of 1.29 and showed the lowest increase in bimolecular recombination coefficient with an increase in charge carrier density as seen in Figure 10.12c.

Because bimolecular recombination is shown to occur at the donor–acceptor interface [44–46], pristine solar cells showed enhanced bimolecular recombination due to excess donor–acceptor interfaces caused by intermixing of amorphous polymer with fullerene. Bimolecular recombination is lower for DIO- and CN-processed cells as the interface states are reduced due to higher crystallinity of the polymer. However, in ODT-processed cells, a higher donor–donor interface led to suppressed bimolecular recombination but, at the same time, led to slow charge carrier transport due to lack of bicontinuous charge transport pathways.

Engineering of Active Layer Nanomorphology

Correlating the nanomorphological and structural characterization to the device results, we can emphasize that the domain purity plays a significant role in J_{sc} because the high purity donor and acceptor domains, even at different domain sizes, in the DIO, CN, and ODT cells led to a comparable J_{sc} (9.5–10.75 mA/cm^2), which is much higher than that (5.55 mA/cm^2) of the pristine cells in which the donor and acceptor are intermixed. As discussed in Section 10.3.5, the high purity donor domain sizes are within the exciton diffusion lengths and seem to have negligible effects on J_{sc}. However, the smaller high purity donor domains in the ODT cells can affect the V_{oc} and FF as a result of charge transport and recombination. Fullerene intercalation between polymer chains (shown in Figure 10.13a for pristine films) leads to higher impure polymer domains as well as highest interdomain interfaces and/or isolated regions giving rise to higher bimolecular recombination as charge recombination is enhanced by the trap states present at the interdomain interfaces. However, lower PCBM intercalation in the additives-added films (Figure 10.13a and b) led to lower bimolecular recombination due to the formation of high purity donor domains and acceptor domains that reduced donor–acceptor interfaces and improved electron/hole transport pathways. This caused a higher J_{sc} in the additive-added (DIO, CN, and ODT) cells than in pristine devices. However, the smaller donor domains in ODT cells led to higher series resistance and

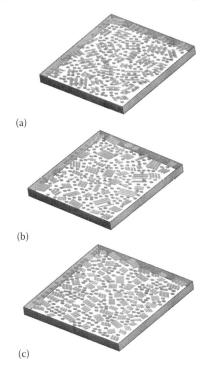

FIGURE 10.13 Nanomorphology when (a) processed without additives, showing highly intermixed domains, (b) processed with additive CN or DIO and domain purity is higher and having adequate interdomain interfaces, and (c) processed with additive ODT and smaller domain of pure domains may lead to lower interdomain interface. (Reproduced from Venkatesan, S. et al., *Nanoscale*, 6, 2, 1011–1019, 2014. With permission.)

lower shunt resistance, causing a lower FF than the DIO and CN devices. The donor and acceptor domain purity, distribution, and size affect charge collection to the electrodes, bimolecular recombination at the domain interfaces, and charge carrier lifetime. In order to attain high photovoltaic device performance, it is necessary to both suppress bimolecular recombination and reduce charge transport/collection time. Hence, it is important to select appropriate solvents, additives, and active layer processing conditions to create ideal morphology in bulk heterojunction devices consisting of optimum polymer domain purity, size (within exciton diffusion length), and distribution.

10.4 SUMMARY

Effects of PCBM weight ratios on the optical and electrical characteristics of a bulk heterojunction 2,1,3-benzothiadiazole-based polymer, namely PBT-T1:PCBM film, were investigated and correlated with changes to nanomorphology. Increasing PCBM loading in active layer blends led to more amorphous polymer chains with smaller domain sizes. The 1:2 mixing ratio was found to be optimum for device performance due to low series and shunt resistance. The addition of small amounts of additive DIO in parent solvent for the 1:2 mixing ratio led to enhanced photovoltaic performance due to a reduction in series resistance, larger shunt resistance, and higher charge carrier mobility caused by improved crystallinity of the polymer. Photo-CELIV measurements with varying delay time showed that 1:2 cells exhibited higher density of extracted charge carriers and was further enhanced when processed from a DCB + DIO solvent mixture. Power conversion efficiency for single junction devices at the 1:2 blending weight ratio with additive and $PC_{70}BM$ reached up to 5.65% with a high open circuit voltage of 0.9 V. In addition, charge transport and bimolecular recombination dynamics were correlated with nanomorphology in polymer solar cells by studying the role of domain purity and domain size on charge transport and recombination kinetics in inverted bulk heterojunction solar cells. Domain purity was found to play an important role in short-circuit current density, and the fill factor is mainly dependent upon domain sizes. Further work to reduce charge recombination in polymer solar cells should be directed toward attaining high domain purity and optimal domain sizes with lower density of trap states.

ACKNOWLEDGMENTS

This research was benefited from NASA EPSCoR (NNX13AD31A), NSF CAREER (ECCS-0950731), and NSF EPSCoR (grant no. 0903804). The XRD measurement was performed using a NSF MRI (grant no. 1229577)-supported Rigaku SmartLab diffractometer.

REFERENCES

1. Darling, S.B., and F. You, The case for organic photovoltaics. *RSC Advances*, 2013. 3(39): pp. 17633–17648.
2. Li, J., M. Yan, Y. Xie, and Q. Qiao, Linker effects on optoelectronic properties of alternate donor–acceptor conjugated polymers. *Energy and Environmental Science*, 2011. 4(10): pp. 4276–4283.

3. Chen, C.-Y., C.-S. Tsao, Y.-C. Huang, H.-W. Liu, W.-Y. Chiu, C.-M. Chuang, U.S. Jeng, C.-J. Su, W.-R. Wu, W.-F. Su, and L. Wang, Mechanism and control of the structural evolution of a polymer solar cell from a bulk heterojunction to a thermally unstable hierarchical structure. *Nanoscale*, 2013. **5**(16): pp. 7629–7638.
4. Jayawardena, K.D.G.I., L.J. Rozanski, C.A. Mills, M.J. Beliatis, N.A. Nismy, and S.R.P. Silva, "Inorganics-in-Organics": Recent developments and outlook for 4G polymer solar cells. *Nanoscale*, 2013. **5**(18): pp. 8411–8427.
5. Jose da Silva, W., H.P. Kim, A. Rashid bin Mohd Yusoff, and J. Jang, Transparent flexible organic solar cells with 6.87% efficiency manufactured by an all-solution process. *Nanoscale*, 2013. **5**(19): pp. 9324–9329.
6. Park, H., Y. Shi, and J. Kong, Application of solvent modified PEDOT:PSS to graphene electrodes in organic solar cells. *Nanoscale*, 2013. **5**(19): pp. 8934–8939.
7. Ho, C.-R., M.-L. Tsai, H.-J. Jhuo, D.-H. Lien, C.-A. Lin, S.-H. Tsai, T.-C. Wei, K.-P. Huang, S.-A. Chen, and J.-H. He, An energy-harvesting scheme employing CuGaSe2 quantum dot-modified ZnO buffer layers for drastic conversion efficiency enhancement in inorganic–organic hybrid solar cells. *Nanoscale*, 2013. **5**(14): pp. 6350–6355.
8. Rathnayake, H., J. Binion, A. McKee, D.J. Scardino, and N.I. Hammer, Perylenediimide functionalized bridged-siloxane nanoparticles for bulk heterojunction organic photovoltaics. *Nanoscale*, 2012. **4**(15): pp. 4631–4640.
9. Adhikary, P., S. Venkatesan, N. Adhikari, P.P. Maharjan, O. Adebanjo, J. Chen, and Q. Qiao, Enhanced charge transport and photovoltaic performance of PBDTTT-C-T/PC70BM solar cells via UV-ozone treatment. *Nanoscale*, 2013. **5**(20): pp. 10007–10013.
10. Adebanjo, O., P.P. Maharjan, P. Adhikary, M. Wang, S. Yang, and Q. Qiao, Triple junction polymer solar cells. *Energy and Environmental Science*, 2013. **6**(11): p. 3150.
11. Kang, D.J., H. Kang, C. Cho, K.-H. Kim, S. Jeong, J.-Y. Lee, and B.J. Kim, Efficient light trapping in inverted polymer solar cells by a randomly nanostructured electrode using monodispersed polymer nanoparticles. *Nanoscale*, 2013. **5**(5): pp. 1858–1863.
12. Chen, W., T. Xu, F. He, W. Wang, C. Wang, J. Strzalka, Y. Liu, J. Wen, D.J. Miller, J. Chen, K. Hong, L. Yu, and S.B. Darling, Hierarchical nanomorphologies promote exciton dissociation in polymer/fullerene bulk heterojunction solar cells. *Nano Letters*, 2011. **11**(9): pp. 3707–3713.
13. Chen, W., M.P. Nikiforov, and S.B. Darling, Morphology characterization in organic and hybrid solar cells. *Energy and Environmental Science*, 2012. **5**(8): pp. 8045–8074.
14. Maharjan, P.P., Q. Chen, L. Zhang, O. Adebanjo, N. Adhikari, S. Venkatesan, P. Adhikary, B. Vaagensmith, and Q. Qiao, Photovoltaic devices and characterization of a dodecyloxybenzothiadiazole-based copolymer. *Physical Chemistry Chemical Physics*, 2013. **15**(18): pp. 6856–6863.
15. Xie, Y., Y. Bao, J. Du, C. Jiang, and Q. Qiao, Understanding of morphology evolution in local aggregates and neighboring regions for organic photovoltaics. *Physical Chemistry Chemical Physics*, 2012. **14**(29): pp. 10168–10177.
16. Dutta, P., Y. Xie, M. Kumar, M. Rathi, P. Ahrenkiel, D. Galipeau, Q. Qiao, and V. Bommisetty, Connecting physical properties of spin-casting solvents with morphology, nanoscale charge transport, and device performance of poly(3-hexylthiophene):phenyl-C61-butyric acid methyl ester bulk heterojunction solar cells. *Journal of Photonics for Energy*, 2011. **1**(1): p. 011124-17.
17. Gao, Y., and J.K. Grey, Resonance chemical imaging of polythiophene/fullerene photovoltaic thin films: Mapping morphology-dependent aggregated and unaggregated C=C species. *Journal of the American Chemical Society*, 2009. **131**(28): pp. 9654–9662.
18. Albrecht, S., W. Schindler, J. Kurpiers, J. Kniepert, J.C. Blakesley, I. Dumsch, S. Allard, K. Fostiropoulos, U. Scherf, and D. Neher, On the field dependence of free charge carrier generation and recombination in blends of PCPDTBT/PC70BM: Influence of solvent additives. *The Journal of Physical Chemistry Letters*, 2012. **3**(5): pp. 640–645.

19. Guo, C., D.R. Kozub, S. Vajjala Kesava, C. Wang, A. Hexemer, and E.D. Gomez, Signatures of multiphase formation in the active layer of organic solar cells from resonant soft x-ray scattering. *ACS Macro Letters*, 2013. **2**(3): pp. 185–189.
20. Lyons, B.P., N. Clarke, and C. Groves, The relative importance of domain size, domain purity and domain interfaces to the performance of bulk-heterojunction organic photovoltaics. *Energy and Environmental Science*, 2012. **5**(6): pp. 7657–7663.
21. Yang, L., J.R. Tumbleston, H. Zhou, H. Ade, and W. You, Disentangling the impact of side chains and fluorine substituents of conjugated donor polymers on the performance of photovoltaic blends. *Energy and Environmental Science*, 2013. **6**(1): pp. 316–326.
22. Karagiannidis, P.G., D. Georgiou, C. Pitsalidis, A. Laskarakis, and S. Logothetidis, Evolution of vertical phase separation in P3HT:PCBM thin films induced by thermal annealing. *Materials Chemistry and Physics*, 2011. **129**(3): pp. 1207–1213.
23. Shuttle, C.G., B. O'Regan, A.M. Ballantyne, J. Nelson, D.D.C. Bradley, and J.R. Durrant, Bimolecular recombination losses in polythiophene: Fullerene solar cells. *Physical Review B*, 2008. **78**(11): p. 113201.
24. Spoltore, D., W.D. Oosterbaan, S. Khelifi, J.N. Clifford, A. Viterisi, E. Palomares, M. Burgelman, L. Lutsen, D. Vanderzande, and J. Manca, Effect of polymer crystallinity in P3HT:PCBM solar cells on band gap trap states and apparent recombination order. *Advanced Energy Materials*, 2013. **3**(4): pp. 466–471.
25. Helgesen, M., S.A. Gevorgyan, F.C. Krebs, and R.A.J. Janssen, Substituted 2,1,3-benzothiadiazole- and thiophene-based polymers for solar cells—Introducing a new thermocleavable precursor. *Chemistry of Materials*, 2009. **21**(19): pp. 4669–4675.
26. Venkatesan, S., S. Ngo, Q. Chen, A. Dubey, L. Mohammad, N. Adhikari, A.Q.Q.S.E. Mitul, and Q. Qiao, Benzothiadiazole based polymer for single and double junction solar cells with high open circuit voltage. *Nanoscale*, 2014. **6**(12): pp. 7093–7100.
27. Yao, Y., J. Hou, Z. Xu, G. Li, and Y. Yang, Effects of solvent mixtures on the nanoscale phase separation in polymer solar cells. *Advanced Functional Materials*, 2008. **18**(12): pp. 1783–1789.
28. Li, G., V. Shrotriya, Y. Yao, J. Huang, and Y. Yang, Manipulating regioregular poly(3-hexylthiophene): [6,6]-phenyl-C$_{61}$-butyric acid methyl ester blends-route towards high efficiency polymer solar cells. *Journal of Materials Chemistry*, 2007. **17**(30): pp. 3126–3140.
29. Kim, H., S. Ok, H. Chae, and Y. Choe, Performance characteristics of polymer photovoltaic solar cells with an additive-incorporated active layer. *Nanoscale Research Letters*, 2012. **7**(1): pp. 1–5.
30. Bhattacharya, J., R.W. Mayer, M. Samiee, and V.L. Dalal, Photo-induced changes in fundamental properties of organic solar cells. *Applied Physics Letters*, 2012. **100**(19): p. 193501.
31. Salim, T., L.H. Wong, B. Brauer, R. Kukreja, Y.L. Foo, Z. Bao, and Y.M. Lam, Solvent additives and their effects on blend morphologies of bulk heterojunctions. *Journal of Materials Chemistry*, 2011. **21**(1): pp. 242–250.
32. Eom, S.H., H. Park, S.H. Mujawar, S.C. Yoon, S.-S. Kim, S.-I. Na, S.-J. Kang, D. Khim, D.-Y. Kim, and S.-H. Lee, High efficiency polymer solar cells via sequential inkjet-printing of PEDOT:PSS and P3HT:PCBM inks with additives. *Organic Electronics*, 2010. **11**(9): pp. 1516–1522.
33. He, Y., and Y. Li, Fullerene derivative acceptors for high performance polymer solar cells. *Physical Chemistry Chemical Physics*, 2011. **13**(6): pp. 1970–1983.
34. Baumann, A., T.J. Savenije, D.H.K. Murthy, M. Heeney, V. Dyakonov, and C. Deibel, Influence of phase segregation on recombination dynamics in organic bulk-heterojunction solar cells. *Advanced Functional Materials*, 2011. **21**(9): pp. 1687–1692.

35. Venkatesan, S., N. Adhikari, J. Chen, E.C. Ngo, A. Dubey, D.W. Galipeau, and Q. Qiao, Interplay of nanoscale domain purity and size on charge transport and recombination dynamics in polymer solar cells. *Nanoscale*, 2014. **6**(2): pp. 1011–1019.
36. Bartelt, J.A., Z.M. Beiley, E.T. Hoke, W.R. Mateker, J.D. Douglas, B.A. Collins, J.R. Tumbleston, K.R. Graham, A. Amassian, H. Ade, J.M.J. Fréchet, M.F. Toney, and M.D. McGehee, The importance of fullerene percolation in the mixed regions of polymer–fullerene bulk heterojunction solar cells. *Advanced Energy Materials*, 2013. **3**(3): pp. 364–374.
37. Miller, N.C., E. Cho, R. Gysel, C. Risko, V. Coropceanu, C.E. Miller, S. Sweetnam, A. Sellinger, M. Heeney, I. McCulloch, J.-L. Brédas, M.F. Toney, and M.D. McGehee, Factors governing intercalation of fullerenes and other small molecules between the side chains of semiconducting polymers used in solar cells. *Advanced Energy Materials*, 2012. **2**(10): pp. 1208–1217.
38. Yang, X., J.K.J. van Duren, M.T. Rispens, J.C. Hummelen, R.A.J. Janssen, M.A.J. Michels, and J. Loos, Crystalline organization of a methanofullerene as used for plastic solar-cell applications. *Advanced Materials*, 2004. **16**(9–10): pp. 802–806.
39. Yun, J.-J., J. Peet, N.-S. Cho, G.C. Bazan, S.J. Lee, and M. Moskovits, Insight into the Raman shifts and optical absorption changes upon annealing polymer/fullerene solar cells. *Applied Physics Letters*, 2008. **92**(25): pp. 251912.
40. Tsoi, W.C., D.T. James, J.S. Kim, P.G. Nicholson, C.E. Murphy, D.D.C. Bradley, J. Nelson, and J.-S. Kim, The nature of in-plane skeleton raman modes of P3HT and their correlation to the degree of molecular order in P3HT:PCBM blend thin films. *Journal of the American Chemical Society*, 2011. **133**(25): pp. 9834–9843.
41. Roelofs, W.S.C., S.G.J. Mathijssen, J.C. Bijleveld, D. Raiteri, T.C.T. Geuns, M. Kemerink, E. Cantatore, R.A.J. Janssen, and D.M. de Leeuw, Fast ambipolar integrated circuits with poly(diketopyrrolopyrrole-terthiophene). *Applied Physics Letters*, 2011. **98**(20): p. 203301.
42. Bijleveld, J.C., A.P. Zoombelt, S.G.J. Mathijssen, M.M. Wienk, M. Turbiez, D.M. de Leeuw, and R.A.J. Janssen, Poly(diketopyrrolopyrrole-terthiophene) for ambipolar logic and photovoltaics. *Journal of the American Chemical Society*, 2009. **131**(46): pp. 16616–16617.
43. Kastner, C., D.K. Susarova, R. Jadhav, C. Ulbricht, D.A.M. Egbe, S. Rathgeber, P.A. Troshin, and H. Hoppe, Morphology evaluation of a polymer–fullerene bulk heterojunction ensemble generated by the fullerene derivatization. *Journal of Materials Chemistry*, 2012. **22**(31): pp. 15987–15997.
44. Monestier, F., J.-J. Simon, P. Torchio, L. Escoubas, F. Flory, S. Bailly, R. de Bettignies, S. Guillerez, and C. Defranoux, Modeling the short-circuit current density of polymer solar cells based on P3HT:PCBM blend. *Solar Energy Materials and Solar Cells*, 2007. **91**(5): pp. 405–410.
45. Mandoc, M.M., F.B. Kooistra, J.C. Hummelen, B. de Boer, and P.W.M. Blom, Effect of traps on the performance of bulk heterojunction organic solar cells. *Applied Physics Letters*, 2007. **91**(26): p. 263505.
46. Szmytkowski, J.D., Analysis of the image force effects on the recombination at the donor–acceptor interface in organic bulk heterojunction solar cells. *Chemical Physics Letters*, 2009. **470**(1–3): pp. 123–125.

Section IV

Devices

11 Inorganic–Organic Nanocomposites and Their Assemblies for Solar Energy Conversion

Jaehan Jung, Ming He, and Zhiqun Lin

CONTENTS

- 11.1 Introduction 308
- 11.2 Grafting Conjugated Polymers onto Semiconductor Inorganic Nanoparticles 309
  - 11.2.1 Introduction 309
  - 11.2.2 Ligand Exchange 310
  - 11.2.3 Direct Grafting of Conjugated Polymers with Nanoparticles 312
  - 11.2.4 *In Situ* Growth of Nanoparticles within the CP Matrix 315
- 11.3 Self-Assembly of Nanoparticles and Conjugated Polymers 318
  - 11.3.1 Introduction 318
  - 11.3.2 Conjugated Polymer Nanofiber 318
  - 11.3.3 Hybrid Nanofibrils 321
- 11.4 Alignment of Nanorods 324
  - 11.4.1 Introduction 324
  - 11.4.2 Alignment of Nanorods 324
    - 11.4.2.1 Temperature 326
    - 11.4.2.2 Aspect Ratio 326
    - 11.4.2.3 Substrates 326
    - 11.4.2.4 Concentration 327
    - 11.4.2.5 Solvent 327
    - 11.4.2.6 Rate of Evaporation 330
    - 11.4.2.7 Surface Ligands 330
    - 11.4.2.8 External Fields 331
  - 11.4.3 Alignment of NRs in the Polymer Matrix 333
- 11.5 Conclusion and Outlook 334
- Acknowledgment 334
- References 334

11.1 INTRODUCTION

Unlike a conventional inorganic semiconductor in which the charge carriers are immediately generated upon the exposure to light, an organic semiconductor forms a spatially localized electron–hole pair (that is, a Frenkel-type exciton).[1,2] These excitons typically have large binding energy of around 0.5 eV, which means that the generated excitons need to migrate to the donor–acceptor interface for their dissociation. The excitons diffuse randomly and are not influenced by an electric field as they are electrically neutral. Hence, the length scale of organic phases must be comparable to the exciton diffusion length ($L = (D \cdot \tau)^{1/2}$ = ~10 nm, where τ is the lifetime of the exciton, and D is the diffusion coefficient).[3–5] To understand the exciton migration process, the energy transfer mechanism should be invoked and can be written as

$$^*D + A \rightarrow D + {}^*A$$

where D and A are the donor and the acceptor, respectively. The asterisk represents the excited state of molecules. The energy transfer may occur either by the dipole–dipole interaction or by the electron exchange interactions. The energy transfer through electron exchange interactions requires an orbital overlap between molecules and is sometimes referred to as "Dexter energy transfer" or "orbital overlap mechanism."[6,7] In this case, the electron is transferred from the LUMO of the donor to that of the acceptor, and the hole is also simultaneously transferred from the HOMO of the donor to that of the acceptor. As the Dexter transfer is governed by the orbital overlap between the electron density of both the excited donor (*D) and the nearby ground state acceptor (A), the rate of the Dexter transfer can be expressed as

$$k_{\text{Dexter}} \propto <\Psi(^*D)\Psi(A)|\ H_{\text{ex}}\ |\Psi(D)\Psi(^*A)>^2$$

where H_{ex} is the electron exchange operator. The form of the H_{ex} operator is $\exp(-R_{\text{DA}})$, where R_{DA} is the distance between the donor (D) and the acceptor (A).

The energy transfer can also occur via the dipole–dipole interaction, and it is sometimes termed as "Förster energy transfer."[8,9] In this mechanism, the energy is transferred by the overlap of dipolar electric fields of *D with A. Similarly, the rate of Förster energy transfer can be also calculated and written as

$$k_{\text{Förster}} \propto <\Psi(^*D)\Psi(A)|\ H_{\text{DD}}\ |\Psi(D)\Psi(^*A)>^2$$

where H_{DD} is the dipole–dipole energy transfer operator and has the form of $\mu_{*D}\mu_{*A}/R_{\text{DA}}^3$. Here, μ_{*D} and μ_{*A} are the strength of an oscillating dipole by *D and *A, respectively. By considering the degree of contribution from Förster and Dexter energy transfers, the overall rate of energy transfer can be expressed as

$$k_{\text{ET}} \propto \alpha <\Psi(^*D)\Psi(A)|\ H_{\text{DD}}\ |\Psi(D)\Psi(^*A)>^2 + \beta <\Psi(^*D)\Psi(A)|\ H_{\text{ex}}\ |\Psi(D)\Psi(^*A)>^2$$

where α and β are the degree of contribution from Förster and Dexter energy transfers, respectively. Because the matrix element is squared, the rate constant is simply proportional to the exp(−2R_{DA}) and 1/R_{DA}^6 for Dexter and Förster energy transfers, respectively. The falloff in the rate of energy transfer from the exchange mechanism is steeper than that from the dipole–dipole interactions. Consequently, the range of distance between *D and *A for Förster energy transfer is relatively larger (that is, 1 ~ 10 nm) than that of Dexter transfer (that is, with a typical length scale of 0.1 ~ 1 nm). Nonetheless, the rate constant (k_{ET}) drops to the negligibly small value compared to the lifetime of *D as the intermolecular separation increases.

The energy transfer rate constant has a connection with the exciton diffusion coefficient, D, and their relationship can be written as[10,11]

$$D = \frac{A}{6}\sum_N R_{DA}^2 k_{ET}(R_{DA})$$

where A is a factor for disorder in the thin film. Here, the energy transfer rate constant, k_{ET} is defined by the Dexter or Förster energy transfer rate depending on the system. Because the diffusion length (L) is expressed as $L = (D(R_{DA})\cdot\tau)^{1/2}$, the interfacial engineering between the donor and the acceptor materials would be a key factor for the efficient charge separation in solar cells. In the following, the methodologies for engineering the donor–acceptor interface are elaborated.

11.2 GRAFTING CONJUGATED POLYMERS ONTO SEMICONDUCTOR INORGANIC NANOPARTICLES

11.2.1 Introduction

Conjugated polymers (CPs) have garnered increasing attention due to their advantageous properties, such as light weight, flexibility, roll-to-roll production, low cost, and large area. After the discovery of CPs such as poly(3-hexylthiophene) (P3HT), the photovoltaic devices utilizing CPs as electron donors and fullerene derivatives as electron acceptors offer a promising route to organic light-harvesting devices. However, the fullerene derivatives are expensive and cannot be synthesized on a large scale. Moreover, their limited charge mobility leads to many studies in searching for other promising materials as electron acceptors.[11] It is worth noting that the size and shape of semiconductor inorganic nanoparticles (that is, CdS, CdSe,[12] CdTe,[13] PbSe,[14,15] and ZnO[16,17]) can be precisely controlled.[18] They possess high intrinsic carrier mobility and their bandgap can be tuned by controlling the nanoparticle size. Moreover, a large surface area originated from the nanosize of particles ensures the effective charge separation as the exciton diffusion length is typically around 10 nm for organic solar cells. In this context, semiconductor nanoparticles have been used as electron-accepting materials to reduce current losses from the recombination process, thus improving photovoltaic performance.[19–21]

In the early study, hybrid solar cells utilizing inorganic nanoparticles (that is, CdSe nanorods or NRs) as electron donors and P3HT as electron acceptors were fabricated.[22] The device was composed of ITO/PEDOT:PSS/P3HT:CdSe NRs/Al as shown in Figure 11.1. The power conversion efficiency (PCE) was 1.7% with an open circuit voltage (V_{oc}) of 0.7 V, fill factor (FF) of 0.4, and current density (J_{sc}) of 5.7 mA/cm².

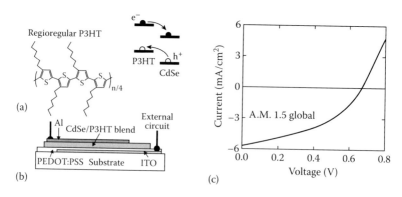

FIGURE 11.1 (a) Structure of regioregular P3HT and the schematic representation of energy level for CdSe nanorods and P3HT. (b) Illustration of the device structure. (c) *I–V* characteristic curve. (From W. U. Huynh, J. J. Dittmer, A. P. Alivisatos, *Science*, 295, 2425–2427, 2002. Reprinted with permission of AAAS.)

However, the overall performance of hybrid devices was not satisfactory as expected when considering the high intrinsic mobility of inorganic materials. Several reasons are speculated as follows:[23–29] First, the use of insulating ligands (surfactant) was unavoidable during the nanoparticle synthesis to prevent their aggregation as well as to obtain the mono-disperse size distribution. Such ligands form a barrier of several Angstrom thickness surrounding nanoparticles, thereby impeding the transport of electrons to adjacent nanoparticles or from CPs. Moreover, the poor miscibility of surfactant-capped nanoparticles within the CP matrix results in the micro-scale phase segregation during the film spin casting, leading to the formation of isolated domains. These micro-sized domains inhibit the charge transport and separation as they trap charge carriers within domains, and the size of domains is bigger than the diffusion length of excitons. As a result, efforts have been made to increase the interfacial area between CPs and inorganic nanoparticles through the surface modification of inorganic nanoparticles.

11.2.2 Ligand Exchange

Ligand exchange is the chemical reaction that replaces one ligand with another via adsorption–desorption dynamics, thus permitting the binding of a broad range of functional groups on the surface of nanoparticles. Briefly, small amounts of nanoparticles with a given ligand coverage are injected into a large volume of selected solvent to dissociate the certain amount of the original capping ligands. The desired ligands are then introduced to replace with original ligands. The reaction processes and rate can be described as follows:[30]

$$\text{Adsorption: M + L} \rightarrow \text{ML}, \ (d[\text{ML}]/dt)_a = k_a[\text{M}][\text{L}] \quad (11.1)$$

$$\text{Desorption: ML} \rightarrow \text{M + L}, \ (d[\text{ML}]/dt)_d = -k_d[\text{ML}] \quad (11.2)$$

Inorganic–Organic Nanocomposites and Their Assemblies

where M and L refer to the surface bonding sites of nanoparticles and the free-state ligand concentration, respectively. The [ML] is the concentration of ligands in bonded state, and k_a and k_d are the absorption and desorption reaction rate constant, respectively. The average surface ligand coverage (θ) can be expressed as

$$\theta = [ML]/([M]+[L])$$

As the number of bonding sites ([M] + [ML]) is constant for a given system, Equations 11.1 and 11.2 can be rewritten as

$$\text{Adsorption: } M + L \rightarrow ML, \ (d[\theta]/dt)_a = k_a(1-\theta)[L]$$

$$\text{Desorption: } ML \rightarrow M + L, \ (d[\theta]/dt)_d = -k_d[\theta]$$

Clearly, the average surface ligand coverage depends on the adsorption rate constant, k_a; the desorption rate constant, k_d; and the concentration of ligands. The adsorption rate and desorption rate are usually determined by the chemical nature of the ligands, which is correlated with the binding ability of functional groups to the nanoparticle surface as well as the steric effects resulting from adjacent chains capped on nanoparticles. Thiols, amines, phosphonic acids, carboxyl acids, and phosphine oxides are the most commonly used ligands. When designing the ligands for photovoltaic applications, several aspects of ligands need to be considered: (1) the head group should have a high affinity for nanoparticles, (2) the energy level of ligands (for example, in case of CP as ligands) need to be well matched with nanoparticles in light of their applications, and (3) the end group must provide nanoparticles with good solubility in the solvents of choice.

To this end, novel CP ligands named pentathiophene phosphonic acid (T5) and terthiophene phosphonic acid (T3) were designed and synthesized (Figure 11.2).[31] The phosphonic acid moiety serving as the head group renders its strong anchoring to the surface of nanoparticles, and the oligohexylthiophene chains as the end group provide good solubility in most of the organic solvents as well as the excellent

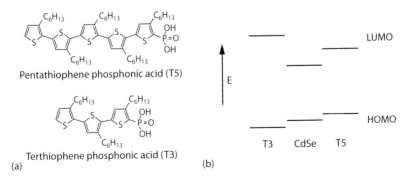

FIGURE 11.2 (a) The molecular structure of ligands. (b) Schematics of the proposed energy level alignment in the CdSe-oligothiophene complexes. (D. J. Milliron, A. P. Alivisatos, C. Pitois, C. Edder, J. M. J. Fréchet: *Advanced Materials*, 2003, 15, 58–61. Copyright Wiley-VCH Verlag GmbH & Co. KGaA. Reproduced with permission.)

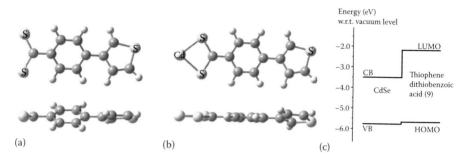

FIGURE 11.3 (a) Optimized structure of thiophenedithiobenzoate in its negative ion form. (b) Optimized structure after the complexation of a Cd^{2+} ion. (c) Electronic energy level alignment for CdSe QDs and carbodithioate oligothiophene. (Reprinted with permission from Querner, C. et al., *Chemistry of Materials*, 18, 4817–4826. Copyright 2006, American Chemical Society.)

miscibility with CPs, especially P3HT, due to their similar structure. The energy level of synthesized CP ligands compared with inorganic nanoparticles (for example, CdSe quantum dots or QDs) is illustrated in Figure 11.2b.

The energy level of T3 molecules and CdSe QDs is type I, and the energy level T5 molecules and CdSe QDs are classified as type II. At the type I interface, the energy transfer occurs while the charges are transferred at the type II interface. Generally, the type II interface is preferred for photovoltaic devices due to the charge transport characteristics whereas the type I is suitable for LED applications as charges can be confined in the core (for example, CdSe QDs), thus facilitating the charge recombination. In this regard, the energy level of CP ligands must be carefully designed to meet the need for specific applications. Indeed, the luminescence quenching in T5-CdSe QDs was observed, and the nanoparticle fluorescence increased after the complexation with T3 ligands. The carbodithioate oligothiphenes were also rationally designed and grafted onto the nanoparticle surfaces (for example, CdSe QDs) for solar cells via the ligand exchange process as shown in Figure 11.3.[32] A staggered (type II) alignment ensures the charge separation: The electron transfers from the CP ligand to nanoparticles, and the hole transfers from nanoparticles to the ligand. Although the ligand was sophisticatedly designed to fulfill the conditions noted earlier in (1) and (2), it failed to satisfy the condition (3), thereby resulting in a poor miscibility with CPs. Consequently, its application in hybrid solar cells based on a spin-coating technique is rather difficult.

11.2.3 Direct Grafting of Conjugated Polymers with Nanoparticles

Although direct ligand exchange is simple, its efficiency is usually low as the targeted CP chains are longer than the passivated ligands.[33] As a result, inorganic nanoparticles prepared with simple ligand exchange sometimes exhibit the phase separation when mixing with CPs, which is detrimental for solar cell applications. To address this problem, a two-step process utilizing bifunctional ligands for coupling was introduced.[34] First, the original ligands of nanoparticles were exchanged with designed bifunctional ligands. The bifunctional ligand is usually short to increase the efficiency of ligand exchange. One end of the bifunctional ligand is tethered to the surface of nanoparticles, and the other end undergoes the coupling reaction with CPs. For example, P3HT was grafted onto CdSe NRs, utilizing bifunctinoal ligands, such as arylbromide phosphine oxide and thiols. Arylbromide-functionalized phosphine oxides or thiols were first introduced to CdSe NRs by ligand exchange. The vinyl-terminated P3HT was then grafted onto the CdSe NR surface via Heck coupling of arylbromide functionalities and vinyl end groups as depicted in Figure 11.4. The grafting density was 250 and 400 P3HT chains per NR for phosphine oxide and thiol functionalities, respectively. This result is consistent with the fact that the thiol ligands coordinate more effectively than the phosphine oxide ligands.

In the approach noted in Section 11.2.3, it is necessary to remove all fatty insulating ligands covering nanoparticles (that is, alkyl phosphonic acid) through the ligand exchange process by utilizing small molecules, such as pyridine. After that, the designed bifunctional ligands were introduced. However, this process is usually time-consuming. In this context, to circumvent the previously mentioned problem (that is, a two-step ligand exchange), CPs (that is, vinyl-terminated P3HT) were grafted onto bifunctional ligand-capped nanoparticles via Heck coupling in the absence of the ligand exchange as illustrated in Figure 11.5.[35] Specifically, monodispersed CdSe QDs were prepared with [(4-bromopheynyl)methyl]dioctylphosphine oxide (DOPO-Br), yielding DOPO-Br-capped CdSe QDs. The as-prepared DOPO-Br capped CdSe QDs were grafted with vinyl-terminated P3HT via Pd-catalyzed Heck coupling without further treatment. However, the weak binding strength of phosphine oxide with nanoparticles and the difficulty in controlling the shape of nanoparticles are the limitations of this approach.

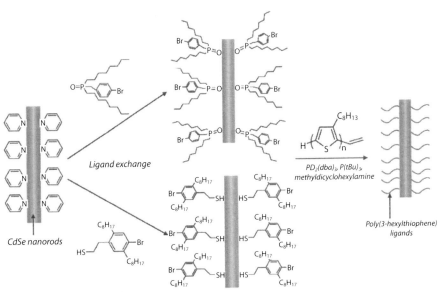

FIGURE 11.4 Schematic illustration of grafting technique. (Reprinted with permission from Zhang, Q. et al., *Chemistry of Materials*, 19, 3712–3716. Copyright 2007, American Chemical Society.)

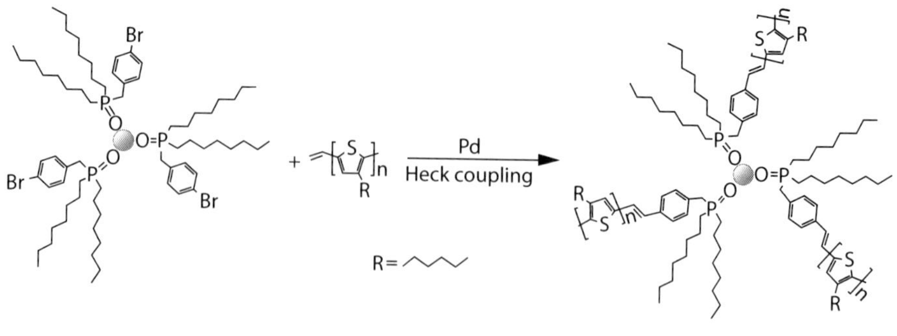

FIGURE 11.5 Illustration of grafting vinyl-terminated P3HT onto CdSe QDs utilizing bifunctional ligands (DOPO-Br). (Reprinted with permission from Xu, J. et al., *Journal of the American Chemical Society*, *129*, 12828–12833. Copyright 2007, American Chemical Society.)

By far, the intimate contact between the donor and acceptor materials can be achieved via the grafting technique mentioned above. Although an intimate contact facilitates the exciton dissociation, which solely happens at the donor–acceptor interface, the efficient charge transport is not determined only by the intimate contact of donor and acceptor. The charge transport properties can be tuned by controlling the size and shape of nanoparticles. For example, anisotropic NRs carry the advantage over QDs because the one-dimensional structures easily form continuous pathways for charge transport, and QDs require more hopping processes. However, the anisotropic shape of nanoparticles cannot be achieved with a previous direct grafting method because of the characteristic of DOPO-Br ligands that do not bind selectively to the certain facets of nanoparticles. To render unidirectional growth of CdSe nanoparticles, the ligands should bind selectively to the surface of CdSe NRs (that is, [01$\bar{1}$0] and [11$\bar{2}$0] facets), which highly depends on the moiety of ligands. The phosphonic acid is one of the most widely used ligands for the growth of the elongated shaped CdSe nanoparticles due to its selectivity. In this context, CP-grafted CdSe NRs were prepared by introducing a new bifunctional ligand, 4-bromobenzyl phosphonic acid (BBPA).[36] The phosphonic acid group of BBPA can strongly coordinate with (01$\bar{1}$0) and (11$\bar{2}$0) facets of CdSe NRs and promote elongated growth of CdSe NRs, yielding BBPA-capped CdSe NRs. Subsequently, the end-functionalized P3HT was grafted with a bromide group of BBPA-capped CdSe NRs via Heck coupling or click chemistry.[36] For the Heck coupling strategy, the procedure was similar to the reaction between DOPO-Br-capped CdSe QDs and vinyl-terminated P3HT. For the click reaction, the bromide group of BBPA was converted into an azide group, forming N_3-BPA-capped CdSe NRs. The ethynyl-terminated P3HT was then grafted with the N_3-functionalized CdSe NRs via the catalyst free Huigen 1,3-dipolar cycloaddition, producing P3HT-CdSe NR nanocomposites (Figure 11.6). It is worth noting that the P3HT-CdSe NR nanocomposites prepared by click chemistry are based on the catalyst-free reaction, and the Heck coupling method requires a Pd catalyst. As a result, nanocomposites produced by click chemistry are desirable for solar cells as there is no need for further purification.

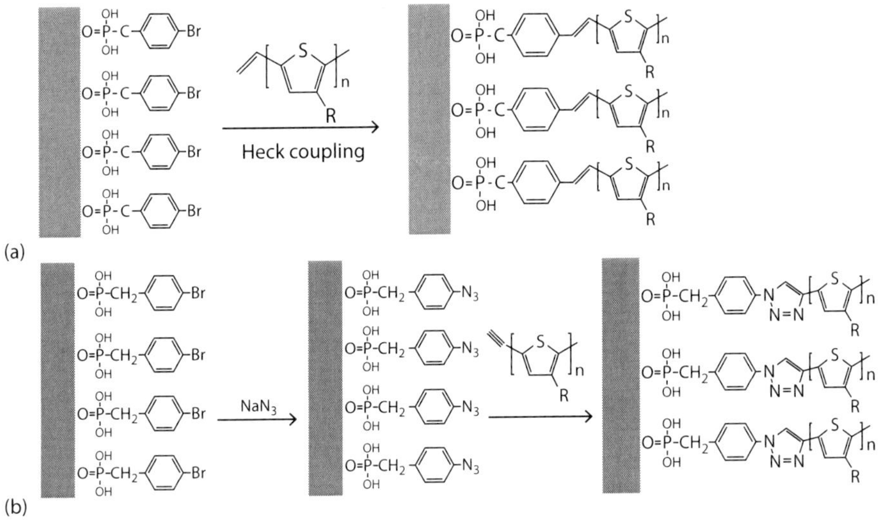

Inorganic–Organic Nanocomposites and Their Assemblies

FIGURE 11.6 (a) Direct grafting of vinyl-terminated P3HT onto CdSe NRs by Heck coupling. (b) Grafting of ethynyl-terminated P3HT onto CdSe NRs by click reaction. (L. Zhao et al.: *Advanced Materials*, 2011, 23, 2844–2849. Copyright Wiley-VCH Verlag GmbH & Co. KGaA. Reproduced with permission.)

In spite of the intriguing properties of anisotropic structure, inorganic NRs usually lie down parallel on the substrates after spin casting. Thus, it does not allow for the exploitation of the full potential of the charge transport property of NRs. In this regard, tetrapods carry advantages over NRs and QDs due to their three-dimensional structures with four arms emanating from the core. The tetrapod shape allows a continuous charge transport pathway regardless of their orientation and ensures the large interfacial area for charge separation. Recently, CdTe tetrapods were prepared with BBPA ligands and then grafted with prepared ethynyl-terminated P3HT via click chemistry, yielding P3HT-CdTe tetrapod nanocomposites as shown in Figure 11.7.[37] Clearly, the nearly complete fluorescence quenching of P3HT-CdTe tetrapod nanocomposites was observed in Figure 11.7f, indicating the efficient charge transfer from the electron-donating P3HT onto the electron-accepting CdTe tetrapods. As we discussed in Section 11.1, the interfacial engineering of nanocomposites is important to increase the efficiency of charge separation. Therefore, these semiconductor CPs–grafted inorganic nanoparticles (QDs, NRs, and tetrapods) can possibly be promising materials as building blocks for solar cell applications due to their inherent intimate contact.

11.2.4 IN SITU GROWTH OF NANOPARTICLES WITHIN THE CP MATRIX

An ideal route to organic–inorganic hybrid nanocomposites is to synthesize uncapped inorganic nanoparticles within the CP matrix, thereby bypassing any further treatment, such as ligand exchange.[38] To avoid the use of insulating ligands, the *in situ* synthesis of inorganic nanoparticles in the polymer matrix was developed. In this method, a critical condition needs to be met, that is, the solvent of choice should dissolve both

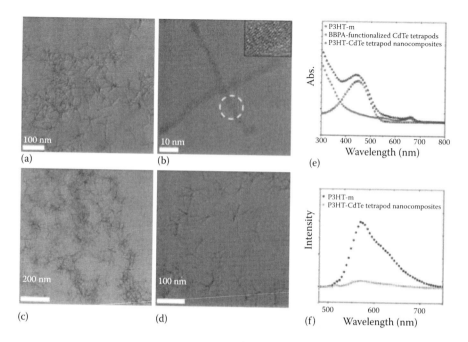

FIGURE 11.7 TEM images of (a) BBPA-functionalized CdTe tetrapods; (b) close-up of individual BBPA-functionalized CdTe tetrapods; the close-up of an arm marked with a dashed circle is shown as the inset; (c, d) P3HT-CdTe tetrapod nanocomposites; and (e, f) absorption and emission spectra. (Reprinted with permission from Jung, J. et al., *Langmuir*, 29, 8086–8092. Copyright 2013, American Chemical Society.)

CPs and precursors. Recently, CdSe nanoparticles have been synthesized with no need for surfactant by utilizing the steric hindrance stemming from hexyl side chains of P3HT, yielding a simple mixture of CdSe/P3HT.[39] In this work, pyrophoric and explosive dimethylcadmium was used as a precursor due to the insolubility of other precursors in organic solvents, that is, octadecene (ODE) and 1,2,3-trichlorobenzene (TCB), chosen for P3HT. Although the size and shape of nanoparticles based on the conventional method can be controlled by tuning the concentration of precursors and temperature, their control in the *in situ* growth approach was still challenging (Figure 11.8). However, the elimination of the need for ligand exchange and grafting procedures as well as the naturally intimate contact between nanoparticles and CPs make the *in situ* growth method a promising route to high-efficiency hybrid solar cells.

Interestingly, anisotropic CdS nanocrystals in the CP matrix (for example, P3HT) based on the *in situ* growth method can also be prepared. Specifically, cadmium acetate was chosen as the precursor, and the mixture of 1,2-dichlorobenzene (DCB) and dimethyl sulfoxide (DMSO) was used as the solvent.[38] In this work, the binary solvent was required to dissolve both precursors and polymers (for example, DCB for P3HT and DSMO for cadmium acetate). The shape evolution from a QD to a NR can be clearly observed with an increased Cd-precursor concentration (Figure 11.9). The growth mechanism of NRs was elucidated by FTIR measurement. As shown in Figure 11.9e, the 1106 cm^{-1} peak that is assigned to S-C stretching of P3HT

Inorganic–Organic Nanocomposites and Their Assemblies 317

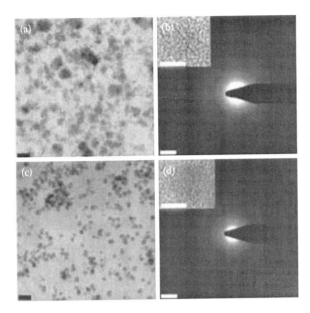

FIGURE 11.8 (a, c) TEM images of CdSe QDs synthesized at 200 and 260°C, respectively. (scale bar = 20 nm). (b, d) Corresponding selected area electron diffraction patterns. Insets show the high-resolution TEM images (scale bar = 5 nm). (Reprinted with permission from Dayal, S. et al., *Journal of the American Chemical Society*, 131, 17726–17727. Copyright 2009, American Chemical Society.)

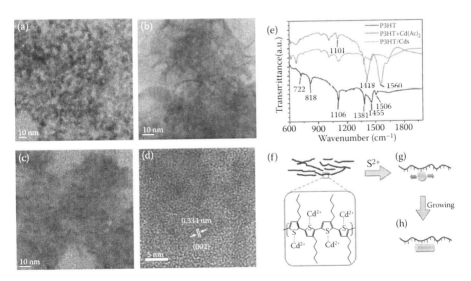

FIGURE 11.9 TEM images of CdS NRs within the P3HT matrix at the cadmium acetate concentrations of (a) 2.5, (b) 8.3, and (c) 12.45 mg/ml. (d) HRTEM image of NR. (e) FTIR spectra before and after adding cadmium acetate in the reaction and after the reaction. (f, g, and h) Proposed synthesis mechanism of CdS/P3HT nanocomposites. (Reprinted with permission from Liao, H.-C. et al., *Macromolecules*, 42, 6558–6563. Copyright 2009, American Chemical Society.)

shifts to 1101 cm^{-1} after mixing with cadmium acetate, indicates the reduction in bond energy of S-C. It suggests that the system experiences the additional intermolecular interaction at the expense of S-C bond energy, probably resulting from a strong dipole–dipole or ion–dipole interaction between the Cd^{2+} ions and S atoms. Based on this study, it was found that CdS NRs were grown along the backbone of P3HT chains after Cd^{2+} ions anchored on the sulfur atoms of P3HT as illustrated in Figure 11.9f, g, and h.

11.3 SELF-ASSEMBLY OF NANOPARTICLES AND CONJUGATED POLYMERS

11.3.1 INTRODUCTION

Hybrid nanocomposites composed of CPs as the electron donor and inorganic nanoparticles as the electron acceptor have received much attention for photovoltaic applications due to the efficient electron transfer, high electron affinity, and large surface area.[40,41] However, the performance of such hybrid solar cells has not yet been successful due probably to the microscale segregation of nanoparticles rather than the nanoscale phase separation, thereby hindering the charge separation and transport. The challenge in hybrid devices is to achieve a large interfacial area between electron donors (that is, CPs) and acceptors (that is, nanoparticles) in order to facilitate exciton dissociation while providing simultaneous continuous pathways for electrons and holes, respectively. The electron transport can be optimized through controlling the shape and size of semiconducting inorganic nanoparticles. For example, compared to QDs, NRs offer more efficient electron transport to the electrode due to the reduced number of hopping processes as in QDs. While semiconductor inorganic nanoparticles have high electron mobility, the hole mobility of most CPs is three orders of magnitude less than that of nanoparticles, thereby resulting in charge accumulation in the film and thus increasing the possibility of recombination between holes and electrons within the device. In this regard, several studies have been performed toward increasing the charge mobility of CPs. It is well known that the mobility of CPs (for example, P3HT) is highly dependent on its crystallinity, the orientation of the crystallites with respect to the charge transport direction, and on its molecular weight.[42–46] Therefore, careful control over the film morphology by the choice of solvent and the deposition conditions is required.

11.3.2 CONJUGATED POLYMER NANOFIBER

There have been many studies in controlling the film morphology of CPs, especially for P3HT. P3HT exhibits high hole mobility (up to 0.1 cm^2/V·s) owing to the strong π–π stacking when the film morphology is precisely controlled. P3HT can form fibrils when polymer chains lie parallel to each other, thus offering good π–π stacking along the length of the fibril. It has been found that P3HT is readily crystallized from poor solvent in the form of nanowires (NWs) with the length of micrometer and the width of nanometer. P3HT chains are packed normal to the NW axis as illustrated in Figure 11.10.[47]

Inorganic–Organic Nanocomposites and Their Assemblies   319

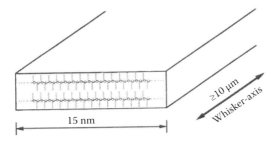

FIGURE 11.10 Illustration of the molecular arrangement of P3HT NWs. (Ihn, K. J.; Moulton, J.; Smith, P.: *Journal of Polymer Science Part B: Polymer Physics*, 1993, *31*, 735–742, Copyright Wiley-VCH Verlag GmbH & Co. KGaA. Reproduced with permission.)

Greenham et al. fabricated devices utilizing CdSe NRs and P3HT nanofibers by careful choice of high boiling solvent, that is, 1,2,3-trichlorobenzene (TCB), thiophene, and chloroform for comparison.[48] The devices prepared by using TCB and thiophene have higher PCE than those by chloroform as the higher boiling point solvent increases the drying time, thereby allowing the large-scale self-organization of P3HT as shown in Figure 11.11. Indeed, P3HT films deposited using high boiling point solvent (that is, TCB) clearly shows the fibrilar structure whereas P3HT does not form the fibrilar structure in case of thiophene and chloroform.

The photovoltaic devices have also been fabricated utilizing P3HT fibrilar structures as electron donors and CdSe NRs as electron acceptors. Clearly, the perfor-

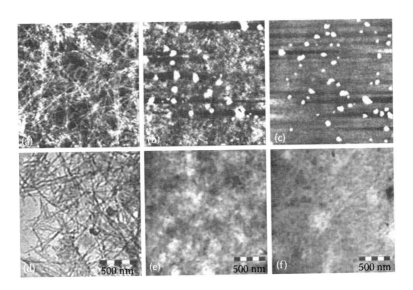

FIGURE 11.11 Tapping-mode AFM images of P3HT films deposited from (a) TCB, (b) thiophene, and (c) chloroform. TEM images of P3HT films deposited from (d) TCB, (e) thiophene, and (f) chloroform. (Sun, B.; Greenham, N. C., *Physical Chemistry Chemical Physics*, 2006, *8* (30), 3557–3560, Reproduced by permission of The Royal Society of Chemistry.)

mance of devices fabricated using high boiling point solvent TCB is the best as shown in Table 11.1.

The increase in the current density of devices fabricated by thiophene and TCB shows the clear evidence of self-organization of P3HT, which leads to higher hole mobility due to the electron delocalization via the π–π interaction. Interestingly, the external quantum efficiency (EQE) of the device fabricated using chloroform shows a steady decrease whereas the device fabricated using TCB retains the EQE up to 100 mW/cm^2 (Figure 11.12). The higher EQE of the device fabricated using TCB even at high light intensity is consistent with the higher hole mobility of P3HT as the poor charge carrier transport leads to high carrier density in the film, which results in enhanced bimolecular recombination and redistributes the electric field within the

TABLE 11.1
The Open-Circuit Voltage, V_{oc}, Short-Circuit Current Density, J_{sc}, Fill Factor, FF, and Power Conversion Efficiency, η, of Solar Cells under AM 1.5 Light Condition

Solvent	V_{oc}/V	J_{sc}/mA cm^{-2}	FF	η
Chloroform	0.66	−6.12	0.41	1.8%
Thiophene	0.60	−8.44	0.45	2.4%
TCB	0.62	−8.79	0.50	2.9%

Source: Sun, B.; Greenham, N. C., *Physical Chemistry Chemical Physics*, 2006, *8* (30), 3557–3560, Reproduced by permission of The Royal Society of Chemistry.

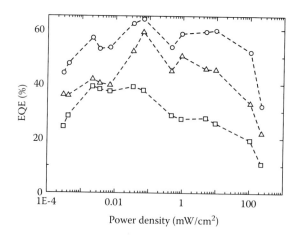

FIGURE 11.12 The EQE as a function of the light intensity for devices prepared by using TCB (○), thiophene (△), and chloroform (□). (Sun, B.; Greenham, N. C., *Physical Chemistry Chemical Physics*, 2006, *8* (30), 3557–3560, Reproduced by permission of The Royal Society of Chemistry.)

Inorganic–Organic Nanocomposites and Their Assemblies

device, thus suppressing the field-assisted charge separation. However, the focus of this work noted above is to achieve the higher crystallinity of P3HT, and the location of inorganic nanoparticles has not yet been considered.

11.3.3 Hybrid Nanofibrils

P-type/n-type integrated one-dimensional nanostructures are very promising structures for solar cells as they possess an extremely efficient donor–acceptor interface, a nanometer-scale phase separation, and an anisotropic continuous charge transport pathway for both electrons and holes. In this context, CdSe QD-decorated P3HT NWs were prepared by a two-step self-assembly process. P3HT NWs were first prepared in a poor solvent, followed by the deposition of CdSe QDs on the surface of P3HT NWs by noncovalent interactions between P3HT and CdSe QDs as illustrated in Figure 11.13.[49]

With this simple two-step assembly process, CdSe QD-decorated P3HT NWs can be prepared. The CdSe QDs were compactly deposited on the surface of P3HT NWs as clearly evidenced in TEM images (Figure 11.14). Such compact alignment of CdSe QDs along P3HT NWs may increase the electron mobility by forming the continuous route for electron transport.[50]

Notably, the measurements on photophysical properties also support the intimate contact between donors and acceptors (Figure 11.15). The maximum absorption peak of hybrid nanowires was shifted from 450 nm to 470 nm, reflecting a coordinate

FIGURE 11.13 Schematic illustration of the stepwise self-assembly of CdSe QD-decorated P3HT NWs. (Reprinted with permission from Xu, J. et al., *Macromolecular Rapid Communications*, 30, 1419–1423. Copyright 2009, American Chemical Society.)

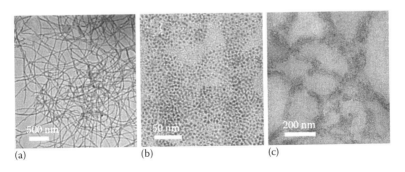

FIGURE 11.14 TEM images of (a) P3HT nanowires, (b) CdSe QDs, and (c) CdSe QDs/P3HT coaxial nanowires. (Reprinted with permission from Xu, J. et al., *Macromolecular Rapid Communications*, 30, 1419–1423. Copyright 2009, American Chemical Society.)

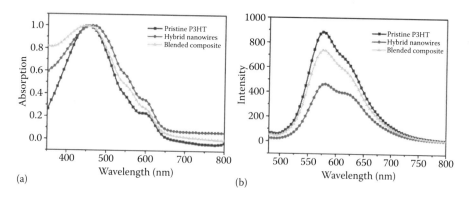

FIGURE 11.15 (a) Absorption spectra, and (b) emission spectra of pristine P3HT, simple blends of P3HT/CdSe QDs, and P3HT/CdSe QD coaxial nanowires. (Reprinted with permission from Xu, J. et al., *Macromolecular Rapid Communications, 30*, 1419–1423. Copyright 2009, American Chemical Society.)

interaction between CdSe QDs and P3HT due to the partial electron transfer from P3HT to CdSe QDs. The intimate contact of inorganic particles with P3HT facilitates the exciton dissociation and the charge generation, thus inducing the photoluminescence quenching. As a result, the efficiency of exciton dissociation and charge generation can be measured by monitoring the photoluminescence. The quenching in the coaxial CdSe QD-incorporated P3HT NWs was more obvious as compared to that in simply blended composites. The absorption spectra and emission spectra both signified the intimate contact of CdSe QDs and P3HT NWs.

It is noteworthy that although the investigation of photophysical properties can provide information on the binding characteristic, it is indirect evidence. In order to fully understand the binding structure of nanoprticle-decorated CP NWs (that is, CdS-decorated P3HT NWs), further studies were conducted using x-ray photoelectron spectroscopy (XPS) and time-resolved photolumninescence (PL) spectroscopy (Figure 11.16).[51] The S_{2p} peaks of XPS are precisely analyzed to identify the binding characteristic between CdS QDs and P3HT during the assembly. The XPS spectra of CdS-grafted P3HT NWs show the appearance of new intermediated peaks, which can be assigned to the C-S-Cd bond. This observation supports the formation of intimate contact between CdS QDs and P3HT. Indeed, the transient PL measurement also shows a shorter lifetime (that is, faster exciton dissociation) in the case of CdS QDs-grafted P3HT NWs compared to the simple mixtures of CdS QDs/P3HT NWs.

The aforementioned methods can achieve a one-dimensional coaxial structure at nanoscale via the coordination of thiophene units to QDs. However, the steric hindrance of the alkyl side chains limits the close packing of QDs. To solve this problem, P3HT was end-functionalized to accommodate the assembly of QDs as shown in Figure 11.17.[52] For example, the ligating moieties, such as thiols and phosphonic acids, were chosen as a functional group to modify polymer chain ends (that is, forming P3HT-SH and P3HT-PO_3H_2). Chain-end–functionalized P3HT were found to both assemble rapidly and form NWs. They can be served as templates for the organization of CdSe/P3HT nanofibrils. It

Inorganic–Organic Nanocomposites and Their Assemblies   323

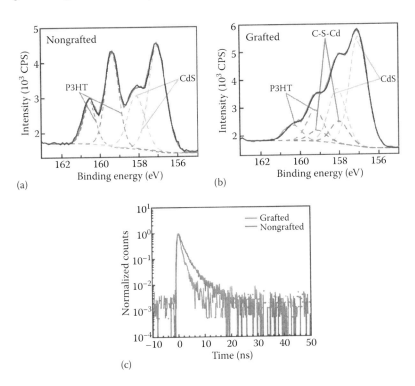

FIGURE 11.16 XPS spectra of S$_{2p}$ in (a) a simple mixture of CdS/P3HT NWs, and (b) CdS-grafted P3HT NWs. (c) Time-resolved photoluminescence spectra. (Reprinted with permission from Ren, S. et al., *Nano Letters*, 11, 3998–4002. Copyright 2011, American Chemical Society.)

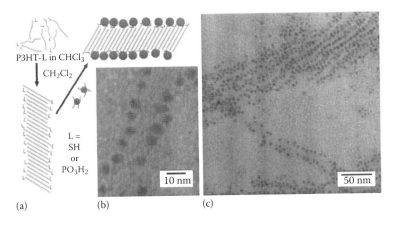

FIGURE 11.17 (a) Schematic representation of fibril formation from end-functionalized P3HT. (b, c) TEM images of CdSe QDs/P3HT NWs. (Pentzer, E. B.; Bokel, F. A.; Hayward, R. C.; Emrick, T.: *Advanced Materials*, 2012, 24 (17), 2254–2258. Copyright Wiley-VCH Verlag GmbH & Co. KGaA. Reproduced with permission.)

is worth noting that the p-type/n-type integrated one-dimensional nanostructures can be promising building materials for solar cells, as such structures maximize the interfacial area for charge dissociation, and ensure an efficient charge transport by forming a high crystalline structure along with the aligned QDs.

11.4 ALIGNMENT OF NANORODS

11.4.1 Introduction

Up to now, there has been much research focused on the interface between inorganic nanoparticle acceptors and conjugated-polymer donors to facilitate the exciton dissociation. The study of controlling the film morphology of photoactive layers should also be highlighted. Generally, the solar cell devices are fabricated using roll-to-roll or spin-casting processes to minimize the cost owing to both the solution processable properties of nanoparticles and conjugated polymers. During these processes, micro-sized phase separation usually occurs, and thus it is not easy to control the film morphology. For example, CdSe NRs lie parallel to the substrate and thus fail to maximize the efficient charge transport that benefits from the anisotropic structure of NRs. An ideal device geometry would consist of a bicontinuous pathway for both holes and electrons, thereby reducing the electron hopping processes. In this context, vertically oriented NRs spanning the full thickness of the film is considered as one of the ideal morphologies in hybrid photovoltaic cells.[53] Within this structure, the bandgap of NRs can be optimized by tuning the diameter, and the light absorption can also be tuned by the NR length. A vertically oriented assembly of inorganic NRs can be prepared by the following methods: (1) self-evaporative drying, and (2) external fields, such as magnetic field or electric field.[54–58]

11.4.2 Alignment of Nanorods

Basically, the total coupling energy between NRs is expressed as follows:[30]

$$V(ij) = V^{vdw}(ij) + V^{steric}(ij) + V^{D}(ij)$$

where $V^{vdw}(ij)$ is the van der Waals interaction energy, $V^{steric}(ij)$ is the energy from steric repulsion, and $V^{D}(ij)$ is the dipole–dipole potential energy between the ith and jth NRs. Although the van der Waals force between NRs is relatively weak during the self-assembly process in solvent, it becomes essential when the concentration of NRs increases. The van der Waals interaction combined with the dipole–dipole interaction would lead to the formation of assembly when they overcome the repulsive force. In this sense, it is of importance to control the distance of nanoparticles to manipulate forces.

Vertical assembly of NRs has been prepared via a solvent-assisted evaporation technique, in which a pinned edge of a drying droplet led to the capillary flow of materials from within the droplet volume to the pinning site. This causes NRs to form a smectic or nemectic superstructure. One more important force is the entropically driven depletion force regardless of its weak strength. The physical origin of the depletion force is coming from the overlap of the restricted volume of the NRs, and this increases the volume accessible to small particles (Figure 11.18a).[59] Consequently, the entropy increases and thus decreases the Gibbs free energy, resulting in an attractive interaction between NRs.

The depletion force between hydrophobic colloidal semiconducting NRs can be controlled by the use of additives, yielding vertically aligned NR arrays. The additive molecules that dissolved in NR solutions drive the solvent away from the surface of NRs and lead to the appearance of depletion force as illustrated in Figure 11.19.

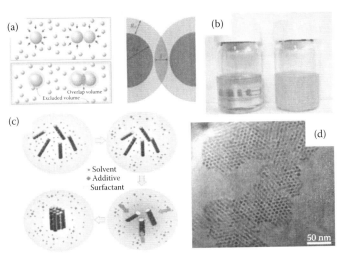

FIGURE 11.18 (a) Schematic illustration of the physical origin of depletion force. (b) Photographs of NR solution before and after the assembly by depletion force. (c) Sketch of the assembly mechanism. (d) TEM images of vertically aligned NRs. (Reprinted with permission from Baranov, D. et al., *Nano Letters*, 10, 743–749. Copyright 2010, American Chemical Society.)

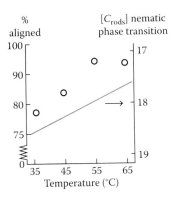

FIGURE 11.19 An elevated temperature increases the percentage of vertically aligned NRs (marked as O) and decreases the required NR concentration for an ordered phase transition (line). (Reprinted with permission from Baker, J.L. et al., *Nano Letters*, 10, 195–201. Copyright 2009, American Chemical Society.)

As a result, the depletion force induces the NR aggregation. In general, the required concentration of additives for the assembly decreases with the increased molecular weight. It was found that anisotropic NRs were preorganized in clusters of ordered particles directly in solution and followed by deposition of monolayer on substrates. Thus, to further understand the mechanism of alignment, several factors influencing the degree of assembly will be discussed in the following.

11.4.2.1 Temperature

The rotational and translational diffusion constants (D) depend on the temperature as expressed in the following:

$$D \propto T/\eta_s$$

where η_s is the shear viscosity. The shear viscosity decreases as the temperature (T) increases. As noted in the equation, the speed of self-assembly would be faster upon heating, which means that the system can have more time for lowering its free energy. Therefore, the increasing number of particularly aligned NRs at an elevated temperature indicates that the aligned state is thermodynamically favorable. However, even further higher temperature led to degradation of the aligned state as the entropic effects on alignment are against by enthalpic interaction (Figure 11.19).[56] The entropically driven ordering may become less important at a higher temperature.

11.4.2.2 Aspect Ratio

Although the depletion force becomes stronger between NRs with a higher aspect ratio, it was found that the low aspect ratio NRs aligned more rapidly than those with high aspect ratio. This is possibly because the kinetic effect outweighs the effect from the depletion force. For example, the diffusion constant for rotation, D^{rot}, can be given as

$$D^{rot} \propto \ln(L)/L^3$$

where L is the length of NRs. Hence, as the length of the NRs increases, the rotational diffusion constant decreases. The reduction in rotational diffusion constant makes the vertical assembly of NRs difficult. The experimental results also show that the increase in the aspect ratio decreases the degree of alignment (Figure 11.20). For NRs with the aspect ratio (AR) approaching 25, they cannot be aligned as supported by the D^{rot} calculation (that is, a decrease in the degree of alignment of ~97% when AR goes from 5 to 25). The kinetic consideration itself is not sufficient to conclude the general influence of AR on alignment. However, little changes in thermodynamics with the change of AR are expected.

11.4.2.3 Substrates

The interaction between substrates and NRs in solution can also influence the assembly of NRs. For example, the favorable interactions between NRs and substrates can lead to a parallel orientation of rods on the substrate rather than a vertical alignment to maximize the contact area, thereby minimizing the energy. At a certain threshold

Inorganic–Organic Nanocomposites and Their Assemblies

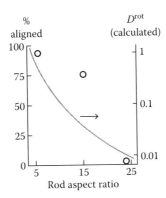

FIGURE 11.20 The degree of assembly (marked as O) of NRs depending on the AR of NRs. The line represents the calculated rotational diffusion constant. (Reprinted with permission from Baker, J.L. et al., *Nano Letters*, *10*, 195–201. Copyright 2009, American Chemical Society.)

(that is, weaker attractive interaction between NRs and substrates), a strong tendency of vertical alignment of NRs can be observed (Figure 11.21). Ligands on NRs are repelled by the substrate if the Hamaker constant of the substrate is greater than that of the solvent. In general, the NR-substrate interactions are relatively weak compared to other factors, enabling the alignment of rods on a wide range of substrates.

11.4.2.4 Concentration

The concentration of NRs determines the nearest neighbor distance in solution, which is of importance for the superlattice formation. For example, NRs were randomly distributed at a low concentration (that is, 8×10^{-7} mol/L), and optimal concentration (that is, $\sim 1 \times 10^{-6}$ mol/L) resulted in perpendicular orientation of rods (Figure 11.22).[60] This phenomenon could be explained by the energy concept. As described in Section 11.7.2, the dipole–dipole interactions that depend on the distance between NRs play an important role in vertical assembly. The optimal concentration ensures that the attractive forces (dipole–dipole interaction) can outweigh the repulsive forces, thereby resulting in assembly.

11.4.2.5 Solvent

The nature of a solvent, such as solubility, dielectric permittivity, and volatility, influences the assembly. The solubility directly affects the ability of dispersion of NRs, which usually have hydrophobic ligands. Random aggregation of NRs was observed in poor solvent as shown in Figure 11.23b. The dielectric constant of solvents is also a key factor for the assembly of NRs. It influences both the dipole–dipole and the Coulomb interactions, which would be screened in a solvent with high dielectric constant, thereby resulting in random distribution of NRs. Indeed, the assembly of NRs in a solvent with high dielectric constant, that is, chloroform ($\varepsilon_r = 5$), was not observed (Figure 11.23a) in spite of the excellent dispersion of NRs in it. The volatility of solvents directly determines the evaporation rate, which plays a very crucial role in the self-assembly of NRs. For example, cyclohexane has a comparable dielectric constant (that is, $\varepsilon_r = 2.02$) to that of toluene (that is, $\varepsilon_r = 2.34$), but the degree of

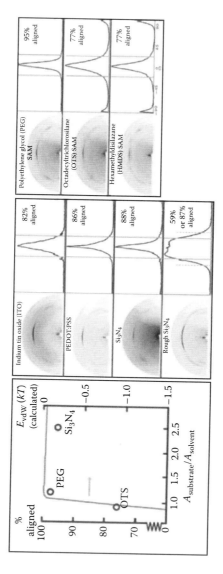

FIGURE 11.21 The degree of alignment (marked as O) depending on the Hamaker constant. The line represents the estimated van der Waals interaction between the NRs and the substrate. Diffraction patterns and corresponding orientation distribution functions for each substrate are shown on the right panels. (Reprinted with permission from Baker, J.L. et al., *Nano Letters*, 10, 195–201. Copyright 2009, American Chemical Society.)

Inorganic–Organic Nanocomposites and Their Assemblies

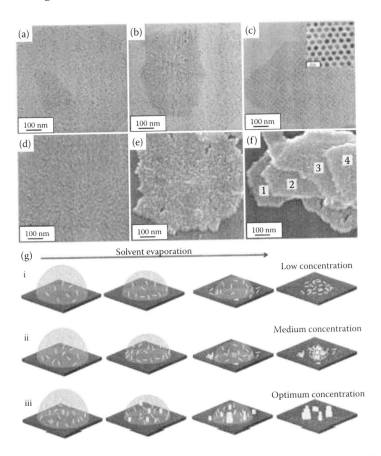

FIGURE 11.22 TEM images of CdS NR arrays with a concentration of (a) 8×10^{-7} mol/L, (b) 1.4×10^{-6} mol/L, (c) 2.1×10^{-6} mol/L, and (d) 3.2×10^{-6} mol/L. (e–f) SEM images showing the monolayer and multilayer of CdS NRs. (g) Schematic representation shows the dependence of the NR concentration on assembly. (Singh, A. et al., *Journal of Materials Chemistry*, 2012, 22 (4), 1562–1569. Reproduced by permission of The Royal Society of Chemistry.)

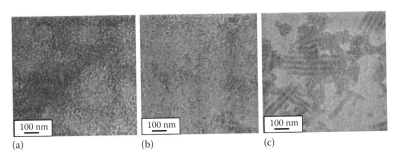

FIGURE 11.23 TEM images of CdSe NRs in (a) dichloromethane, (b) chloroform, and (c) cyclohexane. (Singh, A. et al., *Journal of Materials Chemistry*, 2012, 22 (4), 1562–1569. Reproduced by permission of The Royal Society of Chemistry.)

NR alignment prepared from cyclohexane is not good as that prepared from toluene due to the lower boiling point of cyclohexane as shown in Figure 11.23c.

11.4.2.6 Rate of Evaporation

The evaporation rate of solvents affects the degree of NR assembly as it controls the total time for the dipole–dipole attraction. The toluene NR solution with the optimized concentration with different drying time has been studied. The degree of vertical assembly indeed reduced with the increase in the evaporation rate as seen in Figure 11.24.

11.4.2.7 Surface Ligands

The chemical compositions of the surface passivation of NRs also play a significant role in the assembly of NRs. Surface modification would change the dispersibility of NRs in solvent, which influences the organization of NRs by controlling the time for dipole–dipole interactions between NRs. Moreover, the length of ligands is also of importance as it is directly related to the excluded volume for depletion force to occur as well as the repulsion force from the steric hindrance. When the excluded volume decreases, the region for depletion interactions will be reduced, thereby resulting in random orientation of NRs. For example, pyridine-capped NRs could not be aligned vertically, yet fatty alkyl chain-capped NRs could form vertical alignment as shown in Figure 11.25d. Even though the chain length of surfactant is similar (that is, MUA and TDPA), the degree of assembly can be totally different. For example, the assembly of 11-mercaptoundenoic acid (MUA)-capped NRs was

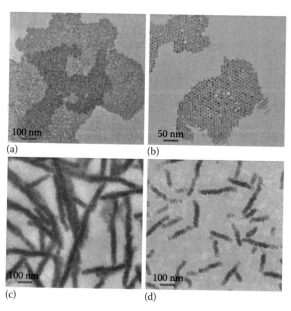

FIGURE 11.24 The effect of the evaporation rate; TME images of NRs with (a, c) the higher evaporation time (3–4 min for evaporation) and (b, d) slow evaporation time (7–8 min for evaporation). (Singh, A. et al., *Journal of Materials Chemistry*, 2012, 22 (4), 1562–1569. Reproduced by permission of The Royal Society of Chemistry.)

Inorganic–Organic Nanocomposites and Their Assemblies

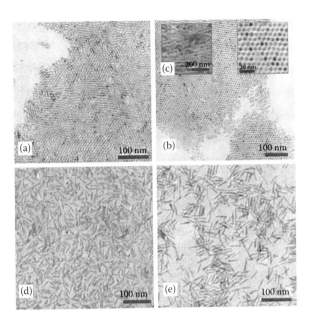

FIGURE 11.25 TEM images of TDPA-capped CdSe NR assembly in PVK (a) and P3HT (b) matrix. The inset (c) is an AFM image showing the fibrilar structure of P3HT at the surface of the composites. Ligands such as (d) pyridine and (e) MUA prevent the formation of ordered NR arrays. (Reprinted with permission from Modestino, M.A. et al., *Macromolecules*, 44, 7364–7371. Copyright 2011, American Chemical Society.)

not successful despite the fact that MUA-capped NRs have almost similar excluded volume with TDPA-capped NRs. This is because of the chemical nature (functionality) of passivation ligands. The electric repulsion between NRs would increase due to the carboxyl group of MUA, preventing the formation of NR arrays as shown in Figure 11.25e.[61]

11.4.2.8 External Fields

The upscaling of a vertically aligned NR monolayer is restricted in the case of solvent assisted self-assembly. In this regard, external fields are employed to facilitate the NR alignment. When the torque of the NRs induced by an electric field is greater than the thermal excitation energy (kT), anisotropic structures with a dipole moment can be aligned parallel along the electric field. The torque of NRs in the electric field can be expressed as $\mathbf{T} = \mathbf{P} \times \mathbf{E}$, where $\mathbf{P}$ and $\mathbf{E}$ are dipole moment and electric field, respectively. The calculated dipole moment of CdSe NRs was ~1450 D, and the electric field was 10^7 V/m. The strength of the torque applied on the NRs is then 4.872×10^{-20} Nm based on the assumption that the NRs and electric field are orthogonal.[62] This value is one order of magnitude larger than thermal energy at room temperature. Therefore, vertical alignment of NRs can be facilitated with the help of an electric field.

In general, the combination of a DC electric field and solvent-assisted evaporation can be used for assembly as illustrated in Figure 11.26.[62] The electric field will rotate

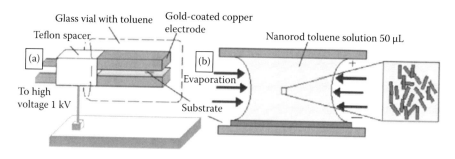

FIGURE 11.26 (a) Schematic of electric field-assisted assembly. (b) Magnified image. (Reprinted with permission from Ryan, K.M. et al., *Nano Letters*, 6, 1479–1482. Copyright 2006, American Chemical Society.)

NRs perpendicular to the substrate, and the evaporation of solvent confines NRs laterally and assemble them into an array.

Moreover, the electrophoresis method uses an electric field on charged particles in solution to cause electromigration of particles toward the oppositely biased electrode for deposition (Figure 11.27).[63] The vertical orientation of NRs (that is, CdS and CdSe) is caused by both inherent overall NR charge and its constituent dipole moment from its noncentrosymmetric wurtzite structure. The inherent charge in NRs causes the NRs to migrate to the oppositely biased electrode, and the dipole moment that arose from noncentrosymmetry leads to orientation of NRs parallel to the electric field, producing perpendicular assembly of NRs.

FIGURE 11.27 (a) Schematic of electrophoresis setup. (b) Photographs of ITO glass before and after CdS NRs deposition. (c) TEM image of vertically aligned CdS NRs. (d) A cross-section SEM image. (S. Ahmed, K. M. Ryan, *Chemical Communications*, 2009, 6421–6423. Reproduced by permission of The Royal Society of Chemistry.)

11.4.3 ALIGNMENT OF NRS IN THE POLYMER MATRIX

In order to fabricate photovoltaic devices utilizing a vertically aligned semiconducting NR assembly, it is necessary to incorporate electron donor materials, such as CPs in the film. To this end, several research groups have reported vertical alignment of NRs within polymer matrices. The vertically aligned NRs have been prepared in the presence of a polymer matrix such as P3HT, poly(9-vinylcarbazole) (PVK), and polystyrene (PS) (Figures 11.25a and c and 11.28a).[61] Vertical NR arrays were formed throughout the film regardless of the polymer morphology. This observation implies that polymer–NR interactions are relatively weak compared with other interactions, such as the NR–NR interaction, thus enabling perpendicular assembly of NRs over a variety of systems. However, NRs aligned vertically exclude polymers from the NR array, which is the limitation of this study, considering that hybrid solar cells need electron-donating materials to be covered on electron-accepting NRs to facilitate exciton dissociation as well as to maximize the charge transport characteristic though a bicontinuous pathway.

Electric fields were then applied to mediate the assembly of inorganic NRs in polymer matrices. The desired nanostructure would be vertically aligned CP-covered NRs to maximize the exciton dissociation though the increased interfacial area between donors and acceptors and the charge transport benefiting from the anisotropic structure. Several groups aligned NRs in the polymer matrix by applying an electric field. However, they observed corralled NRs due to the unfavorable polymer–ligand interaction (Figure 11.28).

The highly unfavorable interaction between polymer and ligands forces a phase separation between them to minimize the interfacial tension, leading to densely packed NRs.[64] When miscible ligands with the surrounding matrix were used, no aggregation was observed as shown in Figure 11.29. This fact demonstrates that the interfacial energy plays a significant role in corralling the NRs. However, even with the combination of an external field (that is, the electric field), vertical assembly of NRs covered by CPs has not yet been achieved. In this sense, CP-grafted semiconducting NRs have garnered much attention to be used for assembly in photovoltaic applications.

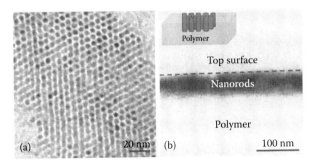

FIGURE 11.28 (a) TME image of the CdSe NR assembly in polystyrene. (b) The cross-section TEM view of P3HT/CdSe NR nanocomposites. (Reprinted with permission from Modestino, M.A. et al., *Macromolecules*, 44, 7364–7371. Copyright 2011, American Chemical Society.)

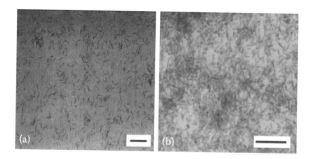

FIGURE 11.29 TEM images of (a) CdSe NRs capped with PEO in a PMMA matrix and (b) CdSe NRs capped with PS in the PMMA matrix. (Reprinted with permission from Gupta, S. et al., *Nano Letters*, 6, 2066–2069. Copyright 2006, American Chemical Society.)

11.5 CONCLUSION AND OUTLOOK

In this chapter, an overview is given on the methodologies in the interfacial engineering of organic–inorganic semiconductor materials based on the ligand exchange approach, the direct grafting technique that utilizes bifunctional ligands, and the direct nanoparticle growth method in the presence of CPs. The intimate contact between electron-donating CPs and electron-accepting inorganic nanoparticles can be successfully achieved by these techniques as confirmed by the time-resolved PL measurement. Although the charge separation can be facilitated due to their inherent intimate contact, the optimization of charge transport via the control over the morphology is of key importance. In order to maximize the charge carrier mobility of CPs, their crystallinity can be controlled by self-assembly via the utilization of unfavorable solvent or thermal treatment, yielding CP NWs. Moreover, inorganic nanoparticles are incorporated on the surface of CP NWs to create both n-type and p-type channels. An ideal device geometry is composed of a bicontinuous pathway for both holes and electrons with a maximized interfacial area between donors and acceptors, thereby reducing the electron hopping processes and electron–hole recombination. Vertical alignment of NRs has been studied to achieve these goals by a solvent evaporation process and/or by the application of external fields. However, the vertical alignment of NRs capped by CPs has not yet been obtained. In this regard, vertically assembled CP-grafted NR nanocomposites could possibly be promising materials as building blocks for photoactive layers in solar cells. The inherent intimate contact between CPs and NRs would facilitate the charge separation, and the bicontinuous geometry would ensure the efficient charge transport to the electrodes.

ACKNOWLEDGMENT

We gratefully acknowledge funding support from NSF (ECCS-1305087).

REFERENCES

1. Agranovich, V.; Benisty, H.; Weisbuch, C. *Solid State Communications* **1997**, *102*, 631.
2. Malyshev, V.; Moreno, P. *Physical Review B* **1995**, *51*, 14587.

3. Lunt, R. R.; Giebink, N. C.; Belak, A. A.; Benziger, J. B.; Forrest, S. R. *Journal of Applied Physics* **2009**, *105*, 053711.
4. Mikhnenko, O. V.; Cordella, F.; Sieval, A. B.; Hummelen, J. C.; Blom, P. W. M.; Loi, M. A. *The Journal of Physical Chemistry B* **2008**, *112*, 11601.
5. Yun-Yue, L.; Chun-Wei, C.; Chang, J.; Lin, T. Y.; Liu, I. S.; Wei-Fang, S. *Nanotechnology* **2006**, *17*, 1260.
6. Baldo, M. A.; Thompson, M. E.; Forrest, S. R. *Nature* **2000**, *403*, 750.
7. Campbell, I. H.; Smith, D. L.; Tretiak, S.; Martin, R. L.; Neef, C. J.; Ferraris, J. P. *Physical Review B* **2002**, *65*, 085210.
8. Wolber, P. K.; Hudson, B. S. *Biophysical Journal* **1979**, *28*, 197.
9. Kasha, M. *Radiation Research* **1963**, *20*, 55.
10. Menke, S. M.; Holmes, R. J. *Energy & Environmental Science* **2014**, *7*, 499.
11. Ahn, T.-S.; Wright, N.; Bardeen, C. J. *Chemical Physics Letters* **2007**, *446*, 43.
12. Peng, X.; Manna, L.; Yang, W.; Wickham, J.; Scher, E.; Kadavanich, A.; Alivisatos, A. P. *Nature* **2000**, *404*, 59.
13. Yu, W. W.; Wang, Y. A.; Peng, X. *Chemistry of Materials* **2003**, *15*, 4300.
14. Schaller, R. D.; Klimov, V. I. *Physical Review Letters* **2004**, *92*, 186601.
15. Ellingson, R. J.; Beard, M. C.; Johnson, J. C.; Yu, P.; Micic, O. I.; Nozik, A. J.; Shabaev, A.; Efros, A. L. *Nano Letters* **2005**, *5*, 865.
16. Vayssieres, L. *Advanced Materials* **2003**, *15*, 464.
17. Özgür, Ü.; Alivov, Y. I.; Liu, C.; Teke, A.; Reshchikov, M. A.; Doğan, S.; Avrutin, V.; Cho, S.-J.; Morkoç, H. *Journal of Applied Physics* **2005**, *98*, 041301.
18. Brandenburg, J. E.; Jin, X.; Kruszynska, M.; Ohland, J.; Kolny-Olesiak, J.; Riedel, I.; Borchert, H.; Parisi, J. *Journal of Applied Physics* **2011**, *110*, 064509.
19. Britt, J.; Ferekides, C. *Applied Physics Letters* **1993**, *62*, 2851.
20. Huynh, W. U.; Dittmer, J. J.; Teclemariam, N.; Milliron, D. J.; Alivisatos, A. P.; Barnham, K. W. J. *Physical Review B* **2003**, *67*, 115326.
21. Li, Z.; Gao, F.; Greenham, N. C.; McNeill, C. R. *Advanced Functional Materials* **2011**, *21*, 1419.
22. Huynh, W. U.; Dittmer, J. J.; Alivisatos, A. P. *Science* **2002**, *295*, 2425.
23. Kalyuzhny, G.; Murray, R. W. *The Journal of Physical Chemistry B* **2005**, *109*, 7012.
24. Puzder, A.; Williamson, A. J.; Zaitseva, N.; Galli, G.; Manna, L.; Alivisatos, A. P. *Nano Letters* **2004**, *4*, 2361.
25. Munro, A. M.; Jen-La Plante, I.; Ng, M. S.; Ginger, D. S. *The Journal of Physical Chemistry C* **2007**, *111*, 6220.
26. Janssen, R. A. J.; Nelson, J. *Advanced Materials* **2013**, *25*, 1847.
27. Reiss, P.; Couderc, E.; De Girolamo, J.; Pron, A. *Nanoscale* **2011**, *3*, 446.
28. Glatthaar, M.; Riede, M.; Keegan, N.; Sylvester-Hvid, K.; Zimmermann, B.; Niggemann, M.; Hinsch, A.; Gombert, A. *Solar Energy Materials and Solar Cells* **2007**, *91*, 390.
29. Heinemann, M. D.; von Maydell, K.; Zutz, F.; Kolny-Olesiak, J.; Borchert, H.; Riedel, I.; Parisi, J. *Advanced Functional Materials* **2009**, *19*, 3788.
30. Ji, X.; Copenhaver, D.; Sichmeller, C.; Peng, X. *Journal of the American Chemical Society* **2008**, *130*, 5726.
31. Milliron, D. J.; Alivisatos, A. P.; Pitois, C.; Edder, C.; Fréchet, J. M. J. *Advanced Materials* **2003**, *15*, 58.
32. Querner, C.; Benedetto, A.; Demadrille, R.; Rannou, P.; Reiss, P. *Chemistry of Materials* **2006**, *18*, 4817.
33. Zhao, L.; Lin, Z. *Advanced Materials* **2012**, *24*, 4353.
34. Zhang, Q.; Russell, T. P.; Emrick, T. *Chemistry of Materials* **2007**, *19*, 3712.
35. Xu, J.; Wang, J.; Mitchell, M.; Mukherjee, P.; Jeffries-El, M.; Petrich, J. W.; Lin, Z. *Journal of the American Chemical Society* **2007**, *129*, 12828.

36. Zhao, L.; Pang, X.; Adhikary, R.; Petrich, J. W.; Jeffries-El, M.; Lin, Z. *Advanced Materials* **2011**, *23*, 2844.
37. Jung, J.; Pang, X.; Feng, C.; Lin, Z. *Langmuir* **2013**, *29*, 8086.
38. Liao, H.-C.; Chen, S.-Y.; Liu, D.-M. *Macromolecules* **2009**, *42*, 6558.
39. Dayal, S.; Kopidakis, N.; Olson, D. C.; Ginley, D. S.; Rumbles, G. *Journal of the American Chemical Society* **2009**, *131*, 17726.
40. Günes, S.; Sariciftci, N. S. *Inorganica Chimica Acta* **2008**, *361*, 581.
41. Wright, M.; Uddin, A. *Solar Energy Materials and Solar Cells* **2012**, *107*, 87.
42. von Hauff, E.; Dyakonov, V.; Parisi, J. *Solar Energy Materials and Solar Cells* **2005**, *87*, 149.
43. Goh, C.; Kline, R. J.; McGehee, M. D.; Kadnikova, E. N.; Fréchet, J. M. J. *Applied Physics Letters* **2005**, *86*, 122110.
44. Schafferhans, J.; Baumann, A.; Wagenpfahl, A.; Deibel, C.; Dyakonov, V. *Organic Electronics* **2010**, *11*, 1693.
45. Assadi, A.; Svensson, C.; Willander, M.; Inganäs, O. *Applied Physics Letters* **1988**, *53*, 195.
46. Kline, R. J.; McGehee, M. D.; Kadnikova, E. N.; Liu, J.; Fréchet, J. M. J. *Advanced Materials* **2003**, *15*, 1519.
47. Ihn, K. J.; Moulton, J.; Smith, P. *Journal of Polymer Science Part B: Polymer Physics* **1993**, *31*, 735.
48. Sun, B.; Greenham, N. C. *Physical Chemistry Chemical Physics* **2006**, *8*, 3557.
49. Xu, J.; Hu, J.; Liu, X.; Qiu, X.; Wei, Z. *Macromolecular Rapid Communications* **2009**, *30*, 1419.
50. Nie, Z.; Petukhova, A.; Kumacheva, E. *Nature Nanotechnology* **2010**, *5*, 15.
51. Ren, S.; Chang, L.-Y.; Lim, S.-K.; Zhao, J.; Smith, M.; Zhao, N.; Bulović, V.; Bawendi, M.; Gradečak, S. *Nano Letters* **2011**, *11*, 3998.
52. Pentzer, E. B.; Bokel, F. A.; Hayward, R. C.; Emrick, T. *Advanced Materials* **2012**, *24*, 2254.
54. Hu, Z.; Fischbein, M. D.; Querner, C.; Drndić, M. *Nano Letters* **2006**, *6*, 2585.
53. Ma, W.; Yang, C.; Gong, X.; Lee, K.; Heeger, A. J. *Advanced Functional Materials* **2005**, *15*, 1617.
55. Artemyev, M.; Möller, B.; Woggon, U. *Nano Letters* **2003**, *3*, 509.
56. Baker, J. L.; Widmer-Cooper, A.; Toney, M. F.; Geissler, P. L.; Alivisatos, A. P. *Nano Letters* **2009**, *10*, 195.
57. Wang, W.; Summers, C. J.; Wang, Z. L. *Nano Letters* **2004**, *4*, 423.
58. Ahmed, W.; Kooij, E. S.; van Silfhout, A.; Poelsema, B. *Nano Letters* **2009**, *9*, 3786.
59. Baranov, D.; Fiore, A.; van Huis, M.; Giannini, C.; Falqui, A.; Lafont, U.; Zandbergen, H.; Zanella, M.; Cingolani, R.; Manna, L. *Nano Letters* **2010**, *10*, 743.
60. Singh, A.; Gunning, R. D.; Ahmed, S.; Barrett, C. A.; English, N. J.; Garate, J.-A.; Ryan, K. M. *Journal of Materials Chemistry* **2012**, *22*, 1562.
61. Modestino, M. A.; Chan, E. R.; Hexemer, A.; Urban, J. J.; Segalman, R. A. *Macromolecules* **2011**, *44*, 7364.
62. Ryan, K. M.; Mastroianni, A.; Stancil, K. A.; Liu, H.; Alivisatos, A. P. *Nano Letters* **2006**, *6*, 1479.
63. Ahmed, S.; Ryan, K. M. *Chemical Communications* **2009**, *42*, 6421.
64. Gupta, S.; Zhang, Q.; Emrick, T.; Russell, T. P. *Nano Letters* **2006**, *6*, 2066.

12 Organic Tandem Solar Cells*

Ning Li, Tayebeh Ameri, and Christoph J. Brabec

CONTENTS

12.1 Introduction 337
12.2 Fundamental Limitations..340
    12.2.1 Single-Junction Solar Cells..340
    12.2.2 Tandem Solar Cells..342
12.3 Working Principles of the Intermediate Layers..346
12.4 Optical Simulations of the Tandem Solar Cells..349
12.5 Review of Experimental Results 352
    12.5.1 Highly Efficient Tandem Organic Solar Cells Based on Novel Materials...352
        12.5.1.1 Polymer-Based Tandem Organic Solar Cells.....................352
        12.5.1.2 Small Molecule-Based Tandem Organic Solar Cells 360
    12.5.2 Highly Efficient Tandem Organic Solar Cells Based on Various Structures..363
        12.5.2.1 Inorganic–Organic Hybrid Tandem Solar Cells.................363
        12.5.2.2 Semitransparent Tandem Organic Solar Cells....................364
12.6 Tandem Solar Cell Measurement...365
12.7 Summary and Outlook 370
References...371

12.1 INTRODUCTION

Addressing climate change, pollution, and energy insecurity problems all at once requires major changes in our energy infrastructure. Over the past decade, a number of studies have proposed large-scale renewable energy plans, mainly based on wind, water, and sunlight resources.[1–4] As shown in Figure 12.1,[5] the solar energy resource potentially dwarfs all other renewable and fossil-based energy resources combined. The yearly sustainable renewable supply of solar energy received by the emerged continents alone is more than 30 times larger than the total planetary reserves of coal and 1500 times larger than the current planetary energy consumption.[1] Therefore,

---

* This chapter is based on the published review paper: Tayebeh Ameri, Ning Li, and Christoph J. Brabec. 2013. Highly efficient organic tandem solar cells: A follow up review. *Energy Environ. Sci.* 6:2390–2413. Reproduced by permission of The Royal Society of Chemistry.

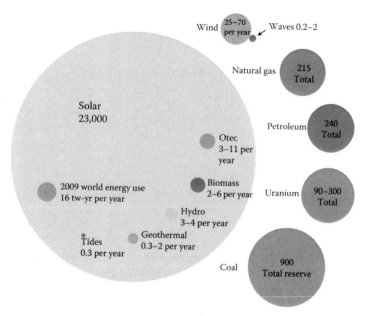

FIGURE 12.1 Comparing finite and renewable planetary energy reserves (Terawatt years). Total recoverable reserves are shown for the finite resources. Yearly potential is shown for the renewables. (From Perez, R., and M. Perez, *The IEA SHC Solar Update* 50:2–3, 2009.)

photovoltaic (PV) technology is the object of steadily growing interest from both academic and industrial protagonists.

After the worldwide PV market more than doubled in 2010, the market grew again by almost 30% in 2011 despite difficult economic conditions. In 2011, the PV industry production reached a worldwide production volume of around 35 GW, and another moderate increase is expected for 2012. Yearly growth rates over the last decade were, on average, more than 40%, which makes the PV industry one of the fastest growing industries at present.[6,7] The most rapid growth in annual production over the last five years could be observed in Asia, where China and Taiwan together now account for almost 65% of worldwide production (Figure 12.2a). With a cumulative installed capacity of more than 66 GW, the European Union is leading in PV installations with two thirds of the total worldwide almost 100 GW of solar PV electricity generation capacity at the end of 2012 (Figure 12.2b). The wafer-based silicon solar cells are still the main technology and have around 85% of the market share.[6] The challenge facing the PV industry is cost-effectiveness through much lower embodied energy. Although the cost of PV systems, especially in the last two years, has significantly reduced, solar power in comparison with fossil energy sources is still expensive. To bridge this difference in costs and make solar power a primary energy source in our society, the solar industry is working on the consistent further development of existing technologies and on the exploration of new solar technologies.

Organic Tandem Solar Cells

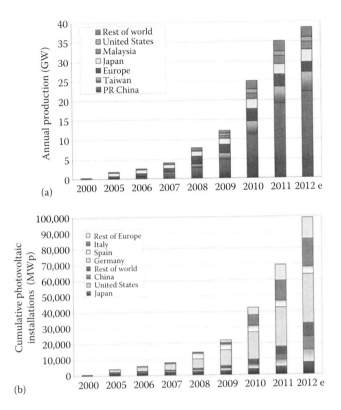

FIGURE 12.2 (a) World PV cell/module production from 2000 to 2012. (b) Cumulative PV installations from 2000 to 2012. (From Jäger-Waldau, A. PV Status Report 2012.)

Among all the alternative technologies, the thin-film approach deserves to gain more attention and is likely to represent about 26% of the overall production by 2015.[6] All the attractive and beneficial features of organic materials, such as their strong absorption coefficient, ease of processing, mechanical flexibility, and ability to tailor the bandgap, have brought them into the focus of attention for thin-film PV applications. Organic solar cells (OSCs) represent a new technology that could lead to the most significant cost reduction in the mid/long term.[8–11] Various printing and coating techniques, such as inkjet, slot die, screen, gravure, spray, and others, are established and already demonstrated for OPV production. The attractive feature of these printing and coating methods is their high production speed of up to several meters per second. With a single production line some 100,000 m² of film per day can be produced, which, at an efficiency of 5% (that is, 50 W m^{-2}), corresponds to a daily capacity of 5 MW. Therefore, a single printing or coating facility could produce modules with the production energy volume of over 1 GW per year.[12,13] Obviously, the printing and coating processes are orders of magnitude more productive than the currently largest production facilities for silicon-based PVs. In addition, the investment in the printing equipment is many times cheaper than for conventional inorganic solar technologies. This illustrates the potential of printed solar power, which can also be easily set up at different locations.

Recently, a large number of OSCs with power conversion efficiencies (PCEs) beyond 6% and up to 10% were reported, making this topic, more than ever, promising.[14–16] However, there is still significant room with respect to research and development necessary to make high-performance OSCs ready for the market. In this regard, various strategies are currently in the exploration phase at different levels of maturity. Among them, the well-known concept of tandem solar cells is one of the most promising approaches to tackle the main losses in single-junction OPV devices and improve the device performance.

Based on our previous work,[17] we aim to review the most important and novel developments that have been recently reported on organic tandem solar cells. In the first part, we introduce some brief theoretical considerations addressing the potential of single and tandem solar cells. The working principles and importance of the intermediate layer are explained as an important subject for fully functional tandem solar cells. Furthermore, the benefits of optical simulations to verify the ultimate efficiency potential of novel materials are discussed. Then, we present and discuss the experimental achievements with considerably high performance recently reported in the literature. Finally, we discuss the intricacies of the precise performance measurement of bulk heterojunction (BHJ) organic tandem solar cells based on complementary absorber materials and describe the applicable method in this regard.

12.2 FUNDAMENTAL LIMITATIONS

Understanding the fundamental limitations of PVs will lead to the achievement of a systematic approach toward further developments of the solar cell technologies. Although the theoretical considerations of PVs have already been described in detail in our previous review, their importance encourages us to address them briefly here as well.

12.2.1 Single-Junction Solar Cells

The fundamental limitations of the energy conversion efficiency of a solar cell are mainly based on the thermodynamical losses. In PVs, only the photons having energy larger than the bandgap of a photoactive material can be absorbed and contribute to the energy conversion. On the other hand, thermalization of the hot charge carriers is another important contributor to losses. Considering a light source of AM 1.5 G, the maximum efficiency (η) achievable with a material having a bandgap energy, E_g, is calculated by

$$\eta(E_g) = J_{sc}(E_g) \times V_{oc}(E_g) \times FF \tag{12.1}$$

where FF is the fill factor (ideally equal to 1), V_{oc} is the open circuit voltage, and J_{sc} is the short-circuit current density, which is calculated by

$$J_{sc}(E_g) = e \int_{E_g}^{\infty} I_{ph}(E) \times \text{EQE}(E) \times dE \quad (12.2)$$

where $I_{ph}(E)$ given in s^{-1} m^{-2} J^{-1} contains the spectrum of the light source, and EQE is the external quantum efficiency (ideally 1 for a photon absorbed). According to the basic investigation of Shockley and Queisser, and considering the spectral losses alone, a solar cell has a peak theoretical efficiency of 48% for a material with E_g close to 1.1 eV.[18] Materials with a bandgap smaller than this suffer from reduction in the open circuit voltage (V_{oc}), and those with a bandgap larger than 1.1 eV are limited by the decrease of the short-circuit current (J_{sc}). Including the effects of blackbody radiation and the radiative/nonradiative recombination of charge carriers, the theoretical maximum efficiency reduces down to 30.1% for a single device under 1 sun illumination.[18] However, this basic calculation describes the fundamental limitations of a solar cell based on inorganic semiconductors. OSCs face further restrictions due to the natural specifications of organic semiconductors.

In organic materials with a small dielectric constant, absorption is governed by Frenkel excitons. The binding energy of Frenkel excitons is around 0.3 eV, and the exciton lifetime is typically in the order of ns.[19,20] Therefore, to achieve substantial energy conversion efficiencies, these excited electron–hole pairs need to be dissociated into free charges with a high yield. Excitons can be dissociated at interfaces of materials with different electron affinities. Blending conjugated polymers (donor) with high electron affinity molecules (acceptor), such as fullerene derivatives ($PC_{60}BM$ and $PC_{70}BM$), has proven to be an efficient way for rapid exciton dissociation. It is shown experimentally that a minimum offset of 0.3 eV either between the donor and acceptor's lowest unoccupied molecular orbital (LUMO) or highest occupied molecular orbital (HOMO) levels is required to achieve an efficient charge transfer between the two components.[21] Correspondingly, the V_{oc} is limited by the difference between the acceptor LUMO and the donor HOMO levels, which also defines the built-in field (V_{bi}). Considering a contact loss of another 0.3 eV,[22–24] the V_{oc} obeys the following empirical equation:[25]

$$V_{oc} = \frac{1}{e}\left(\left|E_{HOMO}^{Donor}\right| - \left|E_{LUMO}^{Acceptor}\right|\right) - 0.3 \quad (12.3)$$

For single-junction OSCs with a donor having a bandgap energy of 1.5 eV, a practical efficiency of nearly 11% can be expected based on the following assumptions: a 0.6 eV loss in V_{oc} (accounting for the minimum required LUMO level offset and contact losses), taking into account absorption/internal quantum efficiency (IQE) limitations (assuming EQE = 65%), and charge carrier transport losses (assuming FF = 65%). In early April 2011, a new efficiency record of >10% from a solution-processed small molecule was reported by Mitsubishi Chemical and certified by Newport.[26] This breakthrough did narrow significantly the performance gap between OPV technology and competitive thin film PV technologies as shown in Figure 12.3.

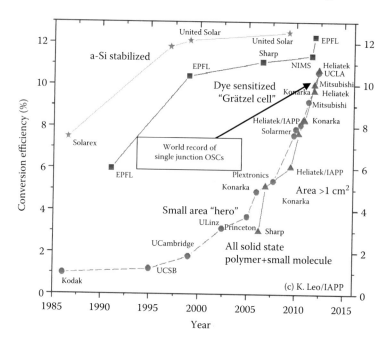

FIGURE 12.3 Comparison of the world records between thin-film PV technologies of a-Si (star), dye-sensitized (square), and OSCs (circle and triangle) from 1985 to present. (From http://www.orgworld.de.)

Figure 12.4a summarizes the efficiency of donor–acceptor organic solar cells versus the bandgap energy of the donor and the LUMO offset between the two materials, assuming an EQE and FF of 65%.[25] This model can be used as a guideline for the material selection and material development for BHJ solar cells. Obviously, the LUMO offset of the materials has an even more pronounced influence on device efficiency than the bandgap of the absorber. To go significantly beyond 10% efficiency, both EQE and FF need to be improved. This would shift the efficiencies shown in Figure 12.4a to higher values but does not change the shape of the contour plot.

12.2.2 Tandem Solar Cells

The performance limitation of single-junction devices is released by employing tandem devices. Tandem or multijunction devices simultaneously tackle absorption and thermalization losses by absorbing the higher energy photons in a wide bandgap cell (higher voltage and lower photocurrent) and the lower energy photons in a smaller bandgap cell (lower voltage and higher photocurrent). The detailed balance limit of the performance of a tandem structure was investigated by De Vos in 1980.[27] The optimal combination of bandgaps for multijunction structures reaches 42% to 53% efficiencies by increasing the number of stacked cells from two to four subcells. Currently, a device based on InGaP/GaAs/InGaAs with an efficiency of

Organic Tandem Solar Cells

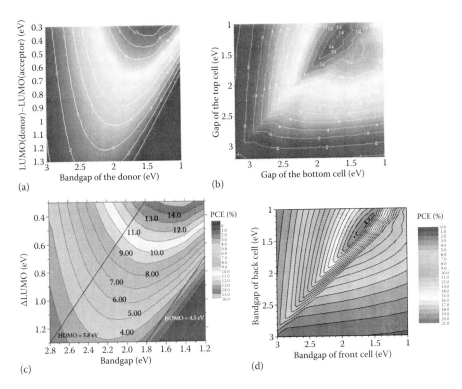

FIGURE 12.4 (a) Efficiency of a donor–acceptor OSC versus the bandgap energy of the donor, and the LUMO offset between the two materials. (Scharber, M. C., D. Mühlbacher, M. Koppe et al.: Design rules for donors in bulk-heterojunction solar cells—Towards 10% energy-conversion efficiency. *Advanced Materials*, 2006, 18, 6, 789–794. Copyright Wiley-VCH Verlag GmbH & Co. KGaA. Reproduced with permission.) (b) Efficiency of a donor–acceptor tandem OSC made of two subcells stacked in series versus the bandgap energy of the top and bottom donors; both top and bottom donor–acceptor couples are assumed to have a LUMO offset of 0.3 eV. (Dennler, G., M. C. Scharber, T. Ameri et al.: Design rules for donors in bulk-heterojunction tandem solar cells-towards 15% energy-conversion efficiency. *Advanced Materials*, 2008, 20, 3, 579–583. Copyright Wiley-VCH Verlag GmbH & Co. KGaA. Reproduced with permission.) Updated efficiencies predicted for (c) single-junction. (Reprinted from *Progress in Polymer Science*, 38, 12, Scharber, M., N. Serdar Sariciftci, Efficiency of bulk-heterojunction organic solar cells, 1929–1940. Copyright 2013, with permission from Elsevier.) and (d) tandem solar cells based on EQE = 80 and FF = 0.75. (Li, N., D. Baran, G. D. Spyropoulos et al.: Environmentally printing efficient organic tandem solar cells with high fill factors: A guideline towards 20% power conversion efficiency. *Advanced Energy Materials*, doi: 10.1002/aenm.201400084, 2014. Copyright Wiley-VCH Verlag GmbH & Co. KGaA. Reproduced with permission.)

37.7% ± 1.2%, and an active area of 1.047 cm², possesses the certified world record of multijunction solar cells under AM 1.5 G.[26]

In an approach similar to the one applied for single-junction cells, Dennler et al.[28] investigated the maximum efficiency possible for organic tandem solar cells. In this study, both top and bottom donor–acceptor couples were assumed to have a LUMO offset of 0.3 eV, and the LUMOs of the first donor (ELUMO1) and the second donor (ELUMO2) were fixed to −4 eV, which is the optimum value for BHJ composites with $PC_{60}BM$ as the acceptor. An ideal fully transparent intermediate layer placed between the subcells was assumed to ensure a loss-free recombination of charge carriers in the intermediate layer. The V_{oc} of the tandem cell is then calculated as the sum of the respective V_{oc} of both subcells.

$$V_{ocTandem} = V_{ocBottom} + V_{ocTop} \qquad (12.4)$$

An EQE of 65% and an IQE of 85% were considered constant over the absorption region of both subcells. As a result, the J_{sc} of bottom and top subcells is calculated as

$$J_{scBottom}(E_{g,1}) = e \int_{E_{g,1}}^{\infty} I_{ph}(E) \times EQE_1(E) \times (1 - Mirror_{loss}) \times dE \qquad (12.5)$$

$$J_{scBottom}(E_{g,2}) = e \int_{E_{g,2}}^{\infty} I_{ph}(E) \times EQE_2(E) \times \left[1 - \frac{EQE_1(E)}{IQE_1(E)} \times (1 - Mirror_{loss})\right] \times dE \qquad (12.6)$$

where $E_{g,1}$ is the bandgap energy of the donor used in the bottom cell, $E_{g,2}$ the bandgap energy of the donor used in the top cell, I_{ph} the flux of photons in AM 1.5 G, and EQE_1/IQE_1 absorption of the bottom cell. $Mirror_{loss}$ represents the loss of absorption in the cell implemented in the tandem versus the standalone single cell that would have a perfect mirror as a top electrode. This loss is induced by the absence of significant reflection at the interface between the bottom cell and the intermediate layer. Depending on the thickness of the bottom cell, these losses were estimated to be about 15%, on average, according to detailed optical simulations based on the transfer matrix formalism (TMF).[28] Tandem operation requires the current densities of both subcells to be matched at the maximum power point. Assuming further that the FFs of the single subcells are identical, as well as their shunt resistances, Kirchoff's law predicts that the J_{sc} of a tandem cell is equal to the smallest J_{sc} of the single cells.

$$J_{scTandem} = Min[J_{scBottom}, J_{scTop}] \qquad (12.7)$$

Based on this consideration, the maximum efficiency of an organic tandem solar cell can be calculated as a function of the bandgaps as presented in Figure 12.4b. This calculation suggests that tandem cells have a 30% higher efficiency potential as compared to single-junction devices. Figure 12.4b shows that a maximum efficiency of up to 15% can be achieved by combining a bottom donor having a bandgap of 1.6 eV with a top donor having a bandgap of 1.3 eV. Under the given assumptions, the most efficient tandem cells require materials with a bandgap energy difference of only 0.3 eV.

Organic Tandem Solar Cells

Owing to the rapid development of novel active materials and device optimization methods, state-of-the-art single-junction solar cells achieved more efficient PV parameters. Recently, Scharber et al.[29] updated the efficiency prediction for single-junction OSCs based on EQE = 0.80 and FF = 0.75. According to the updated prediction, a maximum PCE of ~15%, as shown in Figure 12.4c, is achievable for single-junction solar cells based on a donor material with a bandgap of 1.45 eV. Thus, the theoretical efficiency of organic tandem solar cells should also be correspondingly improved. Based on the updated assumptions (EQE = 0.80 and FF = 0.75), as shown in Figure 12.4d, a maximum PCE of 21% is suggested for organic tandem solar cells incorporating a bottom cell with a bandgap of ~1.6 eV and a top cell with a bandgap of ~1.2 eV.[30]

In previous studies,[28,31–33] the narrow absorption window, characteristic of organic materials, is not considered. When calculating the potential of organic (single and multijunction) solar cells, Minnaert and Veelaert[34] investigated the influence of different spectral absorption widths on the efficiency potential of organic tandem and triple-junction solar cells. According to their simplified model, the efficiency decreases rapidly as soon as one subcell has a low absorption window. The maximum efficiency of an organic tandem cell as a function of the absorber's spectral width of the subcells is presented in Figure 12.5. Interestingly, this study suggests that it would not pay off to try developing organic materials with an absorption window broader than 400 nm. More than 90% of the absolute maximum efficiency can be achieved by having absorbers with a spectral width of 400 nm for both subcells. In addition, the optimum bandgap of the cells shifts toward higher energies for narrower absorption windows. Similar conclusions were drawn for the efficiency potential of triple-junction solar cells as a function of the different absorption windows of the three subcells.

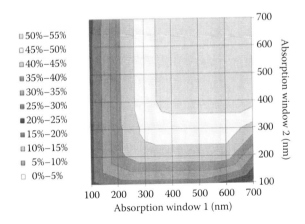

FIGURE 12.5 The maximum efficiency in the ideal scenario for an organic tandem cell as a function of the absorption windows of the subcells. "Absorption window 1" refers to the bottom subcell with the highest absorber bandgap, directed at the sun. (From Minnaert, B., and P. Veelaert. 2012. Guidelines for the bandgap combinations and absorption windows for organic tandem and triple-junction solar cells. *Materials* 5 (10):1933–1953.)

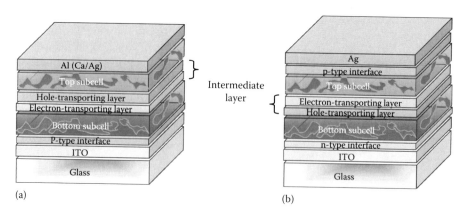

FIGURE 12.6 Schematic representation of an organic tandem device comprised of two subcells having different, complementary absorption spectra in a normal (a) and an inverted configuration (b). (Ameri, T., N. Li, and C. J. Brabec, 2013, Highly efficient organic tandem solar cells: A follow up review, *Energy Environ. Sci.*, 6, 2390–2413. Reproduced by permission of The Royal Society of Chemistry.)

Figure 12.6 shows the most frequently reported configurations of organic tandem solar cells. Indeed, the two cells involved in the device can be connected either in series (two-terminal) or in parallel (three-terminal), depending on the nature of the intermediate layer, and on the way, the intermediate layer and the two electrodes are connected. In the vast majority of reports, however, the series connection is used. In this case, the device can be fabricated in the conventional (or normal) architecture or in the inverted one[35] as depicted in Figure 12.6a and b, respectively. The intermediate layer has to ensure the alignment of the quasi-Fermi level of the acceptor of the bottom cell with the quasi-Fermi level of the donor of the top cell (or vice versa for an inverted architecture). Following the same principle, an infinite number of devices can be theoretically piled up this way.

12.3 WORKING PRINCIPLES OF THE INTERMEDIATE LAYERS

In the context of photovoltaic operation, free charge carriers generated in the active layer are selectively extracted through an electron-transporting layer (ETL) at the cathode or a hole-transporting layer (HTL) at the anode. Depending on the tandem architecture, a semitransparent electrode (in parallel connection) or a semitransparent recombination layer consisting of an ETL and HTL (in series connection) is required to connect the two subcells.

Owing to the lack of a printed semitransparent electrode, highly efficient parallel-connected tandem devices are more difficult to realize than the series-connected ones.[17,36,37] Recently, several solution-processed semitransparent electrodes, such as highly conductive poly(3,4-ethylenedioxythiophene):poly(styrenesulfonate) (PEDOT:PSS)[38,39] and silver-nanowires (AgNWs),[40–42] have been reported to substitute ITO, exhibiting promising electrical and optical properties.

Different from the parallel-connected tandem device, the intermediate layer (IML) of the series-connected tandem device functions as a recombination layer

Organic Tandem Solar Cells

between the electrons and holes selectively collected from different subcells. In this case, the energetic diagram of the device can be represented as in Figure 12.7. When compared to the complex semitransparent electrode for the parallel-connected device, the solution-processed recombination layer for the series-connected device is straightforward to engineer under the following requirements:

- The IML should be highly transparent to minimize the optical losses.
- A quasi-ohmic contact should be formed between the ETL and HTL to ease bipolar recombination.
- The recombination of the charge carriers in the IML needs to be balanced to maximize the short-circuit current.
- The processing of the IML should match the requirements for mass production, that is, solution-processing, low-temperature treatment, thickness requirement, etc.
- The IML should be robust enough to protect the underlying active layer from damage during processing of the upper layers.
- The IML should be environmentally stable to enhance the reliability and lifetime of the tandem device.
- The optical spacer effect of the IML on the light propagation inside the tandem device should be considered for further optimization.

In the past several years, multiple efficient IMLs were developed for organic tandem solar cells.[17] Due to the lack of conductivity, ultrathin thermally evaporated metal layers were frequently evaporated between ETL and HTL to form an

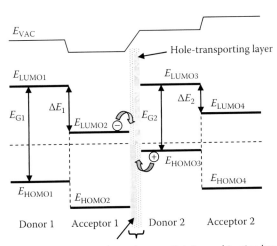

FIGURE 12.7 Simplified band diagram of a tandem cell made of two subcells connected in series via a recombination layer. (Ameri, T., N. Li, and C. J. Brabec, 2013, Highly efficient organic tandem solar cells: A follow up review, *Energy Environ. Sci.*, 6, 2390–2413. Reproduced by permission of The Royal Society of Chemistry.)

equivalent ohmic contact. An IML with nonohmic contact properties typically has s-shaped J–V characteristics.[43–50]

However, to keep the tandem architecture as simple as possible, a directly formed ohmic contact between the ETL and HTL is preferred. Most fundamental solution-processed IMLs are based on combinations of metal oxides (ETL) with a highly conductive PEDOT:PSS (HTL), such as TiOx/PEDOT:PSS (PH500),[51,52] ZnO/PEDOT:PSS (PH500),[53] (Orgacon),[54,55] or (Clevios F CPP105D),[56] etc. In these combinations, the highly conductive PEDOT:PSS plays a crucial role in forming the quasi-ohmic contact. We have reported on an efficient IML consisting of aluminum-doped ZnO (AZO) and low-conductive PEDOT:PSS Al4083.[57] Solution-processed ZnO is not well defined in terms of its electrical and semiconducting properties (density of states and density of charge carriers) and may differ for various processes and routes. Due to the promising electrical properties of AZO versus ZnO, the quasi-ohmic contact that formed between the ETL and HTL is not limited by the conductivity of PEDOT:PSS. Consequently, the thickness of this intermediate layer could be extended from ~50 to 160 nm without compromising the efficiency of the tandem device.

However, owing to the thermal-annealing requirements of most solution-processed IMLs, efficient tandem solar cells reported until now were mainly based on a thermally stable donor material for the bottom cell, such as poly(3-hexylthiophene-2,5-diyl) (P3HT).[53] Very recently, we demonstrated solution-processed, efficient IMLs that were deposited by doctor-blading in air and treated at fairly low temperatures (≤80°C) and still exhibited promising reliability and reproducibility.[58–60] These IMLs facilitate the use of novel and high-efficiency donor materials for the bottom cell in tandem architecture as well as the use of flexible substrates for roll-to-roll mass production.

As a novel approach, applications of CVD-grown graphene[61] as well as solution-processed graphene oxide[62] in the IML of organic tandem solar cells were reported in 2011. Tung et al. reported an IML based on water-soluble graphene oxide (GO)/PEDOT:PSS gel and ZnO for tandem solar cells in a normal architecture by a direct adhesive lamination process.[62] Later, the authors were able to further modify the properties of the GO-based interfaces by adding a small amount of single-walled carbon nanotubes (SWCNTs) into GO, which did not affect their solution processability but improved the vertical resistance of the resulting thin film. This allowed the use of thicker 3–4 nm GO thin films rather than, for example, 1–2 monolayers.[63] Based on a rather similar fabrication procedure, GO–SWCNT composites in combination with ZnO were employed as the IML in inverted as well as normal architecture tandem solar cells.[64]

Moreover, Zhou et al.[65] demonstrated that ethoxylated polyethylenimine (PEIE) can be used to universally modify the work function of electrodes. PEIE was demonstrated to serve as an efficient IML with highly conductive PEDOT:PSS (PH1000) although the interface properties between PEIE and PEDOT:PSS were not well understood.[66] Liu et al. demonstrated a novel IML consisting of conjugated polyelectrolyte (CPE) bilayer and modified PEDOT:PSS for solution-processed small-molecule tandem solar cells in a normal architecture.[67] A notable PCE of 10.1% was obtained for the tandem solar cell based on two identical subcells.

In addition to its function as a recombination layer, the IML of a tandem device may have further functionality. One example is the optical spacer effect, improving

Organic Tandem Solar Cells

the tandem device performance by adjustment of the electrical field distribution within each subcell's active layer. Using PEDOT:PSS Al4083/AZO as an IML, optical simulations suggested a slight tuning of the photon absorption, resulting in an enhancement of J_{sc} for P3HT:PCBM/Si-PCPDTBT:PCBM subcells as a function of the IML thickness. This effect was also confirmed through experimental results in which, for a given thickness of the active layers, the J_{sc} of the tandem device increased from 7.24 to 7.68 mA cm^{-2} as the IML was increased from ~50 to 160 nm.[57] A different but also attractive approach was demonstrated by J. Yang et al.[68] who investigated plasmonic effects in an inverted tandem cell configuration by blending Au nanoparticles (NPs) with the intermediate layer. A 20% improvement in PCE was suggested due to plasmonic near-field enhancement.

12.4 OPTICAL SIMULATIONS OF THE TANDEM SOLAR CELLS

In order to fully exploit the efficiency potential of promising materials, the combinations of absorbers versus layer sequence have to be understood. In addition, even for donor materials with attractive absorption properties, suitable acceptor molecules have to be chosen to ensure balanced photogeneration. Moreover, the thicknesses of the top and the bottom cells must be carefully optimized to match the photocurrent from each subcell. In this regard, optical simulations are frequently employed to exploit the ultimate efficiency potential for different material combinations of the materials as a function of layer thickness.

Optical simulations of tandem solar cells were carried out with different methodologies, such as the Rigorous Coupled Wave Analysis (RCWA)[69–71] or the TMF.[72–74] Using these methods, the number of absorbed photons in each layer is calculated by considering the reflection, the transmission, and the electric field distribution for periodic structures. A multilayer OSC can be treated in the simplest case as a one-dimensional system, and the amplitude of the electromagnetic field vector is coherently calculated for all layers. Prior to the simulations, the complex refractive indices ($n + ik$) of all the layers have to be determined. The imaginary part of the complex index of refraction (k) is directly measured by UV-vis spectrometry, and the real part (n) by variable angle spectroscopic ellipsometry (VASE).

All the aforementioned concepts were successfully demonstrated by Dennler et al.,[75] performing detailed optical simulations of tandem solar cells based on the following organic semiconductors: poly(3-hexylthiophene) (P3HT), poly[2,6-(4,4-bis-(2-ethylhexyl)-4H-cyclopenta[2,1-b;3,4-b′]dithiophene)-alt-4,7-(2,1,3-benzothiadiazole)] (PCPDTBT), 1-(3-methoxycarbonyl)propyl-1-phenyl[6,6] C_{61} (PC$_{60}$BM), and 1-(3-methoxycarbonyl)propyl-1-phenyl[6,6] C_{71} (PC$_{70}$BM). The structure of the tandem cells is depicted in Figure 12.8a. An 8-nm-thick TiOx and 25-nm-thick PEDOT:PSS were employed as the IMLs. For each active layer pair, the thickness of both the bottom (d_{bottom} between 0 and 300 nm) and the top (d_{top} between 0 and 600 nm) active layers was varied and the number of photons absorbed (N_{ph}) in those layers under AM 1.5 G was calculated. An example of such a calculation with PCPDTBT:PC$_{60}$BM as the bottom cell and P3HT:PC$_{70}$BM as the top cell is shown in Figure 12.8b. N_{ph} for the bottom (N_{ph} [bottom]) and the top (N_{ph} [top]) cells are represented in the three-dimensional space as surfaces, blue and red, respectively.

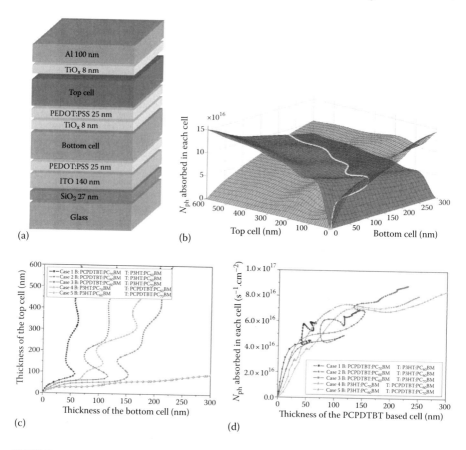

FIGURE 12.8 (a) Schematic diagram of the structure of the tandem cells simulated by Dennler et al. for different material combinations as bottom and top subcells. (b) Number of photons absorbed in the bottom active layer and in the top active layer versus the thickness of the bottom and the top active layers in a 3-D plot. The yellow line indicates the isoline 1, where N_{ph} (bottom)/N_{ph} (top) = 1. (c) Thickness relationship for reaching an equal number of photons absorbed in the top and the bottom cell, for each material combination. (d) Number of photons absorbed in both active layers (isoline 1) versus the thickness of the PCPDTBT-based active layer for various material combinations. (Reprinted with permission from Dennler, G., K. Forberich, T. Ameri et al., 2007, Design of efficient organic tandem cells: On the interplay between molecular absorption and layer sequence, *Journal of Applied Physics*, 102, 12, 123109, Copyright 2007, American Institute of Physics. From Yang, J. et al., *ACS Nano* 5 (8):6210–6217, 2011.)

The two surfaces N_{ph} (bottom) and N_{ph} (top) cross each other along a tortuous line defining the pair (d_{bottom}, d_{top}) for which N_{ph} (bottom) = N_{ph} (top). This line is called "isoline 1." The isoline 1 reports the optimum thicknesses of the active layers under the assumption that the top and bottom cells have the same IQE and FF, independent of the active layer thickness. In contrast, the current of the whole tandem device would be optimized along another isoline for which N_{ph} (bottom)/N_{ph} (top) = const ≠ 1.

Organic Tandem Solar Cells

Figure 12.8c shows the isoline 1 versus the thickness of the top and the bottom active layers for five different material combinations and layer sequences. This plot gives a quantitative understanding of the impact of layer design upon the tandem cell performance. Obviously, PCPDTBT:PC$_{70}$BM for the bottom cell combined with P3HT:PC$_{60}$BM for the bottom cell (case 1) and in vice versa sequence (case 5) have such different absorption properties that tuning of the active layers' thickness and achieving a balanced N_{ph} are very challenging. Replacing PC$_{70}$BM by PC$_{60}$BM in the bottom cell (case 2) allows one to utilize a thicker bottom cell compared to case 1. Interestingly, cases 3 and 4, using both the same blends but in a reversed layer sequence, show quite similar behavior with no obvious saturation for the thickness range investigated.

Valuable information can be deducted from the N_{ph}, plotted versus the thickness of the single layers as shown in Figure 12.8d. This one displays the N_{ph} along the isoline 1 versus the thickness of the PCPDTBT-based cell. This way of plotting was chosen because PCPDTBT cannot be processed in thick layers without losing its performance. As such, it is more critical to optimize the thickness of this cell. The optical interference effects cause quite unregularly shaped curves. N_{ph} for a single film thickness of the PCPDTBT layer can be matched by multiple layer thicknesses of the P3HT layer. Specifically, the case 3, in which PCPDTBT:PC$_{60}$BM is used as the bottom layer and P3HT:PC$_{70}$BM as the top layer, appears to be the most promising one for a reasonable thickness range of PCPDTBT between 130 and 230 nm. Under these conditions, the number of absorbed photons becomes maximized, which is reflected in higher current densities compared to the other scenarios. Despite intuition, the most efficient configuration puts the low bandgap polymer as the front absorber and the wide bandgap polymer as the back absorber. However, due to the specific optical absorption profiles of the two polymers and the thickness limitation of PCPDTBT, a configuration with the low bandgap polymer as the front or the bottom absorber allows higher performance as compared to other layer sequences.

Interestingly, Kim et al.[51] experimentally identified the same configuration as the most promising one. The diagram was approximately a 135 nm PCPDTBT:PC$_{60}$BM layer for the bottom cell and 155 nm P3HT:PC$_{70}$BM layer for the top cell, resulting in efficiencies of 6.5%. However, beyond the fact that these results demonstrated excellent agreement between simulation and the experiment, it became very clear that the choice of the right acceptor plays a major role in the current balance of the tandem cells.

A direct comparison between the calculated photocurrent (based on optical simulations) and the experimental results allows identification of the current limiting subcell. For instance, according to the simulation results presented in Figure 12.8, around 7.3×10^{16} photons per s per cm^2 are absorbed in each of the 135-nm-thick PCPDTBT and 155-nm-thick P3HT layers. Assuming an IQE of 100%, this N_{ph} would yield a short-circuit current density of 11.7 mA cm^{-2}, and the current reported for this device is about 7.8 mA cm^{-2}. This indicates that the limiting subcell driving the J_{sc} in the tandem has an IQE of 67%, a value in good accordance with the IQE values that have been determined for PCPDTBT:PC$_{60}$BM single devices. Hence, this suggests that the performance of the device is limited by the PCPDTBT subcell for this combination.

12.5 REVIEW OF EXPERIMENTAL RESULTS

From the very first organic tandem solar cell based on evaporated small molecules reported by Hiramoto et al. in 1990[76] until the breakthrough of a 6.5% solution-processed tandem cell reported by Kim et al. in 2007,[51] a detailed review of the reported organic multijunction solar cells is presented in our previous work.[17] Since then, a growing number of records (around or higher than 5%) have been reported in which the tandem/multijunction structure improves the performance of the individual single subcells. Fabrication and reproducibility of highly efficient organic tandem solar cells have been reported to be very challenging. Currently, researchers have mainly focused on the exploration of novel materials that are suitable for tandem devices. Thus, the PCEs of organic tandem solar cells have been rapidly boosted in the last several years by up to 10%–12%. In addition, new approaches on the design of IMLs, device architecture, and reliability of tandem solar cells have attracted considerable attention in recent years. In this section, the recent results of different types of highly efficient organic tandem solar cells obtained by various groups are reviewed.

12.5.1 Highly Efficient Tandem Organic Solar Cells Based on Novel Materials

Recently, a great deal of effort is made to improve the PCE of organic tandem solar cells based on novel materials, including polymers as well as vacuum- and solution-processed small molecules. In the following, we present a review of highly efficient tandem solar cells based on both material categories.

12.5.1.1 Polymer-Based Tandem Organic Solar Cells

In 2012, Gevaerts et al.[55] reported efficiencies of more than 7% for a normal structured tandem device. In this case, the large bandgap polymer of poly[N-9′-heptadecanyl-2,7-carbazole-alt-5,5-(4′,7′-di-2-thienyl-2′,1′,3′-benzothiadiazole)] (PCDTBT)[77–80] mixed with PC$_{70}$BM and the novel low bandgap diketopyrrolopyrrole-quintetthiophene copolymer (PDPP5T)[81–84] blended with PCBM were employed as bottom and top cells, respectively. Two different optimal layer thicknesses (80 and 210 nm) were identified, providing power PCEs of 5.6%–5.8% for single PCDTBT:PC$_{70}$BM cells. In these devices, there is a tradeoff between FF and J_{sc} while the V_{oc} is constant. The single cell of PDPP5T:PCBM with an optimized thickness of 120 nm led to efficiencies of 5.3%. Optical simulations, assuming a spectrally averaged IQE of 70% for PCDTBT:PC$_{70}$BM and 76% for PDPP5T:PCBM, predicted the maximum efficiency for a 170-nm PCDTBT:PC$_{70}$BM bottom cell combined with a 120-nm PDPP5T:PCBM top cell. The intermediate layer consisted of a 30-nm-thick layer of 4–5 nm ZnO NPs and a 20-nm layer of pH neutral PEDOT:PSS. Experimentally, the authors demonstrated a 7.0% efficient tandem device with a V_{oc} of 1.44 V, FF of 0.54, and J_{sc} of 9 mA cm^{-2}. The experimental results closely match the optical prediction as well as calculated efficiency extracted from the reconstructed J–V characteristics of the tandem device. Based on the same device architecture, Li et al. improved the PCE of a tandem solar cell by up to 8.9% by replacing PDPP5T with a novel low bandgap polymer, poly[[2,5-bis(2-hexyldecyl-2,3,5,6-tetrahydro-3,6-dioxopyrrolo[3,4-c]-pyrrole-1,4-diyl]-alt-[3′,3″-dimethyl-2,2′:5′,2″-terthiophene]-5,5″-diyl] (PMDPP3T) for the top cell.[85] Single-junction solar cells based on PMDPP3T mixed with PC$_{70}$BM and PCBM reached PCEs of 7.0% and 6.2%, respectively. The optimized thicknesses for bottom and top cells were determined by optical simulations to be 155 and 150 nm, respectively. Compared to the corresponding subcells' PCEs, the optimized tandem cell with the same layers and thickness achieved an unparalleled increase of about 50%. Moreover, by integrating the EQE spectra of subcells with a AM 1.5 G solar spectrum, they found that photocurrents generated from the bottom and top cells were not matched. The PMDPP3T:PCBM-based top cell delivered a J_{sc} of more than 11 mA cm^{-2} in a tandem solar cell, and the PCDTBT:PC$_{70}$BM-based bottom cell reached only a J_{sc} of 9.2 mA cm^{-2}. Therefore, they suggested an additional photoactive layer of the same low bandgap material, creating a triple-junction cell with a 1 + 2 configuration, that is, one junction based on a wide bandgap polymer and two junctions based on a low bandgap polymer, to further improve the efficiency of organic tandem solar cells. Experimentally, they demonstrated a triple-junction cell based on PCDTBT:PC$_{70}$BM (125 nm) as a bottom cell, PMDPP3T:PCBM (95 nm) as a middle cell, and PMDPP3T:PCBM (215 nm) as a top cell with V_{oc} = 2.09 V, J_{sc} = 7.34 mA cm^{-2}, and FF = 0.63, resulting in a PCE of 9.64%.

Zhou et al.[66] reported on inverted polymer tandem solar cells wherein highly conductive PEDOT:PSS (PH1000) in combination with ethoxylated polyethylenimine (PEIE) acted as an efficient charge recombination layer. This surface modification with PEIE led to a high work function contrast between the two opposite interfaces of the charge recombination layer in contact with the bottom and top cells in a tandem cell configuration where the subcells are connected in series. The bottom surface of the PEDOT:PSS displays a high work function of 4.9 eV whereas the top surface modified by PEIE displays a low work function of 3.6 eV. A 200-nm-thick P3HT:ICBA and 80-nm-thick poly[(4,8-bis-(2-ethylhexyloxy)-benzo[1,2-b:4,5-b′]dithiophene)-2,6-diyl-alt-(4-(2-ethylhexanoyl)-thieno[3,4-b]thiophene)-2,6-diyl] (PBDTTT-C):PCBM blend were employed as the bottom and top photoactive layers, respectively. With this combination, tandem cells displayed excellent performance: V_{oc} = 1.48 V, J_{sc} = 7.4 mA cm^{-2}, FF = 0.68, and PCE = 7.5%, averaged over 25 devices among which the champion cell showed V_{oc} = 1.5 V, J_{sc} = 7.7 mA cm^{-2}, FF = 0.68, and PCE = 8.2% under 100 mW cm^{-2} AM 1.5 illumination. It is worth mentioning that no posttreatment, such as thermal annealing and/or UV illumination was required to activate the PEIE buffer layer.[86] The authors also tested PH1000/branched polyethylenimine (PEI) as the recombination layer in tandem solar cells. The cells with PH1000/PEI yielded very similar results to those of PH1000/PEIE. This is consistent with the fact that the reduction of the work function of PH1000 by coating PEIE or PEI is similar.[65] Based on the similar IML, PEDOT:PSS/PEIE, bin Mohd Yusoff et al. demonstrated a tandem solar cell with an averaged V_{oc} of 1.52 V, J_{sc} of 8.73 mA cm^{-2}, and FF of 0.67, resulting in a PCE of 8.91%.[87] A wide bandgap polymer poly(4,4-dioctyldithieno(3,2-b:2′,3′-d)silole)-2,6-diyl-alt-(2,1,3-benzothiadiazole)-4,7-diyl] (PSEHTT) mixed with ICBA (120 nm) and a low bandgap polymer poly[(4,40-bis(2-ethylhexyl)dithieno[3,2-b:20,30-d]silole)-2,6-diyl-alt-(2,1,3-benzothiadiazole)-4,7-diyl] (PSBTBT/Si-PCPDBT) mixed with PC$_{70}$BM (220 nm) were used in bottom and top cells, respectively. Compared to the PCEs of optimized subcells, the corresponding tandem solar cells reached an improvement in PCE by ~50%. Moreover, they investigated the influence of different IMLs on the performance of tandem solar cells. Based on the same active materials and layer thicknesses, they demonstrated that tandem solar cells based on PEDOT:PSS/ZnO neutralized PEDOT:PSS/ZnO, and Graphene/TiO$_x$ IMLs obtained V_{oc} of 1.40, 1.20, and 1.14 V and PCE of 8.30%, 6.04%, and 5.11%, respectively. This study revealed that the IML not only serves as the charge recombination electrode between subcells, but ensures the presence of suitable energy alignment for efficiently recombining the charges from the subcells. The generation of a reverse built-in potential in IML might directly influence the V_{oc} of tandem solar cells. Based on the same subcells, the authors demonstrated elsewhere efficient organic tandem solar cells incorporating a TiO$_2$ and graphene oxide (GO) based IML.[88] The architecture of tandem solar cells was ITO/GO (10 nm)/PSEHTT:ICBA (80 nm)/TiO$_2$ (40 nm)/GO (10 nm)/PEDOT:PSS/PSBTBT:PC$_{70}$BM (110 nm)/ZnO (10 nm)/Al (150 nm). The tandem solar cells with an active area of 0.04 cm^2 obtained an averaged V_{oc} of 1.62 V, J_{sc} of 8.23 mA cm^{-2}, and FF of 0.63, resulting in an averaged PCE of 8.40% over 117 devices. They observed that inserting a thin layer GO (10 nm) between TiO$_2$ and PEDOT:PSS allows for drastically improved device stability for devices under continuous illumination. The efficiency of the tandem solar cell dropped to about 8.3% of the initial PCE after 720 hr and remained at 80% initial PCE after 2880 hr of continuous illumination at 48°C–50°C with 30% relative humidity.

Chang et al. reported on a tandem solar cell based on a fluoro-containing low bandgap polymer, poly[2,6-(4,4-bis(2-ethylhexyl)-4H-cyclopenta[2,1-b;3,4-b′]dithiophene)-alt-4,7-(5-fluoro-2,1,3-benzothia-diazole)] (PCPDTFBT) mixed with PC$_{70}$BM as the top cell.[89] By modifying the active layer morphology with a high boiling point 1,3,5-trichlorobenzene and modifying the ZnO buffer layer with a fullerene self-assembled monolayer (C$_{60}$-SAM), single-junction solar cells based on PCPDFBT:PC$_{70}$BM obtained a PCE of 6.6%, which was among the highest PCE values for cyclopenta[2,1-b;3,4-b′]dithiophene (CPDT)-based organic solar cells. The tandem solar cells based on P3HT:ICBA (150 nm) as the bottom cell and PCPDTFBT:PC$_{70}$BM (100 nm) as the top cell with an inverted architecture reached a V_{oc} of 1.57 V, a J_{sc} of 7.83 mA cm^{-2}, and a FF of 66.5%, resulting in a PCE of 8.2%. The combination of modified PEDOT:PSS (Clevious P VP A1 4083 diluted with equal volume of isopropyl alcohol and 0.2 wt% of Zonyl FSO fluorosurfactant), PEDOT:PSS (PH1000) and ZnO was used as an IML to connect the bottom cell, P3HT:ICBA 150 nm, and the top cell, PCPDTFBT:PC$_{70}$BM 100 nm. They noted that the use of PEDOT:PSS (PH1000) ensures the ohmic contact between modified PEDOT:PSS and ZnO. The tandem solar cells without PEDOT:PSS (PH1000) usually showed a s-shaped kink in J–V characteristics and therefore a deteriorated PCE.

Jo et al. demonstrated an efficient inverted architecture single-junction solar cell based on a low bandgap polymer, poly[{2,5-bis(2-octyldodecyl)-2,3,5,6-tetrahydro-3,6-dioxopyrrolo[3,4-c]pyrrole-1,4-diyl}-alt-{[2,2′-(1,4-phenylene)bisthiophene]-5,5′-diyl}] (PDPPTPT), ($E_g \approx$ 1.49 eV).[90] The PCE of this DPP-based single-junction solar cell was improved from ≈5.5%[91] up to 7% by incorporating an additive mixture consisting of 3 vol.% 1,8-diiodooctane (DIO) and 3 vol.% 1-chloronaphthalene (CN) into the chlorobenzene (CB) host solvent to optimize the BHJ nanomorphology and thus to improve the cell performance. Moreover, an ultrathin conjugated polyelectrolyte (CPE) layer, poly(9,9'-bis(6''-N,N,N-trimethylammoniumhexyl) fluorene-coalt-phenylene), with bromide counterions (FPQ-Br) was employed between the ZnO and BHJ layers to improve the electron transport and interfacial contact. The tandem solar cells based on a wide bandgap thieno[3,4-c]pyrrole-4,6-dione (TPD)-terthiophene copolymer ($E_g \approx 1.8$ eV) mixed with $PC_{70}BM$ as the bottom cell and PDPPTPT:$PC_{70}BM$ as the top cell reached a V_{oc} of 1.60 V, a J_{sc} of 8.4 mA cm^{-2}, and a FF of 0.64, resulting in a champion PCE of 8.58%. The IML consisted of a conductive PEDOT:PSS (Clevios CPP105D) layer and ZnO NPs. The use of ZnO NPs as the ETL in the IML is due to its relatively low annealing temperature ($\approx 80°C$), which does not affect a change of BHJ nanomorphology in the bottom cell. The CPE, which modified the ZnO layer in the bottom and top cells, played an important role in optimizing the tandem solar cells. Without the CPE layers, the tandem solar cell based on the same active layers obtained a PCE of 7.23% along with a J_{sc} of 7.5 mA cm^{-2}, and a FF of 0.60.

Li et al.[92] demonstrated two efficient conjugated copolymers, PBDTFBZO and PBDTFBZS, consisting of dialkylthiol substituted benzo[1,2-b:4,5-b']dithiophene donor and monofluorinated benzotriazole acceptor blocks. The optical bandgaps of both polymers were determined to be ~1.8 eV. Single-junction solar cells based on PBDTFBZS:$PC_{70}BM$ with 3% DIO obtained a high PCE of 7.74% along with a V_{oc} of 0.88 V, a J_{sc} of 12.36 mA cm^{-2}, and a promising FF of 0.712. Tandem solar cells with an inverted architecture based on PBDTFBZS:$PC_{70}BM$ as the bottom cell were constructed. A diketopyrrolopyrrole (DPP)-based copolymer (PNDTDPP)[93] mixed with $PC_{70}BM$ was employed as the top cell in the tandem architecture, which is ITO/ZnO (40 nm)/PBDTFBZS:$PC_{70}BM$ (150 nm)/PEDOT:PSS/ZnO/PNDTDPP:$PC_{70}BM$ (120 nm)/MoO_3 (10 nm)/Ag (100 nm). The tandem solar cells obtained a V_{oc} of 1.59 V, a J_{sc} of 9.10 mA cm^{-2}, and a FF of 0.65, resulting in a PCE of 9.40%, exhibiting an improvement by 48% compared to the corresponding reference single-junction solar cells.

Based on two commercially available conjugated polymers, GEN-2 (Merck)[60,94] and pDPP5T-2 (BASF),[58,60] we demonstrated efficient organic tandem solar cells with inverted architecture incorporating a low-temperature, solution-processed IML.[30,59] The functionality of the IML was fully investigated and reported previously.[59] It is notable to mention that all layers of the tandem solar cells, excluding two electrodes, were fabricated by doctor blading at a fairly low temperature (<80°C) in air, exhibiting an excellent compatibility with the roll-to-roll mass production of organic tandem solar cells. The architecture of tandem solar cells was ITO/ZnO/GEN-2:PCBM/PEDOT HIL3.3/ZnO/Ba(OH)$_2$/pDPP5T-2:$PC_{70}BM$/MoO_x/Ag. Zhang et al.[95] found that modifying the AZO layer with an ultrathin solution-processed Ba(OH)$_2$ layer could dramatically enhance the PCE of the single-junction solar cells with an inverted architecture, especially for the DPP-based solar cells. In this work, we successfully used the same recipe to modify the ZnO for pDPP5T-2:$PC_{70}BM$-based solar cells. The tandem solar cells on glass and flexible substrate obtained a PCE of 7.66% and 5.56%, respectively, which were the highest efficiencies reported until now for organic tandem solar cells fabricated by a mass production–compatible method in air.

The breakthrough in organic tandem solar cell performance was made by Yang's group. In 2012, Dou et al.[53] demonstrated the rational design of a novel low-bandgap conjugated polymer, poly{2,6'-4,8-di(5-ethylhexylthienyl)benzo[1,2-b;3,4-b]dithiophene-alt-5-dibutyloctyl-3,6-bis(5-bromothiophen-2-yl)pyrrolo[3,4-c]pyrrole-1,4-dione} (PBDTT-DPP) with an optical bandgap of 1.44 eV for tandem solar cells. PBDTT-DPP blended with PC$_{70}$BM gave PCEs greater than 6% for single-junction solar cells in both normal and inverted architectures. Among more than 300 devices processed, the best devices gave the following values: $V_{oc} \approx 0.74$ V, $J_{sc} \approx 13.5$ mA cm^{-2}, and FF $\approx 65\%$. The high V_{oc} can be attributed to the deep HOMO level of -5.3 eV, and the high J_{sc} and FF can be attributed to the high hole mobility of PBDTT-DPP (the hole mobility of PBDTT-DPP:PC$_{70}$BM with a 1:2 weight ratio was found to be 2.9×10^{-4} cm^2 V^{-1} s^{-1}). Finally, a PCE of 8.62% was certified by the National Renewable Energy Laboratory (NREL) for an inverted architecture tandem solar cell comprised of P3HT:ICBA and PBDTT-DPP:PC$_{70}$BM as the bottom and top cells, respectively. A combination of m-PEDOT:PSS/ZnO ensured the efficient charge carriers recombination in the IML. The resulting tandem device showed a FF of 0.66, V_{oc} of 1.56 V, and J_{sc} of 8.26 mA cm^{-2}.

The substitution of S by Se on the DPP unit (PBDTT-SeDPP) led to a reduced bandgap (optical bandgap of 1.38 eV) and an enhanced hole mobility (6.9×10^{-4} cm^2 V^{-1} s^{-1}) compared to S-based PBDTT-DPP.[96] A higher photocurrent of 16.8 mA cm^{-2} and a PCE of 7.2% were obtained in a single junction of PBDTT-SeDPP:PC$_{70}$BM. Even more importantly, this new Se-containing polymer significantly enhanced the tandem device performance by up to 9.5%. Tandem solar cells with an inverted architecture were fabricated using PBDTT-SeDPP:PC$_{70}$BM as the top cell and P3HT:ICBA as the bottom cell. The averaged PCE from 20 tandem devices was 9.5% with a V_{oc} of 1.52 V, a J_{sc} of 9.44 mA cm^{-2}, and a FF of 0.66. The V_{oc} of the tandem device is almost the sum of two subcells (0.84 V and 0.69 V), indicating the effectiveness of the high-performance IML of PEDOT:PSS/ZnO. The FF of the tandem device is around the average of the two subcells (70% for the bottom cell and 62% for the top cell). The major improvement compared to the previously reported tandem cells based on PBDTT-DPP is the J_{sc} (from 8.3 mA cm^{-2} to 9.4 mA cm^{-2}).

Finally, efficiencies of 10.6% were achieved by the same group based on a developed structure of the reported low-bandgap polymer of PCPDTBT.[97] First, a significantly enhanced V_{oc} was obtained by introducing two strong electron-withdrawing fluorine atoms on the benzothiadiazole (BT) unit, forming the difluorobenzothiadiazole (DFBT) unit to lower the HOMO level. Second, the bandgap was further decreased by inserting a strong electron-donating oxygen atom into the cyclopentadithiophene (CPDT) unit, creating the dithienopyran (DTP) unit. These two strategies led to the novel polymer poly[2,7-(5,5-bis-(3,7-dimethyl octyl)-5H-dithieno[3,2-b:20,30-d]pyran)-alt-4,7-(5,6-difluoro-2,1,3-benzothiadiazole)] (PDTP-DFBT) with a bandgap of 1.38 eV, which also showed a high hole mobility (3.2×10^{-3} cm^2 V^{-1} s^{-1}) and deep HOMO level (-5.26 eV). Single-junction devices based on PDTP-DFBT showed a high quantum efficiency of >60% from 710 to 820 nm, and the spectral response extends to 900 nm. This led to a PCE of 7.1% and 7.9% when blended with PCBM and PC$_{70}$BM, respectively. Tandem devices based on the bottom cell of P3HT:ICBA and the top cell of PDTP-DFBT:PCBM obtained the certified PCE

Organic Tandem Solar Cells

of 10.6% with a J_{sc} of 10.1 mA cm^{-2}, FF of 68.5%, and V_{oc} of 1.53 V under standard reporting conditions (25°C, 100 mW cm^{-2}, AM 1.5 G). The device architecture is ITO/ZnO/P3HT:ICBA (220 nm)/PEDOT:PSS/ZnO/PDTP-DFBT:PCBM (100 nm)/MoO$_3$/Ag (Figure 12.9a). The J–V characteristics of this tandem device, certified by NREL, are shown in Figure 12.9b. The efficiency of 10.6% ranks at the world's top level among polymer PV cells that are available at the moment. In addition, You et al. constructed the tandem solar cells based on two identical PDTP-DFBT:PC$_{70}$BM subcells.[98] They found that the modified PEDOT:PSS in the IML doesn't work well with the PDPTP-DFBT:PC$_{70}$BM; a thin MoOx layer was therefore thermally evaporated on top of the bottom cell to overcome the contact problem. Finally, the tandem solar cells based on an 80-nm-thick bottom cell and a 100-nm top cell obtained a V_{oc} of 1.36 V along with a J_{sc} of 11.5 mA cm^{-2} and a FF of 0.65, resulting in a PCE of 10.2%, leading to an improvement by ~26% compared to the corresponding champion single-junction solar cells. However, compared to the traditional tandem solar cells with complementary absorption, the identical subcell-based tandem solar cells cannot reduce the thermalization losses. PCE of this kind of tandem solar cells is still limited by the S-Q limit for single-junction solar cells[18] due to relatively low V_{oc}. Nevertheless, considering the thickness limit for most organic active materials, this 1 + 1 approach, that is, tandem solar cells based on two identical subcells, stands a chance to enhance both the IQE and EQE of organic solar cells and, as a result, to further enhance their PCEs.

A brief summary of the reports dealing with highly efficient tandem organic solar cells based on polymeric donors and their corresponding single subcells is presented in Table 12.1.

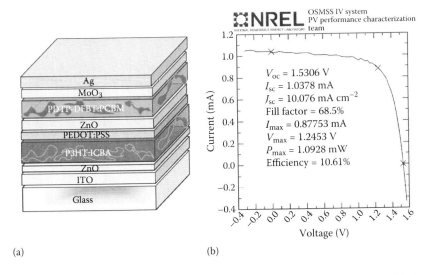

FIGURE 12.9 (a) Schematic of the 10.6% efficient tandem solar cell, demonstrated by You et al. (b) J–V characteristics of the device as measured by NREL under AM 1.5 G illumination (100 mW/cm²). (Reprinted by permission from Macmillan Publishers Ltd., You, J., L. Dou, K. Yoshimura et al., A polymer tandem solar cell with 10.6% power conversion efficiency. *Nat Commun*, 4, 1446. Copyright 2013.)

TABLE 12.1
Nonexhaustive Survey of Reports Dealing with High Efficient Tandem Organic Solar Cells Based on Polymeric Donors

Year	Recombination Layer	Bottom Cell Active Materials	V_{oc} (V)	FF	J_{sc} (mA cm^{-2}) (mW cm^{-2})	Eff (%)	Top Cell Active Materials	V_{oc} (V)	FF	J_{sc} (mA cm^{-2}) (mW cm^{-2})	Eff (%)	Tandem Cell V_{oc} (V)	FF	J_{sc} (mA cm^{-2}) (mW cm^{-2})	Eff (%)	Ref.
2012	30 nm ZnO + 20 nm PEDOT:PSS	PCDTBT:PC$_{70}$BM	0.9	0.52	12.3 (100)	5.8	PDPP5T:PCBM	0.58	0.65	14.0 (100)	5.3	1.44	0.54	9.0 (100)	7.0	[55]
2013	30 nm ZnO + 20 nm PEDOT:PSS	PCDTBT:PC$_{70}$BM	0.87	0.56	9.76 (100)	4.7	PMDPP3T:PCBM	0.61	0.65	15.30 (100)	6.0	1.49	0.62	9.58 (100)	8.9	[85]
2012	40 nm PEDOT:PSS PH1000 + 10 nm PEIE	P3HT:ICBA	0.82	0.63	8.8 (100)	4.5	PBDTTT-C:PCBM	0.67	0.57	15.2 (100)	5.9	1.5	0.72	7.7 (100)	8.2	[66]
2013	40 nm PEDOT:PSS PH1000 + 10 nm PEIE	PSEHTT:ICBA	0.91	0.66	9.98 (100)	5.99	PSBTBT:PC$_{70}$BM	0.64	0.58	13.36 (100)	4.99	1.52	0.67	8.73 (100)	8.91	[87]
2013	35 nm TiO$_2$ + GO	PSEHTT:ICBA	0.94	0.65	10.39 (100)	6.42	PSBTBT:PC$_{70}$BM	0.68	0.61	13.14 (100)	5.51	1.62	0.62	8.23 (100)	8.4	[88]

Organic Tandem Solar Cells

2013	Modified PEDOT:PSS + PEDOT:PSS	P3HT:ICBA	0.83	0.68	11.59 (100)	6.5	PCPDTFBT:PC$_{70}$BM	0.74	0.61	14.24 (100)	6.6	1.57	0.66	7.83 (100)	8.2	[89]
2013	PH1000 + ZnO 60 nm	P2:PC$_{70}$BM	0.88	0.73	9.7 (100)	6.18	PDPPTPT:PC$_{70}$BM	0.78	0.62	14.5 (100)	7.04	1.60	0.64	8.4 (100)	8.58	[90]
2013	PEDOT:PSS + 35 nm ZnO	PBDTFBZS:PC$_{70}$BM	0.88	0.67	10.44 (100)	6.18	PNDTDPP:PC$_{70}$BM	0.74	0.61	13.86 (100)	6.35	1.59	0.65	9.1 (100)	9.40	[92]
2013	PEDOT:PSS + ZnO 30 nm	GEN-2:PCBM	0.77	0.67	10.74 (100)	5.54	pDPP5T-2:PC$_{70}$BM	0.54	0.66	13.00 (100)	4.63	1.30	0.72	8.18 (100)	7.66	[30]
2012	PEDOT:PSS + 30 nm ZnO	P3HT:ICBA	0.85	0.7	9.56 (100)	5.7	PBDTT-DPP:PC$_{70}$BM	0.74	0.65	13.5 (100)	6.5	1.56	0.66	8.26 (100)	8.6	[53]
2012	Modified PEDOT:PSS + ZnO 30 nm	P3HT:ICBA	0.84	0.7	9.5 (100)	5.6	PBDTT-SeDPP:PC$_{70}$BM	0.69	0.62	16.8 (100)	7.2	1.52	0.66	9.44 (100)	9.5	[96]
2013	PEDOT:PSS + 30 nm ZnO	P3HT:ICBA	0.84	0.71	10.3 (100)	6.1	PDTP-DFBT:PCBM	0.7	0.66	15.4 (100)	7.1	1.53	0.68	10.1 (100)	10.6	[97]
2013	PEDOT:PSS + ZnO	PDTP-DFBT:PC$_{70}$BM	0.69	0.65	16.9 (100)	7.6	PDTP-DFBT:PC$_{70}$BM	0.69	0.63	18.6 (100)	8.1	1.36	0.65	11.5 (100)	10.2	[98]
2013	15 nm MoO$_3$ + modified PEDOT:PSS + ZnO															

359

12.5.1.2 Small Molecule-Based Tandem Organic Solar Cells

Recently, Riede et al. presented a certified efficiency of 6.07% for a small molecule-based tandem solar cell.[99] The absorption-complementary materials of fluorinated zinc phthalocyanine derivative (F4-ZnPc) and dicyanovinyl-capped sexithiophene derivative (DCV6T) blended with C_{60} were evaporated as bottom and top active layers,

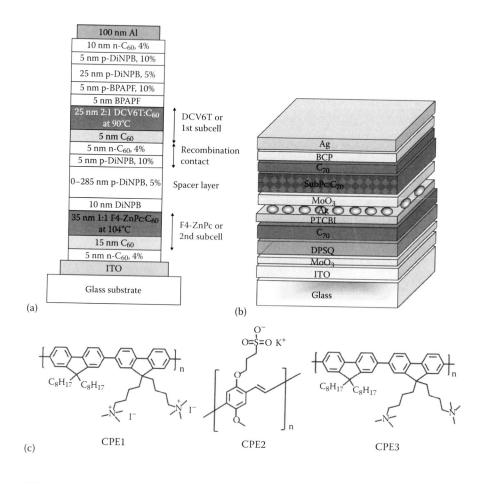

FIGURE 12.10 Schematic of (a) A vacuum-processed small molecule-based tandem solar cell, demonstrated by Riede et al. (Riede, M., C. Uhrich, J. Widmer et al.: Efficient Organic Tandem Solar Cells based on Small Molecules. *Advanced Functional Materials*. 2011. 21. 16. 3019–3028. Copyright Wiley-VCH Verlag GmbH & Co. KGaA. Reproduced with permission.) and (b) A vacuum- and solution-processed small molecule-based tandem solar cell, demonstrated by Lassiter et al. (Reprinted with permission from Lassiter, B. E, J. D. Zimmerman, and S. R. Forrest, *Applied Physics Letters*, 103, 123305, 2013, Copyright 2013, American Institute of Physics.) (c) Chemical structures of CPEs used in ref. 67 for solution-processed small molecule based tandem solar cells. (Reprinted by permission from Macmillan Publishers Ltd., Yongsheng, C.-C. C., Z. Hong et al. 2013. Solution-processed small-molecule solar cells: breaking the 10% power conversion efficiency, *Scientific Reports*, 3, 33, 3356, Copyright 2013.)

Organic Tandem Solar Cells

respectively. As shown in Figure 12.10a, a complicated IML of N,N'-diphenyl-N,N'-bis(4'-(N,N-bis(naphth-1-yl)-amino)-biphenyl-4-yl)-benzidine (DiNPB), 5% p-doped DiNPB (p-DiNPB), 10% p-DiNPB, 4% n-doped C_{60} (n-C_{60}), and C_{60} ensured the recombination of holes and electrons extracted from the bottom and top cells. Thickness variations of the transparent optical spacer p-DiNPB between both subcells in the tandem solar cell were shown to cause a significant change in short-circuit current density J_{sc} due to optical interference effects whereas V_{oc} and FF were hardly affected. The maximum efficiency of about 6.07% under standard test conditions on a module-relevant area of nearly 2 cm² with V_{oc} = 1.59 V, FF = 0.61, and J_{sc} = 6.18 mA cm⁻² was reported for a spacer thickness of 150–165 nm.

Lassiter et al.[100] demonstrated an organic tandem solar cell incorporating solution- and vacuum-deposited small molecules as the active layers. In this work, the authors incorporated a solution-processed and solvent-vapor-annealed 2,4-bis[4-(N,N-diphenylamino)-2,6-dihydroxyphenyl] squaraine[101] (DPSQ)/C_{70} bilayer HJ for the bottom cell and a boron subphthalocyanine chloride (SubPc):C_{70} graded HJ[102] for the top cell. The tandem devices were fabricated with the following layer thicknesses: glass substrate/100 nm, ITO/20 nm, MoO_3/13 nm, DPSQ/10 nm, C_{70}/5 nm, PTCBI/0.1 nm, Ag/5 nm, MoO_3/29 nm, SubPc:C_{70}/3 nm, C_{70}/7 nm, BCP/100 nm Ag (depicted in Figure 12.10b). The IML consisting of PTCBI/Ag/MoO_3 was vacuum-processed. The resulting tandem device was demonstrated with a PCE of 6.66%, FF = 0.54, J_{sc} = 6.2 mA cm⁻², and V_{oc} = 1.97 V, indicating a nearly loss-free IML. The obtained efficiency for the tandem devices was at least 35% improved as compared to the most efficient single cell.

Liu et al.[67] designed and synthesized a two-dimensional conjugated small molecule, SMPV1, based on alkylthienyl-substituted benzo[1,2-b:4,5-b9]dithiophene (BDT-T) as the core unit and 3-octylrodanine as the electron-withdrawing end group for high performance solution-processed OSCs. The single-junction solar cells based on SMPV1 mixed with $PC_{70}BM$ (1:0.8 w/w) and 0.5 mg mL⁻¹ PDMS as additive obtained a PCE of 8.02% along with a V_{oc} of 0.937, a J_{sc} of 12.17 mA cm⁻², and FF of 0.704, which was certificated by Newport Cooperation under AM 1.5 G irradiation (100 mW cm⁻²). The single-junction solar cells with PDMS showed an averaged IQE over 80%, indicating that the PDMS additive plays a positive role in free carrier generation in the BHJ films, which may result from a more effective donor–acceptor interface due to the decreased domain size. To further enhance the absorption of the SMPV1-based solar cells, they demonstrated the tandem solar cells based on two identical subcells incorporating a conjugated polyelectrolyte (CPE)-based, thermal-annealing free IML. Depositing a CPE thin layer on top of the active layer normally leads to a surface dipole at the interface between the photoactive layer and metal electrode, which can suppress the interfacial recombination, and thus improve the device performance.[103] The authors developed a multilayered film based on bilayer CPE and a modified PEDOT:PSS, having a well-defined molecular organization in a smooth geometrical arrangement, to serve as the IML. The architecture of the tandem solar cells was ITO/PEDOT:PSS/SMPV1:$PC_{71}BM$/CPE1/CPE2/M-PEDOT:PSS/SMPV1:$PC_{71}BM$/CPE3/Al. The chemical structures of CPEs, CPE1, CPE2, and CPE3, are depicted in Figure 12.10c. The optimized tandem solar cell achieved a PCE of 10.1% along with a V_{oc} of 1.82 V, a J_{sc} of 7.70 mA cm⁻², and a notable FF of 72%, corresponding to a 25% enhancement

TABLE 12.2
Nonexhaustive Survey of Reports Dealing with High-Efficiency Tandem Organic Solar Cells Based on Small Molecules

Year	Recombination Layer	Bottom Cell Active Materials	V_{oc} (V)	FF	J_{sc} (mA cm^{-2}) (mW cm^{-2})	Eff (%)	Top Cell Active Materials	V_{oc} (V)	FF	J_{sc} (mA cm^{-2}) (mW cm^{-2})	Eff (%)	Tandem Cell V_{oc} (V)	FF	J_{sc} (mA cm^{-2}) (mW cm^{-2})	Eff (%)	Ref.
2011	10 nm DiNPB + 165 nm p-DiNPB, 5% + 5 nm p-DiNPB, 10% + 5 nm n-C$_{60}$, 4% + 5 nm C$_{60}$	F4-ZnPc: C$_{60}$	0.66	0.64	8.3 (100)	3.9	DCV6T:C$_{60}$	0.88	0.66	8.3 (111)	4.3	1.59	0.61	6.18 (100)	6.07	[99]
2012	5 nm PTCBI + 0.1 nm Ag + 5 nm MoO$_3$	SubPc:C$_{70}$	1.04	0.48	8.5 (100)	4.3	DPSQ/C$_{70}$	0.94	0.71	6.1 (100)	4.1	1.97	0.54	6.2 (100)	6.6	[100]
2013	5 nm CPE1 + 5 nm CPE2 + modified PEDOT:PSS	SMPV1: PC$_{70}$BM	0.94	0.69	12.5 (100)	8.1	SMPV1: PC$_{70}$BM	0.94	0.69	12.5 (100)	8.1	1.82	0.72	7.7 (100)	10.1	[67]
2013	NR[a]	NR	NR	NR	NR	NR	NR	NR	NR	NR	NR	NR	NR	NR	12	[104]

[a] Not reported.

362 Organic Solar Cells

Organic Tandem Solar Cells

compared to the best single-junction solar cell. This study demonstrated the first solution-processed small molecule-based solar cell exceeded the 10% milestone, which was among the highest efficiencies of OPV devices.

Recently, a new record of 12% on vacuum-processed organic tandem solar cells has been reported by Heliatek GmbH.[104] With Yang's group (2011) and Heliatek's new record-breaking efficiencies for OPV cells,[97,104] the organics are now matching the a-Si:H. So far, no technical details on this record efficient tandem solar cell have been reported.

A summary of the small molecule-based tandem solar cells and their corresponding single subcells is provided in Table 12.2.

12.5.2 Highly Efficient Tandem Organic Solar Cells Based on Various Structures

In addition to the novel organic materials, various device architectures have been applied to enhance the efficiencies of tandem OPVs. A review of these structures, including inorganic–organic hybrid and semitransparent tandem organic solar cells is presented in the following.

12.5.2.1 Inorganic–Organic Hybrid Tandem Solar Cells

In 2012, Seo and his colleagues[105] reported the record efficiency of inorganic–organic hybrid tandem solar cells. Figure 12.14 shows the device structure of the inorganic–organic hybrid tandem solar cells designed in their work. Thin films of a-Si (p-type/intrinsic/n-type) were deposited as the bottom inorganic subcell using plasma-enhanced chemical vapor deposition. The active layer of the organic top subcell consisted of a blend of PBDTTT-C:PCBM. The organic and inorganic solar cells were connected by a transparent ITO layer (50 nm) and a highly conductive PEDOT:PSS layer (70 nm). The ITO, acting as an ETL, was deposited using magnetron sputtering. The PEDOT:PSS layer was used as the HTL. Consequently, electrons extracted from the inorganic solar cells combined with holes extracted from the OSCs at the ITO/PEDOT:PSS interface. The maximum PCE of 5.72% with a V_{oc} of 1.42 V, J_{sc} of 6.84 mA cm^{-2}, and FF of 0.58 was achieved, which is higher than that of either of the two subcells individually. The V_{oc} of the hybrid cell was 1.42 V, reaching 92% of the sum of the subcell V_{oc} values. This provided evidence that the two subcells were indeed connected in series.

Beiley et al.[106] demonstrated a novel device architecture, hybrid tandem photovoltaic (HTPV), which is composed of an inexpensive and low-temperature processed solar cell, such as an organic or dye-sensitized solar cell, that can be printed on top of one of a variety of more traditional inorganic solar cells. Based on their modeling, an OSC may be added on top of a commercial CIGS cell, as a HTPV cell, to improve its efficiency from 15.1% to 21.4%, thereby reducing the cost of the modules by ~15% to 20% and the cost of installation by up to 30%. In detail, the authors assumed light is first incident on the organic top cell, which absorbs only light above its bandgap, and allows lower energy photons to pass through to be absorbed by the inorganic bottom cell. The organic top cells were modeled assuming a 90% IQE, a 0.70 FF, and

0.6 V loss in V_{oc} compared to the bandgap of the absorber. In ideal cases, in which the top organic solar cells' bandgap can be finely tuned and absorb 100% of the above-bandgap light, the top cell can improve a commercial HelioVolt CIGS cell from 15.1% to 21.4% and the NREL CIGS cell from 19.9% to 23.5% in a two-terminal device, in which the photocurrent generated by both subcells should be matched. However, based on current OPV technology, in which the losses in V_{oc} are typically higher than 0.6 V due to contact loss and the energy required for charge carriers dissociation, an organic cell absorbing 75% of the light above 1.8 eV that achieves a V_{oc} of 945 mV can only improve a HelioVolt CIGS cell by from 15.1% to 16.5% PCE. Nevertheless, this novel approach has the potential to decrease module costs by ~15% to 20% and installation costs by up to 30%, which would make solar power economically competitive in many more places. Later, the authors[107] experimentally demonstrated an HTPV incorporating a semitransparent organic top cell based on PBDTTPD:PC$_{70}$BM and an inorganic bottom CIGS cell. The PCEs of the HTPV cell were estimated to be 11.7% and 14.5% when the light comes from a different side of the semitransparent organic top cell. Although both efficiencies were lower than that of the single-junction CIGS cell (15.7%), the semitransparent cell architecture paved the way for HTPV technology approaching and exceeding the 20% milestone.

12.5.2.2 Semitransparent Tandem Organic Solar Cells

Chen et al.[40] reported on a normal architecture, semitransparent OSC based on an efficient low-bandgap polymer PBDTT-DPP, which was used in the top cell of an efficient organic tandem solar cell in previous work.[53] The architecture of the semi-transparent solar cell was ITO/PEDOT:PSS/PBDTT-DPP:PCBM (~100 nm)/TiO$_2$/AgNW. To form the AgNW-based top electrode, the AgNW dispersion in isopropyl alcohol was spin coated (2 mg/ml dispersion, 2500 rpm, 10 drops) or spray coated (0.05 mg/ml dispersion) onto the TiO$_2$ layer to form AgNW conducting networks.[108] The fusing process of the AgNW network was then carried out by applying diluted TiO$_2$ sol–gel solution in ethanol at 3000 rpm and baking at 100°C for 10 s. The ITO nanoparticle dispersion (10 wt%) was used as transparent conductive filler and was spin coated onto the fused AgNW matrix to form the composite electrode. Mild heating at 80°C for 1 min removed the residual solvent. The thickness of the transparent composite electrode is around 400 nm. The semitransparent solar cell with a maximum transmission of 66% at 550 nm achieved PCE of 4.02% and 3.82% under the illumination from the ITO and AgNW sides, respectively, and the control device based on the same active layer with an opaque reflective electrode obtained a PCE of 6.03%.

Based on the same processing recipe for the top AgNW electrode, the authors fabricated the semitransparent tandem solar cells based on PBDTT-FDPP-C$_{12}$ and PBDTT-SeDPP as active materials.[109] The combination of PFN/TiO2/modified PEDOT:PSS (85 vol.% PH1000, 10 vol.% IPA, and 5 vol.% DMF)/PEDOT:PSS (Al4083). The tandem solar cells with an device architecture of ITO/PEDOT:PSS/PBDTT-FDPP-C12:PCBM/PFN/TiO2/m-PEDOT:PSS/PEDOT:PSS/PBDTT-SeDPP:PCBM/TiO$_2$/AgNW obtained a PCE of 6.4% along with a V_{oc} of 1.46 V, a J_{sc} of 7.2 mA cm^{-2}, a FF of 0.61, and an averaged transmission of 43%. Replacing the PCBM with PC$_{70}$BM in the top cell led to an increase in PCE up to 7.3%, responding

Organic Tandem Solar Cells

to an improvement in PCE by more than 30% compared to the optimized single-junction semitransparent solar cells based on PBDTT-SeDPP:PC$_{70}$BM, and the averaged transmission decreased to 30%. The authors observed that these semitransparent tandem solar cells can be used in either top- or bottom-illumination modes without any major differences in their device performances.

Jen's group used a different approach to fabricating the semitransparent OSCs, where a thin Ag film was thermally evaporated to form the semitransparent top electrode.[110,111] The realized semitransparent solar cells based on poly(indacenodithiophene-co-phenanthro[9,10-b]quinoxaline) (PIDT-PhanQ):PC$_{70}$BM exhibited the PCEs in the range of 2.6% to 5.63% and the averaged visible transmittance in the range of ~30% to 13%.[110] Chen et al. demonstrated the inverted architecture semitransparent OSCs by employing a thin thermally evaporated MoO$_x$/Ag electrode.[111] This device architecture was ITO/ZnO/C$_{60}$-SAM/PBDTTT-C-T:PC$_{70}$BM/MoO$_x$/Ag, and the thin C$_{60}$-SAM was utilized to modify the ZnO buffer layer, thus enhancing the device performance. The synthesis and characterization of the low bandgap polymer based on benzo[1,2-b:4,5-b′]dithiophene (BDT) and thieno[3,4-b]thiophene units (PBDTTT-C-T) was demonstrated elsewhere.[112] By tuning the thickness of the reflective Ag electrode, the averaged visible transmittance accordingly varied from ~35.9% to ~2.0%, and the corresponding PCE of the semitransparent solar cells increased from 4.25% up to 7.56%. It is notable to mention that the optimized semitransparent OSCs obtained a PCE of ~6% along with an averaged visible transmittance of 25% and a color rendering index close to 100. Later, Chang et al. demonstrated semitransparent organic tandem solar cells incorporating this thermally evaporated thin MoO$_x$/Ag electrode.[113] The architecture of the tandem solar cells was ITO/ZnO/C$_{60}$-SAM/PIDT-PhanQ:PCBM/m-PEDOT:PSS/PEDOT:PSS (PH1000)/ZnO/C$_{60}$-SAM/PIDT-PhanQ:PC$_{70}$BM/MoO$_x$/Ag. The IML consisting of bilayer PEDOT:PSS and ZnO/C$_{60}$-SAM was demonstrated and investigated in a previous publication.[89] Based on the thin active layers (total thickness of 155 nm), semitransparent tandem solar cells incorporating a thin top electrode (~10 nm Ag) obtained a PCE of 6.7% (under illumination at 100 mW cm^{-2}) along with a V_{oc} of 1.70 V, a J_{sc} of 5.81 mA cm^{-2}, a FF of 0.67, and an averaged visible transmittance of ~40% in the range between 380 and 700 nm. The authors mentioned that an improved PCE of 7.4% was obtained under a low light intensity of 80 mW cm^{-2}, representing the highest value reported for a semitransparent solar cell.

A summary of the semitransparent organic tandem solar cells and their corresponding single subcells is provided in Table 12.3.

12.6 TANDEM SOLAR CELL MEASUREMENT

For a better insight into the working mechanisms of organic tandem solar cells and the accurate determination of the PCE, it is important to assess the performance of each individual subcell in a tandem device. By knowing the EQE of the subcells, it is possible to determine which of the subcells is limiting the performance. Nowadays, much effort is devoted to the development of multijunction solar cells for record efficiency, and hence, reporting the accurate performance has become an increasingly important subject for the OPV community. EQE measurements are essential to report precise PCEs of the tandem devices. Therefore, there is a strong need for

TABLE 12.3
Nonexhaustive Survey of Reports Dealing with High-Efficiency Semitransparent Tandem Organic Solar Cells

Year	Recombination Layer	Bottom Cell Active Materials	V_{oc} (V)	FF	J_{sc} (mA cm^{-2}) (mW cm^{-2})	Eff (%)	Top Cell Active Materials	V_{oc} (V)	FF	J_{sc} (mA cm^{-2}) (mW cm^{-2})	Eff (%)	Tandem Cell V_{oc} (V)	FF	J_{sc} (mA cm^{-2}) (mW cm^{-2})	Eff (%)	Ref.
2013	PFN + TiO2 + modified	PBDTT-FDPP-C12:PCBM	0.76	0.61	9.1 (100)	4.3	PBDTT-SeDPP:PCBM	0.73	0.58	10.9 (111)	4.6	1.46	0.61	7.2 (100)	6.4	[109]
	PEDOT:PSS + PEDOT:PSS	PBDTT-FDPP-C12:PCBM	0.76	0.61	9.1 (100)	4.3	PBDTT-SeDPP:PC$_{70}$BM	0.73	0.58	13.0 (111)	5.5	1.47	0.59	8.4 (100)	7.3	
2013	Modified PEDOT:PSS + PEDOT:PSS PH1000 + ZnO	PIDT-phanQ:PCBM	0.85	0.68	8.56 (100)	5.0	PIDT-phanQ:PC$_{70}$BM	0.87	0.65	11.45 (100)	6.5	1.63	0.68	5.23 (80)	7.4	[113]

Organic Tandem Solar Cells

a customized protocol to measure the EQE of these cells correctly. In contrast to single-junction solar cells consisting of just one absorber layer, measuring the EQE for a tandem cell is significantly more challenging, especially for the technologically attractive two-terminal configuration in which the intermediate contact is buried and subcells cannot be addressed individually. Although some detailed protocols for EQE measurement of two-terminal tandem and triple-junction inorganic solar cells were developed,[114–117] specific characteristics, such as a field-assisted charge collection and sublinear light intensity dependence, limit their use for organic tandem solar cells. These properties necessitate that EQE experiments are carried out under representative illumination conditions and electrical bias to maintain short-circuit conditions for the addressed subcell.

A tandem cell is only able to pass current when both the wide and small bandgap subcells are simultaneously excited with light. Therefore, the EQE of organic tandem cells requires selective background illumination to optically bias one subcell such that the other subcell, which is under investigation, is current-limiting over the entire wavelength range.[43] Due to frequently observed sublinear light intensity dependence of OSCs, the EQE under reduced light intensity, for example, under monochromatized white light, is often overestimated compared to the response under standard operating conditions (AM 1.5 G with a light intensity of 100 mW cm^{-2}).[118] This consideration needs to be taken into account for EQE measurement as explained in detail in the following.

In addition, under background illumination at short-circuit conditions, the current-limiting subcell actually operates under reverse bias because the optically biased subcell operates close to V_{oc}. This effect is illustrated in Figure 12.11 for a 4.9% organic tandem solar cell comprising wide and small bandgap polymers and a ZnO/PEDOT:PSS recombination layer.[119] Therefore, a forward bias voltage on the

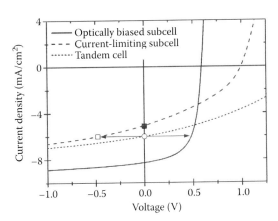

FIGURE 12.11 Influence of the bias voltage of the optically biased subcell on the current-limiting subcell. ■ is the target value, and □ is the obtained value when measuring the tandem cell under the short-circuit conditions. (Gilot, J., M. M. Wienk, and R. A. J. Janssen: Measuring the External Quantum Efficiency of Two-Terminal Polymer Tandem Solar Cells. *Advanced Functional Materials*. 2010. 20. 22. 3904–3911. Copyright Wiley-VCH Verlag GmbH & Co. KGaA. Reproduced with permission.)

tandem cell is required to compensate for this effect and to avoid the overestimation of the EQE. The EQE overestimation depends on the magnitude of the slope of the J–V curve around $V = 0$ (for example, in Figure 12.11, the error induced is about 16%).

Gilot et al. developed a method to determine the magnitudes of the background illumination and bias voltage during EQE measurements, based on the behavior of single-junction cells and optical modeling.[119] Figure 12.12a depicts a simplified schematic diagram of the EQE set-up presented in their work. It consists of a tungsten–halogen white light that is modulated with a mechanical chopper and then dispersed by a monochromator. A further two semiconductor lasers are used for bias illumination with wavelengths tuned to excite the photoactive layers in the different subcells. Their intensity can be varied with circular variable neutral density filters. A lock-in amplifier referenced to the mechanical chopper is used as a measuring unit. The device whose EQE measurement is presented here has been described elsewhere[120] and is based on PFTBT as a wide bandgap polymer and pBBTDPP2 as a small bandgap polymer, both in combination with PCBM. The normalized absorption spectra of individual films of the active layers of the subcells are displayed in Figure 12.12b and show the wavelengths used for the bias illumination. Because the PFTBT:PCBM layer is insensitive to light above 750 nm, bias illumination with a 780 nm laser is well suited to generate excess charges in the small bandgap top cell and measure the current-limiting wide bandgap bottom cell in the tandem device. On the other hand, the low absorption of pBBTDPP2:PCBM in the 500 to 550 nm region enables selective excitation of the small bandgap top cell in the EQE measurement of the tandem cell with bias illumination from a 532 nm laser.

To simulate the 1-sun operating conditions for each subcell, the following procedure was proposed by the authors: for the small-bandgap top cell, sufficient floodlight is provided by the 532-nm bias illumination. This can be explained by considering

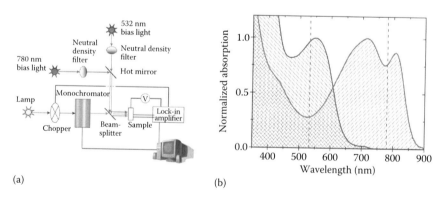

FIGURE 12.12 (a) Setup for the EQE measurement of tandem solar cells. (b) Normalized absorption spectra of both individual active layers (PFTBT:PCBM and pBBTDPP2:PCBM). The dotted lines represent the wavelengths of the bias illumination. ([a] Gilot, J., M. M. Wienk, and R. A. J. Janssen. Measuring the External Quantum Efficiency of Two-Terminal Polymer Tandem Solar Cells. *Advanced Functional Materials*. 2010. 20. 22. 3904–3911, Copyright Wiley-VCH Verlag GmbH & Co. KGaA. Reproduced with permission. [b] From Burdick, J., and T. Glatfelter, *Solar Cells* 18 (3–4):301–314, 1986.)

Organic Tandem Solar Cells

that although the 532 nm light is primarily absorbed by the wide bandgap bottom cell (creating the optical bias), the small bandgap top cell also absorbs a substantial amount at this wavelength, and the intensity of the 532-nm light can be tuned to generate 1-sun operating conditions in the small bandgap top subcell. In this regard, the appropriate AM 1.5 G equivalent intensity for 532-nm laser floodlight was determined from the absorption profile of the tandem subcells obtained by calculating the number of absorbed photons using optical modeling. For the example described here, the authors had found that, at this intensity and wavelength (532 nm), the wide bandgap bottom cell absorbs approximately three times more light, which ensures that the small bandgap top cell remains current limiting. For the EQE measurement of the wide bandgap bottom cell, the 780-nm laser light could only be used as a bias light for the small bandgap top cell, but could not act as a floodlight to have the wide bandgap bottom cell operating under 1-sun conditions. Hence, the intensity of the 780-nm bias light was just tuned to AM 1.5 G equivalent intensity for the small bandgap single-junction cell, which ensures enough charge generation in the small bandgap top cell, exceeding the charge generation in the wide bandgap bottom cell at any wavelength of the modulated probe light. Afterward, a mathematical correction of the measured EQE is applied to correct for the sublinear light intensity dependence of the current. This correction factor is determined as the average ratio between the EQE measurements with and without a 1-sun intensity floodlight of a wide bandgap single-junction device.

In the next step, the magnitude of the electrical bias was determined. The correct electrical bias could be determined by making use of the J–V characteristics of the single junction solar cells under illumination conditions representative for the subcells in the tandem device. The number of absorbed photons and, subsequently, J_{sc} was again calculated by optical modeling. However, during the EQE measurement, the tandem cell also experienced the modulated probe light on top of the bias illumination, generating additional current in the subcells. Because the variation of probe light intensity for different wavelengths caused a variable electrical bias, an average value of the electrical bias at wavelengths of minimal and maximal additional current generation of the modulated probe light could be applied. In this work, the wavelengths of 350 nm and 550 nm, respectively, for the wide bandgap bottom cell and 350 nm and 800 nm for the small bandgap top cell were selected. As a result, the average electrical bias needed for the EQE measurements was 0.60 and 0.90 V for the wide bandgap bottom cell and the small bandgap top cell, respectively.

As illustrated in Figure 12.13, the applied optical and electrical bias influences the EQE of the subcells of the organic tandem solar cell significantly. This clearly demonstrates that careful determination of the magnitude of the optical and electrical bias is essential for accurate EQE measurements.

Finally, the calculated J_{sc} based on the subcells' EQE spectra and accordingly modified J–V characteristics of the single-junction solar cells, representative of bottom and top cells, can be used to construct the J–V characteristics of the tandem device.[119–121] The J_{sc} of a tandem device estimated via convolution of the EQE with the AM 1.5 G solar spectrum has shown a comparable result to that obtained with a class-A solar simulator. This demonstrates that the EQE was accurately determined with the protocol developed by Gilot et al. The same approach can be employed to

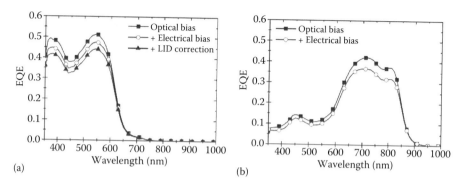

FIGURE 12.13 The EQE of the (a) wide bandgap bottom subcell and (b) small bandgap top subcell after applying relevant optical and electrical biases. For the wide bandgap bottom subcell an extra light intensity dependence (LID) correction is required. (From Burdick, J., and T. Glatfelter, *Solar Cells* 18 (3–4):301–314, 1986. Gilot, J., M. M. Wienk, and R. A. J. Janssen. Measuring the External Quantum Efficiency of Two-Terminal Polymer Tandem Solar Cells. *Advanced Functional Materials*. 2010. 20. 22. 3904–3911. Copyright Wiley-VCH Verlag GmbH & Co. KGaA. Reproduced with permission.)

measure the J–V characteristics of individual subcells in two-terminal organic tandem solar cells, in which the EQE spectra of each subcell are obtained by varying the electrical bias over the tandem cell.[122] Integration of the resulting EQE with the AM 1.5 G solar spectrum for each value of V corresponds to measurement of the J–V characteristics of the subcells under a solar simulator. In an EQE measurement, indeed, the response of the cell to modulated light is measured, which means only the photocurrent (J_{photo}) is determined. Therefore, just the J_{photo}–V curve obtained from $J_{photo} = J_{light} - J_{dark}$ of a single reference cell measured under a solar simulator can be compared to the curve from the EQE measurement. However, the lower temperature in the dark measurement may cause an overestimation of J_{photo}.[122,123] To determine the V_{oc} of the subcells, the photocurrent curve can be summed with a measured dark curve of a single reference cell identical to the subcell. It is worth mentioning that the long and intense illumination during the bias-dependent EQE measurements may degrade the tandem cell as observed by Gilot et al.[122]

12.7 SUMMARY AND OUTLOOK

Organic tandem solar cells have shown tremendous progress in recent years. As illustrated in Figure 12.14, PCEs over 10% were achieved. Although vacuum-processed tandem solar cells have currently demonstrated higher record performance than solution-processed tandem solar cells, most efforts have been concentrated on the latter one. This explains the importance of and attraction toward large-area roll-to-roll printing of OSCs. According to the reports, the large bandgap subcell is dominantly based on P3HT, which is well known for its high crystallinity, but also for its availability at reasonable quality and costs. As a big advantage, P3HT is exclusively functional with many different multiabduct fullerene derivatives, leading to the higher V_{oc}. On the low bandgap absorber side, more work needs to be done with special importance on

Organic Tandem Solar Cells

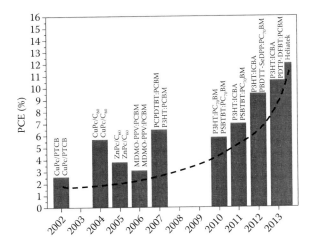

FIGURE 12.14 Recent progress in PCE of organic tandem solar cells. (Ameri, T., N. Li, and C. J. Brabec, 2013, Highly efficient organic tandem solar cells: A follow up review, *Energy Environ. Sci.*, 6, 2390–2413. Reproduced by permission of The Royal Society of Chemistry.)

absorbers outreaching into the 900 nm region. Appropriate low bandgap couples need to be designed and synthesized based on the design rules for organic tandem solar cells. Understanding the detailed operation mechanisms of the device, designing novel high potential materials and improving the robustness of the intermediate layer will open the door to further enhance the efficiency of organic tandem solar cells toward 20%.

REFERENCES

1. Jacobson, M. Z., and G. M. Masters. 2001. Exploiting wind versus coal. *Science* 293 (5534):1438.
2. Pacala, S., and R. Socolow. 2004. Stabilization wedges: Solving the climate problem for the next 50 years with current technologies. *Science* 305 (5686):968–972.
3. Fthenakis, V., J. E. Mason, and K. Zweibel. 2009. The technical, geographical, and economic feasibility for solar energy to supply the energy needs of the US. *Energy Policy* 37 (2):387–399.
4. Jacobson, M. Z., and M. A. Delucchi. 2009. A path to sustainable energy by 2030. *Scientific American* 301 (5):58–65.
5. Perez, R., and M. Perez. 2009. A fundamental look at energy reserves for the planet. *The IEA SHC Solar Update* 50:2–3.
6. Jäger-Waldau, A. 2012. PV Status Report 2012.
7. Jäger-Waldau, A. 2011. PV Status Report 2012.
8. Brabec, C. J. 2004. Organic photovoltaics: Technology and market. *Solar Energy Materials and Solar Cells* 83:273–292.
9. Brabec, C. J., N. S. Sariciftci, and J. C. Hummelen. 2001. Plastic solar cells. *Advanced Functional Materials* 11 (1):15–26.
10. Hoppe, H., and N. S. Sariciftci. 2004. Organic solar cells: An overview. *Journal of Materials Research* 19 (7):1924–1945.
11. Thompson, B. C., and J. M. J. Fréchet. 2008. Polymer–fullerene composite solar cells. *Angewandte Chemie International Edition* 47 (1):58–77.

12. Brabec, C. J., J. A. Hauch, P. Schilinsky, and C. Waldauf. 2005. Production aspects of organic photovoltaics and their impact on the commercialization of devices. *MRS Bulletin* 30 (1):50–52.
13. Brabec, C. J., and J. R. Durrant. 2008. Solution-processed organic solar cells. *MRS Bulletin* 33 (7):670–675.
14. Chen, H.-Y., J. Hou, S. Zhang et al. 2009. Polymer solar cells with enhanced open-circuit voltage and efficiency. *Nature Photonics* 3 (11):649–653.
15. Chu, T.-Y., J. Lu, S. Beaupré et al. 2011. Bulk heterojunction solar cells using thieno [3,4-c]pyrrole-4,6-dione and dithieno[3,2-b:2′,3′-d]silole copolymer with a power conversion efficiency of 7.3%. *Journal of the American Chemical Society* 133 (12): 4250–4253.
16. Zhou, H., L. Yang, A. C. Stuart, S. C. Price, S. Liu, and W. You. 2011. Development of fluorinated benzothiadiazole as a structural unit for a polymer solar cell of 7% efficiency. *Angewandte Chemie International Edition* 50 (13):2995–2998.
17. Ameri, T., N. Li, and C. J. Brabec. 2013. Highly efficient organic tandem solar cells: A follow up review. *Energy and Environmental Science* 6:2390–2413.
18. Shockley, W., and H. J. Queisser. 1961. Detailed balance limit of efficiency of p-n junction solar cells. *Journal of Applied Physics* 32 (3):510–519.
19. Frenkel, J. 1931. On the transformation of light into heat in solids. I. *Physical Review* 37 (1):17.
20. Liang, W. Y. 1970. Excitons. *Physics Education* 5 (4):226.
21. Brédas, J.-L., D. Beljonne, V. Coropceanu, and J. Cornil. 2004. Charge-transfer and energy-transfer processes in π-conjugated oligomers and polymers: A molecular picture. *Chemical Reviews* 104 (11):4971–5004.
22. Cravino, A. 2007. Origin of the open circuit voltage of donor-acceptor solar cells: Do polaronic energy levels play a role? *Applied Physics Letters* 91 (24):243502.
23. Vandewal, K., A. Gadisa, W. D. Oosterbaan et al. 2008. The relation between open-circuit voltage and the onset of photocurrent generation by charge-transfer absorption in polymer: Fullerene bulk heterojunction solar cells. *Advanced Functional Materials* 18 (14):2064–2070.
24. Gruber, M., J. Wagner, K. Klein et al. 2012. Thermodynamic efficiency limit of molecular donor-acceptor solar cells and its application to diindenoperylene/C60-based planar heterojunction devices. *Advanced Energy Materials* 2 (9):1100–1108.
25. Scharber, M. C., D. Mühlbacher, M. Koppe et al. 2006. Design rules for donors in bulk-heterojunction solar cells—Towards 10% energy-conversion efficiency. *Advanced Materials* 18 (6):789–794.
26. Green, M. A., K. Emery, Y. Hishikawa, W. Warta, and E. D. Dunlop. 2013. Solar cell efficiency tables (version 41). *Progress in Photovoltaics: Research and Applications* 21 (1):1–11.
27. Vos, A. De. 1980. Detailed balance limit of the efficiency of tandem solar cells. *Journal of Physics D: Applied Physics* 13 (5):839.
28. Dennler, G., M. C. Scharber, T. Ameri et al. 2008. Design rules for donors in bulk-heterojunction tandem solar cells-towards 15% energy-conversion efficiency. *Advanced Materials* 20 (3):579–583.
29. Scharber, M., and N. S. Sariciftci. 2013. Efficiency of bulk-heterojunction organic solar cells. *Progress in Polymer Science* 38 (12):1929–1940.
30. Li, N., D. Baran, G. D. Spyropoulos et al. 2014. Environmentally printing efficient organic tandem solar cells with high fill factors: A guideline towards 20% power conversion efficiency. *Advanced Energy Materials* 4 (11):1400084.
31. Koster, L. J. A., V. D. Mihailetchi, and P. W. M. Blom. 2006. Ultimate efficiency of polymer/fullerene bulk heterojunction solar cells. *Applied Physics Letters* 88 (9): 093511–093511.

Organic Tandem Solar Cells

32. Moliton, A., and J.-M. Nunzi. 2006. How to model the behaviour of organic photovoltaic cells. *Polymer International* 55 (6):583–600.
33. Lunt, R. R., T. P. Osedach, P. R. Brown, J. A. Rowehl, and V. Bulović. 2011. Practical roadmap and limits to nanostructured photovoltaics. *Advanced Materials* 23 (48):5712–5727.
34. Minnaert, B., and P. Veelaert. 2012. Guidelines for the bandgap combinations and absorption windows for organic tandem and triple-junction solar cells. *Materials* 5 (10):1933–1953.
35. Ameri, T., G. Dennler, C. Waldauf et al. 2008. Realization, characterization, and optical modeling of inverted bulk-heterojunction organic solar cells. *Journal of Applied Physics* 103 (8):084506–084806.
36. Siddiki, M. K., J. Li, D. Galipeau, and Q. Qiao. 2010. A review of polymer multijunction solar cells. *Energy and Environmental Science* 3 (7):867–883.
37. Sista, S., Z. Hong, L.-M. Chen, and Y. Yang. 2011. Tandem polymer photovoltaic cells-current status, challenges and future outlook. *Energy and Environmental Science* 4 (5):1606–1620.
38. Na, S.-I., S.-S. Kim, J. Jo, and D.-Y. Kim. 2008. Efficient and flexible ITO-free organic solar cells using highly conductive polymer anodes. *Advanced Materials* 20 (21):4061–4067.
39. Alemu, D., H.-Y. Wei, K.-C. Ho, and C. W. Chu. 2012. Highly conductive PEDOT:PSS electrode by simple film treatment with methanol for ITO-free polymer solar cells. *Energy and Environmental Science* 5 (11):9662–9671.
40. Chen, C.-C., L. Dou, R. Zhu et al. 2012. Visibly transparent polymer solar cells produced by solution processing. *ACS Nano* 6 (8):7185–7190.
41. Krantz, J., T. Stubhan, M. Richter et al. 2013. Spray-coated silver nanowires as top electrode layer in semitransparent P3HT:PCBM-based organic solar cell devices. *Advanced Functional Materials* 23:1711–1717.
42. Guo, F., X. Zhu, K. Forberich et al. 2013. ITO-free and fully solution-processed semitransparent organic solar cells with high fill factors. *Advanced Energy Materials* 3:1062–1067.
43. Dennler, G., H.-J. Prall, R. Koeppe, M. Egginger, R. Autengruber, and N. S. Sariciftci. 2006. Enhanced spectral coverage in tandem organic solar cells. *Applied Physics Letters* 89 (7):073502–073502.
44. Colsmann, A., J. Junge, C. Kayser, and U. Lemmer. 2006. Organic tandem solar cells comprising polymer and small-molecule subcells. *Applied Physics Letters* 89 (20): 203506.
45. Janssen, A. G. F., T. Riedl, S. Hamwi, H. H. Johannes, and W. Kowalsky. 2007. Highly efficient organic tandem solar cells using an improved connecting architecture. *Applied Physics Letters* 91 (7):073519–073519.
46. Hadipour, A., B. de Boer, J. Wildeman et al. 2006. Solution-processed organic tandem solar cells. *Advanced Functional Materials* 16 (14):1897–1903.
47. Sun, X. W. 2010. Inverted tandem organic solar cells with a MoO$_3$/Ag/Al/Ca intermediate layer. *Applied Physics Letters* 97 (5):053303.
48. Yuan, Y., J. Huang, and G. Li. 2011. Intermediate layers in tandem organic solar cells. *Green* 1 (1):65–80.
49. Ajuria, J., I. Etxebarria, W. Cambarau et al. 2011. Inverted ITO-free organic solar cells based on p and n semiconducting oxides. New designs for integration in tandem cells, top or bottom detecting devices, and photovoltaic windows. *Energy and Environmental Science* 4 (2):453–458.
50. Olthof, S., R. Timmreck, M. Riede, and K. Leo. 2012. Photoelectron spectroscopy investigations of recombination contacts for tandem organic solar cells. *Applied Physics Letters* 100 (11):113302–113302.

51. Kim, J. Y., K. Lee, N. E. Coates et al. 2007. Efficient tandem polymer solar cells fabricated by all-solution processing. *Science* 317 (5835):222–225.
52. Yang, J., R. Zhu, Z. Hong et al. 2011. A robust inter-connecting layer for achieving high performance tandem polymer solar cells. *Advanced Materials* 23 (30):3465–3470.
53. Dou, L., J. You, J. Yang et al. 2012. Tandem polymer solar cells featuring a spectrally matched low-bandgap polymer. *Nature Photonics* 6:180–185.
54. Gilot, J., M. M. Wienk, and R. A. J. Janssen. 2007. Double and triple junction polymer solar cells processed from solution. *Applied Physics Letters* 90 (14):143512.
55. Gevaerts, V. S., A. Furlan, M. M. Wienk, M. Turbiez, and R. A. J. Janssen. 2012. Solution processed polymer tandem solar cell using efficient small and wide bandgap polymer:fullerene blends. *Advanced Materials* 24 (16):2130–2134.
56. Kouijzer, S., S. Esiner, C. H. Frijters, M. Turbiez, M. M. Wienk, and R. A. J. Janssen. 2012. Efficient inverted tandem polymer solar cells with a solution-processed recombination layer. *Advanced Energy Materials* 2 (8):945–949.
57. Li, N., T. Stubhan, D. Baran et al. 2013. Design of the solution-processed intermediate layer by engineering for inverted organic multi junction solar cells. *Advanced Energy Materials* 3:301–307.
58. Li, N., D. Baran, K. Forberich et al. 2013. An efficient solution-processed intermediate layer for facilitating fabrication of organic multi-junction solar cells. *Advanced Energy Materials* 3 (12):1597–1605.
59. Li, N., P. Kubis, K. Forberich, T. Ameri, F. C. Krebs, and C. J. Brabec. 2014. Towards large-scale production of solution-processed organic tandem modules based on ternary composites: Design of the intermediate layer, device optimization and laser based module processing. *Solar Energy Materials and Solar Cells* 120:701–708.
60. Li, N., D. Baran, K. Forberich et al. 2013. Towards 15% energy conversion efficiency: A systematic study of the solution-processed organic tandem solar cells based on commercially available materials. *Energy and Environmental Science* 6:3407–3413.
61. Tong, S. W., Y. Wang, Y. Zheng, M.-F. Ng, and K. P. Loh. 2011. Graphene intermediate layer in tandem organic photovoltaic cells. *Advanced Functional Materials* 21 (23):4430–4435.
62. Tung, V. C., J. Kim, L. J. Cote, and J. Huang. 2011. Sticky interconnect for solution-processed tandem solar cells. *Journal of the American Chemical Society* 133 (24):9262–9265.
63. Kim, J., V. C. Tung, and J. Huang. 2011. Water processable graphene oxide:single walled carbon nanotube composite as anode modifier for polymer solar cells. *Advanced Energy Materials* 1 (6):1052–1057.
64. Tung, V. C., J. Kim, and J. Huang. 2012. Graphene oxide: Single-walled carbon nanotube-based interfacial layer for all-solution-processed multijunction solar cells in both regular and inverted geometries. *Advanced Energy Materials* 2 (3):299–303.
65. Zhou, Y., C. Fuentes-Hernandez, J. Shim et al. 2012. A universal method to produce low–work function electrodes for organic electronics. *Science* 336 (6079):327–332.
66. Zhou, Y., C. Fuentes-Hernandez, J. W. Shim, T. M. Khan, and B. Kippelen. 2012. High performance polymeric charge recombination layer for organic tandem solar cells. *Energy and Environmental Science* 5:9827–9832.
67. Liu, Y., C.-C. Chen, Z. Hong et al. 2013. Solution-processed small-molecule solar cells: Breaking the 10% power conversion efficiency. *Scientific Reports* 3 (33):3356.
68. Yang, J., J. You, C.-C. Chen et al. 2011. Plasmonic polymer tandem solar cell. *ACS Nano* 5 (8):6210–6217.
69. Lalanne, P., and G. M. Morris. 1996. Highly improved convergence of the coupled-wave method for TM polarization. *Journal of the Optical Society of America A* 13 (4):779–784.

70. Hugonin, J. P., and P. Lalanne. 2005. Reticolo software for grating analysis. *Institut d'Optique*, Orsay, France.
71. Persson, N.-K., and O. Inganäs. 2006. Organic tandem solar cells—Modelling and predictions. *Solar Energy Materials and Solar Cells* 90 (20):3491–3507.
72. Knittel, Z. 1976. *Optics of Thin Films*. London: Wiley.
73. Pettersson, L. A. A., L. S. Roman, and O. Inganäs. 1999. Modeling photocurrent action spectra of photovoltaic devices based on organic thin films. *Journal of Applied Physics* 86 (1):487–496.
74. Hoppe, H., N. Arnold, D. Meissner, and N. S. Sariciftci. 2004. Modeling of optical absorption in conjugated polymer/fullerene bulk-heterojunction plastic solar cells. *Thin Solid Films* 451–452:589–592.
75. Dennler, G., K. Forberich, T. Ameri et al. 2007. Design of efficient organic tandem cells: On the interplay between molecular absorption and layer sequence. *Journal of Applied Physics* 102 (12):123109–123109.
76. Hiramoto, M., M. Suezaki, and M. Yokoyama. 1990. Effect of thin gold interstitial-layer on the photovoltaic properties of tandem organic solar cell. *Chemistry Letters* 19 (3):327–330.
77. Blouin, N., A. Michaud, and M. Leclerc. 2007. A low-bandgap poly(2,7-carbazole) derivative for use in high-performance solar cells. *Advanced Materials* 19 (17):2295–2300.
78. Park, S. H., A. Roy, S. Beaupre et al. 2009. Bulk heterojunction solar cells with internal quantum efficiency approaching 100%. *Nature Photonics* 3 (5):297–302.
79. Alem, S., T.-Y. Chu, S. C. Tse et al. 2011. Effect of mixed solvents on PCDTBT:PC$_{70}$BM based solar cells. *Organic Electronics* 12 (11):1788–1793.
80. Sun, Y., C. J. Takacs, S. R. Cowan et al. 2011. Efficient, air-stable bulk heterojunction polymer solar cells using MoOx as the anode interfacial layer. *Advanced Materials* 23 (19):2226–2230.
81. Bürgi, L., M. Turbiez, R. Pfeiffer, F. Bienewald, H.-J. Kirner, and C. Winnewisser. 2008. High-mobility ambipolar near-infrared light-emitting polymer field-effect transistors. *Advanced Materials* 20 (11):2217–2224.
82. Bijleveld, J. C., A. P. Zoombelt, S. G. J. Mathijssen et al. 2009. Poly(diketopyrrolopyrrole–terthiophene) for ambipolar logic and photovoltaics. *Journal of the American Chemical Society* 131 (46):16616–16617.
83. Düggeli, M., E. M. Zahcr, P. Hayoz, O. F Aebischer, T. M. Fonrodona, and M. G. R. Turbiez. 2010. Diketopyrrolopyrrole polymers for use in organic semiconductor devices. Google Patents, WO2010049323 A1.
84. Yi, Z., X. Sun, Y. Zhao et al. 2012. Diketopyrrolopyrrole-based π-conjugated copolymer containing β-unsubstituted quintetthiophene unit: A promising material exhibiting high hole-mobility for organic thin-film transistors. *Chemistry of Materials* 24 (22):4350–4356.
85. Li, W., A. Furlan, K. H. Hendriks, M. M. Wienk, and R. A. J. Janssen. 2013. Efficient tandem and triple junction polymer solar cells. *Journal of the American Chemical Society* 15 (135):5529–5532.
86. Zhou, Y., J. W. Shim, C. Fuentes-Hernandez et al. 2012. Direct correlation between work function of indium-tin-oxide electrodes and solar cell performance influenced by ultraviolet irradiation and air exposure. *Physical Chemistry Chemical Physics* 14 (34):12014–12021.
87. Yusoff, A. R. b. M., S. J. Lee, H. P. Kim, F. K. Shneider, W. J. da Silva, and J. Jang. 2013. 8.91% power conversion efficiency for polymer tandem solar cells. *Advanced Functional Materials* 24 (15):2240–2247.
88. Yusoff, A. R. b. M., W. J. da Silva, H. P. Kim, and J. Jang. 2013. Extremely stable all solution processed organic tandem solar cells with TiO2/GO recombination layer under continuous light illumination. *Nanoscale* 5 (22):11051–11057.

89. Chang, C.-Y., L. Zuo, H.-L. Yip et al. 2013. A versatile fluoro-containing low-bandgap polymer for efficient semitransparent and tandem polymer solar cells. *Advanced Functional Materials* 23:5084–5090.
90. Jo, J., J.-R. Pouliot, D. Wynands et al. 2013. Enhanced efficiency of single and tandem organic solar cells incorporating a diketopyrrolopyrrole-based low-bandgap polymer by utilizing combined ZnO/polyelectrolyte electron-transport layers. *Advanced Materials* 25:4783–4788.
91. Bijleveld, J. C., V. S. Gevaerts, D. Di Nuzzo et al. 2010. Efficient solar cells based on an easily accessible diketopyrrolopyrrole polymer. *Advanced Materials* 22 (35): E242–E246.
92. Li, K., Z. Li, K. Feng, X. Xu, L. Wang, and Q. Peng. 2013. Development of large bandgap conjugated copolymers for efficient regular single and tandem organic solar cells. *Journal of the American Chemical Society* 135 (36):13549–13557.
93. Peng, Q., Q. Huang, X. Hou, P. Chang, J. Xu, and S. Deng. 2012. Enhanced solar cell performance by replacing benzodithiophene with naphthodithiophene in diketopyrrolopyrrole-based copolymers. *Chemical Communications* 48 (93):11452–11454.
94. Larsen-Olsen, T. T., F. Machui, B. Lechene et al. 2012. Round-robin studies as a method for testing and validating high-efficiency ITO-free polymer solar cells based on roll-to-roll-coated highly conductive and transparent flexible substrates. *Advanced Energy Materials* 2:1091–1094.
95. Zhang, H., T. Stuhban, N. Li et al. 2014. A solution-processed barium hydroxide modified aluminum doped zinc oxide layer for highly efficient inverted organic solar cells. *Journal of Materials Chemistry A.* doi: 10.1039/C4TA03421J.
96. Dou, L., W.-H. Chang, J. Gao, C.-C. Chen, J. You, and Y. Yang. 2013. A selenium-substituted low-bandgap polymer with versatile photovoltaic applications. *Advanced Materials* 25:825–831.
97. You, J., L. Dou, K. Yoshimura et al. 2013. A polymer tandem solar cell with 10.6% power conversion efficiency. *Nature Communications* 4:1446.
98. You, J., C.-C. Chen, Z. Hong et al. 2013. 10.2% power conversion efficiency polymer tandem solar cells consisting of two identical sub-cells. *Advanced Materials* 25:3973–3978.
99. Riede, M., C. Uhrich, J. Widmer et al. 2011. Efficient organic tandem solar cells based on small molecules. *Advanced Functional Materials* 21 (16):3019–3028.
100. Lassiter, B. E., J. D. Zimmerman, and S. R. Forrest. 2013. Tandem organic photovoltaics incorporating two solution-processed small molecule donor layers. *Applied Physics Letters* 103 (12):123305–123305.
101. Wei, G., X. Xiao, S. Wang et al. 2011. Functionalized squaraine donors for nanocrystalline organic photovoltaics. *ACS Nano* 6 (1):972–978.
102. Pandey, R., and R. J. Holmes. 2010. Graded donor–acceptor heterojunctions for efficient organic photovoltaic cells. *Advanced Materials* 22 (46):5301–5305.
103. He, Z., C. Zhong, X. Huang et al. 2011. Simultaneous enhancement of open-circuit voltage, short-circuit current density, and fill factor in polymer solar cells. *Advanced Materials* 23 (40):4636–4643.
104. Press-release. 2012. Available at http://www.heliatek.com/wp-content/uploads/2013/01 /130116_PR_Heliatek_achieves_record_cell_effiency_for_OPV.pdf.
105. Seo, J. H., D.-H. Kim, S.-H. Kwon et al. 2012. High efficiency inorganic/organic hybrid tandem solar cells. *Advanced Materials* 24 (33):4523–4527.
106. Beiley, Z. M., and M. D. McGehee. 2012. Modeling low cost hybrid tandem photovoltaics with the potential for efficiencies exceeding 20%. *Energy and Environmental Science* 5 (11):9173–9179.

107. Beiley, Z. M., M. Greyson Christoforo, P. Gratia et al. 2013. Semi-transparent polymer solar cells with excellent sub-bandgap transmission for third generation photovoltaics. *Advanced Materials* 25 (48):7020–7026.
108. Zhu, R., C.-H. Chung, K. C. Cha et al. 2011. Fused silver nanowires with metal oxide nanoparticles and organic polymers for highly transparent conductors. *ACS Nano* 5 (12):9877–9882.
109. Chen, C.-C., L. Dou, J. Gao, W.-H. Chang, G. Li, and Y. Yang. 2013. High-performance semi-transparent polymer solar cells possessing tandem structures. *Energy and Environmental Science* 6 (9):2714–2720.
110. Chueh, C.-C., S.-C. Chien, H.-L. Yip et al. 2013. Toward high-performance semi-transparent polymer solar cells: Optimization of ultra-thin light absorbing layer and transparent cathode architecture. *Advanced Energy Materials* 3 (4):417–423.
111. Chen, K.-S., J.-F. Salinas, H.-L. Yip, L. Huo, J. Hou, and A. K. Y. Jen. 2012. Semi-transparent polymer solar cells with 6% PCE, 25% average visible transmittance and a color rendering index close to 100 for power generating window applications. *Energy and Environmental Science* 5 (11):9551–9557.
112. Huo, L., S. Zhang, X. Guo, F. Xu, Y. Li, and J. Hou. 2011. Replacing alkoxy groups with alkylthienyl groups: A feasible approach to improve the properties of photovoltaic polymers. *Angewandte Chemie International Edition* 123 (41):9871–9876.
113. Chang, C.-Y., L. Zuo, H.-L. Yip et al. 2013. Highly efficient polymer tandem cells and semitransparent cells for solar energy. *Advanced Energy Materials* 4 (7):1301645.
114. Burdick, J., and T. Glatfelter. 1986. Spectral response and I–V measurements of tandem amorphous-silicon alloy solar cells. *Solar Cells* 18 (3–4):301–314.
115. Meusel, M., C. Baur, G. Létay, A. W. Bett, W. Warta, and E. Fernandez. 2003. Spectral response measurements of monolithic GaInP/Ga(In)As/Ge triple-junction solar cells: Measurement artifacts and their explanation. *Progress in Photovoltaics: Research and Applications* 11 (8):499–514.
116. Mueller, R. L. 1993. Spectral response measurements of two-terminal triple-junction a-Si solar cells. *Solar Energy Materials and Solar Cells* 30 (1):37–45.
117. Ran, D., S. Zhang, J. Zhang, X. Zhang, and F. Wan. 2007. Study on voltage biasing technique for spectral response measurement of multi-junction solar cell. Paper read at 3rd International Symposium on Advanced Optical Manufacturing and Testing Technologies: Optical Test and Measurement Technology and Equipment.
118. Peumans, P., S. Uchida, and S. R. Forrest. 2003. Efficient bulk heterojunction photovoltaic cells using small-molecular-weight organic thin films. *Nature* 425 (6954):158–162.
119. Gilot, J., M. M. Wienk, and R. A. J. Janssen. 2010. Measuring the external quantum efficiency of two-terminal polymer tandem solar cells. *Advanced Functional Materials* 20 (22):3904–3911.
120. Gilot, J., M. M. Wienk, and R. A. J. Janssen. 2010. Optimizing polymer tandem solar cells. *Advanced Materials* 22 (8):E67–E71.
121. Kroon, J. M., M. M. Wienk, W. J. H. Verhees, and J. C. Hummelen. 2002. Accurate efficiency determination and stability studies of conjugated polymer/fullerene solar cells. *Thin Solid Films* 403:223–228.
122. Gilot, J., M. M. Wienk, and R. A. J. Janssen. 2011. Measuring the current density–voltage characteristics of individual subcells in two-terminal polymer tandem solar cells. *Organic Electronics* 12 (4):660–665.
123. Ooi, Z. E., R. Jin, J. Huang, Y. F. Loo, A. Sellinger, and J. C. deMello. 2008. On the pseudo-symmetric current-voltage response of bulk heterojunction solar cells. *Journal of Materials Chemistry* 18 (14):1644–1651.

13 Graphene-Based Polymer and Organic Solar Cells

Reg Bauld, Faranak Sharifi, and Giovanni Fanchini

CONTENTS

13.1 Introduction 379
13.2 Methods for Producing Graphene Electrodes..380
    13.2.1 Solution-Based Graphene Exfoliation 380
    13.2.2 Oxidation Reduction Methods...381
    13.2.3 Organic Solvent Exfoliation 382
    13.2.4 Surfactant-Assisted Exfoliation 383
    13.2.5 Superacid Exfoliation 383
13.3 Thin-Film Preparation Methods from Graphene Dispersions......................383
    13.3.1 Vacuum Filtration 383
    13.3.2 Other Preparation Methods 385
13.4 CVD Graphene 385
13.5 Graphene in Organic Solar Cells..385
    13.5.1 Graphene as a Window Electrode 388
    13.5.2 Graphene Inside the Active Layer 390
13.6 Summary 390
References..391

13.1 INTRODUCTION

Graphene is an allotrope of carbon that consists of a sheet of one-atom thick sp^2 bonded carbon atoms as shown in Figure 13.1a. Since its isolation in 2004,[1] graphene has attracted significant attention due to its outstanding electrical and physical properties.[2-5] Single-layer graphene sheets exhibit properties such as high transparency, high electron mobility, excellent mechanical strength, and impressive thermal conductivity. It has been proposed for applications in a variety of different fields and technologies, including solar cells,[6-11] energy storage,[12] batteries,[13] fuel cells,[14] and in biotechnology. Although many applications of graphene have been suggested, its commercial deployment remains limited due to the inherent difficulties of scalability, reproducibility, and cost-effectively preparing graphene samples of high quality.

So far, chemical vapor deposition (CVD) has had the greatest success in producing large-area graphene thin films and boasts impressive characteristics, such as ~97% transparencies at ~200Ω/□ sheet resistances.[15] However, the drawback of this method is that it tends to be very difficult to implement in practice and requires

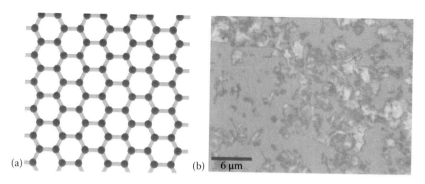

FIGURE 13.1 (a) Ball-stick model of graphene that shows the sp^2 bonding structure responsible for its unique properties. (b) A typical optical image of a solution-processed graphene film prepared by vacuum filtration on a SiO_2 substrate. (Reproduced from R. Bauld et al., *International Journal of Modern Physics B*, 26, 21, 1242004, 2012. With permission.)

extreme control of the deposition conditions in order to prevent the seeding of a second layer. Other methods, including solution processing, end up with lower-quality thin films but with much lower manufacturing costs and at a higher reproducibility. A typical optical image of a solution-processed graphene film is shown in Figure 13.1b.

One particular application for solution-processed graphene thin films that has been widely studied is their use as a flexible replacement for indium tin oxide (ITO) as a transparent and conducting electrode in the construction of solar cells and digital displays. ITO films have excellent properties with transmittances of ~99% and sheet resistances of ~10–30 Ω/□, but its widespread use in computer displays and limited supply of indium has resulted in a high price. The abundant supply of graphitic carbon available in nature, combined with the high conductivity and transparency of graphene, makes it an excellent candidate for ITO replacement.

In this chapter, we will focus on how graphene can be used in organic solar cells and, specifically, how it can be utilized as a cell electrode. In the first part of the chapter, we will review the various methods for the production of solar-grade graphene and graphene thin films, and in the second part, we will discuss the architectures, performances, and perspectives of graphene-based organic photovoltaics.

13.2 METHODS FOR PRODUCING GRAPHENE ELECTRODES

13.2.1 SOLUTION-BASED GRAPHENE EXFOLIATION

The basic mechanism for exfoliation of graphite and the dispersion of graphene in solution relies on a close matching of the surface energy of graphene and the surface energy of a specific solvent. Ultimately, it will take some energy to create a liquid–surface interface so that the minimization of this energy will provide the most stable graphene–liquid dispersions. The theory of stabilization of graphene flakes in solution has been described thoroughly[16–18] and heavily relied on the results previously achieved for the dispersion of carbon nanotubes (CNT). Although theoretical knowledge is certainly important in developing new dispersion methods, trial and error approaches have been more successful in practice due to the complicated nature of graphene dispersions.

The dispersion of graphene in the appropriate solvent is a necessary step for preparing graphene thin films from graphite. The four primarily used methods of dispersion include (i) the dispersion in aqueous solutions by the addition of stabilizing surfactants,[19,20] (ii) the use of organic solvents,[21] (iii) the treatment of graphite by strong acids,[22–24] and (iv) the oxidation of graphene with the formation and subsequent reduction of graphene oxide.[24] To date, none of these methods have proven more advantageous than another.

All of the four methods of dispersion rely on the same principle of surface energy minimization and electrostatic stabilization. Specific advantages and drawbacks arise when they are considered in the context of solar cell fabrication. In addition to the transparency and conductivity of the films, features such as surface roughness, voids, and "wettability" become very important in organic solar cells with graphene contacts as will be discussed later.

13.2.2 Oxidation Reduction Methods

The first solution exfoliation methods that have been developed are based on the dispersion of graphene oxide (GO) in water. GO is most often prepared using Hummer's method.[25] The procedure consists of the oxidation of graphite by refluxing it in the presence of a strong oxidizer, such as potassium permanganate ($KMnO_4$), in concentrated sulfuric acid (H_2SO_4) for a period of up to three days.[25–31] The resulting graphite oxide slurry can be easily dispersed in water by mild ultrasonication.[32] The process ends with a clear yellow solution of GO flakes in water, which can be used to prepare thin films.

GO films are insulating, so they must be reduced to improve their electrical conductivity. Reduction of GO films can be done by exposure to strong reducing agents, such as hydrazine monohydrate, or by heating up to high temperatures of ~500°C–1000°C under an inert atmosphere.[33,34] Such treatments result in a partially restored conductivity of reduced graphene oxide (RGO) films. The reduction removes sp^3-bonded O and OH functional groups, thereby restoring the delocalized π-electron network typical of pristine graphene, which yields a dramatic increase in electrical conductivity. However, the resultant RGO sheets are not infrequently damaged by this process.

The harsh nature of the oxidation reduction scheme introduces a variety of defects into the hexagonal lattice.[32] The presence of defects in the hexagonal structure tends to reduce the mobility of the charge carriers by acting as scattering centers that reduce the mean free path and moves the transport away from the ideal ballistic transport seen in pristine graphene sheets produced by mechanical exfoliation. In addition RGO tends to lose the unique electronic properties of graphene,[26,35] and the sheet resistance of the resulting films is too poor to produce efficient solar cells.

Even with the destructive nature of this method, it is still appealing due to the high concentration dispersions achievable—as high as 7 mg/ml.[36] Oxidation reduction techniques, in combination with a method to repair the defects, could result in a practical production method. Chemical vapor deposition on RGO has shown some promise to this end.[37] Deposition on top of RGO films have resulted in an order of magnitude improvement in conductivity over RGO, although CVD has a limited throughput, and having to resort to CVD for the purposes of improving conductivity essentially eliminates the benefits of solution-processed graphene films.

Less destructive oxidation methods exist that do not require an aggressive oxidation scheme as utilized in Hummer's method. One such method that shows some promise involves the partial oxidation of graphene and sonication in DMF.[38] Exfoliation does not happen spontaneously as in strong oxidation of graphite, so sonication is required to complete the dispersion. This, however, tends to reduce the mean flake size by fracturing the graphene sheets, which then has implications to the resulting film conductivities. Smaller flake sizes mean more flakes are required to cover a given area, so this leads to the presence of more grain boundaries and hence a higher resultant sheet resistance. An ideal solution-based method would require no sonication or oxidation so as to produce pristine graphene.

Research into oxidation processes other than the Hummer's method is one avenue that yielded to some interesting results. Recently, oxidation by dicromate[16] has resulted in an alternate route to GO. The production of GO paper as a protective coating is still an application of interest; hence future research into advanced oxidation methods is certainly to be expected.

Dispersion of GO in solvents other than water have also been investigated by Paredes et al.[29] GO was dispersed in a wide variety of organic solvents by sonication, resulting in GO concentrations far less than what is seen in water. Concentrations are on the order of ~0.1 mg/ml, but the ability to disperse GO in different solvents opens up a range of processing possibilities and reactions that are simply not possible in aqueous solution. With new high-yield methods of pristine graphene exfoliation emerging, and a limited capability to repair RGO films, we suspect this route to graphene films may fall out of favor with regards to the production of high-quality devices.

13.2.3 Organic Solvent Exfoliation

With the inherent limitations of GO exfoliation and reduction, searches have moved to simple solvent-based exfoliation. It has been discovered that graphene can be exfoliated in a number of organic solvents.[38–42] This method shows promise due to its inherent simplicity and the fact that it results in pristine monolayer sheets of graphene with respectable electronic properties. The yield and solution concentrations were once less than those achievable by oxidative methods, but with further development, extremely high concentrations have been achieved. Graphene dispersion concentrations as high as 63 mg/ml have been achieved by dispersion and reprocessing using N-methyl-2-pyrrolidinone.[42] This specific dispersion has yet to be examined to determine if it can produce high-quality conducting films, but the general trend appears to be promising.

Unfortunately, sonication is required to assist in the dispersion, which will result in smaller mean flake sizes and, hence, reduced film conductivities.[39,41] Graphene films created from these solutions tend to contain many multilayer graphene sheets with only a minority of the flakes, ~1%, being monolayer.

13.2.4 Surfactant-Assisted Exfoliation

Surfactants have proven useful in dispersing nanotubes in solutions,[43–47] so it is only natural to try a similar method for the dispersion of graphite. Work by Lotya et al.[16] using water solutions of sulfonates has demonstrated the viability of this method. More recent work[48–50] has resulted in significant improvements in the concentration of flakes in the solution, in the yield of monolayer graphene flakes in the solution, and in the overall conductivity of the films. Recent work by Sharifi et al., focused on biologically inspired surfactants, including ribonucleic acids (RNA), which not only led to improved transport properties, but also opened an avenue toward biocompatible devices.[48–50] A drawback of surfactant-assisted techniques may be in the amount of surfactant that is left behind in the films, which may be detrimental to some specific applications.

13.2.5 Superacid Exfoliation

Chlorosulphonic acid has been proven effective to disperse nanotubes[51] and graphite.[52] Concentrations of graphene flakes up to 20 mg/ml and the formation of a liquid crystal have been reported. Film conductivities of 110,000 S/m have been measured—comparable to that of micromechanical cleavage.[53] This is likely due to the fact this method leaves the graphene without functional groups, and the exfoliation occurs without sonication, resulting in larger mean flake sizes.

The exfoliation is thought to be the result of protonation of carbon.[51,52] The protonated sheets become positively charged, resulting in electrostatic repulsion. Due to the high concentrations, sheet quality, and large flake size obtained by this method, it appears that protonation is the way to go for solution-based exfoliation. However, the resulting graphene is suspended in very corrosive acid, which presents difficulties in creating a film directly from solution. Vacuum filtration can only be done on very resilient filters, making it difficult to transfer the film to another substrate via the vacuum filtration/etching method.

13.3 THIN-FILM PREPARATION METHODS FROM GRAPHENE DISPERSIONS

13.3.1 Vacuum Filtration

The most commonly used solution-based method to fabricate graphene thin films is vacuum filtration. This is, in large part, due to the ease of the method, the consistency of film thickness attainable, and the obvious utility in preparing films from a water-based solution or dispersion. Typically, with this method, a solution of exfoliated graphite or graphite oxide in water or organic solvents is prepared as previously described in Section 13.2. The suspension is left to sediment overnight and is subsequently centrifuged at several thousand rpm. The resulting supernatant is then used to prepare films using a vacuum-filtration apparatus originally proposed by Wu et al.[54] for carbon nanotube networks and adapted by Eda et al.[26] for graphene films. Graphene suspensions are vacuum filtrated through filter membranes with nanometric pore sizes. The filter membrane is chosen specifically to be etchable, typically a nitrocellulose or ester-based membrane is utilized. After filtration, the membrane is transferred film side down to any desired substrate and then dried under load in an oven or vacuum desiccator. Once dry, samples are washed in sequential baths of acetone and methanol to etch the filter membrane leaving behind a graphene film on its substrate. A schematic of the vacuum-filtration method is shown in Figure 13.2.

This method is particularly useful because the film thickness is uniform, and the total thickness of the film can be controlled by altering the concentration and the filtrated volume of the dispersion. The graphene flakes tend to block the solution; hence, the procedure is self-regulating as regions of the filter that are high in flake density will have lower flow rates, and so this mechanism results in a uniform film.[26,27,55] In the case of fabrication of GO "bucky paper," the films may even be thick and flexible enough so they can simply be peeled off[36,56] without the need for filter etching.

Monolayer films can also be created using this method,[26] and the use of a nitrocellulose filter allows the film to be placed on any desired substrate by placing the filter film side down on the substrate with pressure and then etching with acetone. However, this method is not without its limitations. Clearly any graphene solution that would dissolve or destroy the nitrocellulose filter could not be filtrated in this manner as is the case with some organic solvents or superacid exfoliation.[52] Using a filter that is more resilient typically means it will be more difficult to etch. In the case of exfoliation with chlorosulfonic acid, the films are generally examined directly on a nanotube filter without the ability to transfer to another substrate.

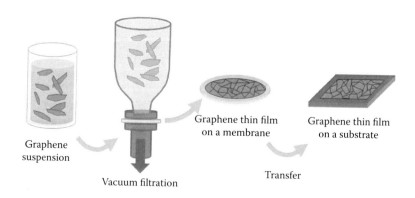

FIGURE 13.2 Schematic of the vacuum filtration method used for depositing solution-processed graphene thin films.[26] This process includes dispersion of graphene or GO in water or organic solvents, filtration on sacrificial filter membranes, transfer of the film onto the requisite substrate, and etching the membrane, leaving behind the graphene thin film on its substrate. (Reproduced from R. Bauld et al., *International Journal of Modern Physics B*, 26, 21, 1242004, 2012. With permission.)

Graphene-Based Polymer and Organic Solar Cells 385

13.3.2 OTHER PREPARATION METHODS

Spin coating from graphene-containing solutions has been widely used as a simple, proof-of-concept method to deposit graphene films onto flat surfaces. Film thickness can be controlled by selecting different rotation speeds and solution concentrations. Although relatively uniform thin GO films have been fabricated using this method,[33] reproducibility tends to be lower than by vacuum filtration.

Other methods include drop casting (that is, placing a drop of the solution on a substrate then drying it either in a controlled environment or in ambient conditions) and dip coating. These are quick methods than can be used to examine the quality and size of individual graphene flakes. Spray coating is also a popular method and can be used to make thicker films. A solution is loaded into an airbrush and sprayed onto a substrate.[27]

13.4 CVD GRAPHENE

High-quality graphene films can be produced by chemical vapor deposition. The most popular method that produces high-quality graphene films involves first growing graphene on copper foil,[57] transferring the graphene to a polymer support, and then moving the graphene thin film to the desired substrate. The polymer film is then etched away, leaving behind a high-quality graphene thin film.[58] Research in this field focuses on the surface structure of the copper foil used and different methods of film transfer. Different surface treatments can highly alter the resulting film morphology and quality.

Graphene films produced by this method are often superior to the solution-processed films described above in terms of mobility, transparency, and overall film conductivity. In addition, these can be produced with a very high percentage of single-layer graphene. This results in a very smooth film, which is ideal for solar cell construction. The benefits of these types of films over solution-processed films are readily apparent; however, they are lacking in some respects. These films are more expensive to produce than solution-processed films due to the nature of the deposition. The process is, however, scalable to large 30-inch films as shown by Bae et al.[15] Films produced in this way have optical and electrical characteristics that are superior to films of ITO with sheet resistances up to 30 $\Omega/\square$ at 90% transmittance in the visible range.

13.5 GRAPHENE IN ORGANIC SOLAR CELLS

While the potential associated with using transparent and conducting graphene platelets in solar cells was recognized quite early,[2-5] the most optimal architecture for exploiting the outstanding properties of graphene in these devices is still unclear. Several reports exist in the literature about the use of graphene-based materials in photovoltaics and offer a variety of different and unrelated approaches about the most appropriate utilization of such materials.[2-5] The vast majority of these reports refer to the use of graphene in solar cells adopting thin film architectures.

In photovoltaics, the thin-film architecture is preferable over a planar architecture only when the solar cell active layers display excessively poor carrier transport properties, which requires them to be sandwiched between two collection electrodes with much higher conductivity: a transparent "window" electrode and a "backing" electrode. Additionally, thin-film solar cells may also benefit from the insertion of additional conducting materials buried within the active layer in order to improve their transport properties for allowing the cell to work in a tandem configuration or, simply, for allowing a more efficient process of collection of photoexcited charge carriers. Examples of thin-film solar cells are amorphous photovoltaic devices, organic photovoltaics (OPV), and dye-sensitized solar cells (DSSC). Typical cell architectures are shown in Figure 13.3.

A major difference between inorganic and organic thin-film photovoltaics rests in the nature of the electron–hole pairs, or "excitons," generated by light absorbed in the active layer of the cell as demonstrated in Figure 13.3. Excitons are the bound electron–hole pair that is created during photoexcitation of an electron in the active layer and are important in understanding how organic solar cells work and specifically why organic solar cells require transparent electrodes, such as graphene or ITO in order to work efficiently. Simply put, the binding energy between an electron and hole pair in an organic material is much stronger than in a solar cell–grade inorganic material, such as silicon. Organic materials can have exciton-binding energies ranging around 0.1 eV. This strong binding energy keeps the pair bound together, making recombination, and hence energy loss, more probable.

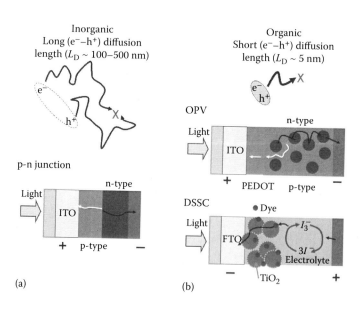

FIGURE 13.3 Various types of thin-film solar cells: (a) Inorganic solar cells in which the exciton diffusion length is relatively large, and a p-n junction architecture is suitable for efficient excitonic dissociation processes. (b) Nanostructured organic solar cell architectures with a high density of interfaces, for materials with short exciton diffusion lengths: bulk heterojunction OPVs and DSSCs. (Reproduced from R. Bauld et al., *International Journal of Modern Physics B*, 26, 21, 1242004, 2012. With permission.)

In inorganic materials, excitonic diffusion lengths are relatively large (typically ~100–500 nm), and the excitons may cover relatively long distances before being dissociated at the p-n interface and generate a hole current drifting toward the ITO window electrode and an electron current drifting toward the backing electrode as in Figure 13.3a. Conversely, excitonic diffusion lengths are very short in organic materials (typically ~5–10 nm) as in Figure 13.3b. Because excitons in such materials may only cover relatively short distances, they need to encounter a p-n interface at distances within their diffusion length for being dissociated into an electron and hole instead of recombining radiatively with the emission of a photon. This means that, in organic solar cells, an extremely high surface area is required between the p-type and n-type components of the photoactive layer for the excitons to encounter a p-n interface as soon as possible and dissociate. Therefore, a suitable architecture for thin-film organic solar cells is formed either by an ultrathin planar p-n junction, or a nanostructured approach must be adopted in order to maximize the density of p-n interfaces. Subsequently, the p-type and n-type constituents of the active layer are frequently mixed at the nanoscale. Typical examples of nanostructured OPVs are represented by bulk heterojunction OPVs and DSSCs as depicted in Figure 13.3b.

In OPVs, the active layer is formed by a p-type organic material (conducting polymers or molecular nanocrystals) blended by an acceptor material, typically fullerene derivatives. In order to prevent the recombination of the electron at the window electrode, a thin electron-blocking, hole-transport layer is generally inserted at the interface between the active layer and such electrode, generally formed by a transparent conducting ITO thin film. The typical material of choice for organic hole transport layers is a water soluble blend of poly-3,4-ethylenedioxythiophene:polystyrene-sulfonate (PEDOT:PSS).

For a transparent electrode ("window") material, the most basic requirement is a good tradeoff between the optical transparency and the sheet resistance. In addition, a work function of the transparent conductor matching the energy level of one of the two species of photoexcited carriers (either electrons or holes) is also essential for the use of a graphene thin film as an electrode material. Graphene thin films can also be treated and functionalized[56] to assist in matching the work function to the materials used in the active layer.

In addition, there is some evidence that solution-based graphene materials may also play a role in the active layer of these devices. These applications include the use of graphene to increase the conductivity of the active layer and to aid in the collection of charges during cell operation.

Applications of graphene thin films for solar cells have a number of common issues. One general issue for assembling organic materials at the top of a graphene thin film is in the poor wettability of graphene. The most popular organic layers used in optoelectronics tend not to adhere well to the graphene surface, resulting in poor coatings and reduced efficiencies or even outright device failure. Some remedies to these problems have been found[9] and will be discussed in Section 13.5.1. This, of course, comes with the drawback of another treatment step that will ultimately increase the net cost of the resulting cell.

Another issue is that, to date, no solution-processing method has been able to produce films primarily consisting of single-layer graphene. Typically, distributions of single-, double-, and multilayer sheets are found.[42,48,59] This introduces an inherent surface roughness in the films that is a source for electrical shorts between the electrodes, which, in turn, reduces the efficiency. A great deal of care must be taken to reduce the occurrence of thick flakes, and this typically requires a "reprocessing" of a dispersed solution. This is normally done by collecting the sediment after a solution-dispersion process and redispersing in a solution. This tends to result in better-quality films with lower mean flake thickness.[48] Nonetheless, the inherent roughness of graphene thin films compared to other electrode materials may also be turned into an opportunity to increase the charge collection area with obvious improvements in term of the efficiency of this process.

13.5.1 Graphene as a Window Electrode

In this section, we will be focusing on the use of graphene as a transparent conductor in OPVs. Transparent graphene anodes synthesized by CVD have been reported by Wang et al.,[11] and polymeric solar cells were assembled on them. The utilized active layers were the "standard" bulk heterojunctions of regioregular poly(3-hexylthiophene) and phenyl-C61-butyric acid methyl ester (P3HT:PCBM). These graphene-based devices demonstrated 0.21% efficiency when assembled on a pristine graphene electrode, but the efficiency could rise up to 1.71% when the graphene film was modified by pyrene butanoic acid succinimidyl ester (PBASE). This was only 55.2% of the efficiency obtained from the ITO control device, which may lead to conclude that P3HT:PCBM is not the best platform for graphene solar cells even when the quality of the graphene films is supposedly high as in the case of CVD-grown graphene.

An interesting alternative to polymer–fullerene blends are OPVs assembled from small polyaromatic molecules. This is a very promising approach because several systems of small polyaromatic molecules are uniquely positioned to self-assemble on graphene and create the most optimal interfaces for efficient charge collection. The PEDOT blocking layer between graphene and the active layer of the cells is sometimes avoided in these cases and can be replaced by the insertion of a hole-blocking, electron-transport layer between the active layer and the backing electrode. A possible choice for the hole-blocking layer is 2,9-Dimethyl-4,7-diphenyl-1,10-phenanthroline[10] (also known as Bathocuproine or BCP), the electron transport material of choice for organic light-emitting devices. For instance, Gomez de Arco et al.[10] compared solar cells constructed from graphene and ITO electrodes fabricated on flexible polyethylene terephthalate (PET) substrates and utilizing copper phtalocyanine-fullerene (CuPc-C_{60}) active bilayers, and they obtained efficiency values of 1.18% and 1.27%, respectively—a figure that indicates that the graphene solar cell possesses an efficiency comparable to the ITO control device within the fabrication uncertainties. The message that can be extracted from this work is that a specific design and specific solar cell architectures are required in order to optimize solar cells assembled on graphene electrodes, and the optimization process cannot simply rely on the results obtained from the optimization of organic solar cells on ITO.

P3HT:PCBM solar cells have also been fabricated[10] by using graphene thin films directly synthesized from organic reagents, which hold a normal-incidence transmittance of ~85% at 4 nm thickness. The advantage of this approach should be in the more accurate control of the roughness of the electrode and in smoother platforms.

The highest external quantum efficiency (EQE) of this type of cells was observed at 520 nm monochromatic light,[10] and at the same conditions, the efficiency of a control ITO device was found to be 47%. Under 510 nm monochromatic light, the efficiency of this graphene-based cell was 1.53%, similar to the ITO-based cell. However, under AM 1.5 sunlight illumination, the graphene-based solar cell showed an efficiency of η = 0.29%, which is, again, much lower than the value reported for the control cell (η = 1.17%). The performance in terms of open circuit voltage of the graphene-based cell and the control were comparable, which may indicate a limited π–π electron conjugation between the graphene electrode and the active layer. The relatively low values of short-circuit current, fill factor, and EQE were attributed by the authors to the high resistance of the prepared graphene films.[6]

Graphene films prepared by spin coating a suspension of reduced graphite oxide (or by reducing a thin layer of spin-coated GO) have been widely used as window electrodes in OPVs.[33,60] AFM characterization revealed that the thickness of these RGO layers is 10 nm or less with a surface roughness lower than 3 nm. Additional analysis demonstrated that the transparency of the film was better than 80% at less than 20 nm film thickness, but the resistance of the films ranged from 1 to 5 MΩ,[60] which does not favorably compare with ITO. In fact, the organic solar cells fabricated using these electrodes held efficiencies of ~0.4%, much inferior to those attained with ITO as an electrode (0.84%). This was most likely due to the very large sheet resistance of the RGO films.

In order to consider the application of graphene as a substitute for ITO anodes in organic solar cells, an OPV was built from a stack of layers of graphene and PEDOT:PSS/CuPc/C_{60}/BCP/Ag.[9] This helped to elucidate the complementary roles of graphene and PEDOT:PSS in graphene-based OPVs and understand the influence of factors, such as the work function and the surface wettability. Graphene sheets with a controlled number of layers were used as transparent electrodes in this case, and different preparation conditions have been explored in order to optimize the solar cell photoconversion efficiency.[9] A challenge that has been identified was the limited surface wettability of graphene electrodes by the aqueous suspensions used to spin coat PEDOT, the hole-transporting layer. Uniform coverage of the graphene surface by PEDOT plays a crucial role in the performance of graphene photovoltaic devices. However, the hydrophobic surface of pristine graphene makes it difficult to uniformly dispense aqueous suspensions of PEDOT:PSS on graphene.

In order to investigate such an issue, different sets of solar cells were built,[9] and graphene layers at different thicknesses were optimized. In one set of solar cells reported,[9] the wettability of the graphene surface by PEDOT:PSS was improved by dissolving PEDOT in an organic solvent (dimethyl sulfoxide, DMSO) and by replacing PSS with poly-(ethylene glycol) (PEG). As a result, the PEDOT/graphene interface improved and the sandwich conductivity increased, which also resulted in an improvement of the short-circuit current. However, an overall decrease in open circuit voltage and fill factor prevented the taking advantage of benefits related to the increase in short-circuit current. This effect was attributed to possible mismatches between the work functions of the electrode and the active layer.[9]

Additional attempts to improve the graphene–polythiophene interface were also reported by Park et al.[9] They reported the functionalization of graphene thin films by AuCl$_3$, which was preceded by an oxygen plasma treatment of the films. Spin coating of PEDOT:PSS on graphene films functionalized by AuCl$_3$ was found to be much easier than on pristine graphene films, which significantly improved the device performance. In these sets of devices, the AM 1.5 efficiencies of graphene-based and ITO-based solar cells were found to be comparable with η = 1.63% and η = 1.77%, respectively.[9]

13.5.2 Graphene Inside the Active Layer

In addition to uses as solar cell electrodes, research has been done into including small amounts of graphene into the active layer of an OSC. Graphene embedded into the active layer can have several positive effects. It can increase the overall conductivity of the active layer, and it can assist in the collection of charge carriers to the electrodes of the device. Work by Yu et al.[61] has demonstrated that incorporating a small (0.005 wt%) concentration of commercially produced graphene powder into the active layer of an OSC can result in an improved solar cell efficiency. The authors find that graphene can work as a good electron acceptor and due to its large surface area and high conductivity, it is an especially effective method of ensuring that electrons reach the cathode of the solar cell. However, this method does come with some drawbacks. Graphene tends to increase the shunt resistance of the cell. A graphene sheet extending through the entire active layer of the cell will provide an alternate current path for the generated photocurrent and will effectively "short" the solar cell. The work by Yu et al., showed that an improvement in power conversion efficiency is possible with the incorporation of graphene; however, this is not an overall improvement. The cells are purposely designed with a deficiency of acceptor material so that graphene can provide the percolating conduction pathways needed for optimal charge collection. It remains to be seen if an optimization of both the graphene content and of the donor–acceptor ratio can result in a net improvement over cells produced without graphene.

13.6 SUMMARY

In summary, we have reviewed some of the methods used for preparing dispersions of graphene in water and organic solvents and examined their success for the fabrication of transparent and conducting films that are more cost-effective than CVD-grown graphene. Device-grade transparent and conducting films have been demonstrated, but the quality of the devices tends to be worse than devices fabricated by CVD graphene albeit with the benefit of cheaper production costs.

We have also examined the use of graphene in solar cell devices and have shown how graphene can be utilized in many different components of organic solar cells. Graphene can be used effectively as a window electrode, and doped graphene can also be used as an electron acceptor in organic photovoltaics. Solution-processed graphene thin films are yet not yet able to effectively compete with ITO; we suspect that further development in solution-processing techniques will result in the more widespread adoption of graphene in several areas of optoelectronics and can potentially reduce the fabrication costs of organic solar cells. CVD deposition of graphene remains a viable option as well; however, it is preferable that solution-processed films can soon provide competitive electrical characteristics.

REFERENCES

1. K. S. Novoselov, A. K. Geim, S. V. Morozov, D. Jiang, Y. Zhang, S. V. Dubonos, I. V. Grigorieva and A. A. Firsov, *Science*, 2004, **306**, 666–669.
2. A. K. Geim and K. S. Novoselov, *Nat Mater*, 2007, **6**, 183–191.
3. T. Seyller, A. Bostwick, K. V. Emtsev, K. Horn, L. Ley, J. L. McChesney, T. Ohta, J. D. Riley, E. Rotenberg and F. Speck, *Phys Status Solidi B*, 2008, **245**, 1436–1446.
4. C. N. R. Rao, A. K. Sood, K. S. Subrahmanyam and A. Govindaraj, *Angew Chem Int Ed*, 2009, **48**, 7752–7777.
5. A. K. Geim, *Science*, 2009, **324**, 1530–1534.
6. X. Wang, L. J. Zhi, N. Tsao, Z. Tomovic, J. L. Li and K. Mullen, *Angew Chem Int Ed*, 2008, **47**, 2990–2992.
7. X. Wang, L. J. Zhi and K. Mullen, *Nano Lett*, 2008, **8**, 323–327.
8. S. R. Sun, L. Gao and Y. Q. Liu, *Appl Phys Lett*, 2010, **96**, 086113.1–086113.3.
9. H. Park, J. A. Rowehl, K. K. Kim, V. Bulovic and J. Kong, *Nanotechnology*, 2010, **21**, 505204.1–505204.6.
10. L. Gomez De Arco, Y. Zhang, C. W. Schlenker, K. Ryu, M. E. Thompson and C. Zhou, *ACS Nano*, 2010, **4**, 2865–2873.
11. Y. Wang, X. H. Chen, Y. L. Zhong, F. R. Zhu and K. P. Loh, *Appl Phys Lett*, 2009, **95**, 063302.1–063302.3.
12. A. Chidembo, S. H. Aboutalebi, K. Konstantinov, M. Salari, B. Winton, S. A. Yamini, I. P. Nevirkovets and H. K. Liu, *Energy Environ Sci*, 2012, **5**, 5236–5240.
13. E. Yoo, J. Kim, E. Hosono, H. Zhou, T. Kudo and I. Honma, *Nano Lett*, 2008, **8**, 2277–2282.
14. J. B. Hou, Y. Y. Shao, M. W. Ellis, R. B. Moore and B. L. Yi, *Phys Chem Chem Phys*, 2011, **13**, 15384–15402.
15. S. Bae, H. Kim, Y. Lee, X. F. Xu, J. S. Park, Y. Zheng, J. Balakrishnan, T. Lei, H. R. Kim, Y. I. Song, Y. J. Kim, K. S. Kim, B. Ozyilmaz, J. H. Ahn, B. H. Hong and S. Iijima, *Nat Nanotechnol*, 2010, **5**, 574–578.
16. M. Lotya, Y. Hernandez, P. J. King, R. J. Smith, V. Nicolosi, L. S. Karlsson, F. M. Blighe, S. De, Z. M. Wang, I. T. McGovern, G. S. Duesberg and J. N. Coleman, *J Am Chem Soc*, 2009, **131**, 3611–3620.
17. Z. Sun, V. Nicolosi, S. D. Bergin and J. N. Coleman, *Nanotechnology*, 2008, **19**, 485702.
18. S. D. Bergin, V. Nicolosi, H. Cathcart, M. Lotya, D. Rickard, Z. Sun, W. J. Blau and J. N. Coleman, *J Phys Chem C*, 2008, **112**, 972–977.
19. B. Z. Tang and H. Y. Xu, *Macromolecules*, 1999, **32**, 2569–2576.
20. L. Vaisman, H. D. Wagner and G. Marom, *Adv Colloid Interface Sci*, 2006, **128**, 37–46.
21. K. D. Ausman, R. Piner, O. Lourie, R. S. Ruoff and M. Korobov, *J Phys Chem B*, 2000, **104**, 8911–8915.
22. H. Hiura, T. W. Ebbesen and K. Tanigaki, *Adv Mater*, 1995, **7**, 275–276.
23. K. Tohji, H. Takahashi, Y. Shinoda, N. Shimizu, B. Jeyadevan, I. Matsuoka, Y. Saito, A. Kasuya, S. Ito and Y. Nishina, *J Phys Chem B*, 1997, **101**, 1974–1978.
24. P. K. Rai, A. N. G. Parra-Vasquez, J. Chattopadhyay, R. A. Pinnick, F. Liang, A. K. Sadana, R. H. Hauge, W. E. Billups and M. Pasquali, *J Nanosci Nanotechnol*, 2007, **7**, 3378–3385.
25. W. S. Hummers and R. E. Offeman, *J Am Chem Soc*, 1958, **80**, 1339–1340.
26. G. Eda, G. Fanchini and M. Chhowalla, *Nat Nanotechnol*, 2008, **3**, 270–274.

27. D. Li, M. B. Muller, S. Gilje, R. B. Kaner and G. G. Wallace, *Nat Nanotechnol*, 2008, **3**, 101–105.
28. H. C. Schniepp, J. L. Li, M. J. McAllister, H. Sai, M. Herrera-Alonso, D. H. Adamson, R. K. Prud'homme, R. Car, D. A. Saville and I. A. Aksay, *J Phys Chem B*, 2006, **110**, 8535–8539.
29. J. I. Paredes, S. Villar-Rodil, A. Martinez-Alonso and J. M. D. Tascon, *Langmuir*, 2008, **24**, 10560–10564.
30. S. Stankovich, D. A. Dikin, R. D. Piner, K. A. Kohlhaas, A. Kleinhammes, Y. Jia, Y. Wu, S. T. Nguyen and R. S. Ruoff, *Carbon*, 2007, **45**, 1558–1565.
31. S. Niyogi, E. Bekyarova, M. E. Itkis, J. L. McWilliams, M. A. Hamon and R. C. Haddon, *J Am Chem Soc*, 2006, **128**, 7720–7721.
32. C. Gómez-Navarro, R. T. Weitz, A. M. Bittner, M. Scolari, A. Mews, M. Burghard and K. Kern, *Nano Lett*, 2007, **7**, 3499–3503.
33. H. A. Becerril, J. Mao, Z. Liu, R. M. Stoltenberg, Z. Bao and Y. Chen, *ACS Nano*, 2008, **2**, 463–470.
34. V. C. Tung, M. J. Allen, Y. Yang and R. B. Kaner, *Nat Nanotechnol*, 2009, **4**, 25–29.
35. S. Stankovich, D. A. Dikin, G. H. B. Dommett, K. M. Kohlhaas, E. J. Zimney, E. A. Stach, R. D. Piner, S. T. Nguyen and R. S. Ruoff, *Nature*, 2006, **442**, 282–286.
36. S. Park, J. H. An, R. D. Piner, I. Jung, D. X. Yang, A. Velamakanni, S. T. Nguyen and R. S. Ruoff, *Chem Mater*, 2008, **20**, 6592–6594.
37. V. Lopez, R. S. Sundaram, C. Gomez-Navarro, D. Olea, M. Burghard, J. Gomez-Herrero, F. Zamora and K. Kern, *Adv Mater*, 2009, **21**, 4683–4685.
38. X. L. Li, G. Y. Zhang, X. D. Bai, X. M. Sun, X. R. Wang, E. Wang and H. J. Dai, *Nat Nanotechnol*, 2008, **3**, 538–542.
39. Y. Hernandez, V. Nicolosi, M. Lotya, F. M. Blighe, Z. Y. Sun, S. De, I. T. McGovern, B. Holland, M. Byrne, Y. K. Gun'ko, J. J. Boland, P. Niraj, G. Duesberg, S. Krishnamurthy, R. Goodhue, J. Hutchison, V. Scardaci, A. C. Ferrari and J. N. Coleman, *Nat Nanotechnol*, 2008, **3**, 563–568.
40. D. Rangappa, K. Sone, M. S. Wang, U. K. Gautam, D. Golberg, H. Itoh, M. Ichihara and I. Honma, *Chem-Eur J*, 2010, **16**, 6488–6494.
41. S. Chandra, S. Sahu and P. Pramanik, *Mater Sci Eng B-Adv*, 2010, **167**, 133–136.
42. U. Khan, H. Porwal, A. O'Neill, K. Nawaz, P. May and J. N. Coleman, *Langmuir*, 2011, **27**, 9077–9082.
43. C. E. Hamilton, J. R. Lomeda, Z. Z. Sun, J. M. Tour and A. R. Barron, *Nano Lett*, 2009, **9**, 3460–3462.
44. V. C. Moore, M. S. Strano, E. H. Haroz, R. H. Hauge, R. E. Smalley, J. Schmidt and Y. Talmon, *Nano Lett*, 2003, **3**, 1379–1382.
45. M. J. O'Connell, P. Boul, L. M. Ericson, C. Huffman, Y. H. Wang, E. Haroz, C. Kuper, J. Tour, K. D. Ausman and R. E. Smalley, *Chem Phys Lett*, 2001, **342**, 265–271.
46. M. J. O'Connell, S. M. Bachilo, C. B. Huffman, V. C. Moore, M. S. Strano, E. H. Haroz, K. L. Rialon, P. J. Boul, W. H. Noon, C. Kittrell, J. P. Ma, R. H. Hauge, R. B. Weisman and R. E. Smalley, *Science*, 2002, **297**, 593–596.
47. M. S. Strano, V. C. Moore, M. K. Miller, M. J. Allen, E. H. Haroz, C. Kittrell, R. H. Hauge and R. E. Smalley, *J Nanosci Nanotechnol*, 2003, **3**, 81–86.
48. M. Lotya, P. J. King, U. Khan, S. De and J. N. Coleman, *ACS Nano*, 2010, **4**, 3155–3162.
49. N. W. Pu, C. A. Wang, Y. M. Liu, Y. Sung, D. S. Wang and M. D. Ger, *J Taiwan Inst Chem E*, 2012, **43**, 140–146.
50. F. Sharifi, R. Bauld, M. S. Ahmed and G. Fanchini, *Small*, 2011, **8**, 699–706.
51. V. A. Davis, A. N. G. Parra-Vasquez, M. J. Green, P. K. Rai, N. Behabtu, V. Prieto, R. D. Booker, J. Schmidt, E. Kesselman, W. Zhou, H. Fan, W. W. Adams, R. H. Hauge, J. E. Fischer, Y. Cohen, Y. Talmon, R. E. Smalley and M. Pasquali, *Nat Nanotechnol*, 2009, **4**, 830–834.

52. N. Behabtu, J. R. Lomeda, M. J. Green, A. L. Higginbotham, A. Sinitskii, D. V. Kosynkin, D. Tsentalovich, A. N. G. Parra-Vasquez, J. Schmidt, E. Kesselman, Y. Cohen, Y. Talmon, J. M. Tour and M. Pasquali, *Nat Nanotechnol*, 2010, **5**, 406–411.
53. K. I. Bolotin, K. J. Sikes, Z. Jiang, M. Klima, G. Fudenberg, J. Hone, P. Kim and H. L. Stormer, *Solid State Commun*, 2008, **146**, 351–355.
54. Z. C. Wu, Z. H. Chen, X. Du, J. M. Logan, J. Sippel, M. Nikolou, K. Kamaras, J. R. Reynolds, D. B. Tanner, A. F. Hebard and A. G. Rinzler, *Science*, 2004, **305**, 1273–1276.
55. D. A. Dikin, S. Stankovich, E. J. Zimney, R. D. Piner, G. H. B. Dommett, G. Evmenenko, S. T. Nguyen and R. S. Ruoff, *Nature*, 2007, **448**, 457–460.
56. J. D. Roy-Mayhew, G. Boschloo, A. Hagfeldt and I. A. Aksay, *ACS Appl Mater Interfaces*, 2012, **4**, 2794–2800.
57. X. S. Li, C. W. Magnuson, A. Venugopal, R. M. Tromp, J. B. Hannon, E. M. Vogel, L. Colombo and R. S. Ruoff, *J Am Chem Soc*, 2011, **133**, 2816–2819.
58. C. Mattevi, H. Kim and M. Chhowalla, *J Mater Chem*, 2011, **21**, 3324–3334.
59. G. Eda and M. Chhowalla, *Nano Lett*, 2009, **9**, 814–818.
60. J. B. Wu, H. A. Becerril, Z. N. Bao, Z. F. Liu, Y. S. Chen and P. Peumans, *Appl Phys Lett*, 2008, **92**, 263302.1–263302.3.
61. F. Yu and V. K. Kuppa, *Mater Lett*, 2013, **99**, 72–75.

Index

Page numbers followed by f and t indicate figures and tables, respectively.

A

ABLs (anode buffer layers), 202–220
   GO, 219–220
   metal oxides, 213–214, 213f
   PEDOT:PSS, 203–213
      additives in, 206–210, 207f, 209t
      advantages, 204
      chemical structure, 203f
      drawbacks, 205
      grades of, 204
      incorporation of, 203
      modified, ITO-free PSCs based on, 210–213, 211t, 212f
      physical treatments, 205–206
      thickness of layer, 204–205, 205f
      types of, 204t
   polymers and small-molecule organic materials, 215–217, 216f, 217t
   self-assembled buffer layers, 218–219, 218f
Absorption, UV-visible
   of PBT-T1, 283–284, 283f
   XRD spectra of PBT-T1:$PC_{60}BM$ films and, 284–285, 284f
Acceptor materials, defect states in, 182–184, 183f, 183t
Acceptor species
   chemical structure, 103–104, 103f
   concentrations of, 96–98, 97f, 99f
   solvents, choice, 98, 100–102, 101f
Active layer, graphene inside, 390
Active layer nanomorphology, engineering in PSCs, 282–300
   fullerene ratio and solvent additives, PBT-T1-based BHJ solar cells and, 283–291
      AFM topography and phase images, PBT-T1:$PC_{60}BM$ films, 285–286, 285f
      current–voltage characteristics, PBT-T1:$PC_{60}BM$ cells, 286–288, 286f, 287t
      current–voltage characteristics, PBT-T1:$PC_{70}BM$ cells, 288–289, 288f, 289t
      photo-CELIV and extracted charge carrier density vs. delay time, PBT-T1, 290–291, 290f, 291t
      UV-visible absorption, PBT-T1, 283–284, 283f
      UV-visible absorption and XRD spectra, PBT-T1:$PC_{60}BM$ films, 284–285, 284f
   overview, 282–283
   solvent additive, PDPP3T:PCBM solar cells and, 291–300, 292f
      additive on charge transport and recombination dynamics, 297–300, 298f, 299f
      additives on surface morphology, 294–296, 295f
      current–voltage characteristics, 296–297, 296f, 296t
      EFTEM images, 292–293, 293f
      selected area electron diffraction and Raman spectra, 293, 294f
Additives
   blending solvent additive, in nanophase separation, 251–252
   in PEDOT:PSS, 206–210, 209t
      alcohols, 206
      DMF, 206–207, 207f
      DMSO, 207, 208
      GO, 208
      gold nanoparticles, 208
      metal oxides, 209–210
      MoO_3, 209–210
      role, 208
      sorbitol, 206
      SWCNTs, 208
      TMS, 208
   processing conditions and, nonequilibrium effects, 118–119
   solvent, on nanomorphology
      PBT-T1-based BHJ solar cells and, 283–291. See also Solvent additives
Admittance spectroscopy, 173
Advanced ternary OPVs, nanophase separation (case study), 271–272
Aerogels, 23
AFM (atomic force microscopy), 94, 252
   phase images of PBT-T1:$PC_{60}BM$ films and, 285–286, 285f
Alcohols, for PEDOT:PSS ABL, 206
Alignment, of NRs, 324–334
   factors influencing, 326–332
      aspect ratio, 326, 327f
      concentration, 327, 329f

395

evaporation rate, 330, 330f
external fields, 331–332, 332f
solvent, 327, 329f, 330
substrates, 326, 327, 328f
surface ligands, 330, 331, 331f
temperature, 326
overview, 324
physical origin of depletion force, 325–326, 325f
in polymer matrix, 333, 333f, 334f
total coupling energy between, 324
vertical assembly, 324
Alkali metal compounds, as CBLs, 223–225, 224t
All atom force field, 94
Alluminium oxide, as CBLs, 222–223
Alq3 (tris-(8-hydroxyquinoline)aluminum), as ABL, 216
Aluminum-doped zinc oxide (AZO), 220, 348
AMBER, force fields for modeling organic systems, 94
Amphiphilic surfactant, as CBL, 227
Anisole, 98, 105
Annealing
in nanophase separation
solvent, 251
thermal, 250–251
treatment, of PEDOT:PSS, 205–206
Anode buffer layers (ABLs), 202–220
GO, 219–220
metal oxides, 213–214, 213f
PEDOT:PSS, 203–213
additives in, 206–210, 207f, 209t
advantages, 204
chemical structure, 203f
drawbacks, 205
grades of, 204
incorporation of, 203
modified, ITO-free PSCs based on, 210–213, 211t, 212f
physical treatments, 205–206
thickness of layer, 204–205, 205f
types of, 204t
polymers and small-molecule organic materials, 215–217, 216f, 217t
self-assembled buffer layers, 218–219, 218f
Area electron diffraction, of PDPP3T:PCBM thin films, 293, 294f
Arylbromide-functionalized phosphine oxides, 313
Aspect ratio, in NR alignment, 326, 327f
Atomic force microscopy (AFM), 94, 252
phase images of PBT-T1:PC$_{60}$BM films and, 285–286, 285f
Attempt-to-escape frequency, 173
Auger recombination, 115
AZO (aluminum-doped zinc oxide), 220, 348

B

Barium (Ba) CBL, 221
Bathocuproine (BCP), as CBL, 225, 226
Benzo[1,2-b:4,5-b']dithiophene (BDT), 7, 8
Benzothiadiazole (BT) unit, 356
BHJ (bulk heterojunctions), 91, 116
charge carrier transport in, 122
charge transport in PBT-T1-based solar cells, fullerene ratio and solvent additives and, 283–291
AFM topography and phase images, PBT-T1:PC$_{60}$BM films, 285–286, 285f
current-voltage characteristics, PBT-T1:PC$_{60}$BM cells, 286–288, 286f, 287t
current-voltage characteristics, PBT-T1:PC$_{70}$BM cells, 288–289, 288f, 289t
photo-CELIV and extracted charge carrier density $vs.$ delay time, PBT-T1, 290–291, 290f, 291t
UV-visible absorption, PBT-T1, 283–284, 283f
UV-visible absorption and XRD spectra, PBT-T1:PC$_{60}$BM films, 284–285, 284f
nanophase separation in OPVs, 248–249
performance, 95
photoactive layer in, 91f, 94
in photocarrier generation regions, 143
PSCs and, 202–203
SM-BHJ, 271
solar cells, 4, 10–13
structure, 143–144, 144f
Bimolecular recombination order, 297–298
Bithiophene, 8
Blending solvent additive, in nanophase separation, 251–252
Blends, D-A
identification and characterization of defects in, 187–195
general depiction, 187–188, 187f
OPV devices, performance, 191–192, 191f
at organic–electrode interfaces, 192–193
origins of defects, 193–195, 194f, 195f
oxygen, structural, and synthesis residuals, 193–195, 194f, 195f
phases and new states, 188–191, 189t
BT (benzothiadiazole) unit, 356
Buckminsterfullerene, 3–4
Bulk heterojunctions (BHJ), 91, 116
charge carrier transport in, 122
charge transport in PBT-T1-based solar cells, fullerene ratio and solvent additives and, 283–291
AFM topography and phase images, PBT-T1:PC$_{60}$BM films, 285–286, 285f
current-voltage characteristics, PBT-T1:PC$_{60}$BM cells, 286–288, 286f, 287t

Index

current-voltage characteristics, PBT-T1:PC$_{70}$BM cells, 288–289, 288f, 289t
photo-CELIV and extracted charge carrier density *vs.* delay time, PBT-T1, 290–291, 290f, 291t
UV-visible absorption, PBT-T1, 283–284, 283f
UV-visible absorption and XRD spectra, PBT-T1:PC$_{60}$BM films, 284–285, 284f
nanophase separation in OPVs, 248–249
organic photovoltaic (OPV) devices, 3
performance, 95
photoactive layer in, 91f, 94
in photocarrier generation regions, 143
PSCs and, 202–203
SM-BHJ, 271
solar cells, 4, 10–13
structure, 143–144, 144f

C

Caesium carbonate, as CBLs, 224–225
Cahn-Hilliard-Cook formalism, 119, 131
Cahn–Hilliard modeling, 282
Calcium (Ca), as CBLs, 220–221
calcium oxide, 223
Capacitance *versus* frequency (CF) method, 173–174, 174f
Capacitance *versus* voltage (CV) method, 172–173, 178–180, 179f, 180f
Capping agents, 67
Carbon-based nanomaterials, 56–64
CNTs, 56–60
application, as transparent electrode of OSCs, 57, 58–60, 59t, 60f
structure and properties, 56–57, 56f, 57f
graphene, 61–64
application as transparent electrode of OSCs, 61, 62–64, 63t
preparation, structure, and properties, 61, 61f, 62f
Carbon nanotubes (CNTs), 56–60
application, as transparent electrode of OSCs, 57, 58–60, 59t, 60f
structure and properties, 56–57, 57f, 58f
Carboxymethyl cellulose (CMC), 60
Carrier dynamics, photoexcited, 143–164
in D-A interfaces, vicinity, 147–153
heterojunction OPV cells, impedance spectroscopy, 148–153, 149f, 150f
overview, 147–148, 148f
films structural effects on, 153–156
heterojunction cells, intermolecular charge-transfer excitons on EQE for, 154–156, 155f
overview, 153, 154f
overview, 143–144, 144f
photocurrent conversion processes, 144–147, 145f
Cathode buffer layers (CBLs), 220–232
alkali metal compounds, 223–225, 224t
GO, 231–232, 231f
low work function metals, 220–221
n-type metal oxides, 221–223, 222f, 223t
organic materials, 225–230
fullerene derivatives, 228, 229–230
nonconjugated polymers, 227–228
small-molecule, 225–227, 225f
water/alcohol–soluble CPs, 228, 229f
SAMs, 230–231, 230f
CB (chlorobenzene), 98, 100, 250
CBLs (cathode buffer layers), 220–232
alkali metal compounds, 223–225, 224t
GO, 231–232, 231f
low work function metals, 220–221
n-type metal oxides, 221–223, 222f, 223t
organic materials, 225–230
fullerene derivatives, 228, 229–230
nonconjugated polymers, 227–228
small-molecule, 225–227, 225f
water/alcohol–soluble conjugated polymers, 228, 229f
SAMs, 230–231, 230f
CF (capacitance *versus* frequency) method, 173–174, 174f
CG (coarse graining), in OPVs, 94, 123
CGMD, 94, 95f
model parameters, 124
for P3HT:PCBM blend, 124
CGMD (coarse grained molecular dynamics), 94, 95f
Characterization, defect states in OPVs, 169
in D-A blends, 187–195
general depiction, 187–188, 187f
OPV devices, performance, 191–192, 191f
at organic–electrode interfaces, 192–193
origins of defects, 193–195, 194f, 195f
oxygen, structural, and synthesis residuals, 193–195, 194f, 195f
phases and new states, 188–191, 189f, 189t
in neat material systems, 175–187
in acceptor materials, 182–184, 183f, 183t
in donor materials, 177–182. *See also* Donor materials, defect states in
general depiction, 175–177, 176f, 177f
origins of defects, 184–187, 185f
oxygen, moisture, structural, and synthesis residuals, 184–187, 185f
Characterization, vertical and planar nanophase separation, 252–264
electrons, techniques using, 253–257, 254f, 255f, 256f, 257f
ions, techniques using, 258–259, 260f

neutrons, techniques using, 263–264
scanning probes, 252–253
x-rays, techniques using, 259, 260–263, 261f, 262f
Charge(s)
accumulation of photogenerated carriers, 157–163, 157f, 158f, 159f, 159t, 160f, 161f, 162f
carrier transport yield, 90
collection yield, 90
generation in OPVs, morphology on, 119–120
transport in PBT-T1-based BHJ solar cells, fullerene ratio and solvent additives and, 283–291
AFM topography and phase images, PBT-T1:PC$_{60}$BM films, 285–286, 285f
current-voltage characteristics, PBT-T1:PC$_{60}$BM cells, 286–288, 286f, 287t
current-voltage characteristics, PBT-T1:PC$_{70}$BM cells, 288–289, 288f, 289t
photo-CELIV and extracted charge carrier density vs. delay time, PBT-T1, 290–291, 290f, 291t
UV-visible absorption, PBT-T1, 283–284, 283f
UV-visible absorption and XRD spectra, PBT-T1:PC$_{60}$BM films, 284–285, 284f
transport in PDPP3T:PCBM solar cells, solvent additives and, 291–300, 292f
additives on surface morphology, 294–296, 295f
current-voltage characteristics, 296–297, 296f, 296t
effect of solvent additives on, 297–300, 298f, 299f
EFTEM images, 292–293, 293f
selected area electron diffraction and Raman spectra, 293, 294f
CHARMM, force fields for modeling organic systems, 94
Chemical vapor deposition (CVD), 56, 379
graphene, 385
on RGO, 382
Chlorobenzene (CB), 7, 98, 100, 250
Chloroform, 98, 100, 102
1-chloronaphthlene (CN), 8
Chlorosulphonic acid, 383
cis-polyacetylene, 265
Clevios P, 47, 48, 49, 50
Clevios PH1000, 47, 48, 49
CL (complex-Langevin) sampling scheme, 131
CMC (carboxymethyl cellulose), 60
CNTs (carbon nanotubes), 56–60
application, as transparent electrode of OSCs, 58–60, 59t, 60f
structure and properties, 56–57, 56f, 57f

Coarse grained molecular dynamics (CGMD), 94, 95f
Coarse graining (CG), in OPVs, 94, 123
CGMD, 94, 95f
model parameters, 124
for P3HT:PCBM blend, 124
Cole–Cole plots, 174–175
Complex-Langevin (CL) sampling scheme, 131
Concentration, of NRs, 327, 329f
Conductivity enhancement, PEDOT:PSS, 47–52, 48f, 50f, 51f, 52f
Conjugated polyelectrolyte (CPE) layer, 355
Conjugated polymers (CPs), 44–45
grafting onto semiconductor inorganic nanoparticles, 309–318
direct grafting, 312–316, 313f, 314f, 315f
ligand exchange, 310–312, 311f, 312f
overview, 309–310, 310f
in situ growth within matrix, 315–318, 317f
LCPs, 27
macromolecular characteristics on morphology, 118–119
equilibrium aspects, 118
nonequilibrium effects, 118–119
nonconjugated polymers, as CBLs, 227–228
self-assembly of nanoparticles and, 318–324
hybrid nanofibrils, 321–324, 321f, 322f, 323f
nanofiber, 318–321, 319f, 320f, 320t
overview, 318
water/alcohol–soluble, as CBLs, 228, 229f
Continuum-based approaches, for OPV modeling, 130–131
Copper oxide ABLs, 214
Copper phthalocyanine (CuPc), as CBL, 226
Coulombic interactions, 117
energy, 121
Covalently bonded atoms, bond potential and, 93
Covalent morphological defects, 168
CPDT (cyclopentadithiophene), 356
CPE (conjugated polyelectrolyte) layer, 355
CPs (conjugated polymers), 44–45
grafting onto semiconductor inorganic nanoparticles, 309–318
direct grafting, 312–316, 313f, 314f, 315f
ligand exchange, 310–312, 311f, 312f
overview, 309–310, 310f
in situ growth within matrix, 315–318, 317f
macromolecular characteristics on morphology, 118–119
equilibrium aspects, 118
nonequilibrium effects, 118–119
nonconjugated polymers, as CBLs, 227–228
self-assembly of nanoparticles and, 318–324

Index

hybrid nanofibrils, 321–324, 321f, 322f, 323f
nanofiber, 318–321, 319f, 320f, 320t
overview, 318
water/alcohol–soluble, as CBLs, 228, 229f
CuPc (copper phthalocyanine), as CBL, 226
Current-voltage characteristics, of solar cells
   PBT-T1:PC$_{60}$BM, 286–288, 286f, 287t
   PBT-T1:PC$_{70}$BM, 288–289, 288f, 289t
   PDPP3T:PCBM, 296–297, 296f, 296t
CVD (chemical vapor deposition), 56, 379
   graphene, 385
   on RGO, 382
CV (capacitance *versus* voltage) method, 172–173, 178–180, 179f, 180f
Cyclic voltammetry, 6
Cyclopentadithiophene (CPDT), 6, 356

D

D-A (donor–acceptor) blends, defects
   identification and characterization, 187–195
     general depiction, 187–188, 187f
     OPV devices, performance, 191–192, 191f
     at organic–electrode interfaces, 192–193
     origins of defects, 193–195, 194f, 195f
     oxygen, structural, and synthesis residuals, 193–195, 194f, 195f
     phases and new states, 188–191, 189f, 189t
D-A (donor–acceptor) functionalized silsesquioxane nanostructures, 19–33
   for OPV devices, 26–32
     overview, 26–27
     PDI nanoparticles, 27–32. *See also* Perylenediimide (PDI)
     P3HT nanoparticles, 27–32. *See also* Poly(3-hexylthiophene) (P3HT), functionalized silsesquioxane nanoparticles
   overview, 19–20
   polysilsesquioxanes, 20–25
     bridging, 22, 22f
     overview, 20–22, 21f, 22f
     POSS, 21–22, 21f
     synthesis by sol-gel method, 23–25, 24f
   silsesquioxanes/bridged silsesquioxanes nanocomposites, 25–26, 25f
D-A (donor–acceptor) interfaces, in OPV cells, 143, 144
   hetero double layered cell, 143–144, 144f
   in photocurrent conversion processes, 146, 147
   photoirradiation, 151, 152, 153, 156–163, 158f, 159f, 159t, 160f, 161f, 162f
   vicinity, photogenerated carrier dynamics in, 147–153

charge accumulation on open-circuit voltage of heterojunction cells, 157–163, 157f, 158f, 159f, 159t, 160f, 161f, 162f
heterojunction cells, impedance spectroscopy, 148–153, 149f, 150f
overview, 147–148, 148f
DCDA (dicyandiamide), as CBL, 226
DCV6T (dicyanovinyl-capped sexithiophene derivative), 360
Debye correlation function, 128
Deep activation energies, defects with, 180–182, 181f, 182t
Deep band, in neat donor materials, 189–190
Deep traps, 172, 173, 176
Defect states, 168–196
   extrinsic, 168
   identification and characterization, in D–A blends, 187–195
     general depiction, 187–188, 187f
     OPV devices, performance, 191–192, 191f
     at organic–electrode interfaces, 192–193
     origins, 193–195, 194f, 195f
     oxygen, structural, and synthesis residuals, 193–195, 194f, 195f
     phases and new states, 188–191, 189f, 189t
   identification and characterization, in neat material systems, 175–187
     in acceptor materials, 182–184, 183f, 183t
     in donor materials, 177–182. *See also* Donor materials, defect states in
     general depiction, 175–177, 176f, 177f
     origins, 184–187, 185f
     oxygen, moisture, structural, and synthesis residuals, 184–187, 185f
   intrinsic, 168
   morphological, 168
   in organic semiconductors, background, 168–169, 168f
   overview, 168
   typical measurement techniques, 169–175
     CF method, 173–174, 174f
     CV method, 172–173
     DLTS, 174
     FTSC, 170–171, 170f
     open-circuit IS, 174–175, 175f
     SCL current modeling, 171–172
     TSC method, 169–170
Delay time, of PBT-T1 solar cells
   photo-CELIV and extracted charge carrier density *vs.*, 290–291, 290f, 291t
Density functional theory (DFT), in OPVs, 108, 118
   electronic structure calculations, 120–122
Density gradient ultracentrifugation (DGU), 60

Devices, OPV
D-A functionalized silsesquioxane
nanostructures for, 26–33
overview, 26–27
PDI nanoparticles, 27–32. See also
Perylenediimide (PDI)
P3HT nanoparticles, 27–32. See also
Poly(3-hexylthiophene) (P3HT),
functionalized silsesquioxane
nanoparticles
polymers in, 114
Dexter energy transfer, 308–309
DFBT (difluorobenzothiadiazole), 356
DFT (density functional theory), in OPVs, 108, 118
electronic structure calculations, 120–122
DGU (density gradient ultracentrifugation), 60
Dicyandiamide (DCDA), as CBL, 226
Dicyanovinyl-capped sexithiophene derivative (DCV6T), 360
Difluorobenzothiadiazole (DFBT), 356
1,8-Diiodooctane (DIO), 7, 12–13, 271, 284, 285, 288–289, 296, 297–300
Diketopyrrolopyrrole (DPP), 7, 269
Diketopyrrolopyrrole-quintetthiophene copolymer (PDPP5T), 352–359, 357f, 358t–359t
N,N-Dimethylformamide (DMF), in PEDOT:PSS ABL, 206–207, 207f
Dimethylquinquethiophene (DM5T) film, 156
Dimethyl sulfoxide (DMSO), 250
for PEDOT:PSS ABL, 207, 208
DiNPB (N,N'-diphenyl-N,N'-bis(4'-(N,N-bis(naphth-1-yl)-amino)-biphenyl-4-yl)-benzidine), 361
DIO (1,8-diiodooctane), 7, 12–13, 271, 284, 285, 288–289, 296, 297–300
2,4-bis[4-(N,Ndiphenylamino)-2,6-dihydroxyphenyl] squaraine (DPSQ), 361
N,N'-Diphenyl-N,N'-bis(4'-(N,N-bis(naphth-1-yl)-amino)-biphenyl-4-yl)-benzidine (DiNPB), 361
4,7-Diphenyl-1,10-phenanthroline-2,9-dicarboxylicacid (DPPA), 226
Direct grafting, of CPs with nanoparticles, 312–316, 313f, 314f, 315f
Dispersions, graphene, 381
thin-film preparation methods from, 383–385
other methods, 385
vacuum filtration, 383–384, 384f
Dissociation, exciton, 117–118
Dithienopyran (DTP), 356
Dithienosilole (DTS), 8, 9, 195, 252, 271
4,7-di(2'-thienyl)-2,1,3-benzothiadiazole (DBT), 6
Dithiols, 6

DL_Poly, 94
DLTS (drive-level transient spectroscopy), 174, 177, 180
DMF (N,N-dimethylformamide), in PEDOT:PSS ABL, 206–207, 207f
DMSO (dimethyl sulfoxide), 250
for PEDOT:PSS ABL, 207, 208
DM5T (dimethylquinquethiophene) film, 156
Donor–acceptor (D-A) blends, defects
identification and characterization, 187–195
general depiction, 187–188, 187f
OPV devices, performance, 191–192, 191f
at organic-electrode interfaces, 192–193
origins of defects, 193–195, 194f, 195f
oxygen, structural, and synthesis residuals, 193–195, 194f, 195f
phases and new states, 188–191, 189f, 189t
Donor–acceptor (D-A) functionalized silsesquioxane nanostructures, 19–33
for OPV devices, 26–32
overview, 26–27
PDI nanoparticles, 27–32. See also
Perylenediimide (PDI)
P3HT nanoparticles, 27–32. See also
Poly(3-hexylthiophene) (P3HT),
functionalized silsesquioxane
nanoparticles
overview, 19–20
polysilsesquioxanes, 20–25
bridging, 22, 22f
overview, 20–22, 21f, 22f
POSS, 21–22, 21f
synthesis by sol-gel method, 23–25, 24f
silsesquioxanes/bridged silsesquioxanes
nanocomposites, 25–26, 25f
Donor–acceptor (D-A) interfaces, in OPV cells, 143, 144
hetero double layered cell, 143–144, 144f
in photocurrent conversion processes, 146, 147
photoirradiation, 151, 152, 153, 156–163, 158f, 159f, 159t, 160f, 161f, 162f
vicinity, photogenerated carrier dynamics in, 147–153
charge accumulation on open-circuit voltage of heterojunction cells, 157–163, 157f, 158f, 159f, 159t, 160f, 161f, 162f
heterojunction cells, impedance spectroscopy, 148–153, 149f, 150f
overview, 147–148, 148f
Donor materials, defect states in, 177–182
electron traps, 182
with relatively deep activation energies, 180–182, 181f, 182t
with relatively shallow activation energies, 177–180, 178f, 179f, 180f

Index

DPP (diketopyrrolopyrrole), 269
DPPA (4,7-diphenyl-1,10-phenanthroline-2,9-dicarboxylicacid), 226
DPP-based polymers, 9
DPP/fullerene-based BHJ solar cells, 9
DPSQ (2,4-bis[4-(N,Ndiphenylamino)-2,6-dihydroxyphenyl] squaraine), 361
Drive-level transient spectroscopy (DLTS), 174, 177, 180
Drop casting, 385
DSCFT (dynamic mean field theory), 131
DSSCs (dye-sensitized solar cells), 70
DTGe, 8
DTP (dithienopyran), 356
DTS (dithienosilole), 8, 9, 195, 252, 271
Dye-sensitized solar cells (DSSCs), 70
Dynamic mean field theory (DSCFT), 131

E

EDOT (ethylenedioxythiophene), 46–47
    as ABL, 215–216
    chemical polymerization, 47
EELS (electron energy-loss spectroscopy), in nanophase separation, 253
EFTEM (energy-filtered TEM)
    images, PDPP3T:PCBM films modified with solvent additives, 292–293, 293f
    in nanophase separation, 253–254, 255, 256f, 257
EG (ethylene glycol), for PEDOT:PSS, 210
Electrode–donor–acceptor interfaces, interfacial properties, 119
Electrodes
    graphene, methods for producing, 380–383
        organic solvent exfoliation, 382–383
        oxidation reduction methods, 381–382
        solution-based exfoliation, 380–381
        superacid exfoliation, 383
        surfactant-assisted exfoliation, 383
    graphene, window, 388–390
    transparent materials, for OSCs, 44–74. *See also* Transparent electrode materials
        carbon-based nanomaterials, 56–64
        CNTs application as transparent electrode, 57, 58–60, 59t, 60f
        conducting polymers, 44–56
        graphene, 61–64. *See also* Graphene(s)
        metal nanostructures, 64–70
        MTFs, 72–74, 73t
        overview, 44
        TCOs to ITOs, 70, 71t
Electron energy-loss spectroscopy (EELS), in nanophase separation, 253
Electronic structure, calculations in OPVs, 120–122, 120f
Electrons, techniques using
    in nanophase separation, 253–257, 254f, 255f, 256f, 257f
Electron-transporting layer (ETL), 346–349, 347f
Electron traps, 182
EMIM-TCB (1-ethyl-3-methylimidazolium tetracyanoborate), 49
Energy-filtered TEM (EFTEM)
    images, PDPP3T:PCBM films modified with solvent additives, 292–293, 293f
    in nanophase separation, 253–254, 255, 256f, 257
Energy transfers
    Dexter, 308–309
    Förster, 308–309
Enhancement, defect states in OPVs, 169
EQE (external quantum efficiency), 90, 117, 341, 344
    defined, 146
    film structure and, 153
    heterojunction OPV cells, intermolecular charge-transfer excitons on, 154–156, 155f
    in photocurrent conversion processes, 146–147
    of tandem organic solar cells, 365, 367–370, 367f, 368f, 370f
Etching method, 383
Ethanol, for PEDOT:PSS, 210
Ethylenedioxythiophene (EDOT), 46–47
    as ABL, 215–216
    chemical polymerization, 47
Ethylene glycol (EG), for PEDOT:PSS, 210
1-Ethyl-3-methylimidazolium tetracyanoborate (EMIM-TCB), 49
ETL (electron-transporting layer), 346–349, 347f
Evaporation rate, in NR alignment, 330, 330f
Excitons, 117–118, 146, 386–387
    categories, 146
    diffusion yield, 90
    dissociation yield, 90
    films structural effects on, 153–156
        heterojunction cells, intermolecular charge-transfer on EQE for, 154–156, 155f
        overview, 153, 154f
    migration process, 308
    in photocurrent conversion processes, 146
Exfoliation
    organic solvent, 382–383
    solution-based graphene, 380–381
    superacid, 383
    surfactant-assisted, 383
Experimental results, tandem solar cells, 352–365
    highly efficient, based on novel materials, 352–363

polymer-based, 352–359, 357f, 358t–359t
small molecule-based, 360–363, 360f, 362t
highly efficient, based on various structures, 363–365
inorganic–organic hybrid cells, 363–364, 371f
semitransparent, 364–365, 366t
External fields, in NR alignment, 331–332, 332f
External quantum efficiency (EQE), 90, 117, 341, 344
defined, 146
film structure and, 153
heterojunction OPV cells, intermolecular charge-transfer excitons on, 154–156, 155f
in photocurrent conversion processes, 146–147
of tandem organic solar cells, 365, 367–370, 367f, 368f, 370f
Extracted charge carrier density
delay time of PBT-T1 solar cells vs. photo-CELIV, 290–291, 290f, 291t
Extrinsic defects, 168

F

Fabrication technique, metal oxides ABLs and, 213
Ferrocene/ ferrocenium standard, 6–7
FF (fill factor), 6
photogenerated carriers and, 157
FIB (focused ion beam), for OPVs, 258
Field theory (FT), 130–131
Fill factor (FF), 6
photogenerated carriers and, 157
Flory-Huggins theory, 132–134, 134f
Fluorinated zinc phthalocyanine derivative (F4-ZnPc), 360
Fluorine-doped tin oxide (FTO), 70
Focused ion beam (FIB), for OPVs, 258
Force fields, in OPVs
all atom, 94
defined, 94
for modeling organic systems, 94
mechanics-based approaches, 123
united-atom, 94
Förster energy transfer, 308–309
Fractional thermally stimulated current (FTSC) method, 170–171, 170f
FT (field theory), 130–131
FTO (fluorine-doped tin oxide), 70
FTSC (fractional thermally stimulated current) method, 170–171, 170f
Fullerene ratio on nanomorphology, charge transport in PBT-T1-based BHJ solar cells and, 283–291

PBT-T1:PC$_{60}$BM solar cells
AFM topography and phase images of films, 285–286, 285f
current-voltage characteristics, 286–288, 286f, 287t
UV-visible absorption and XRD spectra of films, 284–285, 284f
photo-CELIV and extracted charge carrier density vs. delay time, PBT-T1, 290–291, 290f, 291t
UV-visible absorption, PBT-T1, 283–284, 283f
Fullerene(s), 91, 96, 100, 101, 102, 105, 105f
derivatives, as CBLs, 228, 229–230
fullerene-oligothiophene interface, 105, 105f
Functionalized polysilsesquioxanes, 20–25
bridging, 22, 22f
overview, 20–22, 21f, 22f
POSS, 21–22, 21f
synthesis by sol-gel method, 23–25, 24f
Functionalized silsesquioxane nanostructures, D-A, 19–33
for OPV devices, 26–32
overview, 26–27
PDI nanoparticles, 27–32. See also Perylenediimide (PDI)
P3HT nanoparticles, 27–32. See also Poly(3-hexylthiophene) (P3HT), functionalized silsesquioxane nanoparticles
overview, 19–20
polysilsesquioxanes, 20–25
bridging, 22, 22f
overview, 20–22, 21f, 22f
POSS, 21–22, 21f
synthesis by sol-gel method, 23–25, 24f
silsesquioxanes/bridged silsesquioxanes nanocomposites, 25–26, 25f
F4-ZnPc (fluorinated zinc phthalocyanine derivative), 360

G

Generalized gradient approximation (GGA), 121
GISAXS (grazing incidence small angle x-ray scattering), 259, 260, 261f, 266
Glycerol monostearate (GMS), 212
GO (graphene oxides), 62
as ABLs, 219–220
as CBLs, 231–232, 231f
dispersion, 381
oxidation reduction methods, 381–382
for PEDOT:PSS ABL, 208
Gold nanoparticles, for PEDOT:PSS ABL, 208
Grafting CPs, onto semiconductor inorganic nanoparticles, 309–318
direct grafting, 312–316, 313f, 314f, 315f
ligand exchange, 310–312, 311f, 312f

Index   403

overview, 309–310, 310f
in situ growth within matrix, 315–318, 317f
Graphene-based polymers, 379–391
  CVD graphene, 385
  drawback, 379–380
  methods for producing electrodes, 380–383
    organic solvent exfoliation, 382–383
    oxidation reduction methods, 381–382
    solution-based exfoliation, 380–381
    superacid exfoliation, 383
    surfactant-assisted exfoliation, 383
  in OSCs, 385–390
    applications, 387
    cell architectures, 386f
    excitons, 386–387
    inside active layer, 390
    as window electrode, 388–390
  overview, 379–380, 380f
  thin-film preparation methods from dispersions, 383–385
    other methods, 385
    vacuum filtration, 383–384, 384f
Graphene oxides (GO), 62
  as ABLs, 219–220
  as CBLs, 231–232, 231f
  dispersion, 381
  oxidation reduction methods, 381–382
  for PEDOT:PSS ABL, 208
Graphene(s), 61–64
  application as transparent electrode of OSCs, 61, 62–64, 63t
  dispersions, thin-film preparation methods, 383–385
    other methods, 385
    vacuum filtration, 383–384, 384f
  preparation, structure, and properties, 61, 61f, 62f
Grazing incidence small angle x-ray scattering (GISAXS), 259, 260, 261f, 266
Gromacs, 94
Guinier-Porod model, 260

H

Harmonic function, defined, 93
Harmonic potential form, 93
Harvard Clean Energy Project, 121
Heeger, 4
Helium ion microscopy (HIM), 259, 260f
Heterojunctions
  BHJ. *See* Bulk heterojunctions (BHJ)
  defined, 116
  OPV cells
    charge accumulation of photogenerated carriers in D–A interface vicinity, 157–163, 157f, 158f, 159f, 159t, 160f, 161f, 162f

impedance spectroscopy, 148–153, 149f, 150f
intermolecular charge-transfer excitons on EQE, 154–156, 155f
Highest occupied molecular orbital (HOMO), 44, 114, 144, 203, 341
Highly efficient organic tandem solar cells
  based on novel materials, 352–363
    polymer-based, 352–359, 357f, 358t–359t
    small molecule-based, 360–363, 360f, 362t
  based on various structures, 363–365
    inorganic–organic hybrid cells, 363–364, 371f
    semitransparent, 364–365, 366t
High-performance OPVs, 9t
High-pressure carbon monoxide (HiPCO), 57
HIM (helium ion microscopy), 259, 260f
HiPCO (high-pressure carbon monoxide), 57
Hole transporting layers (HTLs), 192–193, 346–349, 347f
Holstein-Peierls polaron model, 122
HOMO (highest occupied molecular orbital), 44, 114, 144, 203, 341
Hopping model, 51
HTLs (hole transporting layers), 192–193, 346–349, 347f
HTPV (hybrid tandem photovoltaics), 363
Hummer's method, 382
Hybrid nanofibrils, 321–324, 321f, 322f, 323f
Hybrid tandem photovoltaics (HTPV), 363
Hydrophilic GO, for PEDOT:PSS ABL, 208

I

Ideal photon-to-current quantum efficiency (IPCE), 3
Identification, defect states in OPVs, 169
  in D-A blends, 187–195
    general depiction, 187–188, 187f
    OPV devices, performance, 191–192, 191f
    at organic-electrode interfaces, 192–193
    origins of defects, 193–195, 194f, 195f
    oxygen, structural, and synthesis residuals, 193–195, 194f, 195f
    phases and new states, 188–191, 189f, 189t
  in neat material systems, 175–187
    in acceptor materials, 182–184, 183f, 183t
    in donor materials, 177–182. *See also* Donor materials, defect states in
    general depiction, 175–177, 176f, 177f
    origins of defects, 184–187, 185f
    oxygen, moisture, structural, and synthesis residuals, 184–187, 185f
Iljima, Sumio, 56
IML (intermediate layers), working principles in tandem solar cells, 346–349, 347f

Impedance spectroscopy (IS)
   heterojunction OPV cells, 148–153, 149f, 150f
   open-circuit, 174–175, 175f
Indane, 98, 100–101
Indium tin oxide (ITO), 44, 119
   ABL and, 202–203
   free PSCs based on modified PEDOT:PSS, 210–213, 211t, 212f
   TCOs and, 70, 71t
   work function, 203
Initial rise method, 171
Inorganic fluorides, as CBLs, 223–224
Inorganic nanoparticles, semiconductor grafting CPs, 309–318
      direct grafting, 312–316, 313f, 314f, 315f
      ligand exchange, 310–312, 311f, 312f
      overview, 309–310, 310f
      in situ growth within matrix, 315–318, 317f
Inorganic–organic hybrid tandem solar cells, 363–364, 371f
Inorganic–organic nanocomposites, 308–334
   alignment of NRs, 324–334. *See also* Alignment, of NRs
      factors influencing, 326–332
      overview, 324
      physical origin of depletion force, 325–326, 325f
      in polymer matrix, 333, 333f, 334f
      total coupling energy between, 324
      vertical assembly, 324
   grafting CPs onto semiconductor inorganic nanoparticles, 309–318
      direct grafting, 312–316, 313f, 314f, 315f
      ligand exchange, 310–312, 311f, 312f
      overview, 309–310, 310f
      in situ growth within matrix, 315–318, 317f
   overview, 308–309
   self-assembly of nanoparticles and CPs, 318–324
      CP nanofiber, 318–321, 319f, 320f, 320t
      hybrid nanofibrils, 321–324, 321f, 322f, 323f
      overview, 318
In situ growth, of nanoparticles within CP matrix, 315–318, 317f
Interfaces
   D-A, in OPV cells, 143, 144. *See also* Donor–acceptor (D-A) interfaces
   hetero double layered cell, 143–144, 144f
   in photocurrent conversion processes, 146, 147
   photoirradiation, 151, 152, 153, 156–163, 158f, 159f, 159t, 160f, 161f, 162f
   vicinity, photogenerated carrier dynamics in, 147–153

   electrode–donor–acceptor interfaces, properties, 119
   fullerene-oligothiophene, 105, 105f
   organic–electrode, defect states at, 192–193
Interfacial materials, for PSCs efficiency enhancement, 202–233
   ABLs, 202–220
      GO, 219–220
      metal oxides, 213–214, 213f
      PEDOT:PSS, 203–213. *See also* PEDOT:PSS, ABLs
      polymers and small-molecule organic materials, 215–217, 216f, 217t
      self-assembled buffer layers, 218–219, 218f
   CBLs, 220–232
      alkali metal compounds, 223–225, 224t
      GO, 231–232, 231f
      low work function metals, 220–221
      n-type metal oxides, 221–223, 222f, 223t
      organic materials, 225–230. *See also* Organic materials, as CBLs
      SAMs, 230–231, 230f
   overview, 202
Interfacial properties, of electrode–donor–acceptor interfaces, 119
Intermediate layers (IML), working principles in tandem solar cells, 346–349, 347f
Intermolecular charge-transfer excitons on EQE for heterojunction OPV cells, 154–156, 155f
Internal quantum efficiency (IQE), 114, 117, 341, 344
Intrinsic defects, 168
Ionized acceptor density, 177–178
Ions, techniques using
   in nanophase separation, 258–259, 260f
IQE (internal quantum efficiency), 114, 117, 341, 344
IS (impedance spectroscopy)
   heterojunction OPV cells, 148–153, 149f, 150f
   open-circuit, 174–175, 175f
Isoindigo-based polymers, 8
Isoline 1, 350–351, 350f
Issues, insights obtained from modeling of OPVs, 116–120
   categories, 117
      charge generation and recombination, morphology on, 119–120
      CPs, macromolecular characteristics on morphology, 118–119
      electrode–donor–acceptor interfaces, interfacial properties, 119
      exciton generation, dissociation, and recombination, 117–118
ITO (indium tin oxide), 44, 119
   ABL and, 202–203

Index

free PSCs based on modified PEDOT:PSS, 210–213, 211t, 212f
TCOs and, 70, 71t
work function, 203

K

Kinetic Monte Carlo (KMC), 129–130
KMC (kinetic Monte Carlo), 129–130
Kohn-Sham ansatz (KS) method, 120
Koopman's condition, 121
KS (Kohn-Sham ansatz) method, 120

L

Laguerre tessellation method, 126
LAMMPS, 94
LBG (low-bandgap) polymers, 268
LCPs (linear conjugated polymers), 27
LDA (local density approximation), 121
LeapFrog algorithm, 92
Lennard-Jones (LJ) pairwise potential, 93
Ligand(s)
 exchange, 310–312, 311f, 312f
 surface, in NR alignment, 330, 331, 331f
Linear conjugated polymers (LCPs), 27
Lithium fluoride, as CBLs, 223–224
LJ (Lennard-Jones) pairwise potential, 93
Local density approximation (LDA), 121
Localized surface plasmon resonance (LSPR) effect, 208
Low-bandgap (LBG) polymers, 268
Lowest unoccupied molecular orbital (LUMO), 44, 90, 144, 203, 341
 defined, 114
Low work function metals, as CBL, 220–221
LSPR (localized surface plasmon resonance) effect, 208
LUMO (lowest unoccupied molecular orbital), 44, 90, 144, 203, 341
 defined, 114

M

Macromolecular characteristics, of CPs on morphology, 118–119
 equilibrium aspects, 118
 nonequilibrium effects, 118–119
Manipulating, nanophase separation in OPVs, 249–252
 blending solvent additive, 251–252
 solvent annealing, 251
 solvent selection, 250
 thermal annealing, 250–251
Maxwell-Boltzmann distribution, 92
MCB (mono-chlorobenzene), 98, 100
MC (Monte Carlo) method

OPV modeling, mechanics-based approaches in, 129–130, 130f
 KMC, 129–130
 MD simulation and, 129
MD (molecular dynamics) simulations, in OPVs
 future directions, 107–108
 mechanics-based approaches in modeling, 123–128, 125f, 126f, 127f, 128f
 MC method and, 129
 of systems relevant, 95–107
 relating synthesis parameters to morphology, 95–107. *See also* Relating synthesis parameters
 theoretical background, 92–95, 95f
Measurement(s)
 Photo-CELIV, extracted charge carrier density *vs.* delay time of PBT-T1 solar cells and, 290–291, 290f, 291t
 tandem solar cells, 365, 367–370, 367f, 368f, 370f
 techniques, for defect states, 169–175
  CF method, 173–174, 174f
  CV method, 172–173
  DLTS, 174
  FTSC, 170–171, 170f
  open-circuit IS, 174–175, 175f
  SCL current modeling, 171–172
  TSC method, 169–170
Mechanics-based approaches, OPV modeling, 122–130
 MC methods, 129–130, 130f
 MD simulations, 123–128, 125f, 126f, 127f, 128f
11-Mercaptoundenoic acid (MUA)-capped NRs, 330, 331
Metal MTFs as transparent electrode, PV performances of OSCs with, 73t
Metal nanostructures, 64–70
 nanomeshes, 64–67, 65t, 66f
 nanowire grids, 67–69, 68t
 ultrathin films, 69–70
Metal oxides
 ABLs, 213–214, 213f
 n-type, as CBLs, 221–223, 222f, 223t
 for PEDOT:PSS ABL, 209–210
Methanesulfonic acid treatment, for PEDOT:PSS films, 52
Methanol, for PEDOT:PSS, 210, 211, 212f
Metropolis algorithm, 129
Mitigation, defect states in OPVs, 169
Modeling
 Cahn–Hilliard, 282
 force fields for organic systems, 94
 Guinier-Porod, 260
 Holstein-Peierls polaron, 122
 Hopping, 51
 Mott-Schottky, 144–145

phase field, 130
reflectivity, 266, 267f
SCLC, 151–152, 171–172
VRH model, 51
Modeling, of OPVs, 113–135
  discussion, 134–135
  insights obtained, issues, 116–120
    charge generation and recombination, morphology on, 119–120
    CPs, macromolecular characteristics on morphology, 118–119
    electrode–donor–acceptor interfaces, interfacial properties, 119
  insights obtained from simulation approaches, 120–134
    continuum-based approaches, 130–131
    electronic structure calculations, 120–122, 120f
    mechanics-based approaches, classical, 122–130. *See also* Mechanics-based approaches
    multiscale methods, 131–134, 132f, 134f
  overview, 114–116, 115f
Moisture, origins of defect states, 184–186, 185f
Molecular dynamics (MD) simulations, in OPVs
  future directions, 107–108
  mechanics-based approaches in modeling, 123–128, 125f, 126f, 127f, 128f
  MC method and, 129
  of systems relevant, 95–107
    relating synthesis parameters to morphology, 95–107. *See also* Relating synthesis parameters
  theoretical background, 92–95, 95f
Molybdenum oxide (MoO$_3$)
  ABLs, 213, 214
  for PEDOT:PSS ABL, 209–210
Molybdenum sulfides ABLs, 214
Mono-chlorobenzene (MCB), 98, 100
Monte Carlo (MC) method
  OPV modeling, mechanics-based approaches in, 129–130, 130f
  KMC, 129–130
  MD simulation and, 129
Morphological defects, 168
Morse potential form, 93
Mott-Gurney law, 172
Mott-Schottky (MS)
  model, 144–145
  relationship, 172
MTFs (multilayer thin films), 72–74, 73t
MUA (11-mercaptoundenoic acid)-capped NRs, 330, 331
Multilayer thin films (MTFs), 72–74, 73t

Multiscale methods, for OPV modeling, 131–134, 132f, 134f
Multi-walled carbon nanotubes (MWNT), 56, 57
MWNT (multi-walled carbon nanotubes), 56, 57

N

Nanofibers, CP, 318–321, 319f, 320f, 320t
Nanofibrils, hybrid, 321–324, 321f, 322f, 323f
Nanomeshes, metal, 64–67, 65t, 66f
Nanomorphology, active layer. *See also* Active layer nanomorphology
  engineering in PSCs, 282–300
    fullerene ratio and solvent additives, charge transport in PBT-T1-based BHJ solar cells, 283–291
    overview, 282–283
    solvent additive, charge transport in PDPP3T:PCBM solar cells and, 291–300, 292f
Nanoparticles
  D-A functionalized silsesquioxane for OPV devices
    PDI, 27–32. *See also* Perylenediimide (PDI)
    P3HT, 27–32. *See also* Poly(3-hexylthiophene) (P3HT), functionalized silsesquioxane nanoparticles
  gold, for PEDOT:PSS ABL, 208
  self-assembly, CPs and, 318–324
    CP nanofiber, 318–321, 319f, 320f, 320t
    hybrid nanofibrils, 321–324, 321f, 322f, 323f
    overview, 318
  semiconductor inorganic, grafting CPs, 309–318
    direct grafting, 312–316, 313f, 314f, 315f
    ligand exchange, 310–312, 311f, 312f
    overview, 309–310, 310f
Nanophase separation, in OPVs, 248–272
  case studies, 264–272
    in advanced ternary OPVs, 271–272
    second-generation semiconducting polymers, 265–268, 265f, 267f
    small-molecule semiconductors, 271
    third generation semiconducting polymers, 268–270
  characterizing vertical and planar, 252–264
    electrons, techniques using, 253–257, 254f, 255f, 256f, 257f
    ions, techniques using, 258–259, 260f
    neutrons, techniques using, 263–264
    scanning probes, 252–253
    x-rays, techniques using, 259, 260–263, 261f, 262f
  manipulation, 249–252

Index

blending solvent additive, 251–252
solvent annealing, 251
solvent selection, 250
thermal annealing, 250–251
overview, 248–249
Nanorods (NRs), alignment, 324–334
  factors influencing, 326–332
    aspect ratio, 326, 327f
    concentration, 327, 329f
    evaporation rate, 330, 330f
    external fields, 331–332, 332f
    solvent, 327, 329f, 330
    substrates, 326, 327, 328f
    surface ligands, 330, 331, 331f
    temperature, 326
  overview, 324
  physical origin of depletion force, 325–326, 325f
  in polymer matrix, 333, 333f, 334f
  total coupling energy between, 324
  vertical assembly, 324
Nanostructures, D-A functionalized silsesquioxane, 19–33
  functionalized polysilsesquioxanes, 20–25
    bridging, 22, 22f
    overview, 20–22, 21f, 22f
    POSS, 21–22, 21f
    synthesis by sol-gel method, 23–25, 24f
  for OPV devices, 26–32
    overview, 26–27
    PDI nanoparticles, 27–32. *See also* Perylenediimide (PDI)
    P3HT nanoparticles, 27–32. *See also* Poly(3-hexylthiophene) (P3HT), functionalized silsesquioxane nanoparticles
  overview, 19–20
  silsesquioxanes/bridged silsesquioxanes nanocomposites, 25–26, 25f
Nanowires (NWs)
  CP, 318, 319f
  metal grids, 67–69, 68t
Naphthalene diimide-containing polymers, 13
National Renewable Energy Laboratory (NREL), 356
NDI-2T, 12
Near edge x-ray absorption fine structure (NEXAFS) spectroscopy, 263
Neat material systems, defects
  identification and characterization, 175–187
    in acceptor materials, 182–184, 183f, 183t
    in donor materials, 177–182. *See also* Donor materials, defect states in
    general depiction, 175–177, 176f, 177f
    origins of defects, 184–187, 185f
    oxygen, moisture, structural, and synthesis residuals, 184–187, 185f

Neutron reflectometry, 263
  reflectivity modeling, 266, 267f
Neutrons, techniques using
  in nanophase separation, 263–264
NEXAFS (near edge x-ray absorption fine structure) spectroscopy, 263
Nonconjugated polymers, as CBLs, 227–228
Noncovalent morphological defects, 168
Nonradiative mechanisms, of recombination, 115
Novel materials, highly efficient organic tandem solar cells based on, 352–363
  polymer-based, 352–359, 357f, 358t–359t
  small molecule-based, 360–363, 360f, 362t
NREL (National Renewable Energy Laboratory), 356
N-type metal oxides, as CBLs, 221–223, 222f, 223t
NWs (nanowires)
  CP, 318, 319f
  metal grids, 67–69, 68t

O

ODCB (o-dichlorobenzene), 98, 100, 250, 252, 286, 287
ODT (1,8 octanedithiol), 292, 293, 294, 296, 297, 298, 299
Oleamide, 227
OLEDs (organic light-emitting diodes), 203
Oligothiophene, 8
Oligothiophene-fullerene interface, 105, 105f
Open-circuit impedance spectroscopy (IS), 174–175, 175f
Open-circuit voltage, photogenerated carrier dynamics
  correlation, 156–163
    charge accumulation in D-A interface vicinity on heterojunction OPV cells, 157–163, 157f, 158f, 159f, 159t, 160f, 161f, 162f
    overview, 156–157
OPLS, force fields for modeling organic systems, 94
Optical simulations, of tandem solar cells, 349–351, 350f
Optical spacer effect, 348–349
OPV. *See* Organic photovoltaic (OPV) solar cells
Orbital overlap mechanism, 308
Organic–electrode interfaces, defect states at, 192–193
Organic films, structural effects on exciton and carrier dynamics, 153–156
  heterojunction cells, intermolecular charge-transfer excitons on EQE for, 154–156, 155f
  overview, 153, 154f
Organic light-emitting diodes (OLEDs), 203

Organic materials
    as CBLs, 225–230
        fullerene derivatives, 228, 229–230
        nonconjugated polymers, 227–228
        small-molecule, 225–227, 225f
        water/alcohol–soluble conjugated polymers, 228, 229f
    small-molecule, as ABLs, 215–217, 216f, 217t
Organic photovoltaic (OPV) solar cells, 3, 90–108
    defect states in materials, thin films, and devices, 168–196. See also Defect states
    devices
        D-A functionalized silsesquioxane nanostructures for, 26–33. See also Functionalized silsesquioxane nanostructures
        polymers in, 114
    future directions, 107–108
    high-performance, 9t
    MD simulations, 92. See also Molecular dynamics (MD) simulations
        CGMD, 94, 95f
        force fields, 94
        LJ pairwise potential, 93
        principle objective, 92
        of systems relevant, 95–107
        theoretical background, 92–95, 95f
    modeling, 113–135. See also Modeling, of OPVs
    nanophase separation in, 248–272. See also Nanophase separation
    overview, 90–92
    PCEs of, 114
    performance, defects on, 191–192, 191f
    photoactive layers, morphology
        in BHJ, 91f, 94, 95
        overview, 91–92
        relating synthesis parameters using MD, 95–107. See also Relating synthesis parameters
        solution-based synthesis, 91
    photoexcited carrier dynamics in, 143–164. See also Photoexcited carrier dynamics
Organic semiconductors, defect states in, 168–169, 168f
Organic solar cells (OSCs)
    graphene-based polymer in, 385–390. See also Graphene-based polymers
    next-generation transparent electrode materials for, 44–74. See also Transparent electrode materials
    OPV. See Organic photovoltaic (OPV) solar cells
    single-junction, 340–342, 342f, 343f
    tandem, 337–371. See also Tandem solar cells
Organic solvent exfoliation, 382–383
Organic vapor phase deposition (OVPD), 70
OSCs (organic solar cells)
    graphene-based polymer in, 385–390. See also Graphene-based polymers
    next-generation transparent electrode materials for, 44–74. See also Transparent electrode materials
    OPV. See Organic photovoltaic (OPV) solar cells
    single-junction, 340–342, 342f, 343f
    tandem, 337–371. See also Tandem solar cells
OVPD (organic vapor phase deposition), 70
Oxidation reduction methods, 381–382
Oxide MTFs as transparent electrode, PV performances of OSCs with, 73t
Oxygen
    origins of defect states
        D-A blends, 193–195, 194f, 195f
        neat material systems, 184–186, 185f
    plasma treatment, of PEDOT:PSS, 206
O-xylene, 12

P

PAH (polycyclic aromatic hydrocarbon), 62
PANI (poly(aniline)), as ABL, 215
Parratt formalism, 266
Particle morphology, PDI/P3HT-functionalized silsesquioxane nanoparticles, 28–31, 28f, 29f, 30f, 31f
PBASE (pyrene butanoic acid succinimidyl ester), 388
PBDTTBT, 7
PBDTT-DPP (poly(2,6′-4,8-di(5-ethylhexylthienyl)benzo[1,2-b;3,4-b]dithiophene-alt-5-dibutyloctyl-3,6-bis(5-bromothiophen-2-yl)pyrrolo[3,4-c]pyrrole-1,4-dione)), 356
PBDTTT-C (poly[(4,8-bis-(2-ethylhexyloxy)-benzo[1,2-b:4,5-b′] dithiophene)-2,6-diyl-alt-(4-(2-ethylhexanoyl)-thieno[3,4-b]thiophene)-2,6-diyl]), 353
PBT-T1 (poly(thiophene-2,5-diyl-alt-[5,6-bis(dodecyloxy)-benzo[c][1,2,5]thiadiazole]-4,7-diyl))-based BHJ solar cells, fullerene ratios and solvent additives and, 283–291
    current-voltage characteristics of PBT-T1:PC$_{70}$BM solar cells, 288–289, 288f, 289t
PBT-T1:PC$_{60}$BM films
    AFM topography and phase images, 285–286, 285f
    current-voltage characteristics of solar cells, 286–288, 286f, 287t

Index

UV-visible absorption and XRD spectra, 284–285, 284f
photo-CELIV and extracted charge carrier density *vs.* delay time of solar cells, 290–291, 290f, 291t
UV-visible absorption, 283–284, 283f
PBT-T1:PC$_{60}$BM films
  AFM topography and phase images, 285–286, 285f
  current-voltage characteristics of solar cells, 286–288, 286f, 287t
  UV-visible absorption and XRD spectra, 284–285, 284f
PBT-T1:PC$_{70}$BM solar cells, current-voltage characteristics, 288–289, 288f, 289t
PBTTT (poly-2,5-bis(3-tetradecylthiophene-2-yl) thieno[3,2-b]-thiophene), 98
PC$_{61}$BM, 7
PC$_{71}$BM, 7
PCBM (phenyl-C61-butyric acid methyl ester), 91, 94, 95f, 250
  blending solvent additive, 252
  charge transport in PDPP3T:PCBM solar cells, solvent additives and, 291–300, 292f
    additive on charge transport and recombination dynamics, 297–300, 298f, 299f
    additives on surface morphology, 294–296, 295f
    current-voltage characteristics, 296–297, 296f, 296t
    EFTEM images, 292–293, 293f
    selected area electron diffraction and Raman spectra, 293, 294f
  chemical structure, 103–104, 103f
  concentration, 96–98, 99f
  in crystalline phases, 105
  identification and characterization of defect states in D–A blends, 187–195. *See also* Defect states
  molecular arrangement, 97, 97f
  in nanophase separation
    second generation semiconducting polymer, 265–268, 265f, 267f
    techniques using electrons, 253–257, 254f, 255f, 256f, 257f
    in thermal annealing, 250–251
  OPV modelling
    MD simulations and, 124–128, 125f, 126f, 127f, 128f
    multiscale methods, 132–134, 132f, 134f
    polymer-based tandem organic solar cells, 352–359, 357f, 358t–359t
  solvents
    choice, 98, 100–102, 101f
    solute concentration, 102–103

PCDTBT (poly[N-9'-heptadecanyl-2,7-carbazole-alt-5,5-(4',7'-di-2-thienyl-2',1',3'-benzothiadiazole)]), 352–359, 357f, 358t–359t
PCEs (power conversion efficiencies), 90, 143
  additives, processing conditions and, 118–119
  defined, 114
  of OPVs, 114
PCPDTBT (poly[2,6-(4,4-bis-(2-ethylhexyl)-4H-cyclopenta[2,1-b;3,4-b']dithiophene)-alt-4,7-(2,1,3-benzothiadiazole)]), 6–7, 268–270, 349–351, 350f
PCPDTFBT (poly[2,6-(4,4-bis(2-ethylhexyl)-4H-cyclopenta[2,1-b;3,4-b'] dithiophene)-alt-4,7-(5-fluoro-2,1,3-benzothia-diazole)]), 354
PDHBT (poly(3,4-dihexyl-2,2'-bithiophene)), 104
PDI (perylenediimide), functionalized silsesquioxane nanoparticles, 27–32
  overview, 27–28
  photovoltaic performances, 31, 32f
  synthesis and particle morphology, 28–31, 28f, 29f, 30f, 31f
PDI-2DTT, 12–13
PDI-T, 13
PDMS*b*-PMMA (poly(dimethylsiloxane)-block-poly(methyl methacrylate)), as CBL, 227
PDPP3T, 7–8
PDPP5T (diketopyrrolopyrrole-quintetthiophene copolymer), 352–359, 357f, 358t–359t
PDPP3T:PCBM solar cells
  charge transport, solvent additives and, 291–300, 292f
    additive on charge transport and recombination dynamics, 297–300, 298f, 299f
    additives on surface morphology, 294–296, 295f
    current-voltage characteristics, 296–297, 296f, 296t
    EFTEM images, 292–293, 293f
    selected area electron diffraction and Raman spectra, 293, 294f
PDS (photothermal deflection spectra), 175–176, 176f
PEDOT (poly(3,4-ethylenedioxythiophene)), 45–47, 119, 389
PEDOT:PSS
  application as transparent electrode of OSCs, 52–54, 53t, 54f, 55t
  chemical structure, 45–46, 46f
  conductivity enhancement, 47–52, 48f, 50f, 51f, 52f
PEDOT:PSS ABLs, 203–213
  additives in, 206–210, 207f, 209t

advantages, 204
chemical structure, 203f
drawbacks, 205
grades of, 204
incorporation of, 203
modified, ITO-free PSCs based on, 210–213, 211t, 212f
physical treatments, 205–206
thickness of layer, 204–205, 205f
types of, 204t
PEDOT doped with tosylate (PEDOT:TsO), 46, 47
PEDOT:PSS, 47
ABLs, 203–213
additives in, 206–210, 207f, 209t
advantages, 204
chemical structure, 203f
drawbacks, 205
grades of, 204
incorporation of, 203
modified, ITO-free PSCs based on, 210–213, 211t, 212f
physical treatments, 205–206
thickness of layer, 204–205, 205f
types of, 204t
application as transparent electrode of OSCs, 52–54, 53t, 54f, 55t
chemical structure, 45–46, 46f
conductivity enhancement, 47–52, 48f, 50f, 51f, 52f
PEDOT:TsO (PEDOT doped with tosylate), 46, 47
PEG (polyethylene glycol)
as CBL, 227
for PEDOT:PSS, 210
PEI (polyethylenimine), 353
PEIE (polyethylenimine), 348, 353
Pentacene, as CBL, 226
PEO (poly(ethylene oxide)) interlayer, 227
Performance(s), PV
defects on, 191–192, 191f
OSCs with CNTs as transparent electrode, 59t
OSCs with graphene as transparent electrode, 63t
OSCs with metal nanomeshes as transparent electrode, 65t
OSCs with metal nanowire grids as transparent electrode, 68t
OSCs with oxide/metal/oxide MTFs as transparent electrode, 73t
OSCs with PEDOT:PSS as transparent electrode, 52–54, 53t, 55t
OSCs with TCOs as transparent electrode, 71t
PDI and P3HT-functionalized silsesquioxane nanoparticles, 31, 32f
Perylenediimide (PDI), functionalized silsesquioxane nanoparticles, 27–32

overview, 27–28
photovoltaic performances, 31, 32f
synthesis and particle morphology, 28–31, 28f, 29f, 30f, 31f
Perylene diimide polymers, 13
Perylenetetracarboxylic diimide, 12
PFBT (polymer (P3HT) and acceptor)), 10, 12
Phase field modeling, 130
[6,6]-phenyl–C_{61}–butyric acid methyl ester ($PC_{61}BM$), 3, 4, 5–6
[6,6]-phenyl–C_{71}–butyric acid methyl ester ($PC_{71}BM$), 3
Phenyl-C61-butyric acid methyl ester (PCBM), 91, 94, 95f, 250
blending solvent additive, 252
charge transport in PDPP3T:PCBM solar cells, solvent additives and, 291–300, 292f
additive on charge transport and recombination dynamics, 297–300, 298f, 299f
additives on surface morphology, 294–296, 295f
current-voltage characteristics, 296–297, 296f, 296t
EFTEM images, 292–293, 293f
selected area electron diffraction and Raman spectra, 293, 294f
chemical structure, 103–104, 103f
concentration, 96–98, 99f
in crystalline phases, 105
identification and characterization of defect states in D-A blends, 187–195. See also Defect states
molecular arrangement, 97, 97f
in nanophase separation
second generation semiconducting polymer, 265–268, 265f, 267f
techniques using electrons, 253–257, 254f, 255f, 256f, 257f
in thermal annealing, 250–251
OPV modelling
MD simulations and, 124–128, 125f, 126f, 127f, 128f
multiscale methods, 132–134, 132f, 134f
polymer-based tandem organic solar cells, 352–359, 357f, 358t–359t
solvents
choice, 98, 100–102, 101f
solute concentration, 102–103
Phosphine oxides, arylbromide-functionalized, 313
Photoactive layers in OPV cells, morphology, 90–108
in BHJ, 91f, 94, 95
future directions, 107–108
overview, 91–92

Index

relating synthesis parameters using MD, 95–107
  other studies, 104–107, 105f, 106f
  polymers and acceptor species, chemical structure, 103–104, 103f
  polymers and acceptor species, concentrations, 96–98, 97f, 99f
  solvents, choice, 98, 100–102, 101f
  solvents, solute concentration, 102–103
  solution-based synthesis, 91
Photo-CELIV measurements, extracted charge carrier density vs. delay time of PBT-T1 solar cells and, 290–291, 290f, 291t
Photocurrent conversion processes, 144–147, 145f
Photoexcited carrier dynamics, in OPVs, 143–164
  in D-A interfaces, vicinity, 147–153
    charge accumulation on open-circuit voltage of heterojunction cells, 157–163, 157f, 158f, 159f, 159t, 160f, 161f, 162f
    heterojunction cells, impedance spectroscopy, 148–153, 149f, 150f
    overview, 147–148, 148f
  films structural effects on, 153–156
    heterojunction cells, intermolecular charge-transfer excitons on EQE for, 154–156, 155f
    overview, 153, 154f
  open-circuit voltage and, correlation, 156–163
    charge accumulation in D-A interface vicinity on heterojunction cells, 157–163, 157f, 158f, 159f, 159t, 160f, 161f, 162f
  overview, 143–144, 144f, 156–157
  photocurrent conversion processes, 144–147, 145f
Photogenerated carrier dynamics, in OPVs
  open-circuit voltage and, correlation, 156–163
    charge accumulation in D-A interface vicinity on heterojunction cells, 157–163, 157f, 158f, 159f, 159t, 160f, 161f, 162f
    overview, 156–157
  in vicinity of D-A interfaces, 147–153
    charge accumulation on open-circuit voltage of heterojunction cells, 157–163, 157f, 158f, 159f, 159t, 160f, 161f, 162f
    heterojunction cells, impedance spectroscopy, 148–153, 149f, 150f
    overview, 147–148, 148f
Photoinduced electron transfer (PET), 3–4
Photoirradiation, of OPV cells, 144, 153
  D-A interface and, 151, 152, 153, 156–163, 158f, 159f, 159t, 160f, 161f, 162f
Photoluminnescence (PL) spectroscopy, 322
Photon absorption efficiency, 90

Photothermal deflection spectra (PDS), 175–176, 176f
Photovoltaics (PV)
  industry
    in 2011, 338
    challenge, 338
    world cell/module production, 339f
  performances
    OSCs with CNTs as transparent electrode, 59t
    OSCs with graphene as transparent electrode, 63t
    OSCs with metal nanomeshes as transparent electrode, 65t
    OSCs with metal nanowire grids as transparent electrode, 68t
    OSCs with oxide/metal/oxide MTFs as transparent electrode, 73t
    OSCs with PEDOT:PSS as transparent electrode, 52–54, 53t, 55t
    OSCs with TCOs as transparent electrode, 71t
    PDI and P3HT-functionalized silsesquioxane nanoparticles, 31, 32f
  solar cells, classification, 90
P3HT (poly(3-hexylthiophene)), 12, 91, 94, 95f
  chemical structure, 103–104, 103f
  concentration, 96–98, 99f
  defect band in, 180–182, 181f
  functionalized silsesquioxane nanoparticles, 27–32
    overview, 27–28
    photovoltaic performances, 31, 32f
    synthesis and particle morphology, 28–31, 28f, 29f, 30f, 31f
  identification and characterization of defect states in D-A blends, 187–195. See also Defect states
  molecular arrangement, 97, 97f
  molecules in anisole, 105, 106, 106f
  in nanophase separation
    second generation semiconducting polymer, 265–268, 265f, 267f
    techniques using electrons, 253–257, 254f, 255f, 256f, 257f
  OPV modelling
    MD simulations and, 124–128, 125f, 126f, 127f, 128f
    multiscale methods, 132–134, 132f, 134f
  PDS of, 176, 176f
  solvents, choice, 98, 100–102, 101f
  TSC signal from, 176, 177f
Physical treatments, of PEDOT:PSS, 205–206
  annealing treatment, 205–206
  O_2 plasma, 206
  thermal, 205
  UV treatment, 206

Planar nanophase separation, characterizing, 252–264
  electrons, techniques using, 253–257, 254f, 255f, 256f, 257f
  ions, techniques using, 258–259, 260f
  neutrons, techniques using, 263–264
  scanning probes, 252–253
  x-rays, techniques using, 259, 260–263, 261f, 262f
PLEDs (polymeric light-emitting diodes), 220
PL (photolumninescence) spectroscopy, 322
PMDPP3T (poly[[2,5-bis(2-hexyldecyl-2,3,5,6-tetrahydro-3,6-dioxopyrrolo[3,4-c]pyrrole-1, 4-diyl]-alt-[3',3''-dimethyl-2,2':5',2''-terthiophene]-5,5''-diyl]), 353–359, 357f, 358t–359t
PMF (potential of mean force), 101
Poly(aniline) (PANI), as ABL, 215
Poly[(4,4'-bis(2-ethylhexyl)dithieno[3,2-b:2,'3'-d]silole)-2,6-diyl-alt-(2,1,3-benzothiadiazole)-4,7-diyl] (PSPTPB), 6, 7, 9t
Poly([4,40-bis(2-ethylhexyl)dithieno(3,2-b;20,30-d)silole]-2,6-diyl-alt-(2,1,3-benzothidiazole)-4,7-diyl) (PSBTBT), 269
Poly[2,6-(4,4-bis(2-ethylhexyl)-4H-cyclopenta[2,1-b;3,4-b']dithiophene)-alt-4,7-(5-fluoro-2,1,3-benzothia-diazole)] (PCPDTFBT), 354
Poly[2,6-(4,4-bis-(2-ethylhexyl)-4H-cyclopenta[2,1-b;3,4-b']dithiophene)-alt-4,7-(2,1,3-benzothiadiazole)] (PCPDTBT), 268–270, 349–351, 350f
Poly[(4,8-bis-(2-ethylhexyloxy)-benzo[1,2-b:4,5-b'] dithiophene)-2,6-diyl-alt-(4-(2-ethylhexanoyl)-thieno[3,4-b]thiophene)-2,6-diyl] (PBDTTT-C), 353
Poly[[2,5-bis(2-hexyldecyl-2,3,5,6-tetrahydro-3,6-dioxopyrrolo[3,4-c]pyrrole-1, 4-diyl]-alt-[3',3''-dimethyl-2,2':5',2''-terthiophene]-5,5''-diyl] (PMDPP3T), 353–359, 357f, 358t–359t
Poly-2,5-bis(3-tetradecylthiophene-2-yl) thieno[3,2-b]-thiophene (PBTTT), 98
Polycyclic aromatic hydrocarbon (PAH), 62
Poly(2,6'-4,8-di(5-ethylhexylthienyl)benzo[1,2-b;3,4-b]dithiophene-alt-5-dibutyloctyl-3,6-bis(5-bromothiophen-2-yl) pyrrolo[3,4-c]pyrrole-1,4-dione) (PBDTT-DPP), 356
Poly(3,4-dihexyl-2,2'-bithiophene) (PDHBT), 104
Poly-(2,2':5',2''-3,3''-dihexylterthiophene) (PTTT), 104

Poly(dimethylsiloxane)-block-poly(methyl methacrylate) (PDMSb-PMMA), as CBL, 227
Poly(4,4-dioctyldithieno(3,2-b:2',3'-d)silole)-2,6-diyl-alt-(2,1,3-benzothiadiazole)-4,7-diyl) (PSEHTT), 353–354
Poly(3,4-ethylenedioxythiophene) (PEDOT), 45–47, 119, 389
PEDOT:PSS
  application as transparent electrode of OSCs, 52–54, 53t, 54f, 55t
  chemical structure, 45–46, 46f
  conductivity enhancement, 47–52, 48f, 50f, 51f, 52f
PEDOT:PSS ABLs, 203–213
  additives in, 206–210, 207f, 209t
  advantages, 204
  chemical structure, 203f
  drawbacks, 205
  grades of, 204
  incorporation of, 203
  modified, ITO-free PSCs based on, 210–213, 211t, 212f
  physical treatments, 205–206
  thickness of layer, 204–205, 205f
  types of, 204t
Polyethylene glycol (PEG)
  as CBL, 227
  for PEDOT:PSS, 210
Poly(ethylene oxide) (PEO) interlayer, 227
Polyethylenimine (PEI), 353
Polyethylenimine (PEIE), 348, 353
Polyhedral oligomeric silsesquioxanes (POSS), 21–22, 21f
Poly(3-hexylthiophene) (P3HT), 91, 94, 95f
  chemical structure, 103–104, 103f
  concentration, 96–98, 99f
  defect band in, 180–182, 181f
  functionalized silsesquioxane nanoparticles, 27–32
    overview, 27–28
    photovoltaic performances, 31, 32f
    synthesis and particle morphology, 28–31, 28f, 29f, 30f, 31f
  identification and characterization of defect states in D–A blends, 187–195. See also Defect states
  molecular arrangement, 97, 97f
  molecules in anisole, 105, 106, 106f
  in nanophase separation
    second generation semiconducting polymer, 265–268, 265f, 267f
    techniques using electrons, 253–257, 254f, 255f, 256f, 257f
OPV modelling

Index

MD simulations and, 124–128, 125f, 126f, 127f, 128f
multiscale methods, 132–134, 132f, 134f
PDS of, 176, 176f
solvents, choice, 98, 100–102, 101f
TSC signal from, 176, 177f
Polymer-based tandem organic solar cells, 352–359, 357f, 358t–359t
Polymeric light-emitting diodes (PLEDs), 220
Polymer/polymer bulk heterojunction solar cells, 10–13
Polymer(s)
as ABLs, 215–217, 216f, 217t
as CBLs
nonconjugated, 227–228
water/alcohol–soluble conjugated, 228, 229f
chemical structure, 103–104, 103f
concentrations of, 96–98, 97f, 99f
conjugated, macromolecular characteristics on morphology, 118–119
equilibrium aspects, 118
nonequilibrium effects, 118–119
CPs. See Conjugated polymers (CPs)
crystallinity of, 105
graphene-based, 379–391. See also Graphene-based polymers
matrix, in NR alignment, 333, 333f, 334f
nanophase separation in OPVs with, case studies
second-generation semiconducting, 265–268, 265f, 267f
third generation semiconducting, 268–270
in OPV devices, 114
solvents, choice, 98, 100–102, 101f
transparent conducting polymers, 44–54
PEDOT:PSS, application as transparent electrode of OSCs, 52–54, 53t, 54f, 55t
PEDOT:PSS, conductivity enhancement, 47–52, 48f, 50f, 51f, 52f
structure, properties, and synthesis, 44–47, 45f, 46f
Polymer (P3HT) and acceptor (PFBT), 10, 12
Polymer solar cells (PSCs)
ABL for, 203. See also Anode buffer layers (ABLs)
active layer nanomorphology via fullerene ratios and solvent additives, 282–300. See also Active layer nanomorphology
BHJ and, 202–203
efficiency enhancement, interfacial materials for, 202–233. See also Interfacial materials
ITO-free, based on modified PEDOT:PSS, 210–213, 211t, 212f
structure, 202

Poly(2-methoxy-5-(2′-ethyl-hexyloxy)-para-phenylenevinylene (MEHPPV)), 3, 4
Poly[N-9′-heptadecanyl-2,7-carbazole-alt-5,5-(4′,7′-di-2-thienyl-2′,1′,3′-benzothiadiazole)] (PCDTBT), 352–359, 357f, 358t–359t
Poly(p-phenylene vinylene) (PPV), 250
Polysilsesquioxanes, functionalized, 20–25
bridging, 22, 22f
overview, 20–22, 21f, 22f
POSS, 21–22, 21f
synthesis by sol-gel method, 23–25, 24f
Polystyrene (PS), 13
Poly(styrene sulphonate) (PSS), 47, 119
PEDOT:PSS
application as transparent electrode of OSCs, 52–56, 53t, 54f, 55t
chemical structure, 45–46, 46f
conductivity enhancement, 47–52, 48f, 50f, 51f, 52f
PEDOT:PSS ABLs, 203–213
additives in, 206–210, 207f, 209t
advantages, 204
chemical structure, 203f
drawbacks, 205
grades of, 204
incorporation of, 203
modified, ITO-free PSCs based on, 210–213, 211t, 212f
physical treatments, 205–206
thickness of layer, 204–205, 205f
types of, 204t
Polythiophene-based donor, 12
Poly(thiophene-2,5-diyl-alt-[5,6-bis(dodecyloxy)-benzo[c][1,2,5]thiadiazole]-4,7-diyl) (PBT-T1)-based BHJ solar cells, fullerene ratios and solvent additives and, 283–291
PBT-T1:PC$_{60}$BM films
AFM topography and phase images, 285–286, 285f
current-voltage characteristics of solar cells, 286–288, 286f, 287t
UV-visible absorption and XRD spectra, 284–285, 284f
PBT-T1:PC$_{70}$BM solar cells, current-voltage characteristics, 288–289, 288f, 289t
photo-CELIV and extracted charge carrier density vs. delay time of solar cells, 290–291, 290f, 291t
UV-visible absorption, 283–284, 283f
Poly(vinylpyrrolidone) (PVP), as CBL, 228
POSS (polyhedral oligomeric silsesquioxanes), 21–22, 21f
Potential of mean force (PMF), 101
Power conversion efficiencies (PCEs), 90, 143
additives, processing conditions and, 118–119

defined, 114
of OPVs, 114
p-type conjugated polymers with, 5f
PPV (poly(*p*-phenylene vinylene)), 250
PSBTBT (poly([4,40-bis(2-ethylhexyl)
  dithieno(3,2-b;20,30-d)silole]-2,6-
  diyl-alt-(2,1,3-benzothidiazole)-4,7-
  diyl)), 269
PSCs (polymer solar cells)
  ABL for, 203. *See also* Anode buffer layers
    (ABLs)
  active layer nanomorphology via fullerene
    ratios and solvent additives,
    282–300. *See also* Active layer
    nanomorphology
  BHJ and, 202–203
  efficiency enhancement, interfacial materials
    for, 202–233. *See also* Interfacial
    materials
  ITO-free, based on modified PEDOT:PSS,
    210–213, 211t, 212f
  structure, 202
PSEHTT (poly(4,4-dioctyldithieno(3,2-
  b:2′,3′-d)silole)-2,6-diyl-alt-(2,1,3-
  benzothiadiazole)-4,7-diyl)), 12,
  353–354
PSiF-DBT solar cells, 6
PSPTPB [(4,4′-bis(2-ethylhexyl)dithieno[3,2-
  b:2,′3′-d]silole)-2,6-diyl-alt-(2,1,3-
  benzothiadiazole)-4,7-diyl], 6, 7
PSS (poly(styrene sulphonate)), 47, 119
PEDOT:PSS
  application as transparent electrode of
    OSCs, 52–54, 53t, 54f, 55t
  chemical structure, 45–46, 46f
  conductivity enhancement, 47–52, 48f,
    50f, 51f, 52f
  PEDOT:PSS ABLs, 203–213
    additives in, 206–210, 207f, 209t
    advantages, 204
    chemical structure, 203f
    drawbacks, 205
    grades of, 204
    incorporation of, 203
    modified, ITO-free PSCs based on,
      210–213, 211t, 212f
    physical treatments, 205–206
    thickness of layer, 204–205, 205f
    types of, 204t
PTB7, 12
PTPD3T, 9
PTPT, 5f, 7, 9t
PTTT (poly-(2,2′:5′,2″-3,3″-dihexylterthiophene)),
  104
P-type conjugated polymers with power
  conversion efficiency, 5f
PV. *See* Photovoltaics (PV)

PVP (poly(vinylpyrrolidone)), 67
  as CBL, 228
Pyrene butanoic acid succinimidyl ester
  (PBASE), 388

R

Radiative mechanisms, of recombination, 115
Radical Voronoi tessellation schemes, 126, 133
Raman spectra, of PDPP3T:PCBM thin films,
  293, 294f
RCWA (rigorous coupled wave analysis), 349
Recombination, in OPVs, 114–116
  Auger, 115
  exciton, 117–118
  morphology on, 119–120
  nonradiative mechanisms, 115
  radiative mechanisms, 115
Recombination dynamics, PDPP3T:PCBM solar
  cells
  solvent additives on, 297–300, 298f, 299f
Reduced graphene oxide (rGO), 62, 381–382
  as CBLs, 232
Reflectivity modeling, 266, 267f
Regioregular poly(3-alkylthiophenes), 4
Regioregular poly(3-hexylthiophene)
  (RR-P3HT), 4, 5–6
Relating synthesis parameters, photoactive layers
  morphology in OPVs, 95–107
  other studies, 104–107, 105f, 106f
  polymers and acceptor species
    chemical structure, 103–104, 103f
    concentrations, 96–98, 97f, 99f
  solvents
    choice, 98, 100–102, 101f
    solute concentration, 102–103
Relatively deep activation energies, defects with,
  180–182, 181f, 182t
Relatively shallow activation energies, defects
  with, 177–180, 178f, 179f, 180f
Resonant soft x-ray scattering (RSoXS), 259,
  260, 262, 262f
RGO (reduced graphene oxide), 62, 381–382
  as CBLs, 232
Rigorous coupled wave analysis (RCWA), 349
RSoXS (resonant soft x-ray scattering), 259, 260,
  262, 262f

S

SAED (selected area electron diffraction), of
  PDPP3T:PCBM thin films, 293, 294f
SAMs (self-assembed monolayers)
  ABLs, 218–219, 218f
  CBLs, 230–231, 230f
SANS (small-angle neutron scattering), 262, 263
Sariciftci, 3–4

Index

Scanning electron microscopy (SEM), in nanophase separation, 253, 259
Scanning probes, in vertical and planar nanophase separation, 252–253
Scanning transmission x-ray microscopy (STXM), 263
Scattering function, 127–128, 128f
  defined, 127
Scattering length density (SLD), 262, 263, 264
SCFT (self-consistent field theory), 130–131
Schottky barrier, 119
SCLC (space charge limited current) modeling, 151–152, 171–172
Secondary ion mass spectroscopy (SIMS), 259
Second-generation semiconducting polymers, nanophase separation (case study), 265–268, 265f, 267f
Selected area electron diffraction (SAED), of PDPP3T:PCBM thin films, 293, 294f
Self-assembed monolayers (SAMs)
  ABLs, 218–219, 218f
  CBLs, 230–231, 230f
Self-assembled buffer layers, 218–219, 218f
Self-assembly, of nanoparticles
  CPs and, 318–324
    hybrid nanofibrils, 321–324, 321f, 322f, 323f
    nanofiber, 318–321, 319f, 320f, 320t
    overview, 318
Self-consistent field theory (SCFT), 130–131
Self-consistent periodic image-charges embedding (SPICE), 122
SEM (scanning electron microscopy), in nanophase separation, 253, 259
Semiconducting polymers, nanophase separation (case studies)
  second-generation, 265–268, 265f, 267f
  third generation, 268–270
Semiconductor inorganic nanoparticles, grafting CPs, 309–318
  direct grafting, 312–316, 313f, 314f, 315f
  ligand exchange, 310–312, 311f, 312f
  overview, 309–310, 310f
  in situ growth within matrix, 315–318, 317f
Semiconductors
  organic, defect states in, 168–169, 168f
  small-molecule, nanophase separation (case study), 271
Semitransparent tandem organic solar cells, 364–365, 366f
Shallow activation energies, defects with, 177–180, 178f, 179f, 180f
Shockley-Read-Hall (SRH) recombination, 221
2,7-silafluorene (SiF), 6
Silsesquioxane nanostructures, D-A functionalized, 19–33
  for OPV devices, 26–32

overview, 26–27
PDI nanoparticles, 27–32. See also Perylenediimide (PDI)
P3HT nanoparticles, 27–32. See also Poly(3-hexylthiophene) (P3HT), functionalized silsesquioxane nanoparticles
overview, 19–20
polysilsesquioxanes, 20–25
  bridging, 22, 22f
  overview, 20–22, 21f, 22f
  POSS, 21–22, 21f
  synthesis by sol-gel method, 23–25, 24f
silsesquioxanes/bridged silsesquioxanes nanocomposites, 25–26, 25f
SIMS (secondary ion mass spectroscopy), 259
Simulation approaches, insights obtained OPV modeling, 120–134
  continuum-based approaches, 130–131
  electronic structure calculations, 120–122, 120f
  mechanics-based approaches, classical, 122–130. See also Mechanics-based approaches
  multiscale methods, 131–134, 132f, 134f
Single-junction solar cells, 340–342, 342f, 343f
Single-walled carbon nanotubes (SWNT), 56, 58, 57f, 60, 60f, 107, 208, 348
SLD (scattering length density), 262, 263, 264
Small-angle neutron scattering (SANS), 262, 263
Small molecule-based BHJ (SM-BHJ), 271. See also Bulk heterojunctions (BHJ)
Small molecule-based tandem organic solar cells, 360–363, 360f, 362t
Small-molecule organic materials
  ABLs, 215–217, 216f, 217t
  CBLs, 225–227, 225f
Small-molecule semiconductors, nanophase separation (case study), 271
SM-BHJ (small molecule-based BHJ), 271. See also Bulk heterojunctions (BHJ)
Solar cells, 3
Solar cells, organic
  OPV. See Organic photovoltaic (OPV) solar cells
Sol-gel method, polysilsesquioxanes synthesis by, 23–25, 24f
Solute concentration, in solvents, 102–103
Solution-based graphene exfoliation, 380–381
Solvent additive on nanomorphology,
  PDPP3T:PCBM solar cells and, 291–300, 292f
  on charge transport and recombination dynamics, 297–300, 298f, 299f
  current-voltage characteristics, 296–297, 296f, 296t
  EFTEM images, 292–293, 293f

selected area electron diffraction and Raman spectra, 293, 294f
on surface morphology, 294–296, 295f
Solvent additives on nanomorphology, PBT-T1-based BHJ solar cells and, 283–291
PBT-T1:PC$_{60}$BM solar cells
AFM topography and phase images of films, 285–286, 285f
UV-visible absorption and XRD spectra of films, 284–285, 284f
PBT-T1:PC$_{70}$BM solar cells, current-voltage characteristics, 288–289, 288f, 289t
photo-CELIV and extracted charge carrier density vs. delay time, PBT-T1, 290–291, 290f, 291t
UV-visible absorption, PBT-T1, 283–284, 283f
Solvent(s), in OPVs
nanophase separation
blending solvent additive, 251–252
selection, 250
NR alignment, 327, 329f, 330
photoactive layers morphology
choice of, 98, 100–102, 101f
solute concentration, 102–103
Sorbitol, in PEDOT:PSS ABL, 206
Space charge limited current (SCLC) modeling, 151–152, 171–172
SPICE (self-consistent periodic image-charges embedding), 122
Spray coating, 385
SRH (Shockley-Read-Hall) recombination, 221
Stöber method, 25–26, 29
Structural defects, origins from residuals
in D-A blends, 193–195, 194f, 195f
in neat material systems, 186–187
STXM (scanning transmission x-ray microscopy), 263
Substrates, NRs and, 326, 327, 328f
Sulfuric acid treatment, for PEDOT:PSS films, 50–52, 50f, 51f, 52f, 54, 54f
Superacid exfoliation, 383
Surface ligands, in NR alignment, 330, 331, 331f
Surface morphology, of PDPP3T:PCBM films solvent additives on, 294–296, 295f
Surfactant-assisted exfoliation, 383
Surfactants, in PEDOT:PSS solution, 49–50
SWNT (single-walled carbon nanotubes), 56, 58, 57f, 60, 60f, 107, 208, 348
Synthesis
PDI and P3HT-functionalized silsesquioxane nanoparticles, 28–31, 28f, 29f, 30f, 31f
polysilsesquioxanes, by sol-gel method, 23–25, 24f
residual

D-A blends, 193–195, 194f, 195f
neat material systems, 186–187
transparent conducting polymers, 44–47, 45f, 46f

T

Tandem solar cells, 337–371
experimental results, review, 352–365. See also Highly efficient organic tandem solar cells
based on novel materials, 352–363
based on various structures, 363–365
fundamental limitations, 340–346
configurations, 346, 346f
EQE and IQE, 344
maximum efficiency of, 344–345, 345f
operation, 344
performance, 342, 343–344
single-junction solar cells, 340–342, 342f, 343f
measurement, 365, 367–370, 367f, 368f, 370f
optical simulations, 349–351, 350f
overview, 337–340, 338f, 339f
working principles of IML, 346–349, 347f
TCB (1,2,3-trichlorobenzene), 319–320, 319f
TCOs (transparent conducting oxides), 44, 70, 71t
TDOS (trap density of states), 169
TEM (transmission electron microscopy), 94
in nanophase separation, 253, 255, 257, 257f, 258, 259
EFTEM, 253–254, 255, 256f, 257
Temperature, in NR alignment, 326
TEOS (tetraethoxysilane), 25, 28–29
Ternary OPVs, advanced
nanophase separation (case study), 271–272
Tessellation schemes, 126–127
Tetraethoxysilane (TEOS), 25, 28–29
Tetramethylene sulfone (TMS), for PEDOT:PSS ABL, 208
Thermal annealing, 4
in nanophase separation, 250–251
Thermally stimulated current (TSC) method, 169–170, 176, 177, 177f
Thermal treatment, of PEDOT:PSS, 205
Thieno[3,4-b]thiophene (TT), 7
Thieno[3,4-c]pyrrole-4,6-dione (TPD), 8
Thin-film preparation methods
from graphene dispersions, 383–385
other methods, 385
vacuum filtration, 383–384, 384f
Thiophene oligomers, 105, 105f
Third generation semiconducting polymers, nanophase separation (case study), 268–270
Tinker, 94

Index

TIPD (titanium (diisopropoxide) bis(2,4-pentanedionate)), as CBL, 226
Titanium (diisopropoxide) bis(2,4-pentanedionate) (TIPD), as CBL, 226
Titanium oxide, as CBLs, 221
TMF (transfer matrix formalism), 344
TMS (tetramethylene sulfone), for PEDOT:PSS ABL, 208
Toluene, 98, 100, 101, 102
Tomography, in OPVs, 255, 257
TPC (transient photocurrent) spectroscopy, 297
TPV (transient photovoltage) spectroscopy, 297
Transfer matrix formalism (TMF), 344
Transient photocurrent (TPC) spectroscopy, 297
Transient photovoltage (TPV) spectroscopy, 297
Transmission electron microscopy (TEM), 94
   in nanophase separation, 253, 255, 257, 257f, 258, 259
      EFTEM, 253–254, 255, 256f, 257
Transmittance, of PEDOT:PSS films, 51–52, 52f
Transparent conducting oxides (TCOs), 44, 70, 71t
Transparent conducting polymers, 44–54
   PEDOT:PSS
      application as transparent electrode of OSCs, 52–54, 53t, 54f, 55t
      conductivity enhancement, 47–52, 48f, 50f, 51f, 52f
      structure, properties, and synthesis, 44–47, 45f, 46f
Transparent electrode materials, for OSCs, 44–74
   carbon-based nanomaterials, 56–64
      CNTs, 56–60. *See also* Carbon nanotubes (CNTs)
      graphene, 61–64. *See also* Graphene(s)
   conducting polymers, 44–54
      PEDOT:PSS, application as transparent electrode, 52–54, 53t, 54f, 55t
      PEDOT:PSS, conductivity enhancement, 47–52, 48f, 50f, 51f, 52f
      structure, properties, and synthesis, 44–47, 45f, 46f
   metal nanostructures, 64–70
      nanomeshes, 64–67, 65f, 66f
      nanowire grids, 67–69, 68t
      ultrathin films, 69–70
   MTFs, 72–74, 73t
   overview, 44
   TCOs to ITOs, 70, 71t
Trans-polyacetylene, 265
Trap bands, in blended devices, 188–189, 189t
Trap density of states (tDOS), 169
1,2,3-Trichlorobenzene (TCB), 319–320, 319f
Tris-(8-hydroxyquinoline)aluminum (Alq3), as ABL, 216
Tris(thienylenevinylene) (TTV), 12

TSC (thermally stimulated current) method, 169–170, 176, 177, 177f

U

UHV (ultrahigh vacuum) chamber, 153
Ultimate efficiency, 114
Ultrahigh vacuum (UHV) chamber, 153
Ultrathin metal films, 69–70
Ultraviolet photoelectron spectroscopy (UPS), 225
Ultraviolet (UV) treatment, of PEDOT:PSS, 206
Ultraviolet (UV)-visible absorption
   of PBT-T1, 283–284, 283f
   XRD spectra of PBT-T1:PC$_{60}$BM films and, 284–285, 284f
United-atom force field, 94
UPS (ultraviolet photoelectron spectroscopy), 225

V

Vacuum filtration, 383–384, 384f
Vacuum level, 6–7
Vanadium oxide ABLs, 213, 214
Vapor-phase polymerization (VPP), 47
Variable angle spectroscopic ellipsometry (VASE), 349
Variable range hopping (VRH) model, 51
VASE (variable angle spectroscopic ellipsometry), 349
Velocity Verlet algorithm, 92
Verlet algorithm, 92
Vertical nanophase separation, characterizing, 252–264
   electrons, techniques using, 253–257, 254f, 255f, 256f, 257f
   ions, techniques using, 258–259, 260f
   neutrons, techniques using, 263–264
   scanning probes, 252–253
   x-rays, techniques using, 259, 260–263, 261f, 262f
Voronoi tessellation schemes, 126, 133
VPP (vapor-phase polymerization), 47
VRH (variable range hopping) model, 51

W

Water, for PEDOT:PSS, 210
Water/alcohol–soluble conjugated polymers, as CBLs, 228, 229f
Window electrode, graphene as, 388–390
Wudl, 4

X

Xerogels, 23
XPS (x-ray photon spectroscopy), 225, 259, 263, 322

X-ray diffraction (XRD) spectra of PBT-T1:PC$_{60}$BM films, 284–285, 284f
X-ray photon spectroscopy (XPS), 225, 259, 263, 322
X-ray reflectivity (XRR), 259, 262
X-rays, techniques using
   in nanophase separation, 259, 260–263, 261f, 262f

XRD (x-ray diffraction) spectra of PBT-T1:PC$_{60}$BM films, 284–285, 284f
XRR (x-ray reflectivity), 259, 262
Xylene, 98

Z

Zinc oxide, 348
   as CBLs, 221, 222, 230–231

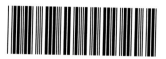